MW00836824

The Geometric Universe

The Geometric Universe

Science, Geometry, and the Work of Roger Penrose

EDITED BY

S. A. HUGGETT

School of Mathematics and Statistics, University of Plymouth

L. J. MASON

K. P. TOD

S. T. TSOU

N. M. J. WOODHOUSE

Mathematical Institute, University of Oxford

OXFORD • NEW YORK • TOKYO

OXFORD UNIVERSITY PRESS

1998

Oxford University Press, Great Clarendon Street, Oxford OX2 6DP
Oxford New York
Athens Auckland Bangkok Bogota Bombay Buenos Aires
Calcutta Cape Town Dar es Salaam Delhi Florence Hong Kong
Istanbul Karachi Kuala Lumpur Madras Madrid Melbourne
Mexico City Nairobi Paris Singapore Taipei Tokyo Toronto Warsaw
and associated companies in
Berlin Ibadan

Oxford is a trade mark of Oxford University Press

Published in the United States by
Oxford University Press Inc., New York

A catalogue record for this book is available from the British Library

Library of Congress Cataloging in Publication Data
(Data available)
ISBN 0 19 850059 9

Typeset by The Author
Printed in Great Britain by
Bookcraft (Bath) Ltd,
Midsomer Norton, Avon

Laudatio

John A. Wheeler
Department of Physics, Princeton University, Princeton, New Jersey[1]

If I had the good fortune to be present at this celebration of the life and work of Roger Penrose, I would recall the shocked sense of scientific-technical inferiority given the Western world when on October 4, 1957 the then Soviet Union became first, with Sputnik, to launch mankind into space. Already in the summer of 1957, I found myself in Paris, chairman of a committee to recommend to the 3rd Annual Conference of NATO Parliamentarians measures to build up Western capabilities in the sciences. The people who were going to have to pay the bill accepted NATO-sponsored international Scientific Conferences and Workshops and support for Fellowships for young people of promise in one NATO country to go to another to broaden their experience. One such NATO fellowship, through no doing of mine, went to young Roger Penrose and brought him to Princeton. Well do I remember the pre-dawn darkness near the end of his stay in Princeton when he reached out over a snowdrift to hand me his Adams Prize essay that I had promised to deliver in two hours to the international mail terminal at New York's Kennedy Airport, so it could make the Cambridge deadline. To the pleasure of us all, he won the Adams Prize. He has been winning prizes ever since.

Over the years, Roger Penrose has won a great prize for us all, a deeper understanding of the structure of spacetime, especially the causality relationship between one point of spacetime and another, probably the most important prediction of general relativity, since it seems to imply that spacetime has a beginning or an end.

Roger Penrose, like all of us, knows that in the year 2000 we will be celebrating the centenary of Max Planck's discovery of the quantum. Will the quantum count as Glory for the wonderful insight it has given us in every branch of physics? Or Shame that we still have not fought our way through to understanding how come the quantum? Glory or Shame? The writings of Roger Penrose direct us again and again to that challenging issue. Three cheers for him and his vision!

[1]This eulogy was received by the editor from Professor Wheeler before the conference.

Preface

The symposium 'Geometric Issues in the Foundations of Science' was held over 5 days in June 1996 in St John's College Oxford in honour of Professor Sir Roger Penrose in his 65th year. The unifying theme guiding the scientific content of the symposium was the impact of Sir Roger's geometric viewpoint on a wide range of fields in basic science and mathematics. The object was to use the opportunity provided by the 65th birthday of Sir Roger to draw a group of distinguished speakers together whose work could broadly be classed as geometrical in order to bring out what was common to these endeavours.

There were 17 plenary lectures held in the auditorium of St John's and 16 shorter lectures delivered in two pairs of parallel sessions in the Mathematical Institute. These were attended by 186 participants from a broad range of backgrounds from the President of the Royal Society on one hand to graduate students on the other.

Sir Michael Atiyah opened with a lecture setting the scene for the symposium, giving an overview of the interaction between geometry and physics and himself and Sir Roger from which many important developments in mathematics and mathematical physics have emerged.

There followed lectures in pure mathematics, including geometry, both classical differential geometry and non-commutative geometry, topology including knot invariants and the applications of gauge theory and developments arising from string theory. Lectures on applied mathematics included integrable systems and general relativity. Lectures on theoretical physics included string theory, quantum gravity and the foundations of quantum mechanics, and in experimental physics there were talks on quasi-crystals and astrophysics. Less easy to classify were the talks on quantum computation, quantum cryptography and the possible role of micro-tubules in a theory of consciousness.

Sir Roger closed the symposium with a review of twistor theory, the problems currently confronting the theory and prospects for their solution.

This volume collects together the contributions of all these lecturers, giving an overview of the many applications of geometrical ideas and techniques across mathematics and the physical sciences.

The organisers wish to thank the Scientific Organising Committee and Professor E Corrigan and also gratefully acknowledge administrative and secretarial support received from the Mathematical Institute, Oxford, particularly from Jill Drake and Brenda Willoughby.

The symposium was supported by a substantial grant from the EPSRC and also by St John's College, Oxford, the London Mathematical Society and Oxford University Press and we are very grateful to them. SAH, LJM, KPT and NMJW benefited from the hospitality of the Erwin Schrödinger Institute, Vienna, while this volume was being prepared and we would like to thank Piotr Kobak for help with the typesetting.

Oxford
August 1997

SAH., LJM., KPT.,
TST., and NMJW.

CONTENTS

Contributors

Jeeva Anandan Max–Planck–Institut für Physik, Föhringer Ring 6, D-80805 Munich, Germany *and* Department of Physics and Astronomy, University of South Carolina, Columbia, SC 20208, USA.

Abhay Ashtekar Center for Gravitational Physics and Geometry, Physics Department, Penn State University, University Park, PA 16802–6300, USA.

Sir Michael Atiyah The Master's Lodge, Trinity College, Cambridge, CB2 1TQ, England.

Toby Bailey Department of Mathematics, University of Edinburgh, James Clerk Maxwell Building, The Kings Buildings, Mayfield Rd, Edinburgh EH9, 3JZ, Scotland.

Dorje C. Brody Blackett Laboratory, Imperial College, South Kensington, London SW7 2BZ, England.

Brandon Carter Department d'Astrophysique Relativiste et de Cosmologie, Observatoire de Paris, Place J. Janssen, F-92195 Meudon Cedex, France.

Alain Connes Department of Mathematics, Collège de France, 11, Place Marcellin Berthelet, 75005, Paris Cedex 05, France.

Simon Donaldson The Mathematical Institute, 24–29 St Giles, Oxford, OX1 3LB, England.

Artur Ekert The Clarendon Laboratory, Parks Road, Oxford OX1 3PU, England.

Helmut Friedrich The Albert Einstein Institute, Max Planck Institute für Gravitationsphysik, Schlaatzweg 1, D-14473, Potsdam, Germany.

Simonetta Frittelli Department of Physics and Astronomy, University of Pittsburgh, Pittsburgh, PA 15260, USA.

Gary Gibbons Department of Applied Mathematics and Theoretical Physics, Silver St, Cambridge CB3 9EW, England.

Simon Gindikin Department of Mathematics, Rutgers University, New Brunswick, NJ 08903, USA.

Stuart Hameroff Department of Anaesthesiology, University of Arizona, Tucson, AZ 8572, USA.

Stephen Hawking Department of Applied Mathematics and Theoretical Physics, Silver St, Cambridge CB3 9EW, England.

Nigel Hitchin Department of Pure Mathematics and Mathematical Statistics, University of Cambridge, 16 Mill Lane, Cambridge, CB2 1SB, England.

Andrew Hodges Wadham College, Oxford OX1 3PN, England.

Lane Hughston Merrill Lynch International, 25 Ropemaker St, London EC2Y 9LY, England.

Richard Josza School of Mathematics and Statistics, University of Plymouth, Plymouth PL4 8AA, England.

Louis Kauffman Department of Mathematics, Statistics, and Computer Science, University of Illinois at Chicago, 851 South Morgan Street, Chicago, Illinois, 60607-7045, USA.

Carlos Kozameh FaMAF, Universidad Nacional de Córdoba, 5000 Córdoba, Argentina.

Claude Lebrun Department of Mathematics, State University of New York at Stony Brook, Stony Brook, NY 11794, USA.

Sergei Merkulov Department of Mathematics, University of Glasgow, Scotland.

Ted Newman Department of Physics and Astronomy, University of Pittsburgh, Pittsburgh, PA 15260, USA.

Sir Roger Penrose The Mathematical Institute, 24–29 St Giles, Oxford, OX1 3LB, England.

L. J. Schwachhöfer Universität Leipzig.

Dennis Sciama SISSA/ISAS, Via Beirut 2-4, 34013 Trieste, Italy.

Graeme Segal Department of Pure Mathematics and Mathematical Statistics, University of Cambridge, 16 Mill Lane, Cambridge, England.

Abner Shimony Department of Physics, Boston University, 590 Commonwealth Avenue, Boston, MA 02215, USA.

Lee Smolin Center for Geometry and Gravity, The Physics Department, Penn State University, USA.

George Sparling Department of Mathematics and Statistics, University of Pittsburgh, Pittsburg, PA 15260, USA.

Paul Steinhardt Department of Physics and Astronomy, David Rittenhouse Laboratory, University of Pennsylvania, Philadelphia, PA 19104-6396, USA.

R. P. Thomas The Mathematical Institute, 24–29 St Giles, Oxford, OX1 3LB, England.

Andrzej Trautman Instytut Fizyki Teoretycznej, Uniwersytet Warszawski, ul. Hoża 69, PL–00681 Warszawa, Poland.

Lev Vaidman School of Physics and Astronomy, Raymond and Beverly Sackler Faculty of Exact Sciences, Tel-Aviv University, Tel-Aviv 69978, Israel.

Gabriele Veneziano Theory Division, CERN, 1211 Geneva 23, Switzerland.

Richard Ward Department of Mathematical Sciences, Univerity of Durham, Durham DH1 3LE, England.

John A. Wheeler Department of Physics, Princeton University, Princeton, New Jersey, USA.

PART I

Plenary Lectures

1

Roger Penrose—A Personal Appreciation

Michael Atiyah
The Master's Lodge, Trinity College, Cambridge CB2 1TQ

1 Personal and historical remarks

Roger Penrose and I were research students together in Cambridge from 1952 to 1955. Moreover, we were both algebraic geometers at the time. I did my research under Hodge, and Roger, after starting with Hodge, moved over to work with Todd. In view of later events it is perhaps interesting to review briefly what areas our research was concerned with.

Under Hodge I was steered towards differential geometry and topology. From Hodge's book I learnt about harmonic forms and their origin in Maxwell's equations, while from the work of the French school (Leray, Cartan, Serre) I learnt about sheaf cohomology. Although I had attended Dirac's course on quantum mechanics it was not until some years later that I became acquainted with spin and the Dirac operator.

Roger's work centred on more classical algebraic geometry, in particular the theory of invariants. In the course of his research he invented a private notation for keeping track of indices in tensorial calculations. Years later this linked up with the technology of Feynman diagrams. It is interesting to note that similar diagrammatic ideas have come to the fore in recent years in the frontier between topology and physics (e.g. knots and Chern–Simons theory).

After our time together as research students, Roger and I pursued quite different routes. While I remained a geometer of sorts, Roger moved into theoretical physics and in due course made his name by his work (jointly with Stephen Hawking) on the generic existence of singularities in general relativity.

Although we met briefly from time to time we only got together on a more permanent basis in Oxford from 1973 to 1990. When I was about to leave the Institute for Advanced Study in Princeton to return to Oxford, I remember a discussion with Freeman Dyson about the possibility of Roger Penrose coming to Oxford as Coulson's successor. Dyson expressed his admiration for Roger's work on black holes but admitted that he was mystified by "twistors". "Perhaps", he said, "you will understand them". This was a percipient remark since I did indeed spend the next decade or so trying to understand and use twistors!

2 Twistors

When Roger and I were once again established as colleagues (this time as professors), the subject of twistors soon became the subject of attention. Roger explained his ideas and although the physical motivation, in terms of quantum theory, was new to me the underlying geometry was extremely familiar. The Klein representation of lines in 3-space by points of a quadric in 5-space I had learnt from Todd's book and it had always fascinated me. Roger explained that he was using complex contour integrals to represent solutions of various differential equations. He pointed out that, as with the usual residue calculus, the precise integrands were not the key thing. The singularities really determined the story.

It was not long before it dawned on me that Roger was essentially struggling with sheaf cohomology but did not realize it. Once this was pointed out Roger and his students became fervent converts. After a few private seminars in my study they really took off. Within a short period of time Roger's group were more expert with sheaf cohomology than I had ever been.

I found the whole twistor programme a fascinating story. Its first success was in the beautiful way in which the sheaf cohomology groups $H^1(\mathbb{P}T^+, \mathcal{O}(n))$ corresponded precisely to the solutions of the zero-rest-mass field equations. In retrospect one can view this as a complexification of the Radon transform (now resurrected for application to tomography), but the Penrose version is both richer and more beautiful.

The second success of the twistor programme was the observation by Richard Ward that it could be used to solve the self-dual Yang–Mills equations. Among other things this stimulated work on instantons and in due course led to Donaldson's remarkable work on 4-manifolds.

Finally[1] the twistor programme led to a deep understanding of the self-dual Einstein equations in which the Riemannian geometry gets encoded entirely into the holomorphic geometry of a complex 3-manifold. This is certainly the deepest application of twistor methods and the final result is quite stunning in its simplicity.

While Roger and his group have continued to use the twistor picture as an alternative to the usual space-time picture, mathematicians have found twistor methods a powerful and subtle tool for various geometric problems. In higher dimensions hyper-Kähler manifolds are the natural generalization of self-dual Einstein manifolds and the twistor theory applies in this more general context. Interestingly enough hyper-Kähler geometry arises naturally in super-symmetric gauge theories. Although this is familiar from various examples the general phenomenon probably deserves further investigation, particularly in view of the current interest and activity in super-symmetric theories.

---

[1] Not chronologically: the 'graviton' preceded the 'instanton'.

3 Integrable systems and solitons

There is little doubt that the subject of integrable systems of differential equations, and their soliton solutions, has been one of the most interesting developments of the past decades. It has attracted interest on a very wide front because of its relevance to physics and applied mathematics while also exhibiting a beautiful structure which has appealed to pure mathematicians.

Nearly all the standard work on integrable systems has focused on equations such as the KdV equation or the non-linear Schrödinger equation. In particular these involve just two independent variables (one space, one time). Key features of the theory are the existence of solitons, the inverse scattering method of solution and the derivation of explicit formulae based on Riemann surface theory.

It was pointed out early on, mainly by Richard Ward, that many of the known integrable systems could be obtained by dimensional reduction from the 4-dimensional self-dual Yang–Mills equations. Moreover the twistor methods seemed similar to the standard methods of dealing with the integrable systems. This observation has now been pursued methodically and a good case can be made for saying that the self-dual Yang–Mills equation is the ancestor of all 2-dimensional integrable systems. This point of view has recently been developed in the new book by Mason and Woodhouse (1996). I hope this attracts the attention it deserves. On the whole, physicists are over-impressed by explicit formulae and the more powerful twistor technique (which can generate formulae as required) appears too abstract to many. The book by Mason and Woodhouse should redress the balance.

Given the fact that integrable models play a key role in various quantum field theories, it seems that one should explore further the implications of twistors at the quantum level. Much has been made by Faddeev and others of the quantum inverse scattering. It would be interesting to translate this into twistorial terms.

4 Rival philosophies

As we all know, the fundamental problem that all physicists would like to solve is how to reconcile quantum theory with general relativity. In other words how to produce a unified theory incorporating "quantum gravity".

There are many different approaches to this goal and it is not, I hope, blasphemous to make comparisons with the way rival religions offer alternative approaches to God. To the protagonists the various religions are mutually incompatible and (at times) distinctly hostile. To the sceptic the different religious approaches simply cancel each other out leaving a vacuum. At the other extreme we have mystics who see underlying commonality in all religious experience and search for some fusion. As a side-remark we mathematicians might note, by analogy, that a combination of oscillatory functions can cancel themselves out but yet leave a few delta functions at specific frequencies.

If we turn now to the different approaches to quantum gravity there are several rival philosophies or religions. The orthodox one is string theory and associated quantum field theory, and here the "prophet" is Edward Witten. We then have the twistor approach led by Roger Penrose. There is also a newer approach based on non-commutative geometry pioneered by Alain Connes (and expounded at this symposium). As "sceptics" we have the hard-headed experimentalists who have little time for theories which deal with monstrously small (or large) quantities, beyond the realm of measurement. On the other hand there are "mystics", among whom I include myself, who hope for a synthesis which embraces aspects of all the rival theories. As a mathematician I would find it a pity if God had not found some use for all the beautiful ideas that have been put forward.

Clues, indicating that such a synthesis is not totally hopeless, include the key role of integrable systems, solitons, duality, holomorphic geometry and supersymmetry. These can be found in quantum field theory as well as in twistor theory.

Roger's ideas on quantum gravity include one which has always appealed to me, and that relates to the famous "collapse of the wave-function" in quantum mechanics. Roger speculates that this collapse is a gravitational effect, thus combining quantum theory and gravity at a very fundamental level and altering the whole philosophical basis of quantum theory. This puts Roger very much in Einstein's camp in the famous Bohr–Einstein debate and promises to open up an argument which has lain dormant for decades.

5 Other topics

I have concentrated on Roger's twistor programme, but I should say a few words about some of his other contributions. The "Penrose tilings" are now well-known (and can even be bought as jig-saw puzzles!) The history of these is very interesting and I talked about them in my Anniversary Address to the Royal Society in 1994 (Atiyah 1995). They provide an excellent example of how a pure mathematical curiosity can develop into a theory of importance in the real world. In fact Roger told me that he had developed his ideas while "doodling" in the waiting room of a hospital when he was visiting a sick friend! The applications now involve quasi-crystals, materials which have been put to commercial use in making new style frying pans.

Roger has also with his two "popular" books entered the controversial field where philosophers spar with scientists and consciousness is the centre of attraction. I can claim a modest contribution on this front since I was a delegate of the Oxford University Press when *The Emperor's New Mind* was published.

6 Conclusion

To sum up, it is clear that Roger is in a real sense one of the original thinkers of our time. Although he is aware of the mainstream work in theoretical physics he is continuously branching out on his own. He thinks deeply and when he has an idea that he thinks is worth developing he pursues it tenaciously over many years.

These days most physicists follow the latest band-wagon, usually within microseconds. Roger steers his own path and eschews band-wagons. He may not always be right but it is important that we have individuals who stick to their guns. Future progress with ideas, as in evolutionary genetics, depends on a sufficient stock so that some good ones will survive and prosper. Roger is one of those who are helping to diversify our "gene pool" of ideas.

A close examination of Roger's work shows that he manages to combine genuine physical insight with the development of beautiful mathematical techniques to go alongside. It is this close harmony of the physics and mathematics which persuades him that he is on to something worthwhile. He has been proved right in the past and will, I hope, be proved right in the future.

Bibliography

Atiyah, M.F. (1995). Royal Society Anniversary Address 1994. *Notes and Records of the Royal Society*, **49**, No 1, p 141.

Mason, L.J., and Woodhouse, N.M.J. (1996). *Integrability, Self-Duality, and Twistor Theory.* LMS Monographs New Series. Clarendon Press, Oxford.

2

Hypercomplex Manifolds and the Space of Framings

Nigel Hitchin

Department of Pure Mathematics and Mathematical Statistics
University of Cambridge, 16 Mill Lane, Cambridge, CB2 1SB

1 Introduction

In general relativity, one commonly formulates Einstein's equations as evolution equations for a metric and second fundamental form on a hypersurface. A rather different approach was adopted in the paper of Ashtekar, Jacobson and Smolin (1988), which in particular reduced the equations for a positive definite self-dual spacetime (a 4-dimensional hyperkähler manifold in other language) to Nahm's equations for a triple of volume-preserving vector fields on a 3-manifold M. The initial aim of this paper is to relax the volume-preserving condition and to place this result in a general framework. Here we regard the triple of vector fields as a trivialization of the tangent bundle of M—a *framing.*

The infinite-dimensional manifold of all framings on a given 3-manifold is an object which perhaps deserves closer study. It is naturally defined, has an action of Diff$M$ on it, and is a principal bundle over the more conventional space of metrics, with group $\mathrm{Map}(M; O(3))$. It has itself a natural Riemannian metric, and also a natural functional f on it. This is the Chern–Simons invariant of an $SO(3)$ connection on TM associated with the framing: we think of a framing as a point of $\Omega^1(M) \otimes \mathbf{R}^3$, identify $\mathbf{R}^3$ with the Lie algebra of $SO(3)$ and take the connection which has this as connection matrix, relative to the trivialization of TM given by the framing. In the context of Floer theory, using the analogous space of connections on a 3-manifold, it is natural to consider the gradient flow of this functional. We find that integrating it is equivalent to integrating Nahm's equations for arbitrary vector fields on a 3-manifold.

In this setting, we introduce the problem of integrating the gradient flow for left-invariant framings on $M = SU(2)$, and this leads to the nonlinear equation for a 3×3 matrix-valued function B:

$$\frac{dB}{ds} = 2(\mathrm{tr}\, B)B - 2BB^T - 2\,\mathrm{adj}\, B.$$

The remainder of the paper solves the equation by cutting a path through a Penrosean landscape: self-dual geometry, twistor spaces, and cohomology classes represented by contour integrals. Without the power of twistor theory and the tentacles which it sends out to so many corners of mathematics, it would be difficult to see how such an equation could be solved explicitly.

Our route first interprets Nahm's equations for vector fields as generating a 4-dimensional hypercomplex manifold, sliced by the level sets of a harmonic function. In this context it becomes relatively easy to see how the volume-preserving condition in Ashtekar *et al.* (1988) yields a hyperkähler metric. The equation for B in this setting now represents an $SU(2)$-invariant hypercomplex manifold. The hyperkähler case then consists of the situation where B is diagonal, in which case our equations reduce to those solved by Halphen (in 1881!) and used in Atiyah and Hitchin (1988) to construct the natural hyperkähler metric on the moduli space of two monopoles.

The next stage is to invoke twistor theory and isomonodromic deformations as in Hitchin (1995) and Maszczyk *et al.* (1994). It turns out that the isomonodromy problem for an invariant hypercomplex manifold involves monodromy contained in the group of upper-triangular matrices in $SL(2,\mathbf{C})$, and this can be described by elements in the cohomology group $H^1(\Sigma, \mathcal{L}^2)$ for a local coefficient system on a 4-times punctured sphere Σ. The contour integral description of these classes means ultimately that we can solve the equations for B using hypergeometric functions.

The concrete outcome of this work is an analytical description of all $SU(2)$-invariant hypercomplex manifolds which parallels in some way the description in Hitchin (1995) of all $SU(2)$-invariant self-dual Einstein manifolds, though we have neither space nor motivation here to discuss global questions of completeness.

2 Framings

Let M be a compact, oriented 3-manifold. It is well-known that the tangent bundle TM is trivial. We denote by $\mathcal{F}$ the space of all trivializations of TM, also known in the literature as parallelizations or *framings*. We adopt the latter terminology, so that a framing, a point $x \in \mathcal{F}$, is a triple

$$(X_1, X_2, X_3)$$

of vector fields on M which are linearly independent at each point. Equivalently we can think of x as the dual basis, a triple $(\theta_1, \theta_2, \theta_3)$ of 1-forms. The space $\mathcal{F}$ of framings has a number of features:

- $\mathcal{F}$ is acted on naturally by $\mathrm{Diff}\, M$.
- Any framing (Y_1, Y_2, Y_3) is related to a fixed framing (X_1, X_2, X_3) by a uniquely defined invertible matrix-valued function B_{ij}. We have $Y_i = \sum_j B_{ji} X_j$, and hence

$$\mathcal{F} \cong \mathrm{Map}(M; GL(3, \mathbf{R}))\,.$$

- A framing determines a Riemannian metric g on M:
$$g = \sum_{i=1}^{3} \theta_i^2$$
so that (X_1, X_2, X_3) is an *orthonormal* basis at each point.
- $\mathcal{F}$ is a principal $\mathrm{Map}(M; O(3))$-bundle over the space of metrics $\mathcal{M}$ on M.
- $\mathcal{F}$ itself has a natural metric: if $\dot{X}_i$ are vector fields on M, then $V = (\dot{X}_1, \dot{X}_2, \dot{X}_3)$ is a tangent vector to $\mathcal{F}$ at the framing (X_1, X_2, X_3) and we define
$$g(V, V) = \int_M \Big(\sum_{i,j} \theta_i(\dot{X}_j)^2\Big)\, \theta_1 \wedge \theta_2 \wedge \theta_3 \,.$$
This is just the $\mathcal{L}^2$ norm of V with respect to the metric on M defined by the framing.

We see then that framings can be used to define all sorts of geometrical entities: metrics, volumes and for our purposes more importantly, *connections.* The standard way to get a connection on TM from a framing is to take the flat connection defined by
$$\nabla_{X_i} X_j = 0$$
but there are other, equally natural, connections. If we regard a framing as a triple of 1-forms, then this is an element of $\Omega^1(M) \otimes \mathbf{R}^3$. Identifying $\mathbf{R}^3$ with the Lie algebra $\mathfrak{so}(3)$ gives the connection
$$\begin{aligned} \nabla_{X_1} X_2 &= X_3 \\ \nabla_{X_1} X_1 &= 0 \\ \nabla_{X_1} X_3 &= -X_2 \end{aligned}$$
and relations obtained by cyclically symmetrizing. This connection preserves the metric g, but in general has torsion. It has the property that an integral curve of any linear combination of X_1, X_2, X_3 is a geodesic. In fact, if $M = SU(2)$ and (X_1, X_2, X_3) is the standard left-invariant trivialization, this is the flat connection giving the right-invariant trivialization. The connection form, relative to the trivialization (X_1, X_2, X_3), is
$$A = \begin{pmatrix} 0 & \theta_3 & -\theta_2 \\ -\theta_3 & 0 & \theta_1 \\ \theta_2 & -\theta_1 & 0 \end{pmatrix}.$$
From this formula we can calculate the basic invariant of a connection over a 3-manifold: the *Chern–Simons invariant*
$$\int_M \mathrm{tr}(A \wedge dA + \frac{2}{3} A \wedge A \wedge A)\,.$$

We obtain (a multiple of) the function

$$f = \frac{1}{2}\int_M \theta_1 \wedge d\theta_1 + \theta_2 \wedge d\theta_2 + \theta_3 \wedge d\theta_3 - 2\theta_1 \wedge \theta_2 \wedge \theta_3 \,. \tag{2.1}$$

Consider now the functional $f : \mathcal{F} \to \mathbf{R}$, and its gradient flow. We have (using $*\theta_1 = \theta_2 \wedge \theta_3$ etc.)

$$df(\dot\theta_1, \dot\theta_2, \dot\theta_3) = \int_M \sum_i (\dot\theta_i \wedge d\theta_i - \dot\theta_i \wedge *\theta_i)\,.$$

Now considering framings as triples of 1-forms, we have for $V = (\dot\theta_1, \dot\theta_2, \dot\theta_3), W = (\dot\varphi_1, \dot\varphi_2, \dot\varphi_3)$

$$g(V, W) = \int_M \sum_{i,j} \dot\theta_j(X_i)\dot\varphi_j(X_i)\theta_1 \wedge \theta_2 \wedge \theta_3 = \int_M \sum_j \dot\theta_j \wedge *\dot\varphi_j \,.$$

Thus the gradient flow of the functional f is defined by the differential equation

$$\frac{d\theta_j}{dt} = *d\theta_j - \theta_j \tag{2.2}$$

or, reverting to vector fields,

$$\begin{aligned} \frac{dX_1}{dt} &= X_1 - [X_2, X_3] \\ \frac{dX_2}{dt} &= X_2 - [X_3, X_1] \\ \frac{dX_3}{dt} &= X_3 - [X_1, X_2]\,. \end{aligned}$$

Remark Note that the critical points of f occur when the framing defines an $\mathfrak{so}(3)$ subalgebra of the algebra of vector fields. From Milnor (1984), this integrates to an action of $SU(2)$, and since the vector fields are linearly independent it has finite stabilizers and so M is a quotient of $SU(2)$ with a left-invariant framing. Note also that in general the flow may be incomplete—in particular the vector fields may become linearly dependent in finite time.

These differential equations can be put in a more familiar form by setting $s = e^t$ and $Y_i = -e^{-t}X_i$. We then obtain

$$\begin{aligned} \frac{dY_1}{ds} &= [Y_2, Y_3] \\ \frac{dY_2}{ds} &= [Y_3, Y_1] \\ \frac{dY_3}{ds} &= [Y_1, Y_2]\,. \end{aligned} \tag{2.3}$$

These are *Nahm's equations*, dimensional reductions of the self-dual Yang–Mills equations, but in our case the gauge group is not a finite-dimensional Lie group, but is instead the group of diffeomorphisms $\mathrm{Diff}\, M$ of a 3-manifold.

3 $SU(2)$-invariance

The specific problem which concerns us here is that of solving the equations (2.3) in the case that M is the 3-sphere $S^3 = SU(2)$ and the triple (Y_1, Y_2, Y_3) is left-invariant. There are actually two cases here. Thinking of the framing as an element of $\Omega^1 \otimes \mathbf{R}^3$, we can have either a trivial action of $SU(2)$ on the $\mathbf{R}^3$ factor, or the adjoint action. If E_1, E_2, E_3 are vector fields on M defining the action, the two cases give either

$$\mathcal{L}_{E_i} Y_j = 0 \qquad \text{or} \qquad \mathcal{L}_{E_1} Y_2 = 2Y_3 \quad \text{etc.}$$

The first case means that the Y_i are themselves left-invariant vector fields, elements of $\mathfrak{su}(2)$, and we obtain the original Nahm's equations for 2×2 matrices. These actually reduce to Euler's top, and are solvable by elliptic functions. It is the other invariance which leads to a more challenging equation.

First we write

$$Y_i = \sum_j B_{ji} E_j$$

for matrix-valued functions B_{ji}. Now using the summation convention,

$$\begin{aligned}
\mathcal{L}_{E_i} Y_j &= \mathcal{L}_{E_i}(B_{kj} E_k) \\
&= (E_i \cdot B_{kj}) E_k + B_{kj} \mathcal{L}_{E_i} E_k \\
&= (E_i \cdot B_{kj}) E_k + 2 B_{kj} \epsilon_{ikl} E_l \,.
\end{aligned}$$

But $\mathcal{L}_{E_i} Y_j = 2\epsilon_{ijl} Y_l = 2\epsilon_{ijl} B_{kl} E_k$ and hence

$$E_i \cdot B_{kj} = 2\epsilon_{ijl} B_{kl} - 2\epsilon_{ilk} B_{lj} \,. \tag{3.1}$$

This equation tells us the representation space in $C^\infty(SU(2))$ that the nine functions B_{ij} lie in.

Nahm's equations (2.3) are

$$\frac{dY_i}{ds} = \frac{1}{2} \epsilon_{ijk} [Y_j, Y_k]$$

so

$$\begin{aligned}
\frac{dB_{ki}}{ds} E_k &= \frac{1}{2} \epsilon_{ijl} [Y_j, Y_l] \\
&= \frac{1}{2} \epsilon_{ijl} \mathcal{L}_{B_{mj} E_m} Y_l \\
&= \frac{1}{2} \epsilon_{ijl} (B_{mj} \mathcal{L}_{E_m} Y_l - (Y_l \cdot B_{mj}) E_m) \\
&= \frac{1}{2} \epsilon_{ijl} (2 B_{mj} \epsilon_{mln} B_{kn} E_k - B_{pl} (E_p \cdot B_{kj}) E_k)
\end{aligned}$$

and using (3.1), this is

$$\epsilon_{ijl}\epsilon_{mln}B_{mj}B_{kn}E_k - \epsilon_{ijl}\epsilon_{pjq}B_{pl}B_{kq}E_k + \epsilon_{ijl}\epsilon_{pqk}B_{pl}B_{qj}E_k\,.$$

Expanding this out gives us the following invariantly-defined equation for the matrix B:

$$\frac{dB}{ds} = 2(\operatorname{tr} B)B - 2BB^T - 2\operatorname{adj} B \tag{3.2}$$

where $\operatorname{adj} A$ is the usual matrix of cofactors satisfying $B\operatorname{adj} B = (\det B)I$.

A solution of the equation (3.2) represents an integral curve of a vector field on the space of 3×3 matrices. Note that if B is symmetric so is $2(\operatorname{tr} B)B - 2BB^T - 2\operatorname{adj} B$ and so the flow is tangential to the space of symmetric matrices. Moreover, if B is diagonal, so is this expression. Thus for a symmetric matrix we can reduce the equations to diagonal form: if B has eigenvalues $u_1/2, u_2/2, u_3/2$ then (3.2) becomes

$$\begin{aligned} u_1' + u_2' &= u_1u_2 \\ u_2' + u_3' &= u_2u_3 \\ u_3' + u_1' &= u_3u_1\,. \end{aligned}$$

Curiously, these equations were solved long ago in Halphen's 1881 paper (Halphen 1881), using complete elliptic integrals. We shall attack the general case here, but what we need to do is to adopt a more geometrical approach.

4 Hypercomplex manifolds

The standard approach to Nahm's equations is to rewrite them in the Lax form

$$\left[\frac{d}{ds} - iY_1, Y_2 + iY_3\right] = 0$$

and the two equations obtained by cyclic permutation of the indices. In our case, where we replace the finite-dimensional Lie group by $\operatorname{Diff} M$, we see that

$$-\frac{\partial}{\partial s} + iY_1 \qquad \text{and} \qquad Y_2 + iY_3$$

are two commuting complex vector fields on $U \times M$ for some interval $U \subseteq \mathbf{R}$. They span the space of vector fields of type $(1,0)$ for an integrable complex structure I. Similarly

$$(-\frac{\partial}{\partial s} + iY_2, Y_3 + iY_1) \qquad \text{and} \qquad (-\frac{\partial}{\partial s} + iY_3, Y_1 + iY_2)$$

define integrable complex structures J and K and I, J, K satisfy the algebraic identities of the quaternions i, j, k. A manifold with this structure is a *hypercomplex manifold.* Geometrically, then, the gradient flow of the Chern–Simons invariant f on the space of framings $\mathcal{F}$ generates a hypercomplex manifold.

It is well-known that a hypercomplex manifold, although not being naturally a Riemannian manifold, has a natural connection—the Obata connection (Obata 1956)—which is torsion-free and preserves I, J, K. We shall derive the basic features of the hypercomplex manifold, including properties of this connection, from the evolution equation for a frame.

First let η_1, η_2, η_3 be the basis of 1-forms dual to the vector fields Y_1, Y_2, Y_3. Then the complex structures acting on 1-forms give

$$\eta_1 = Ids, \quad \eta_2 = Jds, \quad \eta_3 = Kds\,.$$

The hypercomplex structure defines a conformal structure (the structure group of the tangent bundle is reduced to the group of quaternions $\mathbf{H}^* \subset SO(4) \cdot \mathbf{R}^*$) which in explicit terms is represented by the metric

$$ds^2 + \eta_1^2 + \eta_2^2 + \eta_3^2\,.$$

In this formalism the evolution of the framing (X_1, X_2, X_3) as t varies has a natural interpretation. Firstly t is a function on the product manifold $U \times M$. The conformal structure defines a normal distribution to the level sets of t and the dual of dt on the integral curves defines a normal vector field N. Since $I\partial/\partial s = Y_1$ and $s = e^t$, we have

$$I\frac{\partial}{\partial t} = e^t Y_1 = -X_1$$

thus we can say that the framing (X_1, X_2, X_3) is simply $-(IN, JN, KN)$. It is the hypercomplex structure which turns the normal vector field into an evolving frame on the slices $t = const.$

It will be convenient to reinterpret Nahm's equations (2.3) in a dual formalism:

Lemma 4.1 *The triple of vector fields (Y_1, Y_2, Y_3) satisfies Nahm's equations if and only if the 2-forms $d\eta_1, d\eta_2, d\eta_3$ are anti-self-dual with respect to the metric $ds^2 + \eta_1^2 + \eta_2^2 + \eta_3^2$.*

Proof We write d_0 for the exterior derivative on the 3-manifold M^3. Then

$$d\eta_i = ds \wedge \frac{d\eta_i}{ds} + d_0\eta_i\,.$$

But since $\eta_i(Y_j)$ is a constant,

$$d_0\eta_i(Y_2, Y_3) = \eta_i([Y_2, Y_3]) \qquad \text{and} \qquad \frac{d\eta_i}{ds}(Y_j) = -\eta_i(\frac{dY_j}{ds})\,.$$

Thus if (Y_1, Y_2, Y_3) satisfies Nahm's equations,

$$\frac{d\eta_i}{ds}(Y_1) = -\eta_i(\frac{dY_1}{ds}) = -\eta_i([Y_2, Y_3]) = -d_0\eta_i(Y_2, Y_3)$$

and this, with similar relations, is the anti-self-duality of $d\eta_i$. The converse is established in the same way. Note that the gradient flow equation (2.2), which uses the frame dual to (X_1, X_2, X_3), is essentially this result. □

We need now to discover more about the Obata connection. Consider

$$\nabla(ds) = ds \otimes \alpha_0 + \eta_1 \otimes \alpha_1 + \eta_2 \otimes \alpha_2 + \eta_3 \otimes \alpha_3 \tag{4.1}$$

where α_i are connection 1-forms. Since the connection commutes with I we have

$$\nabla(\eta_1) = \nabla(Ids) = \eta_1 \otimes \alpha_0 - ds \otimes \alpha_1 + \eta_3 \otimes \alpha_2 - \eta_2 \otimes \alpha_3 \tag{4.2}$$

and similar terms for η_2 and η_3. The 1-forms α_i satisfy a number of relations. Firstly, since the Obata connection is torsion free,

$$0 = d^2 s = \alpha_0 \wedge ds + \alpha_1 \wedge \eta_1 + \alpha_2 \wedge \eta_2 + \alpha_3 \wedge \eta_3$$

and so, setting $Y_0 = \partial/\partial s$ we have

$$\alpha_i(Y_j) = \alpha_j(Y_i)\,. \tag{4.3}$$

Similarly, from (4.2) we have

$$d\eta_1 = \alpha_0 \wedge \eta_1 - \alpha_1 \wedge ds + \alpha_2 \wedge \eta_3 - \alpha_3 \wedge \eta_2\,. \tag{4.4}$$

The condition that $d\eta_i$ is anti-self-dual now gives the extra relation

$$\alpha_0(Y_0) + \alpha_i(Y_i) = 0\,. \tag{4.5}$$

We can use these facts to highlight some features of the geometry of the hypercomplex manifold. First note that the connection defines a divergence operator on vector fields and 1-forms. For a vector field Y, we have $\nabla Y \in C^\infty(T \otimes T^*)$ and so we define $\operatorname{div} Y \in C^\infty$ by contracting this. With a torsion-free connection and a top-degree form ω, the divergence of a vector field satisfies:

$$\mathcal{L}_Y \omega = d(\iota(Y)\omega) = \nabla_Y \omega - \operatorname{div}(Y)\omega\,. \tag{4.6}$$

For a 1-form, the conformal structure defines an isomorphism $C : T^* \cong T \otimes L$ where L is the line bundle of half-densities, so we can define $\operatorname{div} \alpha \in C^\infty(L)$ by taking $\nabla\alpha \in C^\infty(T^* \otimes T^*) \cong C^\infty(T^* \otimes T \otimes L)$ and contracting. We define the Laplacian of a function f by $\Delta f = \operatorname{div}(df) \in C^\infty(L)$

Lemma 4.2 *$\operatorname{div} \eta_i = 0$ and $\Delta s = 0$ and $\operatorname{div} Y_i = -2\alpha_0(Y_i)$.*

Proof Since the conformal structure is represented by the metric

$$ds^2 + \eta_1^2 + \eta_2^2 + \eta_3^2$$

then if L is trivialized by $(ds \wedge \eta_1 \wedge \eta_2 \wedge \eta_3)^{1/2}$, using $C(\eta_i) = Y_i$ and $C(ds) = Y_0$ we have from (4.2),

$$\operatorname{div} \eta_1 = \alpha_0(Y_1) - \alpha_1(Y_0) + \alpha_2(Y_3) - \alpha_3(Y_2)$$

and this vanishes from (4.3). Similarly from (4.1),

$$\Delta s = \alpha_0(Y_0) + \alpha_1(Y_1) + \alpha_2(Y_2) + \alpha_3(Y_3)$$

which vanishes from (4.5). From (4.1) and (4.2) and similar expressions we evaluate

$$\nabla Y_1 = -Y_0 \otimes \alpha_1 - Y_1 \otimes \alpha_0 - Y_2 \otimes \alpha_3 + Y_3 \otimes \alpha_2$$

and then similarly

$$\operatorname{div} Y_1 = -\alpha_1(Y_0) - \alpha_0(Y_1) - \alpha_3(Y_2) + \alpha_2(Y_3) = -2\alpha_1(Y_0)$$

from (4.3). □

Remark Note that evolution equations where time is harmonic have appeared in many parts of the literature, for example in Hoppe (1996), or in the work of Krichever and Novikov (1990) on Riemann surfaces.

We now reverse this process so as to derive the evolution equations from the harmonic function:

Theorem 4.3 *Let X be a hypercomplex 4-manifold and s a harmonic function on X. Defining the vector field $\partial/\partial s$ normal to the level sets of s, decompose $X = U \times M$ where U is an interval and M a 3-manifold. Then the frame (Y_1, Y_2, Y_3) dual to Ids, Jds, Kds on M evolves according to Nahm's equations.*

Proof Setting $\eta_1 = Ids$, etc., the conformal structure determined by the hypercomplex structure is represented by the metric

$$ds^2 + \eta_1^2 + \eta_2^2 + \eta_3^2 \,.$$

Now consider $d\eta_1 = d(Ids)$. Since I is integrable, this is a $(1,1)$-form and is anti-self-dual if and only if it is orthogonal to the 2-form $\omega_1 = ds \wedge \eta_1 + \eta_2 \wedge \eta_3$ corresponding to the hermitian metric and the complex structure I:

$$d(Ids) \wedge \omega_1 = 0 \,.$$

But this is the statement that s is harmonic. This follows since the connection is torsion-free and I is covariant constant, so that $d(Ids)$ factors through $\nabla(ds)$. The contraction is then linear algebra and just as in the Kähler case. Thus if s is harmonic, $d\eta_i$ is anti-self-dual, and from (4.1) this is equivalent to Nahm's equations. □

A special case of this is the following, which is the result of Ashtekar, Jacobson and Smolin (Ashtekar *et al.* 1988) which was the initial stimulus for this work:

Theorem 4.4 *Let X be a hypercomplex manifold generated by a framing (Y_1, Y_2, Y_3) on a 3-manifold M satisfying Nahm's equations. Then X is hyperkähler if and only if the vector fields Y_i are volume-preserving.*

Proof A hypercomplex manifold is hyperkähler if the Obata connection is the Levi-Civita connection of a metric in the conformal class, equivalently if there exists a covariant constant section of $\Lambda^4 T^*$. From (4.1) and (4.2), the covariant derivative of the 4-form $ds \wedge \eta_1 \wedge \eta_2 \wedge \eta_3$ is given by

$$\nabla(ds \wedge \eta_1 \wedge \eta_2 \wedge \eta_3) = 4(ds \wedge \eta_1 \wedge \eta_2 \wedge \eta_3) \otimes \alpha_0 \,. \tag{4.7}$$

Thus $4\alpha_0$ is the connection form for the induced connection on $\Lambda^4 T^*$. The hypercomplex structure is therefore hyperkähler if and only if there is a function u such that $\alpha_0 = du$. The covariant constant volume form is then $e^{-4u} ds \wedge \eta_1 \wedge \eta_2 \wedge \eta_3$.

Suppose first that X is hyperkähler, so that $\alpha_0 = du$, then define the 3-form $\Omega = e^{-2u}\eta_1 \wedge \eta_2 \wedge \eta_3$ on M. We first show that it is independent of s.

$$\frac{d\Omega}{ds} = -2\frac{du}{ds}\Omega + e^{-2u}\left(\frac{d\eta_1}{ds} \wedge \eta_2 \wedge \eta_3 + \eta_1 \wedge \frac{d\eta_2}{ds} \wedge \eta_3 + \eta_1 \wedge \eta_2 \wedge \frac{d\eta_3}{ds}\right).$$

But from (4.4), we have

$$\frac{d\eta_1}{ds} = \alpha_0(Y_0)\eta_1 + \alpha_1(Y_i)\eta_i + \alpha_2(Y_0)\eta_3 - \alpha_3(Y_0)\eta_2$$

so

$$\frac{d\eta_1}{ds} \wedge \eta_2 \wedge \eta_3 = (\alpha_0(Y_0) + \alpha_1(Y_1))\eta_1 \wedge \eta_2 \wedge \eta_3 \,.$$

Adding the similar terms, we have

$$\frac{d\Omega}{ds} = -2\frac{du}{ds}\Omega + e^{-2u}(3\alpha_0(Y_0) + \alpha_i(Y_i))\eta_1 \wedge \eta_2 \wedge \eta_3 \,.$$

But from (4.5) this gives

$$\frac{d\Omega}{ds} = \left(-2\frac{du}{ds} + 2\alpha_0(Y_0)\right)\Omega$$

and since $\alpha_0 = du$ this yields

$$\frac{d\Omega}{ds} = 0 \,.$$

Using the metric

$$g = e^{-2u}(ds^2 + \eta_1^2 + \eta_2^2 + \eta_3^2) \tag{4.8}$$

which is preserved by the connection, the vector field $e^{2u}Y_i$ (for $i = 1, 2, 3$) is dual to the 1-form η_i. From (4.2), $\operatorname{div}\eta_i = 0$, so

$$\operatorname{div}(e^{2u}Y_i) = 0 \,.$$

On a Riemannian manifold, a divergence-free vector field preserves the volume form, so

$$d(\iota(e^{2u}Y_i)e^{-4u}ds \wedge \eta_1 \wedge \eta_2 \wedge \eta_3) = 0\,.$$

But since, as we have seen, $\Omega = e^{-2u}\eta_1 \wedge \eta_2 \wedge \eta_3$ is independent of s, and Y_i is tangential to $s = const.$, so

$$\mathcal{L}_{Y_i}\Omega = d(\iota(Y_i)\Omega) = 0$$

and the vector fields Y_i preserve the volume form Ω on M.

Conversely, suppose that $\Omega = e^{-2u}\eta_1 \wedge \eta_2 \wedge \eta_3$ is independent of s, then from the calculation above,

$$\alpha_0(Y_0) = \frac{du}{ds}\,. \tag{4.9}$$

If Y_i preserves Ω, and Ω is independent of s then again as above $\mathcal{L}_{Y_i}(e^{-2u}ds \wedge \eta_1 \wedge \eta_2 \wedge \eta_3) = 0$. Since $\operatorname{div} Y_i = -2\alpha_0(Y_i)$, from (4.6) this means that

$$\nabla_{Y_i}(e^{-2u}ds\wedge\eta_1\wedge\eta_2\wedge\eta_3) = e^{-2u}(-2du(Y_i)-2\alpha_0(Y_i)+4\alpha_0(Y_i))ds\wedge\eta_1\wedge\eta_2\wedge\eta_3 = 0$$

and hence

$$\alpha_0(Y_i) = du(Y_i)\,.$$

Hence from (4.9), $\alpha_0 = du$ as required. □

This discussion has seemingly taken us far afield from the differential equation (3.2)

$$\frac{dB}{ds} = 2(\operatorname{tr} B)B - 2BB^T - 2\operatorname{adj} B$$

but Theorem 4.4 has the following consequence:

Proposition 4.5 *The matrix valued function B satisfying (3.2) generates an $SU(2)$-invariant hyperkähler manifold if and only if it is symmetric.*

Proof From Theorem 4.4 we need to show that the vector fields $Y_i = \sum_j B_{ji}E_j$ preserve a volume form on $SU(2)$. By left invariance it is enough to show that the invariant covolume form $E_1 \wedge E_2 \wedge E_3$ is preserved. Now the invariance properties of Y_i are expressed by $\mathcal{L}_{E_1}Y_2 = 2Y_3$, etc., so

$$[Y_1, E_1] = 0 \qquad [Y_1, E_2] = -2Y_3 \qquad [Y_1, E_3] = 2Y_2$$

and so

$$\begin{aligned}\mathcal{L}_{Y_1}(E_1 \wedge E_2 \wedge E_3) &= -2E_1 \wedge Y_3 \wedge E_3 + 2E_1 \wedge E_2 \wedge 2Y_2\\ &= 2(B_{32} - B_{23})E_1 \wedge E_2 \wedge E_3\,.\end{aligned}$$

Invariance for Y_1 thus implies $B_{23} = B_{32}$ and considering Y_2, Y_3 we obtain that B is symmetric. □

As remarked above, the symmetric case reduces to Halphen's equations. In the setting of $SU(2)$-invariant hyperkähler metrics the equations were derived in Gibbons and Pope (1979) and solved in the context of the natural metric on the moduli space of two $SU(2)$ monopoles in Atiyah and Hitchin (1988). We are concerned now with the general case, which requires yet more geometry to handle—this time the use of Penrose's twistor space in the hypercomplex setting.

5 Twistor spaces and isomonodromic deformations

Following the usual treatment of Nahm's equations (e.g. in Hitchin 1983), it is natural to introduce an indeterminate z and to write

$$\begin{aligned} Y &= (Y_1 + iY_2) + 2izY_3 + z^2(Y_1 - iY_2) \\ Y_+ &= iY_3 + z(Y_1 - iY_2)\,. \end{aligned}$$

Then, instead of three Lax equations, we obtain a single equation

$$\left[-\frac{d}{ds} + Y_+, Y\right] = 0\,.$$

This formalism leads immediately to a geometrical structure on the manifold $Z = X \times \mathbf{C}P^1$ where $X = U \times M$ is our hypercomplex manifold. We let ζ be an affine parameter on the projective line and set $z = \bar{\zeta}$, then as a consequence of the Lax equation we have three *commuting* complex vector fields W_i on Z:

$$W_1 = -\frac{\partial}{\partial s} + Y_+ \qquad W_2 = Y \qquad W_3 = \frac{\partial}{\partial \zeta}\,.$$

By the Newlander–Nirenberg theorem, these span the space of $(1,0)$ tangent vectors for an integrable complex structure on Z. This complex 3-manifold is the twistor space of the hypercomplex manifold X. The projection $p : Z \to \mathbf{C}P^1$ is easily seen to be holomorphic. As is well-known (Penrose 1976), the holomorphic geometry of Z encodes the differential geometry of the hypercomplex manifold.

We have seen that the differential equation (3.2) for B represents in geometrical terms a hypercomplex manifold with an $SU(2)$ action which acts non-trivially on the complex structures I, J, K. In the language of twistor theory we have a twistor space Z with a (local) holomorphic $SL(2, \mathbf{C})$ action which commutes with the projection p and acts as the group of Möbius transformations on $\mathbf{C}P^1$. This is part of the more general situation of twistor spaces with symmetry treated in Hitchin (1995), and relates to the study of isomonodromic deformations and Painlevé's sixth equation. We shall solve the equation for B by using this method.

The construction uses the linearized action of $SL(2, \mathbf{C})$ expressed as a homomorphism of holomorphic vector bundles:

$$\alpha : Z \times \mathfrak{sl}(2) \to TZ\,.$$

Its inverse $\alpha^{-1} \in \Omega^{1,0}(Z, \mathfrak{sl}(2))$ is a meromorphic 1-form, which when restricted to a section of $p : Z \to \mathbf{C}P^1$ acquires four simple poles. The matrix-valued 1-form α^{-1} defines a *flat* meromorphic connection on the trivial bundle over Z, and the holonomy on each section is induced from that on Z, and so as the sections vary we obtain an *isomonodromic deformation.* The resulting nonlinear ordinary differential equation will in our case be a substitute for the nonlinear ordinary differential equation (3.2).

Proposition 5.1 *The meromorphic 1-form α^{-1} on the section $(s, x) = const.$ is*

$$-\frac{\sum B_{jk} q_k E_j}{\sum B_{jk} q_k q_j}.$$

Proof To derive the meromorphic connection, define complex vector fields on Z:

$$\begin{aligned} Z_1 &= Y_+ - \bar{Y}_+ \\ Z_2 &= Y \\ Z_3 &= \bar{Y}. \end{aligned}$$

Note that since B and E_i are real,

$$Z_3 = (Y_1 - iY_2) - 2i\zeta Y_3 + \zeta^2 (Y_1 + iY_2) = \sum_{jk} B_{jk} q_k E_j \tag{5.1}$$

where $q(\zeta)$ is the null vector

$$q = (1 + \zeta^2, -i(1 - \zeta^2), -2i\zeta). \tag{5.2}$$

The three vector fields Z_1, Z_2, Z_3 have no $\partial/\partial s$ component and so there is an (invertible) matrix C such that

$$Z_i = \sum_j C_{ji} E_j.$$

The $SU(2)$ action on $M = SU(2)$ is given by the inverse of this. The action on the twistor space $\mathbf{R} \times M \times \mathbf{C}P^1$ incorporates also the Möbius action on the projective line. This is the holomorphic action

$$E_i \mapsto -q_i \frac{\partial}{\partial \zeta}$$

where $q = (q_1, q_2, q_3)$ in (5.2). The total action is holomorphic, so α is defined by the $(1, 0)$ part:

$$\alpha(E_i) = \sum_j (C^{-1})_{ji} Z_j^{(1,0)} - q_i \frac{\partial}{\partial \zeta}.$$

From the definition of the complex structure $Z_3^{(1,0)} = 0$ so relative to the basis E_1, E_2, E_3 of $\mathfrak{sl}(2)$ and $Z_1^{(1,0)}, Z_2^{(1,0)}, \partial/\partial\zeta$ of TZ, α is represented by the matrix M where the first two rows of M are the first two rows of C^{-1} and the last row is the vector $-q$.

The holomorphic sections of the twistor space are given by fixing $x \in M$ and $s \in U$, so the connection form on a section is M^{-1} applied to $\partial/\partial\zeta$, i.e. the third column of M^{-1}. Now since the first two rows of M are those of C^{-1}, we have

$$MC = \begin{pmatrix} 1 & 0 & 0 \\ 0 & 1 & 0 \\ c_1 & c_2 & c_3 \end{pmatrix}$$

where $c_i = -\sum_j q_j C_{ji}$. Thus

$$M^{-1} = c_3^{-1} C \begin{pmatrix} 1 & 0 & 0 \\ 0 & 1 & 0 \\ -c_1 & -c_2 & 1 \end{pmatrix}.$$

Now from (5.1) $C_{j3} = \sum_k B_{jk} q_k$, so $c_3 = -\sum_j q_j C_{j3} = -\sum_{jk} B_{jk} q_j q_k$ and $(M^{-1})_{j3} = c_3^{-1} C_{j3}$ which gives the result. □

The problem has now been reduced to solving the isomonodromic deformation problem for the connection

$$\frac{d}{d\zeta} - \frac{\sum B_{jk} q_k E_j}{\sum B_{jk} q_k q_j}. \tag{5.3}$$

This equation has four regular singular points at the four roots of the quartic

$$\sum B_{jk} q_k(\zeta) q_j(\zeta) = 0.$$

The general problem of this type is only solvable by using Painlevé transcendants, but as we shall see next, the monodromy group for our particular example is simpler and leads to an explicit solution. We simplify the calculations by using 3-dimensional vector notation instead of Lie brackets for $SL(2, \mathbf{C})$. If we write $b_i = \sum B_{ij} q_j$, then a covariant constant section of the adjoint bundle is a vector-valued function $\mathbf{v}$ which satisfies

$$\frac{d\mathbf{v}}{d\zeta} - 2\frac{\mathbf{b} \times \mathbf{v}}{\mathbf{b} \cdot \mathbf{q}} = 0.$$

The quadratic $\mathbf{q}(\zeta)$ is a nonvanishing holomorphic section of $\mathbf{C}^3 \otimes \mathcal{O}(2)$ and so defines a holomorphic subbundle isomorphic to $\mathcal{O}(-2)$ in the trivial bundle $\mathbf{C}P^1 \times \mathbf{C}^3$.

Proposition 5.2 *The line bundle generated by* $\mathbf{q}(\zeta)$ *is preserved by the connection (5.3).*

Proof We have

$$\frac{d\mathbf{q}}{d\zeta} = (2\zeta, 2i\zeta, -2i) = \mathbf{p} \times \mathbf{q} \tag{5.4}$$

where $\mathbf{p} = (1, i, 0)$, and is the unique null vector satisfying (5.4). Note that $\mathbf{p} \cdot \mathbf{q} = 2$. Hence

$$\frac{d\mathbf{q}}{d\zeta} - 2\frac{\mathbf{b} \times \mathbf{q}}{\mathbf{b} \cdot \mathbf{q}} = \left(\mathbf{p} - 2\frac{\mathbf{b}}{\mathbf{b} \cdot \mathbf{q}}\right) \times \mathbf{q}\,.$$

But

$$(\mathbf{b} \times \mathbf{q}) \times \mathbf{q} = (\mathbf{b} \cdot \mathbf{q})\mathbf{q} - (\mathbf{q} \cdot \mathbf{q})\mathbf{b} = (\mathbf{b} \cdot \mathbf{q})\mathbf{q}$$

since $\mathbf{q}$ is a null vector, so

$$\left(\frac{d\mathbf{q}}{d\zeta} - 2\frac{\mathbf{b} \times \mathbf{q}}{\mathbf{b} \cdot \mathbf{q}}\right) \times \mathbf{q} = (\mathbf{p} \cdot \mathbf{q} - 2)\mathbf{q} = 0\,.$$

Thus the holonomy preserves the line bundle. We know therefore that there exists a 1-form θ such that

$$\frac{d\mathbf{q}}{d\zeta} - 2\frac{\mathbf{b} \times \mathbf{q}}{\mathbf{b} \cdot \mathbf{q}} = \theta \mathbf{q}\,.$$

We can find θ by taking the dot product with $\mathbf{p}$, using

$$\mathbf{p} \cdot \mathbf{q} = 2 \qquad \text{and} \qquad \frac{d\mathbf{q}}{d\zeta} \cdot \mathbf{p} = (\mathbf{p} \times \mathbf{q}) \cdot \mathbf{p} = 0\,.$$

We obtain

$$\theta = -\frac{(\mathbf{b} \times \mathbf{q}) \cdot \mathbf{p}}{\mathbf{b} \cdot \mathbf{q}}\,. \tag{5.5}$$

The holonomy preserves a rank 1 null (and hence nilpotent) subbundle in a trivial $\mathfrak{sl}(2)$ bundle. As a representation in $SL(2, \mathbf{C})$ this means that the holonomy is upper triangular—in a fixed Borel subgroup. Near the poles, the holonomy is determined up to conjugation by the residues of the form α^{-1} (at least in the general case where the eigenvalues don't differ by an integer). Since the line bundle is always preserved, and has connection form given by θ, the eigenvalues of the local holonomy are $\exp(\pm 2\pi i \lambda)$ where λ is the residue of θ at the singularity. To find this local holonomy, it is convenient to write $B = A + S$ where A is skew-symmetric and S symmetric. Thus

$$\mathbf{b} = \mathbf{c} + \mathbf{a} \times \mathbf{q}$$

for some vector $\mathbf{a}$ where $c_i = \sum S_{ij} q_j$. It follows that $\mathbf{b} \cdot \mathbf{q} = \mathbf{c} \cdot \mathbf{q}$. We can then write θ from (5.5) as

$$\theta = -\frac{(\mathbf{c} \times \mathbf{q}) \cdot \mathbf{p} + ((\mathbf{a} \times \mathbf{q}) \times \mathbf{q}) \cdot \mathbf{p}}{\mathbf{c} \cdot \mathbf{q}} = \frac{[\mathbf{c}, \mathbf{p}, \mathbf{q}]}{\mathbf{c} \cdot \mathbf{q}} - 2\frac{\mathbf{a} \cdot \mathbf{q}}{\mathbf{c} \cdot \mathbf{q}}\,.$$

The singularities are simple poles and occur where $\mathbf{c}\cdot\mathbf{q} = \sum S_{ij}q_iq_j = 0$ so since

$$(\mathbf{c}\cdot\mathbf{q})' = 2\mathbf{c}\cdot\mathbf{q}' = 2\mathbf{c}\cdot(\mathbf{p}\times\mathbf{q})$$

the residue is

$$\lambda = \frac{1}{2} - \frac{\mathbf{a}\cdot\mathbf{q}}{[\mathbf{c},\mathbf{p},\mathbf{q}]}\,. \tag{5.6}$$

□

One consequence of this is the following:

Proposition 5.3 *The hypercomplex manifold defined by the evolution equations is hyperkähler if and only if the residues of θ are all equal to $1/2$.*

Proof From Proposition 4.5, the structure is hyperkähler if and only if B is symmetric, equivalently $\mathbf{a} = 0$. So first suppose $\mathbf{a} = 0$, then from (5.6) the residues are all $1/2$. If conversely they are all $1/2$, then $\mathbf{a}\cdot\mathbf{q}(\zeta) = 0$ for each pole $\zeta = \zeta_i$. But

$$\det\begin{pmatrix} 1+\zeta_1^2 & -i+i\zeta_1^2 & -2i\zeta_1 \\ 1+\zeta_2^2 & -i+i\zeta_2^2 & -2i\zeta_2 \\ 1+\zeta_3^2 & -i+i\zeta_3^2 & -2i\zeta_3 \end{pmatrix} = -4(\zeta_1-\zeta_2)(\zeta_2-\zeta_3)(\zeta_3-\zeta_1)$$

so if the ζ_i are distinct then $\mathbf{a} = 0$. Note that since we are describing a meromorphic connection form on a line bundle of degree -2 the sum of the residues is always 2, so any three determine the fourth. □

With this result we can see how solving Halphen's equations requires elliptic integrals. Since each residue is $1/2$, the local holonomy is ± 1. This means that if we take the double covering of $\mathbf{C}P^1$ branched over the four singular points and pull back the connection, the local holonomy in the adjoint bundle is trivial, and so the connection has holonomy determined by the abelian fundamental group of the elliptic curve, and can be evaluated explicitly. In Hitchin (1995) this procedure is carried out in a more general context. The calculations there are not far removed from what we shall do next, but it turns out that we need *hypergeometric functions* rather than elliptic functions to solve the isomonodromic problem where B is not necessarily symmetric.

6 Holonomy and hypergeometric functions

Removing four points from the 2-sphere $\mathbf{C}P^1$ leaves a noncompact surface Σ which retracts to a wedge of three circles. Its fundamental group is free on the three generating loops, and so the holonomy of a connection of the form we are considering consists of taking three matrices

$$\begin{pmatrix} u_i & v_i \\ 0 & u_i^{-1} \end{pmatrix} \qquad (i = 1,2,3)\,.$$

The diagonal entry $u_i = \exp(2\pi i \lambda_i)$ where λ_i is the residue calculated in (5.6).

Isomonodromic deformations consist of variations of the connection as a function of the four singular points which fix the holonomy up to conjugation. Now suppose u_1, u_2, u_3 are fixed—this can be read off as an algebraic condition on the connection by calculating the residues. By conjugation we can, in the generic case, make $v_1 = 0$ and $v_2 = 1$. There remains then a single parameter v_3 to determine the holonomy.

There is a more invariant way of seeing this parameter. Fixing the u_i fixes a flat line bundle L on the punctured sphere, associated to the homomorphism

$$\begin{pmatrix} u & v \\ 0 & u^{-1} \end{pmatrix} \mapsto u$$

from the upper-triangular group to $\mathbf{C}^*$. The sheaf of locally flat sections of L defines a local coefficient system $\mathcal{L}$, and the off-diagonal terms in the representation define invariantly an element of the cohomology group $H^1(\Sigma, \mathcal{L}^2)$. If the u_i^2 are nontrivial, then $H^0(\Sigma, \mathcal{L}^2) = 0$ and since the Euler characteristic of Σ is -2, we have

$$\dim H^1(\Sigma, \mathcal{L}^2) = 2\,.$$

A change of basis in $\mathbf{C}^2$ has the effect of multiplying the cohomology class by a scalar, so that the parameter which determines the holonomy is geometrically a point in the projective line:

$$\mathbf{P}(H^1(\Sigma, \mathcal{L}^2))\,.$$

In the language of connections we are trying to describe equivalence classes of flat connections A on an extension of line bundles

$$L \to V \to L^*$$

which preserve L and a skew form. If d_L is the covariant exterior derivative of the induced connection on L, then just as holomorphic extensions are classified by elements of the Dolbeault cohomology group $H^1(\Sigma, \mathcal{O}(L^2))$, flat extensions d_A are classified by de Rham cohomology classes in $H^1(\Sigma, \mathcal{L}^2)$ determined by using the connection d_{2L} on L^2 instead of the Dolbeault operator $\bar{\partial}_{2L}$.

To describe the class, suppose that u is a nonvanishing section of L and v is a section of V such that $\langle u, v\rangle = 1$. Then since L is preserved by the connection

$$d_A u = d_L u = \gamma u \qquad \text{and} \qquad d_A v = \alpha u + \beta v$$

for 1-forms α, β, γ. In fact the line bundle $L^2 \subset \operatorname{End} V$ is preserved by the connection, so $2\gamma = \theta$. Now

$$0 = d\langle u, v\rangle = \gamma + \beta$$

so, knowing the flat connection on L, which is the 1-form γ, it is simply α which determines the connection on V. If we make a different choice $\tilde{v}$ instead of v, then $\tilde{v} = v + fu$, and

$$d_A\tilde{v} = \alpha u - \theta v + dfu + f\gamma u = \alpha u - \gamma(\tilde{v}) + (df + 2f\gamma)u\,.$$

Thus if $d_A\tilde{v} = \tilde{\alpha}u - \gamma\tilde{v}$

$$\tilde{\alpha} = \alpha + d_{2L}f$$

and α or $\tilde{\alpha}$ represent the same flat cohomology class of the extension in $H^1(\Sigma, \mathcal{L}^2)$.

To describe explicitly such elements, take $\alpha \in \Omega^1(\Sigma, L^2)$ with $d_{2L}\alpha = 0$. Restricted to a generating loop Γ_i, $H^0(\Gamma_i, \mathcal{L}^2)$ vanishes and since the Euler characteristic vanishes too, so does $H^1(\Gamma_i, \mathcal{L}^2)$. There is thus a unique $f_i \in \Omega^0(\Sigma, L^2)$ such that $\alpha = d_{2L}f_i$. Fixing a point $p \in \Sigma$ and a trivialization of L^2 at p,

$$(f_1(p) - f_2(p), f_1(p) - f_3(p)) \in \mathbf{C}^2$$

represents the cohomology class. Indeed, clearly replacing α by $\alpha + d_{2L}f$ changes f_i to $f_i + f$ and leaves this unchanged. If we apply a gauge transformation g, then $f_i(p) \mapsto g^2(p)f_i(p)$, so the projective equivalence class is invariant and we can write it explicitly as

$$\frac{f_1(p) - f_2(p)}{f_1(p) - f_3(p)}\,. \tag{6.1}$$

This number precisely describes the monodromy of our connections, and an isomonodromic deformation is one for which this remains constant.

We now put this approach into effect for the connection in the adjoint representation given by

$$d_A\mathbf{v} = \frac{d\mathbf{v}}{d\zeta} - 2\frac{\mathbf{b}\times\mathbf{v}}{\mathbf{b}\cdot\mathbf{q}}\,.$$

In the above notation, $L^2 \subset \mathfrak{sl}(2)$ is the subbundle preserved by the connection, and of which $\mathbf{q}$ is a nonvanishing section on $\mathbf{C}P^1\backslash\{\infty\}$. Thus

$$d_{2L}\mathbf{q} = \frac{d\mathbf{q}}{d\zeta} - 2\frac{\mathbf{b}\times\mathbf{q}}{\mathbf{b}\cdot\mathbf{q}} = 2\gamma\mathbf{q} = \theta\mathbf{q}\,.$$

Now pass to the 2-dimensional spin representation. The adjoint representation of $SL(2, \mathbf{C})$ is the symmetric square of this representation and if ψ_1, ψ_2 form a symplectic basis of $\mathbf{C}^2$ then $((\psi_1^2 + \psi_2^2)/2, i(\psi_1^2 - \psi_2^2)/2, i\psi_1\psi_2)$ forms an orthonormal basis in the adjoint representation. In particular, if $u = (1, \zeta)$ and $v = (0, 1)$, then

$$u^2 = \mathbf{q}, \qquad v^2 = \mathbf{p}, \qquad uv = (\zeta, i\zeta, -i) = \frac{1}{2}\mathbf{p}\times\mathbf{q}\,.$$

From this, u generates the subbundle $L \subset \mathbf{C}P^1 \times \mathbf{C}^2$, and with $v = (0,1)$ forms a basis with $\langle u, v\rangle = 1$ as above. With this relationship between the two representations, we obtain $d_A v^2 = 2\alpha uv - \gamma v^2$ and so

$$d_A \mathbf{p} = -2\frac{\mathbf{b}\times\mathbf{p}}{\mathbf{b}\cdot\mathbf{q}} = \alpha\mathbf{p}\times\mathbf{q} - \gamma\mathbf{p}\,.$$

Taking the dot product with $\mathbf{p}\times\mathbf{q}$ gives

$$\alpha = \frac{\mathbf{b}\cdot\mathbf{p}}{\mathbf{b}\cdot\mathbf{q}}d\zeta\,. \tag{6.2}$$

At this juncture we should pause to understand how the isomonodromic deformation problem is to be described. The connection has simple poles at the four roots of $\mathbf{b}\cdot\mathbf{q} = \mathbf{c}\cdot\mathbf{q} = 0$. Geometrically, if we pass from $\mathbf{C}^3$ to the projective space $\mathbf{C}P^2$, $\mathbf{q}$ traces out the standard conic $z_1^2+z_2^2+z_3^2 = 0$ and then the four points of intersection with the conic $\sum B_{ij}z_iz_j = \sum S_{ij}z_iz_j = 0$ are the four singular points. These determine the pencil of conics, but only determine S up to addition of tI for some t. The residues of θ at the singular points are determined, as in the proof of Proposition 5.3, by the skew part $\mathbf{a}$ of the matrix B. Thus if we fix the singularities and their residues, B is uniquely determined modulo an overall scale and addition of a multiple of the identity. The effect of changing B to $B+tI$ changes $\mathbf{b}$ to $\mathbf{b}+t\mathbf{q}$ and so from (6.2) changes α to

$$\alpha + t\frac{2}{\mathbf{b}\cdot\mathbf{q}}d\zeta\,.$$

For an isomonodromic deformation, fixing the projective cohomology class of α in $\mathbf{P}(H^1(\Sigma,\mathcal{L}^2))$ expresses t as a function of the four singular points. To do this explicitly, however we must depart from the general terminology and normalize our connection.

We therefore make a projective transformation of $\mathbf{C}P^1$ so that the singular points are $0,1,x,\infty$. By a gauge transformation we may still assume that the line bundle L is spanned by $(1,\zeta)$, but now

$$\gamma = \frac{1}{2}\left(\frac{a}{\zeta} + \frac{b}{\zeta-1} + \frac{c}{\zeta-x}\right)d\zeta\,.$$

We have as before $d_A u = \gamma u$ and $d_A v = \alpha u - \gamma v$. Now since $\zeta = \infty$ is a simple pole of the connection, the connection matrix relative to the constant basis $(1,0),(0,1)$ vanishes at infinity. But

$$d_A(1,0) = d_A(u-\zeta v) = (\gamma - \zeta\alpha, 2\zeta\gamma - \zeta^2\alpha - d\zeta)\,.$$

If this vanishes at infinity, then α is of the form

$$\alpha = \frac{A(\zeta-x)+B}{\zeta(\zeta-1)(\zeta-x)}d\zeta \tag{6.3}$$

where $A = a+b+c-1$. Thus B is the parameter to be determined by the holonomy.

To do this, recall that we need to find around each generating loop Γ_i the unique single-valued solution to the equation $d_A f = \alpha$, or in our explicit terms,

$$\frac{df}{d\zeta} + \left(\frac{a}{\zeta} + \frac{b}{\zeta-1} + \frac{c}{\zeta-x}\right) f = \frac{A(\zeta-x)+B}{\zeta(\zeta-1)(\zeta-x)}.$$

Choosing a base-point p this is solved by

$$f(\zeta) = \frac{1}{\zeta^a(\zeta-1)^b(\zeta-x)^c} A \int_p^\zeta t^{a-1}(t-1)^{b-1}(t-x)^c dt$$
$$+ B \int_p^\zeta t^{a-1}(t-1)^{b-1}(t-x)^{c-1} dt + C$$

where $C = f(p)$ is to be determined so that f is single-valued. In our situation, we can take x real and $x > 1$ and $p \in (0,1)$. Then performing the integral around the first loop Γ_1—the contour passing from $\zeta = p$ to $\zeta = 1$ below the real axis, and returning above—we find f_1 is single-valued if

$$C = f_1(p) = -A \int_p^1 t^{a-1}(t-1)^{b-1}(t-x)^c dt - B \int_p^1 t^{a-1}(t-1)^{b-1}(t-x)^{c-1} dt.$$

Evaluating the other integrals similarly, we obtain the constant k which determines the holonomy as in (6.1)

$$k = \frac{f_1(p) - f_2(p)}{f_1(p) - f_3(p)} = \frac{AI_1 + BI_2}{AJ_1 + BJ_2}$$

where the integrals I_1, I_2, J_1, J_2 are

$$I_1 = \int_0^1 t^{a-1}(t-1)^{b-1}(t-x)^c dt, \quad I_2 = \int_0^1 t^{a-1}(t-1)^{b-1}(t-x)^{c-1} dt = -\frac{1}{c}\frac{dI_1}{dx}$$

and

$$J_1 = \int_1^x t^{a-1}(t-1)^{b-1}(t-x)^c dt, \quad J_2 = \int_1^x t^{a-1}(t-1)^{b-1}(t-x)^{c-1} dt = -\frac{1}{c}\frac{dJ_1}{dx}.$$

Thus, fixing the constant k to determine an isomonodromic deformation, the coefficient B is defined by

$$B^{-1} = \frac{1}{c(a+b+c-1)} \frac{d}{dx} \ln(I - kJ)$$

where

$$I = \int_0^1 t^{a-1}(t-1)^{b-1}(t-x)^c dt \qquad \text{and} \qquad J = \int_1^x t^{a-1}(t-1)^{b-1}(t-x)^c dt.$$

Since the hypergeometric function is defined by

$$F(\alpha,\beta;\gamma;z) = \frac{\Gamma(\gamma)}{\Gamma(\beta)\Gamma(\gamma-\beta)}\int_0^1 t^{\beta-1}(1-t)^{\gamma-\beta-1}(1-tz)^{-\alpha}dt$$

it is clear that I (and after a change of variable J) can be expressed in terms of hypergeometric functions. In the general context of self-dual conformal structures defined using Painlevé transcendants (Hitchin (1995), Maszczyk *et al.* (1994)), the solution $y(x)$ to Painlevé VI is determined by the value of ζ at which the connection form has a common eigenvector with the residue at infinity. Since the line bundle $(1,\zeta)$ is preserved by the connection, $v=(0,1)$ is an eigenvector at infinity and $d_A v = \alpha u - \gamma v$ so that we seek the value of ζ at which $\alpha = 0$. From (6.3) this is when $\zeta = x - B/A$ so the solution to the Painlevé equation is

$$y(x) = x + \frac{I_1 - kJ_1}{I_2 - kJ_2} = x + \frac{c}{(\ln(I-kJ))'}\,.$$

Remarks 1. Solutions of Painlevé's equation involving hypergeometric functions have appeared in Okamoto (1987) and elsewhere. In Okamoto's formalism the condition $a+b+c-A=1$ places his set of solutions on the wall of the Weyl chamber. For us, it is the statement that the connection preserves a line bundle of degree -1.

2. In the hyperkähler case we saw in Proposition 5.3 that $a=b=c=1/2$. Thus the integrals I_1 and I_2 are the elliptic integrals

$$I_1 = \int_0^1 \frac{(t-x)}{\sqrt{t(t-1)(t-x)}}dt, \quad I_2 = \int_0^1 \frac{1}{\sqrt{t(t-1)(t-x)}}dt$$

which evaluate in Weierstrassian terms to be $2(-\eta_1 + (1-2x)\omega_1/3)$ and $2\omega_1$ respectively. The integrals J_1, J_2 are obtained by integration around another cycle in the elliptic curve: we replace ω_1, η_1 by ω_3, η_3. In this form the solution coincides with that produced in Hitchin (1995), which is another way of writing Halphen's solution.

Retracing the steps from this solution to find the matrix B_{ij}, or the hypercomplex manifold X, is a task I leave to the reader. What global properties one seeks are unclear. Certainly we are unlikely to find compact examples, since according to Boyer (1988), hypercomplex ones are confined to tori, Hopf surfaces and K3 surfaces.

Bibliography

Ashtekar, A., Jacobson, T., and Smolin, L. (1988). A new characterization of half-flat solutions to Einstein's equations. *Commun. Math. Phys.*, **115**, 631–48.

Atiyah, M. F. and Hitchin, N. J. (1988). *The geometry and dynamics of magnetic monopoles.* Princeton University Press, Princeton (1988).

Boyer, C. P. (1988). A note on hyperHermitian four-manifolds. *Proc. Amer. Math. Soc.*, **102**, 157–64.

Gibbons, G. W. and Pope, C. N. (1979). The positive action conjecture and asymptotically Euclidean metrics in quantum gravity. *Commun. Math. Phys.*, **66**, 267–90.

Halphen, G.-H. (1881). Sur un système d'équations différentielles. *C.R. Acad. Sci. Paris*, **92**, 1101–3.

Hitchin, N. J. (1983). On the construction of monopoles. *Commun. Math. Phys.*, **89**, 145–90.

Hitchin, N. J. (1995). Twistor spaces, Einstein metrics and isomonodromic deformations. *J. Differential Geometry*, **42**, 30–112.

Hoppe, J. (1996). Multilinear evolution equations for time-harmonic flows in conformally flat manifolds. *Class. Quantum Grav.*, **13**, 87–93.

Krichever, I. M. and Novikov, S. P. (1990). Riemann surfaces, operator fields, strings. Analogues of the Fourier-Laurent bases. In: *Physics and mathematics of strings*, 356–88. World Sci. Publishing, Teaneck, NJ.

Maszczyk, R., Mason, L. J., and Woodhouse, N. M. J. (1994). Self-dual Bianchi metrics and the Painlevé transcendents. *Class. Quantum Grav.*, **11**, 65–71.

Milnor, J. (1984). Remarks on infinite-dimensional Lie groups. In: *Relativity, groups and topology, II (Les Houches, 1983)*, 1007–57, North-Holland, Amsterdam-New York.

Obata, M. (1956). Affine connections on manifolds with almost complex, quaternionic or Hermitian structure. *Jap. J. Math.*, **26**, 43–79.

Okamoto, K. (1987). Studies on the Painlevé equations. I. Sixth Painlevé equation P_{VI}. *Ann. Mat. Pura Appl.* (4), **146**, 337–81.

Penrose, R. (1976). Nonlinear gravitons and curved twistor theory, *Gen. Rel. Grav.*, **7**, 31–52.

3

Gauge Theory in Higher Dimensions

S. K. Donaldson and R. P. Thomas
The Mathematical Institute, Oxford

1 Introduction

The interaction between geometry in the adjacent dimensions $2, 3$ and 4 is a theme which runs through a great deal of the work by mathematicians on gauge theory over the past few years. In this article we will examine the possibility of developing this theme in higher dimensions. We will find extensions following two intertwining threads. One thread, which we say more about here, replaces real variables by complex variables, and hence operates in *complex* dimensions $2, 3, 4$. The other thread involves, from one point of view, replacing the quaternions by the octonians, and operates in the realm of manifolds with "exceptional holonomy". The picture we will find pulls together various ideas which have been touched on in the literature but the striking analogies which emerge do not seem to be well-known. Our treatment will be informal throughout this article—our main aim is to advertise the potential for research in these directions. A great deal of technical work is needed to develop these ideas in detail, and a more thorough and wide-ranging account of the Calabi–Yau story will appear in the D.Phil. thesis of the second author. The first author would like to emphasise the debt due to other mathematicians in forming parts of the picture described here; particularly Dominic Joyce, Simon Salamon and Christopher Lewis for lessons on exceptional holonomy. A substantial part of this picture is essentially due to Joyce and Lewis, and again futher details will appear in the doctoral thesis of Lewis.

2 The familiar theory

Let us begin by reviewing very briefly, ignoring many important technicalities, the sort of ideas and structures in gauge theory that we wish to generalise. These involve gauge theory, with structure group a compact Lie group (which we may wish to take to be $SU(2)$ or $SO(3)$ in a detailed development) over differentiable manifolds of dimensions $2, 3, 4$, all of which have definite *orientations*. It is convenient to focus on the intermediate dimension 3, where we have the well-known theories of Casson and Floer. If $Y_{\mathbf{R}}$ is a compact 3-manifold, the Chern–Simons functional gives a map $CS : \mathcal{B}(Y_{\mathbf{R}}) \to \mathbf{R}/\mathbf{Z}$ where $\mathcal{B}(Y_{\mathbf{R}})$ is the infinite-dimensional space of gauge-equivalence classes of connections on a fixed bundle over $Y_{\mathbf{R}}$. The derivative of CS is then given by the curvature of a connection, regarded as a 1-form on $\mathcal{B}(Y_{\mathbf{R}})$, through the formula:

$$\delta CS = \int_{Y_{\mathbf{R}}} \mathrm{Tr}(F \wedge \delta A), \tag{2.1}$$

and so the critical points are the *flat* connections. We assume for simplicity that CS is a generic function, so the critical points are non-degenerate and in particular isolated. The Casson invariant of $Y_{\mathbf{R}}$ is given by counting, with signs, the flat connections. It can be interpreted, formally, as the Euler characterisitic of the infinite-dimenional space $\mathcal{B}(Y_{\mathbf{R}})$. The Floer homology of $Y_{\mathbf{R}}$ is defined by fixing a metric on $Y_{\mathbf{R}}$, and hence on $\mathcal{B}(Y_{\mathbf{R}})$. This allows one to define the gradient vector field of CS; the integral curves of the gradient vector field connecting different critical points give, in Floer's celebrated construction, a chain complex which computes the Floer homology. The homology groups do not depend on the metric and are formally the homology groups of $\mathcal{B}(Y_{\mathbf{R}})$ in "semi-infinite dimensions".

The four-dimensional view-point on these ideas comes from the fact that the gradient flow equation for CS is precisely the Yang–Mills instanton equation on $Y_{\mathbf{R}} \times \mathbf{R}$: thus the pointwise symmetry group $SO(3)$ of the three-dimensional theory has a surprising extension to $SO(4)$ (related to the Lorentzian invariance of Maxwell's equations). More generally, if $Y_{\mathbf{R}}$ is the boundary of a 4-manifold $X_{\mathbf{R}}$ one gets an interaction between the instanton theory on $X_{\mathbf{R}}$, made into a complete manifold by adjoining an infinite cylinder, and the Floer theory on $Y_{\mathbf{R}}$.

To go down to two dimensions we consider a splitting of the 3-manifold, $Y_{\mathbf{R}} = Y_{\mathbf{R}}^+ \cup_{S_{\mathbf{R}}} Y_{\mathbf{R}}^-$, by a surface $S_{\mathbf{R}} \subset Y_{\mathbf{R}}$. The moduli space $M(S_{\mathbf{R}})$ of flat connections over the surface $S_{\mathbf{R}}$ is, roughly speaking, a finite-dimensional manifold with an intrinsic *symplectic structure* induced by the formula:

$$\Omega(a, b) = \int_{S_{\mathbf{R}}} \mathrm{Tr}(a \wedge b), \tag{2.2}$$

where a and b are bundle-valued 1-forms over $S_{\mathbf{R}}$ representing tangent vectors to $M(S_{\mathbf{R}})$. Now we consider the subsets $L^+, L^- \subset M(S_{\mathbf{R}})$ given by the connections which extend over $Y_{\mathbf{R}}^{\pm}$. These are Lagrangian submanifolds of $M(S_{\mathbf{R}})$ and the flat connections on $Y_{\mathbf{R}}$ appear as the intersection points $L^+ \cap L^-$. (This is the point of view Casson took in his original definition.) The instantons in four dimensions are more elusive in this picture but they are related to the *holomorphic discs* in $M(S_{\mathbf{R}})$ with boundary in the $L^{\pm}$. (Here one chooses a metric on $S_{\mathbf{R}}$, which makes $M(S_{\mathbf{R}})$ into a Kähler manifold.) This is the essence of the "Atiyah–Floer conjecture", versions of which have been proved by D. Salamon and others (Dostoglou and Salamon 1994). The main point is that if T is another surface the "adiabatic limit" of the instanton equations on the product $S_{\mathbf{R}} \times T$, as the metric in the $S_{\mathbf{R}}$ direction is scaled down, can be identified with the holomorphic mapping equation for maps from T to $M(S_{\mathbf{R}})$.

3 The complex analogy

In elementary terms, our procedure for extending the ideas sketched above is to replace ordinary derivatives $\frac{\partial}{\partial x_\alpha}$, where x_α are real co-ordinates, by Cauchy–Riemann operators $\frac{\partial}{\partial \overline{z}_\alpha}$, where z_α are complex co-ordinates. The important role played by orientation in the real case leads to the need for a "complex orientation"—a trivialisation of the canonical line bundle. Thus the geometrical setting for our discussion involves Calabi–Yau manifolds. From the point of view of analysis the crucial thing is that the ordinary derivative $\frac{d}{dx}$ on $\mathbf{R}$ and the Cauchy–Riemann operator $\frac{\partial}{\partial \overline{z}}$ on $\mathbf{C}$ are both ellipic operators, so behave in rather similar ways.

We begin then in complex dimension 3 with a compact Calabi–Yau 3-fold Y, so there is a nowhere degenerate holomorphic form $\theta \in \Omega^{3,0}(Y)$. We sometimes want to suppose that Y is Kähler, and so admits a Kähler–Einstein metric with holonomy $SU(3)$. Fix a C^∞ complex vector bundle E over Y and let $\mathcal{A}$ be the space of $\overline{\partial}$-operators on E: that is differential operators:

$$\overline{\partial}_\alpha : \Omega^0(E) \to \Omega^{0,1}(E) \tag{3.1}$$

satisfying the usual Leibnitz rule. Any two elements of $\mathcal{A}$ differ by a tensor in $\Omega^{0,1}(End\, E)$. If E has a fixed Hermitian metric these $\overline{\partial}$-operators may be identified with unitary connections, by projecting to the $(0,1)$ part. We consider the action of the complex gauge group $\mathcal{G}^c$ of general linear automorphisms of E, which act on $\mathcal{A}$ by conjugation. Thus we have a quotient space $\mathcal{C}_E = \mathcal{A}/\mathcal{G}^c$. This definition should not be taken too literally: we know very well from other problems that issues involving "stability" arise in forming such quotients, and we may wish to restrict attention to a suitable set of stable points in $\mathcal{A}$. In particular we may employ the well-known framework involving the comparison of symplectic and complex quotients and, having fixed metrics, work with the quotient of the space of connections whose curvature satisfies a moment map condition $F.\omega = 0$ (where ω is the Kähler form), by the group $\mathcal{G}$ of unitary automorphisms of E (see De Bartolomeis and Tian 1996, for example). However for the present we shall ignore such technicalities and imagine that $\mathcal{C}_E$ is an infinite-dimensional complex manifold. Now any operator $\overline{\partial}_\alpha \in \mathcal{A}$ prolongs to the E-valued $(0,q)$ forms, and the composite $\overline{\partial}_\alpha^2$ defines a tensor in $\Omega^{0,2}(End E)$. If we identify the operators with connections, this is just the $(0,2)$ part of the curvature, so we denote it by $F^{0,2}(\alpha)$. Then we define a complex 1-form U on the space $\mathcal{A}$ by

$$U_\alpha(\delta\alpha) = \int_Y \mathrm{Tr}(\delta\alpha \wedge F^{0,2}(\alpha)) \wedge \theta. \tag{3.2}$$

Here α is a point in $\mathcal{A}$ and $\delta\alpha$ is a tensor in $\Omega^{0,1}(End\, E)$, regarded as a tangent vector to $\mathcal{A}$ at α. The analogy with the case of connections over 3-manifolds will be clear to the reader. Just as in that case one shows that U descends to the quotient space $\mathcal{C}_E$ and defines a closed 1-form, so is locally the derivative of a complex-valued function Φ. The new feature in our present case is that

Φ is a holomorphic function on the complex manifold $\mathcal{C}_E$. We can identify Φ more explicitly: if we regard $\mathcal{A}$ as a space of connections then it is just given by pairing the Chern–Simons invariant with the holomorphic form θ. In any event we get a well-defined function on a covering space of $\mathcal{C}_E$, with covering group at most $H^3(Y; \mathbf{Z})$—for our current exposition we will largely ignore this covering issue. Now the main point is that, just as in the Casson–Floer theory, the critical points of Φ—the zeros of the 1-form U—have a solid geometric meaning. They are just the operators satisfying the integrability condition $\overline{\partial}_\alpha^2 = 0$, that is, those which endow E with a *holomorphic* structure. Clearly then we should hope that "counting" the holomorphic bundles of a fixed topological type over a Calabi–Yau manifold Y will yield an invariant which can be regarded as analogous to the Casson invariant of a 3-manifold.

Some remarks are now in order. First, the point of view above is very close to the discussion by Witten (1995, Sec 4.5), which is aimed more at an analogy with the Chern–Simons theory on 3-manifolds, involving integration of the exponential of the Chern–Simons functional. Second, a lot of work would need to be done to give precision to these ideas. For example one would expect to bring in a suitable notion of stability, as mentioned above. One would have to deal with the problem that the zeros of the 1-form U may be degenerate, i.e. the situation is not generic, and consider suitable perturbations, or appropriate methods of counting degenerate zeros. One would need compactness results to ensure that counting gave a finite answer. However, at least as far as the "local" discussion goes (in the space of connections) one can take over much of the usual Fredholm-theory analysis (Taubes 1990) from the usual 3-manifold case to this complex setting. The setting for the local analysis in the ordinary case is the de Rham complex

$$\Omega^0 \xrightarrow{d} \Omega^1 \xrightarrow{d} \Omega^2 \xrightarrow{d} \Omega^3, \tag{3.3}$$

coupled to the flat connection on the adjoint bundle. In the holomorphic case we get the Dolbeault complex

$$\Omega^{0,0} \xrightarrow{\overline{\partial}} \Omega^{0,1} \xrightarrow{\overline{\partial}} \Omega^{0,2} \xrightarrow{\overline{\partial}} \Omega^{0,3} \tag{3.4}$$

coupled to the endomorphism bundle. The point is that these are both elliptic complexes. As far as compactness goes, one knows at least that the L^2 norm of the curvature of a Hermitian–Yang–Mills connection over a compact Kähler manifold of any dimension m is fixed by the topology of the bundle. This follows from the Chern-Weil theory and the identity:

$$|F|^2\omega^m = -\mathrm{Tr}(F^2) \wedge \omega^{m-2} \tag{3.5}$$

for curvature tensors F of type $(1,1)$ with $F.\omega = 0$. A third remark is that, while the tight analogy with the Casson theory is restricted to Calabi–Yau manifolds, one may try more generally to approach the integrabilty equations $F^{0,2} = 0$ (or what is more or less the same, the Hermitian–Yang–Mills equations for unitary connections) from the point of view of nonlinear Fredholm theory. The problem is that the equations are overdetermined in complex dimension 3 or more. This difficulty is not so serious in a situation where one knows that all the higher dimensional cohomology groups $H^i(End_0 E)$, for all the relevant holomorphic bundles E, vanish for $i \geq 3$. Here End_0 denotes the trace-free endomorphisms. From this perspective the good feature of a Calabi–Yau 3-fold is that the cohomology group $H^3(End_0 E)$ is dual to $H^0(End_0 E)$, which vanishes for stable bundles E.

We now go up to complex dimension 4, beginning with the product $\mathbf{C} \times Y$. Here we fix a Kähler–Einstein metric on Y and a metric on E, so that $\mathcal{A}$ gets an induced Hermitian L^2-metric. Thus the complex cotangent vector U is dual to a tangent vector $\widehat{U}$. We would like to consider a "complex gradient equation", in parallel with Floer's theory. Some subtleties are involved, however, because the complex gauge group $\mathcal{G}^c$ does not preserve the vector field $\widehat{U}$, so there are different ways to proceed. On the one hand we could consider a map $\alpha(z)$ from $\mathbf{C}$ to $\mathcal{A}$ which satisfies the equation

$$\frac{\partial \alpha}{\partial \bar{z}} = \widehat{U}. \tag{3.6}$$

On the other hand we can consider the complex quotient $\mathcal{C} = \mathcal{A}/\mathcal{G}^c$ which (modulo questions of stability) is identified with the symplectic quotient $\{F.\omega = 0\}/\mathcal{G}$, and in this way get an induced Kähler metric. The vector field $\widehat{U}$ on $\mathcal{A}$ induces a vector field $\underset{\sim}{U}$ on $\mathcal{C}$, and we can consider a map $\tilde{\alpha}(z)$ from $\mathbf{C}$ to $\mathcal{C}$ satisfying

$$\frac{\partial \tilde{\alpha}}{\partial \bar{z}} = \widetilde{U}. \tag{3.7}$$

We want to interpret these constructions in terms of gauge theory over $\mathbf{C} \times Y$. Observe that on any n-dimensional Calabi–Yau manifold we have complex *antilinear* maps $*_n : \Omega^{0,q} \to \Omega^{0,n-q}$ defined by the condition that the $(0,n)$-form $\alpha \wedge *_n \alpha$ is $|\alpha|^2$ times the conjugate of the complex volume form. The vector field $\widehat{U}$ can be identified with $*_3 F^{0,2}$, defined by combining the above operation on the form component with the antilinear adjoint map on $End\, E$.

Now move to 4 dimensions. If X is any Calabi–Yau 4-fold we have $*_4 : \Omega^{0,2}_X \to \Omega^{0,2}_X$ with $*_4^2 = 1$, so we get a decomposition $\Omega^{0,2}_X = \Omega^{0,2}_+ \oplus \Omega^{0,2}_-$ into ± 1 eigenspaces. (It is important to realise that these are real, not complex, subspaces. In terms of representations, what we are saying is that the representation $\Lambda^{0,2}$ of $SU(4)$ is a *real* representation. Notice also that multiplying the complex volume form θ by a complex scalar gives a different splitting.) This decomposition of the forms means that there is a "complex anti-self-duality" equation for unitary connections over X: $F^{0,2}_+ = 0$. This equation has been found and studied independently by Lewis. Putting the pieces together, in the first setting of a map from $\mathbf{C}$ to $\mathcal{A}$ we define a connection over $\mathbf{C} \times Y$ with zero components in the $\mathbf{C}$ direction and one sees that solutions of (3.6) correspond to connections over $\mathbf{C} \times Y$ which satisfy the two equations

$$\begin{cases} F_+^{0,2} = 0, \\ F.\omega_{\mathbf{C}} = 0, \end{cases} \tag{3.8}$$

where $F.\omega_{\mathbf{C}}$ is the component of the curvature in the $\mathbf{C}$-variable. In the second setting, of a map from $\mathbf{C}$ to $\mathcal{C}$, one finds that the solutions of (3.7) correspond to solutions of the two equations

$$\begin{cases} F_+^{0,2} = 0, \\ F.\omega_Y = 0. \end{cases} \tag{3.9}$$

On the other hand, over any Calabi–Yau 4-fold X the natural supplement to the anti-self-duality equation, as studied by Lewis, is $F.\omega_X = 0$. We say that a unitary connection over X is an $SU(4)$*-instanton* if it satisfies the two equations $F_+^{0,2} = 0$, $F.\omega_X = 0$. These are elliptic equations, modulo unitary gauge equivalence. Notice that (when X is compact) Hermitian–Yang–Mills connections on stable holomorphic bundles give examples of such instantons. Lewis has shown that if a certain characteristic class condition is satisfied then these are the only solutions. Moreover there is in any case an L^2 bound on the curvature of an $SU(4)$-instanton, coming from an identity like (3.5).

In the case when $X = \mathbf{C} \times Y$ we can write the $SU(4)$-instanton equations as

$$\begin{cases} F_+^{0,2} = 0, \\ F.\omega_{\mathbf{C}} + F.\omega_Y = 0. \end{cases} \tag{3.10}$$

These three equations (3.8), (3.9) and (3.10) all fit into the continuous family $F_+^{0,2} = 0$, $tF.\omega_{\mathbf{C}} + (1-t)F.\omega_Y = 0$ $(0 \leq t \leq 1)$, and, supposing that this family is well-behaved, the solutions of the three equations are more-or-less equivalent (at least in regard to topological "counting" of the solutions). The advantage of the intermediate equation (3.10) (or, more generally, the equation for any parameter t with $0 < t < 1$) is that it is elliptic. Clearly the existence of these different equations has to do with different ways of dealing with the gauge invariance of the problem and we can think of (3.10) as a regularisation of the extremes (3.8), (3.9).

Now we can carry through the discussion above replacing $\mathbf{C}$ by any Riemann surface with trivial cotangent bundle, and in particular by $\mathbf{R} \times S^1$. The conclusion we are finally led to is that the $SU(4)$-instantons over $\mathbf{R} \times S^1 \times Y$ which converge to Hermitian–Yang–Mills connections on holomorphic bundles E^+, E^- at $\pm\infty$ in the $\mathbf{R}$-variable play the role of complex gradient curves for the holomorphic Chern–Simons functional, in the same way that instantons over tubes are viewed as gradient curves in Floer's theory. In Section 8 below we shall return to discuss the topological interpretation of these complex gradient curves; we shall see there that the solutions which are invariant under rotations in the S^1-variable have a particularly simple interpretation and it is worth noting now that for rotation-invariant connections the three equations (3.8), (3.9), (3.10) are all directly equivalent since $F.\omega_{\mathbf{C}}=0$.

4 Exceptional holonomy

There is a remarkable feature of these $SU(4)$-instanton equation, which leads us to the second thread mentioned at the beginning of this article. This thread gives us, in some ways, a more direct generalisation of the familiar 3 and 4-dimensional picture. (Much of what follows was explained to the first author by Dominic Joyce: a general reference for exceptional holonomy is Salamon (1989).) Consider the 8-dimensional, real, spin representation V of $Spin(7)$, and the standard embedding of $Spin(6)$ in $Spin(7)$. There is an exceptional isomorphism between $Spin(6)$ and $SU(4)$, so we get an embedding of $SU(4)$ in $Spin(7)$, and under this embedding V becomes the fundamental representation of $SU(4)$, of complex dimension 4. This means that a Calabi–Yau 4-fold, with holonomy group $SU(4)$, furnishes an example of an 8-manifold with an integrable $Spin(7)$-structure, where $Spin(7)$ acts via the spin representation. Now the second exterior power $\Lambda^2 V$ splits under the $Spin(7)$ action into two irreducible pieces, a copy of the adjoint representation (given by the infinitesimal action on V), and a complement, H say, of dimension 7. Restricting to $SU(4) \subset Spin(7)$, the representation H splits into

$$H = \mathbf{R}\omega \oplus \Lambda^{0,2}_{+}. \tag{4.1}$$

So we conclude that the $SU(4)$-instanton equation $F^{0,2}_{+} = F.\omega = 0$ has a further symmetry under $Spin(7)$: it is just the condition that the H component of the curvature vanishes, which makes sense on any manifold with a $Spin(7)$-structure. (So we will also refer to the equations as the $Spin(7)$-instanton equations.) This is rather similar, although the similarity does not seem to fit into the general pattern of this article, to the way in which the Hermitian–Yang–Mills equation on a Kähler surface (with holonomy $U(2)$) is just the ordinary instanton equation, and as such makes sense on any oriented Riemannian 4-manifold (with holonomy $SO(4)$). Higher dimensional versions of the instanton equation have been discussed by various authors (see Corrigan *et al.* 1983, Ward 1984, Fubini and Nicolai 1985, for example), and these $Spin(7)$-instanton equations are part of a general theory developed by Salamon and Reyes-Carrion (Reyes-Carrion 1997).

We may also take these ideas over to seven real dimensions, and the exceptional holonomy group G_2. Let us start on a different tack, and consider a compact oriented n-manifold N furnished with a closed $(n-3)$-form σ. Then we may consider a functional on the space of unitary connections on a bundle over N defined by the closed 1-form

$$\int_N \mathrm{Tr}(F \wedge \delta A) \wedge \sigma, \tag{4.2}$$

following the notation of (1). The critical points of this functional are the connections with $F \wedge \sigma = 0$. Linearising the theory leads to a complex which is a bundle-valued version of:

$$\Omega^0_N \xrightarrow{d} \Omega^1_N \xrightarrow{\sigma\wedge d} \Omega^{n-1}_N \xrightarrow{d} \Omega^n_N. \tag{4.3}$$

The condition that this complex be *elliptic* is an algebraic condition on σ at each point which (choosing a metric) essentially comes down to the condition that $*\sigma$ defines a non-degenerate skew-symmetric cross-product $TM \times TM \to TM$. As is well-known, this leads to an almost complex structure on the $(n-1)$ sphere, and can only exist in dimensions $n = 3, 7$, where the algebraic models are the cross-product on the imaginary quaternions and octonians. The first case is the ordinary Floer theory. The second case operates best when we have a 7-manifold with holonomy G_2, and $*\sigma$ is the fundamental covariant constant 3-form, which can be taken as the definition of the structure. In this situation one gets an L^2 bound on the curvature, by an identity similar to (3.5). In sum then, on a manifold N with a G_2-structure, we have the basis for a Casson–Floer-type theory, in which the role of the flat connections is played by the solutions of the equation $F \wedge \sigma = 0$. The corresponding gradient lines are just the solutions of the $Spin(7)$ instanton equation on the manifold $N \times \mathbf{R}$ (which has a $Spin(7)$ structure, corresponding to the natural inclusion $G_2 \subset Spin(7)$). This picture interacts with the previous one, when we consider $N = Y \times S^1$, where Y is our Calabi–Yau manifold with holonomy $SU(3)$. For a general G_2 manifold, and a bundle $E \to N$, one could expect to have a Casson invariant, and a Floer homology theory which bears the same relation to the $Spin(7)$-instantons on a $Spin(7)$ manifold which is asymptotic to a tube $N \times [0, \infty)$, as in the ordinary case of three and four dimensions.

5 The two-dimensional picture

We now return to the complex geometry strand of Section 3, and view things from the standpoint of complex dimension 2. Let S be a compact Calabi–Yau surface (so either a torus or a $K3$ surface), and let $M(S)$ be a moduli space of vector bundles over S (as usual, we ignore niceties involving stability). In favourable cases at least, this is a complex manifold with a complex symplectic structure due to Mukai (1984). The tangent space of $M(S)$ at a bundle E is the cohomology group $H^1(End\,E)$ and the symplectic form is given by the cup product $H^1 \otimes H^1 \to H^2$, composed with the trace pairing $End\,E \otimes End\,E \to \mathbf{C}$ and the evaluation map $H^2(\mathcal{O}) = H^2(K_S) \to \mathbf{C}$. More explicitly, if tangent vectors are represented by bundle-valued $(0,1)$ forms, the symplectic form is given by a formula analogous to (2):

$$\langle a, b\rangle = \int_S \mathrm{Tr}(a \wedge b) \wedge \theta_S. \tag{5.1}$$

Now suppose that S is embedded in a compact complex 3-fold Y^+, and that it is cut out as the zero-set of a section of the anticanonical bundle $K^{-1}_{Y^+}$

(i.e. there is a meromorphic 3-form on Y^+ with no zeros and with a simple pole along S—in this situation the adjunction formula forces S to be a Calabi–Yau surface). We consider a moduli space L^+ of holomorphic bundles over Y^+ of the appropriate topological type, and the map from L^+ to $M(S)$ given by restriction of bundles. It is an observation of Tyurin (1990) that the image is (roughly speaking) a complex Lagrangian submanifold of $M(S)$, with respect to the complex symplectic structure. At the level of tangent vectors, the situation is described by a portion of the long exact sequence associated with the restriction map, for a bundle E over Y^+:

$$H^1(Y^+, End\,E \otimes K_{Y^+}) \to H^1(Y^+, End\,E) \to H^1(S, End\,E|_S) \to$$
$$\to H^2(Y^+, End\,E \otimes K_{Y^+}) \to H^2(Y^+, End\,E).$$

The key thing is that this exact sequence is its own transpose with respect to Serre duality. This is a standard general fact about duality in complex geometry: if one works with Dolbeault cohomology it essentially comes down to the fact that $1/z$ is a fundamental solution for the $\overline{\partial}$-operator over $\mathbf{C}$. We suppose that the final term is zero: which is roughly speaking the assumption that L^+ is smooth and of the proper dimension. Then the first term also vanishes and we get a short exact sequence

$$0 \to TL^+ \to TM(S) \to T^*L^+ \to 0,$$

self-dual with respect to the symplectic form on $TM(S)$, and this just expresses the fact that TL^+ maps to a Lagrangian subspace under the restriction map.

We want next to consider the analogue of the Lagrangian intersection picture for the Casson invariant in the real case. Suppose that the same surface S is embedded as an anticanonical divisor in another compact 3-fold Y^-. Then we can form a singular space with a normal crossing singularity by gluing together the two copies of S: $Y_0 = Y^+ \cup_S Y^-$. Then we may consider deformations of Y_0 in a family Y_t such that Y_t is smooth. This is a standard kind of deformation problem, to which an extensive theory can be applied (Friedman 1983). Locally around S, the picture can be modelled within the total space of the bundle $\nu_+ \oplus \nu_-$ over S, where $\nu_\pm$ are the normal bundles of S in $Y^\pm$ (which are the restrictions of the anti-canonical bundles). For any section ϵ of the line bundle $\nu_+ \otimes \nu_-$ over S the equation $s_+ s_- = \epsilon$ cuts out a 3-dimensional subvariety V_ϵ of the total space. Here $s^\pm$ are tautological sections of the lifts of $\nu^\pm$ to the total space. The double-crossing space Y_0 is modelled on V_0. If we can choose ϵ to have transverse zeros (forming a smooth curve Z in S) then V_ϵ will be smooth. If appropriate obstruction spaces vanish we can extend this local model to a 1-parameter family of deformations of the whole space, modelled near S on $V_{t\epsilon}$. Even if Z has simple nodes, so that V_ϵ has double points, we can get a family of smooth manifolds by making small resolutions (which will change the topology). A particular example of this comes if ϵ is the product of sections ϵ^+, ϵ^- of ν^+, ν^-, so that $Z = Z^+ \cup Z^-$ is a reducible curve. Then we can proceed in another way, in which the topology of the construction is rather transparent. We blow up Y^+ along Z^+ and Y^- along Z^- to get new 3-folds $\overline{Y}^+, \overline{Y}^-$. The proper transforms of S, which we denote by the same letter, in these manifolds have trivial normal bundles, and the smooth 3-fold we seek to construct is given, topologically, by cutting out tubular neighbourhoods of S in $\overline{Y}^+, \overline{Y}^-$ and gluing together the resulting boundaries, each of which is a product of S with a circle. It is particularly easy to see in this case that the canonical bundle of the manifold Y we make by deforming will be trivial, and in fact this is true for any ϵ. (A useful example in lower dimensions to have in mind comes by taking the rational elliptic surface—the projective plane blown up in nine points of intersection of two cubics. A fibre of the elliptic fibration is an anticanonical divisor and the fibre sum of two copies of this manifold yields a K3 surface, with trivial canonical bundle.)

Now a holomorphic bundle over the singular space Y_0 is given by a pair of bundles E^+, E^- over Y^+, Y^-, which are isomorphic over S. Thus, ignoring questions of stability, etc., these holomorphic bundles correspond to intersection points of L^+, L^- in $M(S)$. The general idea should now be clear: we want to regard these Lagrangian intersection points as a limit of the holomorphic bundle moduli space on the smooth Calabi–Yau manifolds Y_t as the complex structure degenerates. So, for example, we would hope that the putative "holomorphic Casson invariant" should go over to the intersection number of L^+, L^- in $M(S)$.

6 Adiabatic limits and dimension reduction

Here we consider briefly the analogue in the complex case of the link between holomorphic maps into the real moduli space $M(S_{\mathbf{R}})$ and ordinary instantons in four dimensions. To do this we take the product $X = T \times S$ of two Calabi–Yau surfaces (tori or K3 surfaces). This is a manifold with holonomy $Sp(1) \times Sp(1) \subset SU(4)$. We have simultaneous actions of the quaternions I, J, K on the tangent spaces of T and S. It is well-known that the moduli space $M(S)$—viewed as a moduli space of instantons—has an induced hyperkähler structure, so the quaternions also act on tangent vectors to $M(S)$. Now consider the $SU(4)$-instanton equations over the product, but with the metric on S scaled by a small factor ϵ. We can decompose the curvature F of a connection over the product into three pieces $F_S \in \Lambda^2(T^*S)$, $F_T \in \Lambda^2(T^*T)$, $F_{ST} \in \Lambda^1 T^*S \otimes \Lambda^1 T^*T$. The $SU(4)$-instanton equations then break up into two parts: one part is

$$F_S^+ = \epsilon F_T^+, \tag{6.1}$$

which makes sense since the bundles of self-dual forms on S, T are trivialised, and the other part is

$$\pi(F_{ST}) = 0, \tag{6.2}$$

where π is the projection from $\Lambda^1 T^*S \otimes \Lambda^1 T^*T$ to $\Lambda^{0,2}_+$. The second equation (6.2) does not involve ϵ. If we naively put $\epsilon = 0$, the first equation (6.1) tells us that the connection is an ordinary instanton on each copy of S in the product, so yields a map f from T to $M(S)$. The second equation goes over to a linear condition on the derivative of f. To see what this linear equation is, let U, V be quaternionic vector spaces and consider the real vector space $Hom_R(U,V)$ of **R**-linear maps. This can be decomposed into four subspaces

$$Hom_R(U,V) = H_1 \oplus H_I \oplus H_J \oplus H_K,$$

where H_1 consists of the quaternion linear maps, H_I consists of the maps which are I-linear but J and K antilinear, and so on. So we get four projection maps $\pi_1, \pi_I, \pi_J, \pi_K$ to the different factors. Now we can apply this when U is the tangent space of T and V is the tangent space of $M(S)$: the condition on the map f which arises from the $SU(4)$ instanton equations is $\pi_J(df) = 0$. This is an elliptic equation which is the natural quaternionic analogue of the holomorphic mapping equation in the complex case. (Notice that the choice of this equation breaks the symmetry between I, J, K: this just comes from the fact that the $SU(4)$ equations depend on a particular complex structue on the product, and a particular holomorphic 2-form. We get a family of similar equations by interchanging I, J, K, which corresponds to different choices of $SU(4)$-structure on the product.)

We will now outline analogues of Hitchin's theory (Hitchin 1987), studying translation invariant solutions of the instanton equation. Let S^+ be the positive spin space of $\mathbf{R}^4$. Then $W = \mathbf{R}^4 \times S^+$ is a real 8-dimensional vector space with a $Spin(7)$-structure: that is, the obvious action of $Spin(4)$ on W extends to the spin representaion of $Spin(7)$, under a certain embedding $Spin(4) \subset Spin(7)$. (Another way of saying this is to exhibit a certain natural 4-form on W, Salamon 1989.) Let us consider solutions of the $Spin(7)$ instanton equation on the flat space W which are invariant under translations in the S^+ directions. These connections correspond to pairs consisting of a connection A on a bundle E over $\mathbf{R}^4$ and a Higgs field Φ, which is a section of $S^+ \otimes_{\mathbf{R}} ad(E) = S^+ \otimes_{\mathbf{C}} ad(E)^c$, where $ad(E)^c$ is the complexification of the bundle associated to the adjoint representation. The $Spin(7)$-instanton equations go over to equations of the shape:

$$F^+(A) = [\Phi, \Phi^*], \quad D_A\Phi = 0, \tag{6.3}$$

where D_A is the Dirac operator in four dimensions, coupled to the connection A, and the bracket in the first equation denotes the combination of the bracket on the Lie algebra with the map $S^+ \otimes \overline{S}^+ = S^+ \otimes S^+ \to \Lambda^+$. Having written the equations for flat space $\mathbf{R}^4$, we see that they make sense over any spin 4-manifold.

This is clearly a 4-dimensional analogue of Hitchin's theory, but also has obvious similarities with the renowned Seiberg–Witten equations in four dimensions (in a similar mould to the equations studied recently by Pidstragatch and Tyurin, Okonek and Teleman and others). If we look for reducible solutions, where the bundle E is $L \oplus L^{-1}$, the connection A is induced from a $U(1)$ connection on L, and Φ takes values in $L^2 \otimes S^+$, then we essentially arrive at the standard Seiberg–Witten equations.

We can play the same game with the G_2 equations $F \wedge \sigma = 0$ in seven dimensions. Here there are two standard models. We can either look at $\mathbf{R}^3 \times S$, where S is the spin space of $\mathbf{R}^3$, or $\mathbf{R}^4 \times \Lambda^+$, where Λ^+ is the bundle of self-dual 2-forms. The first model leads to the 3-dimensional Seiberg–Witten equations, and non-abelian versions of these. The second leads to the Vafa–Witten equations for a pair (A, ϕ) consisting of a connection A over a 4-manifold and a section ϕ of $\Lambda^+ \otimes ad(E)$ (Vafa and Witten 1994).

7 An example: quadrics in P^5

We will now discuss an example which illustrates the ideas of Section 4 above. We consider a Calabi–Yau manifold $Y_1 \subset \mathbf{CP}^5$ which is the complete intersection of a quadric Q_0 and a quartic hypersurface V. We can degenerate V into a union of two quadrics $Q_1 \cup Q_2$, and in this way embed Y_1 in a family Y_t with Y_0 the union of two pieces $Y_0 = Y^+ \cup_S Y^-$, where

$$Y^+ = Q_0 \cap Q_1, \; Y^- = Q_0 \cap Q_2, \; S = Q_0 \cap Q_1 \cap Q_2.$$

So S is the intersection of three quadrics in $\mathbf{CP}^5$; a $K3$ surface. There is a wonderful explicit construction of a certain moduli space of holomorphic bundles over such a surface S, an example of Mukai duality. To explain this we must review some basic facts about quadrics in $\mathbf{P}^5$. If $Q \subset \mathbf{P}^5$ is any nonsingular quadric there are two families of planes in Q, the "α-planes" and "β-planes" in the language of twistor theorists. The α-planes through each point are parametrised by a copy of $\mathbf{P}^1$ and similarly for the β-planes. So we get two $\mathbf{P}^1$-bundles P_α, P_β over Q. (These can be lifted to vector bundles, but it easier to work with $\mathbf{P}^1$ bundles here.) Another way to view this is to identify Q with the Klein quadric $Gr_2(\mathbf{C}^4)$. Then P_α and P_β are the projectivisations of the tautological bundle U and the quotient bundle $\mathbf{C}^4/U$. However it is important to realise that in general there is a complete symmetry between P_α and P_β, with no preferred way to choose which one is which. (This happens because the orthogonal group has two components.) If we take instead a singular quadric Q', with one singular point, then there is just one family of planes in Q': so roughly speaking the bundles P_α and P_β coalesce as the quadric becomes singular.

Now consider the sextic curve $B \subset \mathbf{P}^2$ given by the equation

$$\det(\lambda_0 Q_0 + \lambda_1 Q_1 + \lambda_2 Q_2) = 0.$$

Each point $(\lambda_0, \lambda_1, \lambda_2)$ of $\mathbf{P}^2 \setminus B$ defines a nonsingular quadric containing the surface S. We get a double cover of this complement by fixing a choice of which of two bundles is to be P_α. For each point of this double cover we get a bundle over S, the restriction of the bundle P_α, for that quadric, to S. This construction extends over the curve B, where the quadrics become singular, but the cover becomes branched there (because the bundles coalesce, as mentioned above). The upshot is that the moduli space $M(S)$ of bundles of the relevant topological type over the K3 surface S is the double cover of the plane branched along the sextic curve B, which is another $K3$ surface.

Now consider the manifold $Y^+ = Q_0 \cap Q_1$. The quadrics through Y^+ are given by setting $\lambda_2 = 0$, and we can repeat the construction above to get a moduli space $L^+ \subset M(S)$ of bundles over Y^+ which is the double cover of the line $\{\lambda_2 = 0\}$ in $\mathbf{P}^2$, branched along the six intersection points with B. This is the description by Newstead of the moduli space of bundles of odd degree over a curve of genus 2 as the intersection of two quadrics, but seen in the opposite way: the universal bundle can be seen either as a family of bundles over the curve L^+ parametrised by Y^+ or vice-versa. In just the same way the subset L^-, parametrising bundles over Y^-, is the double cover of the line $\{\lambda_1 = 0\}$. The intersection $L^+ \cap L^-$ consists of two points and, following through the definitions, one sees that these just correspond to the restriction to Y_0 of the two bundles P_α and P_β over the original quadric Q_0. This suggests that the only stable bundles of the relevant topological type over the smooth Calabi–Yau manifold Y_1 are the restrictions of P_α, P_β. (Note that P_α, P_β have different topological type over Q_0 since they have different characteristic classes in $H^4(Q_0) = \mathbf{Z} \oplus \mathbf{Z}$, but this difference is not seen over Y_1 since $H^4(Y_1) = \mathbf{Z}$.)

8 Vanishing cycles and pseudoholomorphic curves

Here we return to discuss the analogue of the Floer theory in our complex setting. The starting point for the ordinary Floer theory is the Morse theory of a real valued function, and it is well-known that the Picard–Lefschetz theory of monodromy and vanishing cycles is the complex analogue of the Morse theory, so it is not surprising that these ideas emerge from our discussion. If we have any Kähler manifold Z and a holomorphic function $\phi : Z \to \mathbf{C}$ with nondegenerate critical points we can define the complex gradient flow equation, for a map $\Gamma : S^1 \times \mathbf{R} \to Z$,

$$\overline{\partial}\Gamma = \widehat{d\phi}. \tag{8.1}$$

Here $\widehat{d\phi}$ is the tangent vector obtained from the derivative of ϕ using the metric. This is a deformation of the holomorphic mapping equation, and fits into a class which has been studied quite extensively in that setting. Indeed if H is any real valued function on Z there is a standard deformation of the holomorphic mapping equation given, from one point of view, by adding

$$\int_{S^1} H \circ \gamma$$

to the symplectic action functional for maps $\gamma : S^1 \to Z$. The equation (8.1) is just this standard deformation for the function $H = Re(\phi)$. Pursuing this line, one sees that there will generically be no solutions of our equation (8.1) which interpolate between different critical points of ϕ (which are also the critical points of H), essentially because the Morse indices of H at all these critical points are the same. However we can bring in the fact that H came from a holomorphic function by considering the family of equations like (8.1) parameterised by a circle, just multiplying ϕ by a complex number λ of unit modulus. The simplest solutions are those which are invariant under the rotations acting on $S^1 \times \mathbf{R}$: these are just the ordinary gradient lines of H. Standard Morse theory arguments give the following picture. For each pair of critical points p^+, p^- we count the S^1-invariant solutions of the family of deformed equations, as λ varies over the circle (but dividing of course by the obvious action of translations) to give a number $n(p^+, p^-)$. Now for each critical point p there is a *vanishing cycle* $W(p)$, well-defined up to isotopy, in any nearby fibre. We take the straight line in $\mathbf{C}$ between $\phi(p^+)$ and $\phi(p^-)$, which generically does not contain any other critical value. Parallel transport along this line allows us to regard the vanishing cycles $W(p^+), W(p^-)$ as submanifolds of the same fibre, and $n(p^+, p^-)$ is the intersection number of these two vanishing cycles. (The data from the more general moduli spaces of solutions, not rotation invariant, describes the action of the "quantum multiplication" in the fibres of ϕ on the vanishing cycles.) The set of numbers $n(p^+, p^-)$, as p^+, p^- range over the different critical points, together with knowledge of the location of the critical values in $\mathbf{C}$, and the reflection formula for the monodromy around a single critical value, gives a complete description of the monodromy action of the fundamental group of $\mathbf{C} \setminus \{\text{critical values}\}$ on the part of the homology of the fibre generated by the vanishing cycles. This is the complex analogue of the boundary operator in the familar Morse–Floer picture, defined by gradient lines, which gives a complete description of the homology of the space. Just as in that case, the numbers $n(p^+, p^-)$ themselves can change under continuous deformations of the set up (Z, ϕ). In the complex case this change comes about when a critical point p' moves across the line segment between p^-, p^+, so one is essentially changing the choice of homotopy class of path used to identify fibres around different critical points.

Now we can take these ideas over to the gauge theory case, where Z becomes the space $\mathcal{C}_E$ of equivalence classes of $\bar{\partial}$-operators on a bundle E over our Calabi–Yau manifold Y (or, perhaps better, a suitable covering space), and ϕ becomes our holomorphic Chern–Simons functional Φ. Multiplying the map Φ by a scalar just corresponds to changing the choice of holomorphic volume form on Y, and hence on $X = \mathbf{R} \times S^1 \times Y$. What we expect then is that for each pair of holomorphic bundles E^+, E^- of the same topological type over Y we can define a number $n(E^+, E^-)$ by counting the rotation-invariant solutions of the 1-parameter family of $SU(4)$-instanton equations over X, asymptotic to $E^{\pm}$ at $\pm\infty$, and that these numbers can be interpreted as giving the monodromy action on the semi-infinite dimensional cohomology of the fibres of Φ. The data from other moduli spaces (not rotation-invariant) should describe the quantum multiplication between the semi-infinite and finite dimensional cohomology of the fibres.

We finish this discussion with one more remark. In a finite-dimensional situation the index associated to the Cauchy–Riemann equations for mappings from a closed Riemann surface Σ of genus g to a complex manifold Z is

$$\langle c_1(Z), [\Sigma]\rangle - (2g-2)\dim Z. \tag{8.2}$$

We get the same index for any deformed equation, which merely adds lower order terms. Now in our case we take Σ to be a 2-torus, and Z to be the infinite-dimensional space $\mathcal{C}_E$. For simplicity we take the gauge group of our connections to be $SU(2)$. From what we have seen above, $SU(4)$-instantons on a bundle E' over $\Sigma \times Y$ can be interpreted as solutions of a deformation of the holomorphic mapping equation. The familiar slant product construction gives an isomorphism $\mu : H_2(Y) \to H^2(\mathcal{C})$. Under this isomorphism the homology class of Σ in $\mathcal{C}$ can be identified with the component of $c_2(E')$ in $H^2(Y) = H^2(Y) \otimes H^2(\Sigma) \subset H^4(Y \times \Sigma)$. On the other hand the index associated to the elliptic $SU(4)$-instanton equation in this situation is just $\langle \frac{1}{6}c_2(Y)c_2(E') - \frac{2}{3}c_2(E')^2, \Sigma \times Y\rangle$. The upshot is that the formula (8.2) is true in the infinite dimensional case if we *define*

$$c_1(\mathcal{C}_E) = \tfrac{1}{6}\mu(PD(c_2(Y) - 8c_2(E))),$$

where $PD(c_2(Y) - 8c_2(E)) \in H_2(Y)$ is the Poincaré dual. The point here is that this Chern class is not defined in any conventional sense for general infinite dimensional complex manifolds (since the infinite general linear group is contractible), but our index theory allows us to give a meaning to it. Notice also that whereas in finite dimensions the deformations of the holomorphic mapping equation do not play any role in the index theory, the analogous deformation is crucial to the discussion above. The condition for a genuine holomorphic mapping from Σ to $\mathcal{C}_E$ can be interpreted as a gauge theory problem on $\Sigma \times Y$, but leads to a nonelliptic equation which is not governed by any index theorem.

9 Submanifolds

In this article we have discussed gauge theory over various manifolds of dimensions $4, 6, 7, 8$. We close by pointing out that there are analogues of most of our constructions which bear instead on special *submanifolds*, in the framework of Harvey and Lawson's "calibrations" (Harvey and Lawson 1982). This is a very active research area at the moment, in part because of connections with "mirror symmetry". The viewpoint which we arrive at, by analogy with the gauge theory set-up, is perhaps new.

On the one hand, returning to our Calabi–Yau 3-fold Y, it is natural to expect connections between counting holomorphic bundles (at least of rank 2), and counting complex *curves*. Indeed one of the standard procedures in algebraic geometry is to go from a rank 2 bundle over a 3-dimensional variety to a curve by taking the zero set of a generic section. On the other hand, we can forget about bundles, and mimic the general scheme in this article, replacing connections by suitable submanifolds.

So let us consider our Calabi–Yau 3-fold Y, with a fixed Kähler metric ω with holonomy $SU(3)$. Let $\mathcal{S}$ be the space of (real) 2-dimensional submanifolds

Σ, representing a given homology class in Y, which are symplectic with respect to ω; i.e. such that $\omega|_\Sigma$ is non-degenerate. Fix a base point Σ_0 in $\mathcal{S}$, and let $\tilde{\mathcal{S}}$ be the covering of $\mathcal{S}$ consisting of pairs $(\Sigma, [Z])$ where $[Z]$ is an equivalence class of 3-chains Z in Y with boundary $\Sigma - \Sigma_0$, and $Z \sim Z'$ if $[Z-Z']$ is zero in $H_3(Y)$. Then we can define a functional

$$\Psi : \widetilde{\mathcal{S}} \to \mathbf{C}$$

by

$$\Psi(\Sigma, [Z]) = \int_Z \theta.$$

It is an easy exercise to show that the critical points of Ψ correspond to the *holomorphic curves* in Y. Now we can define a (real) gradient equation associated to this functional as follows. If ξ, η are independent tangent vectors to Σ at a point p, let v be a tangent vector corresponding, under the metric on Y, to the 1-form

$$\frac{1}{\omega(\xi,\eta)}\, \xi \lrcorner \eta \lrcorner \mathrm{Re}(\theta).$$

This does not depend on the choice of basis ξ, η for $T_p\Sigma$, and we denote it by $v_\Sigma(p)$. Thus we get a vector field v_Σ along Σ in Y which can be regarded as a tangent vector in $\mathcal{S}$. Then the gradient equation is $\partial\Sigma/\partial t = v_\Sigma$. Similarly, we can define a complex gradient equation for a map from $\mathbf{C}$ to $\mathcal{S}$:

$$\frac{\partial \Sigma}{\partial t} + i\frac{\partial \Sigma}{\partial s} = v_\Sigma.$$

The surprising thing is that these equations have straightforward interpretation in seven and eight dimensions. Consider the model $\mathbf{R}^7 = \mathbf{R}^4 \times \Lambda^+(\mathbf{R}^4)$ for the imaginary octonians. A 3-dimensional subspace of $\mathbf{R}^7$ is called *associative* if it lies in the G_2-orbit of $\Lambda^+(\mathbf{R}^4)$ in the Grassmannian of 3-planes in $\mathbf{R}^7$. This leads to the notion of an *associative* 3-dimensional submanifold of a manifold N^7 with a G_2-structure. In the same way the model $\mathbf{R}^8 = \mathbf{R}^4 \times S_+(\mathbf{R}^4)$ leads to the notion of a *Cayley submanifold*, of dimension four, in an 8-manifold with a $Spin(7)$ structure.

Suppose Σ_t is a 1-parameter family of surfaces in Y. The family may be considered in an obvious way as a 3-dimensional submanifold Γ of $Y \times \mathbf{R}$, and a little thought shows the real gradient equation precisely goes over to the condition that Γ be an associative submanifold. Similarly for the complex gradient equation and Cayley submanifolds of $Y \times \mathbf{C}$. In the same spirit, we can look at the space of 3-manifolds in a manifold N with a G_2 structure, and get a gradient flow equation whose solutions are the Cayley submanifolds in $N \times \mathbf{R}$. In sum, we might hope that there are Floer theories, etc., involving these special submanifolds in close parallel to the gauge theory structures we have discussed in this article. The "special Lagrangian" submanifolds of Y can also be brought into this picture: they are just the associative submanifolds in $Y \times \mathbf{R}$ which lie in a single fibre $Y \times \{t\}$: they can be interpreted as gradient lines taking the empty set to itself. Similarly for the co-associative submanifolds in a 7-manifold N with a G_2 structure, which appear as the "vertical" Cayley submanifolds in $N \times \mathbf{R}$. In the analogy between the gauge theory and submanifold theory, the L^2 curvature identities like (3.5) go over to the familiar volume identities for calibrated submanifolds of Harvey and Lawson.

Bibliography

Corrigan, E., Devchand, C., Fairlie, D., and Nuyts, J. (1983). First-order equations for gauge fields in spaces of dimension greater than four *Nucl. Phys.*, **B 214**, 452–64.

De Bartolomeis, P. and Tian, G. (1996). Stability of complex vector bundles *Jour. Differential Geometry*, **43**, 231–75.

Dostoglou, S. and Salamon, D. (1994). Self-dual instantons and holomorphic curves *Annals of Math.* **139**, 581–640.

Friedman, R. (1983). Global smoothings of varieties with normal crossings. *Annals of Math.*, **118**, 75–114.

Fubini, S. and Nicolai, H. (1985). The octonionic instanton. *Phys. Lett.*, **155**, 369–72.

Harvey, R. and Lawson, H. B. (1982). Calibrated geometries. *Acta Math.* **148**, 47–157.

Hitchin, N. J. (1987). The self-duality equations on a Riemann surface. *Proc. Lond. Math. Soc.*, **55**, 59–126.

Mukai, S. (1984). Symplectic structure of the moduli space of stable sheaves on an abelian or $K3$ surface. *Inv. Math.*, **77**, 101–116.

Reyes-Carrion, R. (1997). On a generalisation of instantons. To appear in *Diff. Geom. and its Applications.*

Salamon, S. (1989). *Riemannian geometry and holonomy groups.* Pitman Research Notes in Mathematics Series, **201**.

Taubes, C. H. (1990). Casson's invariant and gauge theory. *Jour. Differential Geometry* **31**, 547–99.

Tyurin, A. (1990). *The moduli spaces of vector bundles on threefolds, surfaces and curves.* Math. Inst. Univ. Gottingen Preprint.

Vafa, C. and and Witten, E. (1994). A strong coupling test of S-duality. *Nucl. Phys.* **431B**, 3–77.

Ward, R. S. (1984). Completely solvable gauge-field equations in dimensions greater than 4. *Nucl. Phys.*, **236B**, 381–96.

Witten, E. (1995). Chern-Simons gauge theory as a string theory. In: *The Floer Memorial Volume.* Progr. Math. **133**, Birkhauser, 637–78.

4

Noncommutative Differential Geometry and the Structure of Space–Time

Alain Connes
Collège de France, 3, rue Ulm, 75005 PARIS
and
I.H.E.S., 35, route de Chartres, 91440 BURES-sur-YVETTE

Notes by Olga Kravchenko

Foreword

One of the original motivations of noncommutative geometry is to apply geometric ideas and concepts to spaces which are intractable if considered from the usual set–theoretic ideas of Riemannian geometry. Among the first examples of such spaces are the leaf spaces of foliations or the duals of noncommutative discrete groups. A more recent example comes from number theory and the Riemann zeta function, and allows us to give a spectral interpretation of the critical zeros of that function. One of the difficulties of the subject is to acquire an intuitive idea of the nature of noncommutative spaces. I remember with emotion the talk given by Roger Penrose a few years ago in which he exhibited, showing the superposition of two transparencies representing Penrose tilings, the quantum nature of the space of Penrose tilings of the plane. The point is that even though the space of such tilings is uncountable, any two tilings can be superposed exactly on arbitrarily large portions, thus exhibiting fluctuations very reminiscent of quantum mechanics. The space of Penrose tilings is a prototype of a noncommutative space and it gives me great pleasure to expound the latest developments of noncommutative geometry on the occasion of his birthday.

1 Generalities

The basic data of Riemannian geometry consists of a manifold M whose points are locally labelled by a finite number of real coordinates $\{x^\mu\}$ and a *metric*, which is given by the infinitesimal line element:

$$ds^2 = g_{\mu\nu}\, dx^\mu\, dx^\nu\,. \tag{1.1}$$

The distance between two points $x, y \in M$ is given by

$$d(x,y) = \mathrm{Inf}\{\text{Length } \gamma \,|\, \gamma \text{ is a path between } x \text{ and } y\} \tag{1.2}$$

where

$$\text{Length } \gamma = \int_\gamma ds\,. \tag{1.3}$$

Riemannian geometry is flexible enough to give a good description of space–time in general relativity (up to a sign change). The essential point here is that the differential and integral calculus allows one to go from the local to the global, while simple notions of Euclidean geometry continue to make sense. For instance the idea of a straight line gives rise to the notion of geodesic. The geodesic equation

$$\frac{d^2\, x^\mu}{dt^2} = -\Gamma^\mu_{\nu\rho}\, \frac{dx^\nu}{dt}\, \frac{dx^\rho}{dt} \tag{1.4}$$

where $\Gamma^\mu_{\nu\rho} = \frac{1}{2}\, g^{\mu\alpha}(g_{\alpha\nu,\rho} + g_{\alpha\rho,\nu} - g_{\nu\rho,\alpha})$, gives the Newton equation of the motion of a particle in the Newtonian potential V provided one uses the metric $dx^2 + dy^2 + dz^2 - (1 + 2V(x, y, z))dt^2$ (cf. Weinberg, 1972, for the more precise formulation). Recent experimental data on binary pulsars confirms through general relativity (Damour and Taylor, 1992) that Riemannian geometry works well as a model for space–time on a sufficiently large scale. However, it is not clear (Riemann, 1953) whether this geometry is adequate for the description of the small scale structure of space–time. The Planck length:

$$\ell_p = (G\hbar/c^3)^{1/2} \sim 10^{-33}\text{cm} \tag{1.5}$$

is considered as a natural lower limit for the precision at which coordinates of an event in space–time make sense. (See for example Fröhlich, 1994, or Doplicher *et al.* 1995, for a physical argument establishing this limit using quantum mechanics.)

In these notes we shall present a new notion of a geometric space where *points* do not play the central role, thus giving much more freedom for describing the small scale texture of space–time. The proposed framework is sufficiently general to treat discrete spaces, Riemannian manifolds, configuration spaces of quantum field theory, and the duals of discrete groups which are not necessarily commutative. The main problem is to show that the usual geometrical ideas and the tools of the infinitesimal calculus do adapt to this more general framework. It turns out that the operator formalism of quantum mechanics together with the analysis of logarithmic divergences of traces of operators give the generalization of the differential and integral calculus that we are looking for. Several direct applications of this approach are described in theorems 2.1, 2.2, and 3.1.

We consider a geometric space to be a *spectral triple:*

$$(\mathcal{A}, \mathcal{H}, D) \tag{1.6}$$

where $\mathcal{A}$ is an involutive algebra of operators in a Hilbert space $\mathcal{H}$ and D is a self-adjoint operator on $\mathcal{H}$. The involutive algebra $\mathcal{A}$ corresponds to a given space M like in the classical duality "Space $\leftrightarrow$ Algebra" in algebraic geometry. The operator $D^{-1} = ds$ corresponds to the infinitesimal line element in Riemannian geometry.

One can see the difference between this *spectral geometry* and Riemannian geometry in two ways. Firstly it is very important that one does not assume that the algebra $\mathcal{A}$ is commutative any more. Secondly the infinitesimal ds in spectral geometry becomes an operator and does not commute with elements of $\mathcal{A}$ even if the algebra $\mathcal{A}$ is commutative itself.

As we shall see, simple commutation relations between ds and elements of $\mathcal{A}$, together with Poincaré duality, characterize the spectral triples (1.6) which come from Riemannian manifolds (Theorem 4.1). When the algebra $\mathcal{A}$ is commutative the spectrum of its norm closure $\bar{\mathcal{A}}$ in bounded operators on $\mathcal{H}$ is a compact space M. A point of M is a character of $\bar{\mathcal{A}}$, i.e. a homomorphism from $\bar{\mathcal{A}}$ to $\mathbb{C}$,

$$\begin{aligned} &\chi : \bar{\mathcal{A}} \to \mathbb{C}, \quad \chi(a+b) = \chi(a) + \chi(b), \quad \chi(\lambda a) = \lambda \chi(a), \\ &\chi(ab) = \chi(a)\chi(b) \qquad \forall\, a, b \in \bar{\mathcal{A}},\ \forall\, \lambda \in \mathbb{C}\,. \end{aligned} \tag{1.7}$$

As an example let us take $\mathcal{A}$ to be the group algebra $\mathbb{C}\Gamma$ for a discrete group Γ acting on the Hilbert space $\mathcal{H} = \ell^2(\Gamma)$ by the regular (left) representation of Γ. When the group Γ and hence the algebra $\mathcal{A}$ are commutative then the characters of $\bar{\mathcal{A}}$ are elements of the Pontryagin dual of Γ,

$$\hat{\Gamma} = \{\chi : \Gamma \to U(1)\ ;\ \chi(g_1\, g_2) = \chi(g_1)\,\chi(g_2) \qquad \forall\, g_1, g_2 \in \Gamma\}\,. \tag{1.8}$$

Elementary notions of differential geometry for the space $\hat{\Gamma}$ continue to make sense in the general case when Γ is no longer commutative. The right column in the following dictionary does not use the commutativity of the algebra $\mathcal{A}$:

Space X	**Algebra $\mathcal{A}$**
Vector bundle	Finite projective module
Differential form of degree k	Hochschild cycle of dimension k
De Rham current of dimension k	Hochschild cocycle of dimension k
De Rham homology	Cyclic cohomology of $\mathcal{A}$

The power of this generalization to the noncommutative case is demonstrated for example in the proof of the Novikov conjecture (Novikov, 1965) for hyperbolic groups Γ (Connes and Moscovici, 1990).

In the general case the notion of a point given by (1.7) is not of much interest; but the notion of probability measure keeps its full meaning. Such a measure φ, also called a state, is a normalized positive linear form on $\mathcal{A}$ such that $\varphi(1) = 1$,

$$\varphi : \bar{\mathcal{A}} \to \mathbb{C}\,,\ \varphi(a^* a) \geq 0 \qquad \forall\, a \in \bar{\mathcal{A}}\,,\ \varphi(1) = 1\,. \tag{1.9}$$

Instead of measuring distances between points using the formula (1.2) we measure distances between states φ, ψ on $\bar{A}$ by a dual formula. This dual formula involves *sup* instead of *inf* and does not use paths in the space

$$d(\varphi, \psi) = \text{Sup}\,\{|\varphi(a) - \psi(a)|\,;\;\; a \in \mathcal{A}\;,\;\; \|[D,a]\| \leq 1\}\,. \tag{1.10}$$

Let us show that this formula indeed gives the geodesic distance in the Riemannian case. Let M be a Riemannian compact manifold with a K-orientation, i.e. a spin structure. The associated spectral triple $(\mathcal{A}, \mathcal{H}, D)$ is given by the representation

$$(f\,\xi)(x) = f(x)\,\xi(x) \qquad \forall\, x \in M\;,\;\; f \in \mathcal{A}\;,\;\; \xi \in \mathcal{H} \tag{1.11}$$

of the algebra $\mathcal{A}$ of functions on M in the Hilbert space

$$\mathcal{H} = L^2(M, S) \tag{1.12}$$

of the square integrable sections of the spinor bundle.

The operator D is the Dirac operator (cf. Lawson and Michelson, 1989). One can check that the commutator $[D, f]$, for $f \in \mathcal{A} = C^\infty(M)$ is an operator which is the Clifford multiplication by the gradient ∇f and that its operator norm is:

$$\|[D,f]\| = \;\text{Sup}_{x \in M}\, \|\nabla f(x)\| = \text{Lipschitz norm}\; f\,. \tag{1.13}$$

Let $x, y \in M$ and φ, ψ be the corresponding characters: $\varphi(f) = f(x)$, $\psi(f) = f(y)$ for $\forall\; f \in \mathcal{A}$. Then formula (1.10) gives the same result as formula (1.2), i.e. it gives the geodesic distance between x and y.

Unlike the formula (1.2) its dual formula (1.10) makes sense in general, namely, for example for discrete spaces and even totally disconnected spaces.

The usual notion of *dimension* of a space is replaced by the *dimension spectrum* which is a subset of $\mathbb{C}$ with real part not bigger than some $\alpha > 0$ if

$$\lambda_n^{-1} = O(n^{-\alpha}) \tag{1.14}$$

where λ_n is the nth eigenvalue of $|D|$.

The relation between local and global data is given by the local index formula, Theorem 3.1 (Connes and Moscovici, 1995).

The characteristic property of *differentiable manifolds* which is carried over to the noncommutative case is *Poincaré duality.* Poincaré duality in ordinary homology is not sufficient to describe homotopy type of manifolds (Milnor and Stasheff, 1974) but D. Sullivan (1971) showed (in the simply connected PL case of dimension ≥ 5 ignoring 2-torsion) that it is sufficient to replace ordinary homology by KO-homology.

Moreover K-homology (cf. for example Atiyah, 1966), thanks to the work of Brown, Douglas, Fillmore, Atiyah, and Kasparov, admits a fairly simple definition in algebraic terms, given by

Space X	Algebra $\mathcal{A}$
$K_1(X)$	Stable homotopy class of the spectral triple $(\mathcal{A}, \mathcal{H}, D)$
$K_0(X)$	Stable homotopy class of $\mathbb{Z}/2$ graded spectral triple

(i.e. for K_0 we suppose that $\mathcal{H}$ is $\mathbb{Z}/2$-graded by γ, where $\gamma = \gamma^*$, $\gamma^2 = 1$ and $\gamma a = a\gamma \quad \forall\, a \in \mathcal{A}$, $\gamma D = -D\gamma$).

This description works for the complex K-homology which is 2-periodic.

The *fundamental class* of a noncommutative space is a class μ in the KR-homology of the algebra $\mathcal{A} \otimes \mathcal{A}^0$ equipped with the involution

$$\tau(x \otimes y^0) = y^* \otimes (x^*)^0 \qquad \forall\, x, y \in \mathcal{A} \tag{1.15}$$

where $\mathcal{A}^0$ denotes the algebra opposite to $\mathcal{A}$. The Kasparov intersection product (Kasparov, 1980) allows one to formulate Poincaré duality in terms of the invertibility of μ. In KR-homology besides the above data one has to give an antilinear isometry J on $\mathcal{H}$ which implements the involution τ in as much as

$$JwJ^{-1} = \tau(w) \qquad \forall\, w \in \mathcal{A} \otimes \mathcal{A}^0 \,. \tag{1.16}$$

KR-homology is periodic with period 8 and the dimension modulo 8 is specified by the following commutation rules. One has $J^2 = \varepsilon$, $JD = \varepsilon' DJ$, $J\gamma = \varepsilon'' \gamma J$ where $\varepsilon, \varepsilon', \varepsilon'' \in \{-1, 1\}$ and with n the dimension modulo 8,

n	0	1	2	3	4	5	6	7
ε	1	1	-1	-1	-1	-1	1	1
ε'	1	-1	1	1	1	-1	1	1
ε''	1		-1		1		-1	

The antilinear isometry J is given in Riemannian geometry by the charge conjugation operator and in the noncommutative case by the Tomita operator (Takesaki, 1970). When an operator algebra $\mathcal{A}$ admits a cyclic vector which is cyclic for the commutant $\mathcal{A}'$, the Tomita operator establishes an anti-isomorphism

$$\mathcal{A}'' \to \mathcal{A}' : a \mapsto Ja^* J^{-1} \,. \tag{1.17}$$

The class μ specifies only the stable homotopy class of the spectral triple $(\mathcal{A}, \mathcal{H}, D)$ equipped with the isometry J (and $\mathbb{Z}/2$-grading γ if n is even). The nontriviality of this homotopy class shows up in the intersection form

$$K(\mathcal{A}) \times K(\mathcal{A}) \to \mathbb{Z} \tag{1.18}$$

given by the Fredholm index of D with coefficients in $K(\mathcal{A} \otimes \mathcal{A}^0)$.

In order to compare different spectral triples in the same homotopy class defined by μ we shall use the following spectral functional

$$\mathrm{Trace}\,(\varpi(D))\,, \tag{1.19}$$

where $\varpi : \mathbb{R} \to \mathbb{R}_+$ is a suitable positive function.

Once the algebra $\mathcal{A}$ fixed, a spectral geometry is determined by the unitary equivalence class of a spectral triple $(\mathcal{A}, \mathcal{H}, D)$ together with the isometry J. Denote the representation of $\mathcal{A}$ in $\mathcal{H}$ by π. $(\pi_1, \mathcal{H}_1, D_1, J_1)$ is unitarily equivalent to $(\pi_2, \mathcal{H}_2, D_2, J_2)$ if there exists a unitary $U : \mathcal{H}_1 \to \mathcal{H}_2$ such that

$$U\pi_1(a)\,U^* = \pi_2(a) \quad \forall\, a \in \mathcal{A}\ ,\ UD_1U^* = D_2\ ,\ UJ_1U^* = J_2 \tag{1.20}$$

(and $U\gamma_1 U^* = \gamma_2$ for even n).

The automorphism group $\mathrm{Aut}(\mathcal{A})$ of the involutive algebra $\mathcal{A}$ acts on the variety of spectral geometries by the following composition:

$$\pi'(a) = \pi(\alpha^{-1}(a)) \qquad \forall\, a \in \mathcal{A}\ ,\ \alpha \in \mathrm{Aut}(\mathcal{A})\,. \tag{1.21}$$

Let $\mathrm{Aut}^+(\mathcal{A})$ be the subgroup of automorphisms which preserve the class μ. $\mathrm{Aut}^+(\mathcal{A})$ acts on the stable homotopy class determined by μ and preserves the action functional (1.19). In general this subgroup is not compact. For Riemannian manifolds it coincides with the group of diffeomorphisms preserving the K-orientation, $\mathrm{Diff}^+(M)$, i.e. the subgroup preserving the spin structure on M. On the other hand, the isotropy group of a given geometry is automatically *compact* (for a unital $\mathcal{A}$). This implies that the action functional (1.19) automatically produces the phenomenon of spontaneous symmetry breaking.

We will show that for the right choice of the algebra $\mathcal{A}$ the functional of action (1.19) added to $\langle \xi, D\xi \rangle$, $\xi \in \mathcal{H}$, gives the standard Glashow–Weinberg–Salam model coupled with gravity. The algebra $\mathcal{A}$ is the tensor product of the algebra of functions on a Riemannian space M and a finite dimensional noncommutative algebra whose spectral geometry is specified by the phenomenological data from particle physics.

2 Infinitesimal calculus

Here we shall give a precise meaning to the notion of infinitesimal variable using the operator formalism of quantum mechanics.

The notion of infinitesimal is supposed to have an obvious intuitive meaning. Let us take a particular example (Bernstein and Wattenberg, 1969). Let us consider darts thrown arbitrarily into a target Ω. Let $dp(x)$ be the probability for the dart to hit the point $x \in \Omega$. It is clear that $dp(x) < \varepsilon$ for any $\varepsilon > 0$ but nevertheless saying that $dp(x) = 0$ is unsatisfactory. The usual formalisms of measure theory or of the theory of differential forms avoid the problem by giving sense to the following expression

$$\int f(x)\, dp(x)\ , \qquad f : \Omega \to \mathbb{C} \tag{2.1}$$

but it is insufficient to give any meaning to expressions like $e^{-\frac{1}{dp(x)}}$. The answer given by nonstandard analysis, a so-called nonstandard real, is equally deceiving. From every nonstandard real number one can construct canonically a subset of the interval $[0,1]$, which is not Lebesgue measurable. No such set can be exhibited (Stern, 1985). This implies that not a single nonstandard real number can actually be exhibited. The formalism which we propose below will give a substantial and computable answer to the above problem.

Our framework consists of a separable Hilbert space decomposed into two orthogonal infinite dimensional subspaces. This decomposition can be described by the linear operator F on $\mathcal{H}$ which is the identity, $F\xi = \xi$, on one subspace and minus identity, $F\xi = -\xi$, on the other; so one has

$$F = F^* \,,\ F^2 = 1\,. \tag{2.2}$$

This framework is determined uniquely up to isomorphism. The following is the beginning of a long dictionary which translates classical notions into operator language:

Classical	**Quantum**
Complex variable	Operator in $\mathcal{H}$
Real variable	Self-adjoint operator
Infinitesimal	Compact operator
Infinitesimal of order α	Compact operator with eigenvalues μ_n satisfying $\mu_n = O(n^{-\alpha})\,,\ n \to \infty$
Differential of a real or complex variable	$đ f = [F, f] = Ff - fF$
Integral of an infinitesimal of order 1	$⨍ T =$ Coefficient of logarithmic divergence in the trace of T.

The first two lines of the dictionary are familiar from quantum mechanics. The range of a complex variable corresponds to the *spectrum* of an operator. The holomorphic functional calculus gives a meaning to $f(T)$ for all holomorphic functions f on the spectrum of T. It is only holomorphic functions which operate in this generality which reflect the difference between complex and real analysis. When $T = T^*$ is self-adjoint then $f(T)$ has a meaning for all Borel functions f. Notice that every usual random variable X on the probability space (Ω, P) can be trivially considered as a self-adjoint operator. One takes $\mathcal{H} = L^2(\Omega, P)$ and T to be the operator of multiplication by X:

$$(T\xi)(p) = X(p)\,\xi(p) \qquad \forall\, p \in \Omega\,, \xi \in \mathcal{H}\,. \tag{2.3}$$

Let us consider the third line of the dictionary. We look for a definition of "infinitesimal variables", i.e. operators T in $\mathcal{H}$ such that

$$\|T\| < \varepsilon \qquad \forall \varepsilon > 0\,, \tag{2.4}$$

where $\|T\| = \text{Sup}\,\{\|T\xi\|\,;\ \|\xi\| = 1\}$ is the operator norm. Clearly, if one takes (2.4) literally then one gets $\|T\| = 0$ and $T = 0$ as the unique solution, but one can weaken (2.4) in the following way:

$$\forall \varepsilon > 0\ ,\ \exists \text{ finite–dimensional space } E \subset \mathcal{H} \text{ such that } \|T|_{E^\perp}\| < \varepsilon\,. \tag{2.5}$$

Here $E^\perp$ denotes the orthogonal complement to E in the space $\mathcal{H}$:

$$E^\perp = \{\xi \in \mathcal{H}\,;\ \langle \xi, \eta \rangle = 0 \quad \forall \eta \in E\} \tag{2.6}$$

which is a subspace of finite codimension in $\mathcal{H}$. $T|_{E^\perp}$ denotes the restriction of T to this subspace,

$$T|_{E^\perp} : E^\perp \to \mathcal{H}\,. \tag{2.7}$$

Operators which satisfy the condition (2.5) are the *compact operators*, i.e. those for which the image of the unit ball of $\mathcal{H}$ has compact closure. The operator T is compact if and only if $|T| = \sqrt{T^*T}$ is compact, and this means that the spectrum of $|T|$ is a sequence of eigenvalues $\mu_0 \geq \mu_1 \geq \mu_2 \ldots\ ,\ \mu_n \downarrow 0$.

These eigenvalues are the characteristic values of T and one has

$$\mu_n(T) = \text{Inf}\,\{\|T - R\|\,;\ R \text{ is an operator of rank } \leq n\} \tag{2.8}$$

$$\mu_n(T) = \text{Inf}\,\{\|T/E^\perp\|\,;\ \dim E \leq n\}\,. \tag{2.9}$$

Compact operators form a two–sided ideal $\mathcal{K}$ in the algebra $\mathcal{L}(\mathcal{H})$ of bounded operators in $\mathcal{H}$.

Consider now the fourth entry in our dictionary. The size of the infinitesimal $T \in \mathcal{K}$ is governed by the order of decay of the sequence $\mu_n = \mu_n(T)$ as $n \to \infty$. In particular, for all real positive α the following condition defines infinitesimals of order α:

$$\mu_n(T) = O(n^{-\alpha}) \qquad \text{when } n \to \infty \tag{2.10}$$

(i.e. there exists $C > 0$ such that $\mu_n(T) \leq Cn^{-\alpha} \quad \forall n \geq 1$). Infinitesimals of order α also form a two–sided ideal as can be checked using (2.8), (cf. Connes, 1994) and moreover,

$$T_j \text{ of order } \alpha_j \Rightarrow T_1 T_2 \text{ of order } \alpha_1 + \alpha_2\,. \tag{2.11}$$

(For $\alpha < 1$ this ideal is a normed ideal given by real interpolation between the ideal $\mathcal{L}^1$ of trace class operators and the ideal $\mathcal{K}$, Connes, 1994.)

Hence, apart from commutativity, intuitive properties of the infinitesimal calculus are fulfilled.

Since the size of the infinitesimal is measured by the sequence $\mu_n \to 0$ it might seem that one does not need the operator formalism at all, that it would be enough to replace the ideal $\mathcal{K}$ in $\mathcal{L}(\mathcal{H})$ by the ideal $c_0(\mathbb{N})$ of sequences converging to zero in the algebra $\ell^\infty(\mathbb{N})$ of bounded sequences. A variable would be a bounded sequence, and an infinitesimal sequence μ_n, $\mu_n \downarrow 0$. However, this commutative version does not allow for the existence of variables with range a continuum since all elements of $\ell^\infty(\mathbb{N})$ have a point spectrum and a discrete spectral measure. Only *noncommutativity* of $\mathcal{L}(\mathcal{H})$ makes it possible to include in one framework variables with Lebesgue spectrum together with infinitesimal variables.

Noncommutativity of $\mathcal{L}(\mathcal{H})$ is also crucial for the next line of the dictionary. The differential df of a real or complex variable

$$df = \Sigma \frac{\partial f}{\partial x^\mu} \, dx^\mu \tag{2.12}$$

is replaced by the commutator

$$đ f = [F, f] \,. \tag{2.13}$$

Going from (2.12) to (2.13) is similar to taking the commutator $[f, g] = fg - gf$ of quantum observables instead of the Poisson bracket $\{f, g\}$ of two functions f, g in classical mechanics.

For a given algebra $\mathcal{A}$ of operators in $\mathcal{H}$ the *dimension* of the corresponding space (in the sense of the dictionary on page 51) is governed by the size of the differentials $đ f$, for $f \in \mathcal{A}$. In dimension p one has

$$đ f \quad \text{of order } \frac{1}{p}, \qquad \text{for any } f \in \mathcal{A} \,. \tag{2.14}$$

We shall see below (Theorem 2.1) concrete examples where p is the Hausdorff dimension of Julia sets. Simple algebraic manipulations on the functional

$$\tau(f^0, \ldots, f^n) = \mathrm{Trace}\,(f^0 \, đ f^1 \ldots đ f^n) \,, \qquad n \text{ odd}, \; n > p \tag{2.15}$$

show that τ is a cyclic cocycle and, moreover, that the cocycle τ is *integral*, i.e. that

$$\langle \tau, K_1(\mathcal{A}) \rangle \subset \mathbb{Z}. \tag{2.16}$$

This integrality result is a very powerful tool, see (Connes, 1994).

If the dictionary would have ended here the calculus would lack a vital tool, the *locality*, namely the possibility of neglecting infinitesimals of order > 1. These are contained in the following two–sided ideal

$$\left\{ T \in \mathcal{K} \,; \; \mu_n(T) = o\left(\frac{1}{n}\right) \right\}, \tag{2.17}$$

where the lower case o has the usual meaning, namely, that $n\mu_n(T) \to 0$ when $n \to \infty$.

Hence if we use the trace as in (2.15) for integration we would encounter two problems:

(1) infinitesimals of order 1 are not in the domain of the trace;
(2) the trace of infinitesimals of order > 1 is not zero.

The natural domain of the trace is the two–sided ideal $\mathcal{L}^1(\mathcal{H})$ of trace class operators

$$\mathcal{L}^1 = \left\{ T \in \mathcal{K} \, ; \, \sum_0^\infty \mu_n(T) < \infty \right\} . \tag{2.18}$$

The trace of an operator $T \in \mathcal{L}^1(\mathcal{H})$ is given by the sum

$$\text{Trace}\,(T) = \sum \langle T\xi_i, \xi_i \rangle \tag{2.19}$$

independent of the choice of the orthonormal basis (ξ_i) of $\mathcal{H}$. One has

$$\text{Trace}\,(T) = \sum_0^\infty \mu_n(T) \qquad \forall \, T \geq 0 \, . \tag{2.20}$$

Let $T \geq 0$ be an infinitesimal of order 1. The only condition on $\mu_n(T)$ is

$$\mu_n(T) = O\left(\frac{1}{n}\right) \tag{2.21}$$

which is not sufficient to ensure the finiteness of the sum in (2.20). This shows the nature of both of the problems since the trace is not zero even on the smallest ideal in $\mathcal{L}(\mathcal{H})$, namely, the ideal of finite rank operators, $\mathcal{R}$.

The solution is obtained by employing the Dixmier trace (Dixmier, 1966), i.e. by the following analysis of the logarithmic divergence of the partial traces

$$\text{Trace}_N(T) = \sum_0^{N-1} \mu_n(T) \ , \ T \geq 0 \, . \tag{2.22}$$

In fact, it is useful to define $\text{Trace}_\Lambda(T)$ for any positive real $\Lambda > 0$ by the interpolation formula

$$\text{Trace}_\Lambda(T) = \text{Inf}\, \{ \|A\|_1 + \Lambda \|B\| \, ; \ A + B = T \} \tag{2.23}$$

where $\|A\|_1$ is the $\mathcal{L}^1$ norm of A, $\|A\|_1 = \text{Trace}\,|A|$, and $\|B\|$ is the operator norm of B. This formula coincides with (2.22) for integer Λ and gives its piecewise affine interpolation for noninteger Λ. Then also (Connes, 1994)

$$\text{Trace}_\Lambda(T_1 + T_2) \leq \text{Trace}_\Lambda(T_1) + \text{Trace}_\Lambda(T_2) \qquad \forall \, \Lambda > 0 \tag{2.24}$$

$$\text{Trace}_{\Lambda_1 + \Lambda_2}(T_1 + T_2) \geq \text{Trace}_{\Lambda_1}(T_1) + \text{Trace}_{\Lambda_2}(T_2) \qquad \forall \, \Lambda_1, \Lambda_2 > 0 \tag{2.25}$$

where T_1, T_2 are positive in (2.25).

Define for all order 1 operators $T \geq 0$

$$\tau_\Lambda(T) = \frac{1}{\log \Lambda} \int_e^\Lambda \frac{\mathrm{Trace}_\mu(T)}{\log \mu} \, \frac{d\mu}{\mu} \tag{2.26}$$

which is the Cesaro mean of the function $\frac{\mathrm{Trace}_\mu(T)}{\log \mu}$ over the scaling group $\mathbb{R}_+^*$.

For $T \geq 0$, an infinitesimal of order 1, one has

$$\mathrm{Trace}_\Lambda(T) \leq C \, \log \Lambda \tag{2.27}$$

so that $\tau_\Lambda(T)$ is bounded. The essential property is the following *asymptotic additivity* of the coefficient $\tau_\Lambda(T)$ of the logarithmic divergence (2.27):

$$|\tau_\Lambda(T_1 + T_2) - \tau_\Lambda(T_1) - \tau_\Lambda(T_2)| \leq 3C \, \frac{\log(\log \Lambda)}{\log \Lambda} \tag{2.28}$$

for $T_j \geq 0$.

An easy consequence of (2.28) is that any limit point τ of the nonlinear functionals τ_Λ for $\Lambda \to \infty$ defines a positive and linear trace on the two–sided ideal of infinitesimals of order 1,

$$\begin{aligned} &\tau(\lambda_1 T_1 + \lambda_2 T_2) = \lambda_1 \, \tau(T_1) + \lambda_2 \, \tau(T_2) \qquad \forall \, \lambda_j \in \mathbb{C} \\ &\tau(T) \geq 0 \qquad \forall \, T \geq 0 \\ &\tau(ST) = \tau(TS) \qquad \text{for any bounded } \; S \\ &\tau(T) = 0 \text{ if } \mu_n(T) = o\left(\frac{1}{n}\right). \end{aligned} \tag{2.29}$$

In practice the choice of the limit point τ is irrelevant because in all important examples (in particular as a corollary of the axioms in the general framework cf. Section 4) T is a *measurable* operator, i.e.

$$\tau_\Lambda(T) \text{ converges when } \; \Lambda \to \infty\,. \tag{2.30}$$

Thus the value $\tau(T)$ is independent of the choice of the limit point τ and is denoted

$$⨍ T\,. \tag{2.31}$$

The first interesting example is provided by pseudodifferential operators T on a differentiable manifold M. When T is of order 1 (in the sense of (2.21)) it is measurable and $⨍ T$ is the noncommutative residue of T (Wodzicki, 1987; Kassel, 1989). It has a local expression in terms of the distribution kernel $k(x,y)$, $x, y \in M$. For T of order 1 the kernel $k(x,y)$ diverges logarithmically near the diagonal,

$$k(x,y) = -a(x) \log|x-y| + O(1) \text{ (for } \; y \to x) \tag{2.32}$$

where $a(x)$ is a 1-density independent of the choice of Riemannian distance $|x-y|$. Then one has (up to normalization),

$$⨍ T = \int_M a(x). \tag{2.33}$$

The right hand side of this formula makes sense for all pseudodifferential operators (cf. Wodzicki, 1987) since one can see that the kernel of such an operator is asymptotically of the form

$$k(x,y) = \sum a_k(x, x-y) + a(x) \log |x-y| + O(1) \tag{2.34}$$

where $a_k(x,\xi)$ is homogeneous of degree $-k$ in ξ, and the 1-density $a(x)$ is defined intrinsically.

The same principle of extension of $⨍$ to infinitesimals of order < 1 works for hypoelliptic operators and more generally (cf. Theorem 3.1) for spectral triples whose dimension spectrum is simple.

After this description of the general framework let us discuss examples. The infinitesimal variable $dp(x)$, which is the probability in the darts game, is given by the operator

$$dp = \Delta^{-1} \tag{2.35}$$

where Δ is the Dirichlet Laplacian for the domain Ω, acting on the Hilbert space $L^2(\Omega)$. Also the algebra of functions $f(x_1,x_2)$, $f : \Omega \to \mathbb{C}$, acts by multiplication operators on $L^2(\Omega)$ (cf. (2.3)). From the H. Weyl theorem on the asymptotic behaviour of eigenvalues of Δ it follows that dp is of order 1, $f\,dp$ is measurable, and that

$$⨍ f\,dp = \int_\Omega f(x_1,x_2)\,dx_1 \wedge dx_2 \tag{2.36}$$

gives the usual expression for the probability. Note that $e^{-\frac{1}{dp(x)}}$ has now full meaning as the heat kernel.

Now let us show how this infinitesimal calculus can be used to make sense of some expressions like the area of 4-dimensional manifolds, which do not exist in the usual calculus.

There is a canonical quantization of the infinitesimal calculus over $\mathbb{R}$ which is invariant under translations and dilations. It is given by the representation of the algebra of functions f on $\mathbb{R}$ as multiplication operators in $L^2(\mathbb{R})$ (cf. (2.3)), while the operator F in $\mathcal{H} = L^2(\mathbb{R})$ is the Hilbert transform (Stein, 1970)

$$(f\xi)(s) = f(s)\,\xi(s) \qquad \forall\, s \in \mathbb{R}\ ,\ \xi \in L^2(\mathbb{R})\ ,\ (F\xi)(t) = \frac{1}{\pi i}\int \frac{\xi(s)}{s-t}\,ds\,. \tag{2.37}$$

One has a unitary equivalent description for $S^1 = P_1(\mathbb{R})$ with $\mathcal{H} = L^2(S^1)$ and

$$F\,e_n = \mathrm{Sign}\,(n)\,e_n\ ,\ e_n(\theta) = \exp\,(in\theta) \qquad \forall\,\theta \in S^1\,,\ (\mathrm{Sign}\,(0) = 1)\,. \tag{2.38}$$

The operator $đ f = [F, f]$, for $f \in L^\infty(\mathbb{R})$, is represented by the kernel $\frac{1}{\pi i} k(s,t)$, with

$$k(s,t) = \frac{f(s) - f(t)}{s - t}. \tag{2.39}$$

Since f and F are bounded operators, $đ f = [F, f]$ is also bounded for all bounded measurable f on S^1, and it makes sense to talk about $|đ f|^p$ for all $p > 0$. Let us give an example where one has to use such an expression for a non-differentiable f. Let J be the Julia set associated to the iterations of the map ($c \in \mathbb{C}$)

$$\varphi(z) = z^2 + c\,, \quad J = \partial B\,, \quad B = \{z \in \mathbb{C};\ \sup_{n \in \mathbb{N}} |\varphi^n(z)| < \infty\}. \tag{2.40}$$

For small c, J is a Jordan curve and B is the bounded component of its complement. The Riemann mapping theorem provides us with a conformal equivalence $D \sim B$ of the unit disc $D = \{z \in \mathbb{C}, |z| < 1\}$ with B. By a theorem of Caratheodory it extends continuously to a homeomorphism $Z : S^1 \to J$ where $S^1 = \partial D$. Since (by a result of D. Sullivan) the Hausdorff dimension p of J is strictly bigger than 1 (for $c \neq 0$) the function Z is nowhere of bounded variation on S^1 and $|Z'|$, the absolute value of the derivative of Z, does not make sense as a distribution. However $|đ Z|$ is well defined and one has:

Theorem 2.1

(1) *$|đ Z|$ is an infinitesimal of order $\frac{1}{p}$.*

(2) *For every continuous function h on J, the operator $h(Z)\,|đ Z|^p$ is measurable.*

(3) $\exists\, \lambda > 0$,

$$⨍ h(Z)\,|đ Z|^p = \lambda \int_J h\, d\Lambda_p \qquad \forall\, h \in C(J)$$

where $d\Lambda_p$ denotes the Hausdorff measure on J.

The first statement of the theorem uses a result of V.V. Peller which characterizes functions f for which $\operatorname{Trace}(|đ f|^\alpha) < \infty$. The constant λ is determined by the asymptotic expansion in $n \in \mathbb{N}$ for the distance in $L^\infty(S^1)$ between Z and restrictions to S^1 of rational fractions with at most n poles outside the unit disc. This constant is of order $\sqrt{p-1}$ and so it is zero for $p = 1$. This is related to a specific feature of dimension 1 manifolds, namely, the differential $đ f$ of a function $f \in C^\infty(S^1)$ is not just of order $(\dim S^1)^{-1} = 1$ but is even trace class, with

$$\operatorname{Trace}(f^0\, đ f^1) = \frac{1}{\pi i} \int_{S^1} f^0\, df^1 \qquad \forall\, f^0, f^1 \in C^\infty(S^1). \tag{2.41}$$

In fact, by a classical result of Kronecker, $đ f$ is of finite rank if and only if f is a rational fraction (cf. Power, 1982).

The quantized calculus can be used in the same manner to describe the projective space $P_1(K)$ over any local field K (i.e. non-discrete locally compact). This calculus is invariant under projective transformations. The special cases of $K = \mathbb{C}$ and $K = \mathbb{H}$ (the field of quaternions) are examples of the calculus on oriented conformal even–dimensional compact manifolds, $M = M^{2n}$. The calculus is defined as follows:

$$\mathcal{H} = L^2(M, \Lambda^n T^*)\,,\ (f\xi)(p) = f(p)\,\xi(p) \qquad \forall\, f \in L^\infty(M)\,,\ F = 2P-1, \quad (2.42)$$

where the scalar product on the Hilbert space of differential forms of degree $n = \frac{1}{2}\dim M$ is given by $\langle \omega_1, \omega_2\rangle = \int \omega_1 \wedge * \omega_2$ which only depends on the conformal structure on M. The operator P is the orthogonal projection on the subspace of exact forms.

Consider $n = 1$, i.e. M being a Riemann surface. An easy calculation shows that

$$⨍ đ f\, đ g = -\frac{1}{\pi} \int df \wedge * dg \qquad \forall\, f, g \in C^\infty(M)\,. \quad (2.43)$$

Let X be a smooth map from M to the space $\mathbb{R}^N$ equipped with a Riemannian metric $g_{\mu\nu}\, dx^\mu\, dx^\nu$. The components X^μ of the map X are functions on M. One has

$$⨍ g_{\mu\nu}\, đ X^\mu\, đ X^\nu = -\frac{1}{\pi} \int_M g_{\mu\nu}\, dX^\mu \wedge * dX^\nu \quad (2.44)$$

where the right hand side is the Polyakov action in string theory. However, the equality (2.44) does not hold for $n = 4$: the right hand side is not very interesting because it is not conformally invariant but the left hand side is still conformally invariant, because $đ X^\mu = [F, X^\mu]$ and F are conformally invariant. It defines the natural conformal analogue of the Polyakov action in the 4-dimensional case. A calculation yields:

Theorem 2.2 *Let X be a smooth map from M^4 to $(\mathbb{R}^N, g_{\mu\nu}\, dx^\mu\, dx^\nu)$,*

$$⨍ g_{\mu\nu}(X)\, đ X^\mu\, đ X^\nu = (16\pi^2)^{-1} \int_M g_{\mu\nu}(X)$$
$$\left\{\tfrac{1}{3}\, r\langle dX^\mu, dX^\nu\rangle - \Delta\langle dX^\mu, dX^\nu\rangle + \langle \nabla dX^\mu, \nabla dX^\nu\rangle - \tfrac{1}{2}\,(\Delta X^\mu)\,(\Delta X^\nu)\right\}\, dv$$

where for the right hand side one uses a Riemannian structure η on M, compatible with the given conformal structure. The scalar curvature r, the Laplacian Δ, the Levi–Civita connection ∇, and the measure dv are defined by η but the result is independent of its choice.

Theorem 2.2 is related to the following formula expressing the Hilbert–Einstein action as the area of the four–dimensional manifold, cf. Kastler (1995); Kalau and Walze (1995)

$$⨍ ds^2 = \frac{-1}{96\pi^2} \int_{M_4} r\, \sqrt{g}\; d^4x \quad (2.45)$$

($dv = \sqrt{g}\; d^4x$ is the volume form and $ds = D^{-1}$ the length element, i.e. the inverse of the Dirac operator).

When the metric $g_{\mu\nu}\, dx^\mu\, dx^\nu$ on $\mathbb{R}^N$ is invariant under translations the action functional of Theorem 2.2 is given by the Paneitz operator on M. It is a fourth order operator which plays the role of the Laplacian in conformal geometry (Branson and Ørsted, 1991). Its conformal anomaly was computed by Branson (1996).

Let us go back to the case $n = 2$ and modify the conformal structure on M by a Beltrami differential $\mu(z,\bar z)\, d\bar z/dz$, $|\mu(z,\bar z)| < 1$. Thus if z is a conformal local coordinate, we now measure angles at $z \in M$ by the identification

$$T_z(M) \to \mathbb{C}:\ X \mapsto \langle X, dz + \mu(z,\bar z)\, d\bar z\rangle \tag{2.46}$$

instead of $X \mapsto \langle X, dz\rangle$. The quantized calculus on M associated to the new conformal structure has the same $\mathcal{H}$, $\mathcal{A}$, and representation of $\mathcal{A}$ in $\mathcal{H}$ unchanged but the operator F is replaced by F' with

$$F' = (\alpha F + \beta)(\beta F + \alpha)^{-1}\ ,\ \alpha = (1-m^2)^{-1/2}\ ,\ \beta = m(1-m^2)^{-1/2}. \tag{2.47}$$

Here m is the operator in $\mathcal{H} = L^2(M, \Lambda^1 T^*)$ given by the endomorphism of the vector bundle $\Lambda^1 T^* = \Lambda^{(1,0)} \oplus \Lambda^{(0,1)}$ with matrix:

$$m(z,\bar z) = \begin{bmatrix} 0 & \bar\mu(z,\bar z)\, d\bar z/dz \\ \mu(z,\bar z)\, dz/d\bar z & 0 \end{bmatrix}. \tag{2.48}$$

The crucial properties of the operator $m \in \mathcal{L}(\mathcal{H})$ are as follows:

$$\|m\| < 1\ ,\ m = m^*\ ,\ m\, f = f\, m \qquad \forall\, f \in \mathcal{A} = C^\infty(M) \tag{2.49}$$

and the deformation (2.47) is a particular case of the

Proposition 2.3 *Let $\mathcal{A}$ be an involutive algebra of operators in $\mathcal{H}$ and*

$$N = \mathcal{A}' = \{T \in \mathcal{L}(\mathcal{H})\,;\, Ta = aT \quad \forall a \in \mathcal{A}\}$$

the von Neumann algebra commutant of $\mathcal{A}$.

(1) *The following formula defines an action of the group $G = GL_1(N)$ of invertible elements of N on the operators F, that satisfy $F = F^*$, $F^2 = 1$*

$$g(F) = (\alpha F + \beta)\,(\beta F + \alpha)^{-1} \qquad \forall g \in G$$

where $\alpha = \frac{1}{2}(g - (g^{-1})^*)$, $\beta = \frac{1}{2}(g + (g^{-1})^*)$.

(2) *One has*

$$[g(F), a] = Y[F,a]\,Y^* \quad \forall a \in \mathcal{A}, \quad \textit{where} \qquad Y = (\beta F + \alpha)^{*-1}.$$

The last equality shows that the transformation $F \to g(F)$ preserves the condition

$$[F,a] \in J \tag{2.50}$$

for all two–sided ideals $J \subset \mathcal{L}(\mathcal{H})$. Only *measurability* of the Beltrami differential μ is needed for m to satisfy (2.50), and similarly one only needs that the conformal structure on M is measurable in order to define the associated quantized calculus. Moreover, the second equality in Proposition 2.3 shows that the regularity condition on $a \in L^\infty(M)$ imposed by (2.50) only depends on the quasi-conformal structure on the manifold M (Connes *et al.* 1994). A local homeomorphism φ of $\mathbb{R}^n$ is *quasi-conformal* if and only if there exists $K < \infty$ such that

$$H_\varphi(x) = \operatorname*{Lim\,sup}_{r\to 0} \frac{\max\{|\varphi(x)-\varphi(y)|\,;\,|x-y|=r\}}{\min\{|\varphi(x)-\varphi(y)|\,;\,|x-y|=r\}} \leq K\ ,\ \forall\, x \in \text{Domain}\,\varphi\,. \tag{2.51}$$

A quasi-conformal structure on a topological manifold M^n is given by a quasi-conformal atlas. The discussion above applies to the general case (n even) (Connes *et al.* 1994) and shows that the quantized calculus is well defined on all quasi-conformal manifolds. The result of D. Sullivan (1977) based on Kirbi (1969) shows that all topological manifolds M^n, $n \neq 4$, admit a quasi-conformal structure. Using the quantized calculus and cyclic cohomology instead of the differential calculus and Chern–Weil theory one gets (Connes *et al.* 1994) a local formula for the topological Pontryagin classes of M^n.

3 The local index formula and the transverse fundamental class

In this section we show how the infinitesimal calculus allows us to go from local to global in the general framework of spectral triples $(\mathcal{A},\mathcal{H},D)$. We shall apply the general result to the cross product of a manifold by a group of diffeomorphisms.

Let us make the following regularity hypothesis on $(\mathcal{A},\mathcal{H},D)$

$$a \text{ and } [D,a] \in \bigcap \text{Dom}\,\delta^k,\ \forall\, a \in \mathcal{A} \tag{3.1}$$

where δ is the derivation $\delta(T) = [|D|,T]$ for any operator T.

Let $\mathcal{B}$ denote the algebra generated by $\delta^k(a)$, $\delta^k([D,a])$. The *dimension* of a spectral triple is bounded above by $p > 0$ if and only if $a(D+i)^{-1}$ is an infinitesimal of order $\frac{1}{p}$ for any $a \in \mathcal{A}$. When $\mathcal{A}$ is unital it depends only on the spectrum of D.

The precise notion of dimension is given by the subset $\Sigma \subset \mathbb{C}$ of singularities of the analytic functions

$$\zeta_b(z) = \text{Trace}\,(b|D|^{-z})\,, \qquad \text{Re}\,z > p\ ,\ b \in \mathcal{B}\,. \tag{3.2}$$

We assume that Σ is discrete and simple, i.e. that ζ_b can be extended to $\mathbb{C}/\Sigma$ with simple poles in Σ.

We refer to Connes and Moscovici (1995) for the case of a spectrum with multiplicities.

The Fredholm index of the operator D determines an additive map $K_1(\mathcal{A}) \xrightarrow{\varphi} \mathbb{Z}$ given by the equation

$$\varphi([u]) = \text{Index}\,(PuP)\ ,\ u \in GL_1(\mathcal{A}) \tag{3.3}$$

where P is the projector $P = \frac{1+F}{2}$, $F = \text{Sign}\,(D)$.

This map is calculated by the pairing of $K_1(\mathcal{A})$ with the following cyclic cocycle

$$\tau(a^0, \ldots, a^n) = \text{Trace}\,(a^0[F, a^1] \ldots [F, a^n]) \qquad \forall\, a^j \in \mathcal{A} \tag{3.4}$$

where $F = \text{Sign}\ D$ and $n \geq p$ is an odd integer.

It is difficult to compute τ in general because the formula (3.4) employs the ordinary trace instead of the local trace $⨍$.

This problem is solved by the following general formula:

Theorem 3.1 *(Connes and Moscovici, 1995).*

Let $(\mathcal{A}, \mathcal{H}, D)$ be a spectral triple satisfying the hypothesis (3.1) and (3.2). Then

(1) *The equality*

$$⨍ P = \text{Res}_{z=0}\ \text{Trace}\,(P|D|^{-z})$$

defines a trace on the algebra generated by $\mathcal{A}$, $[D, \mathcal{A}]$ and $|D|^z$, where $z \in \mathbb{C}$.

(2) *There is only a finite number of non-zero terms in the following formula. It defines the odd components $(\varphi_n)_{n=1,3,\ldots}$ of a cocycle in the bicomplex (b, B) of $\mathcal{A}$,*

$$\varphi_n(a^0, \ldots, a^n) = \sum_k c_{n,k} ⨍ a^0[D, a^1]^{(k_1)} \ldots [D, a^n]^{(k_n)}\, |D|^{-n-2|k|}\,,$$

$$\forall\, a^j \in \mathcal{A}$$

where the following notations are used: $T^{(k)} = \nabla^k(T)$ and $\nabla(T) = D^2T - TD^2$, k is a multi-index, $|k| = k_1 + \ldots + k_n$,

$$\begin{aligned} c_{n,k} &= (-1)^{|k|}\sqrt{2i}(k_1! \ldots k_n!)^{-1} \\ &\quad \times((k_1+1)\ldots(k_1+k_2+\ldots+k_n+n))^{-1}\Gamma\left(|k| + \frac{n}{2}\right). \end{aligned}$$

(3) *The pairing of the cyclic cohomology class $(\varphi_n) \in HC^*(\mathcal{A})$ with $K_1(\mathcal{A})$ givcs the Fredholm inder of D with coefficients in $K_1(\mathcal{A})$.*

Let us recall that the bicomplex (b, B) is given by the following operators acting on multi-linear forms on $\mathcal{A}$,

$$\begin{aligned} (b\varphi)(a^0, \ldots, a^{n+1}) &= \sum_0^n (-1)^j\, \varphi(a^0, \ldots, a^j a^{j+1}, \ldots, a^{n+1}) \\ &\quad + (-1)^{n+1}\, \varphi(a^{n+1}a^0, a^1, \ldots, a^n) \end{aligned} \tag{3.5}$$

$$\begin{aligned}
&B = AB_0\,,\\
&B_0\varphi(a^0,\ldots,a^{n-1}) = \varphi(1,a^0,\ldots,a^{n-1}) - (-1)^n\,\varphi(a^0,\ldots,a^{n-1},1)\,,\\
&(A\psi)(a^0,\ldots,a^{n-1}) = \sum_0^{n-1}(-1)^{(n-1)j}\,\psi(a^j,a^{j+1},\ldots,a^{j-1})\,.
\end{aligned} \tag{3.6}$$

For the normalization of the pairing between HC^* and $K(\mathcal{A})$ see Connes (1994).

Remarks

(a) The statement of Theorem 3.1 remains valid when D is replaced by $D|D|^\alpha$, $\alpha \geq 0$.

(b) In the even case, i.e. when $\mathcal{H}$ is $\mathbb{Z}/2$ graded by γ,

$$\gamma = \gamma^*,\ \ \gamma^2 = 1,\ \ \gamma a = a\gamma \quad \forall\, a \in \mathcal{A},\ \gamma D = -D\gamma,$$

there is an analogous formula for a cocycle (φ_n), n even, which gives the Fredholm index of D with coefficients in K_0. However, φ_0 is not expressed in terms of the residue $⨍$ because it is not local for a finite dimensional $\mathcal{H}$ (cf. Connes and Moscovici, 1995).

(c) There exists an analogous formula for the case when the dimension spectrum Σ has multiplicities. There are some correction terms, their number is finite and bounded independently of the multiplicity, cf. Connes and Moscovici (1995).

The dimension spectrum of a manifold M is the set $\{0,1,\ldots,n\}$, $n = \dim M$; it is simple. Multiplicities appear for singular manifolds. Cantor sets provide examples of complex points $z \notin \mathbb{R}$ in the dimension spectrum.

Starting from a manifold M we are going to perform a general geometrical construction which will yield a spectral triple satisfying the above hypothesis (3.1) and (3.2) and give the fundamental class in K-homology of a K-oriented manifold M without breaking the symmetry of $\mathrm{Diff}^+(M)$, the group of diffeomorphisms of M which preserve K-orientation. More precisely, we are looking for a spectral triple, $(C^\infty(M),\mathcal{H},D)$, in the same K-homology class as the Dirac operator associated to a Riemannian metric (cf. (1.12) and (1.13)) but which is equivariant under the action of the group $\mathrm{Diff}^+(M)$ in the sense of Kasparov (1980). This means that one has a unitary representation $\varphi \to U(\varphi)$ of $\mathrm{Diff}^+(M)$ in $\mathcal{H}$ such that

$$U(\varphi)\,f\,U(\varphi)^{-1} = f\circ\varphi^{-1} \qquad \forall\, f \in C^\infty(M)\ ,\ \varphi \in \mathrm{Diff}^+(M) \tag{3.7}$$

and

$$U(\varphi)\,D\,U(\varphi)^{-1} - D \text{ is bounded for all } \varphi \in \mathrm{Diff}^+(M)\,. \tag{3.8}$$

When D is the Dirac operator associated to a Riemannian structure, the principal symbol of D in turn determines the Riemannian metric and hence diffeomorphisms which satisfy (3.8) are isometries.

A solution to this problem is essential in order to define the transverse geometry of foliations. It is obtained in two steps. The first step consists of using the negative curvature metric on $GL(n)/O(n)$ and the "dual Dirac" operator of Miscenko and Kasparov (Connes, 1986) to reduce the problem to the action of $\mathrm{Diff}^+(M)$ on the total space P of the bundle of metrics over M. The second step, following the idea of Hilsum and Skandalis (1987), is to use hypoelliptic operators to construct D on P.

Although the equivariant geometry obtained on P is finite dimensional and satisfies hypothesis (3.1) and (3.2), the geometry obtained for M by using the intersection product with the "dual Dirac" is infinite dimensional and θ-summable,

$$\mathrm{Trace}\,(e^{-\beta D^2}) < \infty \qquad \forall\,\beta > 0\,. \tag{3.9}$$

By construction the fibre of $P \xrightarrow{\pi} M$ is the quotient $F/O(n)$ of the $GL(n)$-principal bundle F of frames on M by the action of the orthogonal group $O(n) \subset GL(n)$. The space P admits a canonical foliation: the vertical foliation $V \subset TP$, $V = \mathrm{Ker}\,\pi_*$ and on the fibres V and on $N = (TP)/V$ the following Euclidean structures. A choice of $GL(n)$-invariant Riemannian metric on $GL(n)/O(n)$ determines a metric on V. The metric on N is defined tautologically: for every $p \in P$ one has a metric on $T_{\pi(p)}(M)$ which is isomorphic to N_p by π_*.

This construction is functorial for diffeomorphisms on M.

The hypoelliptic calculus adapted to this structure is a particular case of the pseudodifferential calculus on Heisenberg manifolds (Beals and Greiner, 1988). One simply modifies the notion of homogeneity of symbols $\sigma(p,\xi)$ by using the following homotheties:

$$\lambda \cdot \xi = (\lambda \xi_v, \lambda^2 \xi_n) \qquad \forall\,\lambda \in \mathbb{R}_+^* \tag{3.10}$$

where ξ_v, ξ_n are vertical and perpendicular to vertical components of the covector ξ. The formula (3.10) depends on local coordinates (x_v, x_n) adapted to the vertical foliation, but the corresponding pseudodifferential calculus does not depend upon this choice. The principal symbol of a hypoelliptic operator of order k is a function, homogeneous of degree k in the sense of (3.10) on the fibre $V^* \oplus N^*$. The distribution kernel $k(x,y)$ of a pseudodifferential operator T in this hypoelliptic calculus has the following behaviour near the diagonal

$$k(x,y) \sim \sum a_j(x, x-y) - a(x) \log |x-y|' + O(1) \tag{3.11}$$

where a_j is homogeneous of degree $(-j)$ in $(x-y)$ in the sense of (3.10) and where the metric $|x-y|'$ is locally of the form

$$|x-y|' = ((x_v - y_v)^4 + (x_n - y_n)^2)^{1/4}\,. \tag{3.12}$$

As in ordinary pseudodifferential calculus the residue is extended to operators of all degrees and is given by the equality

$$⨍ T = \frac{1}{v + 2m} \int a(x) \tag{3.13}$$

where the 1-density $a(x)$ does not depend on the choice of metric $|\ |'$ and where $v = \dim V$, $m = \dim N$ and $v + 2m$ is the Hausdorff dimension of the metric space $(P, |\ |')$.

The operator D is defined by the equation $D|D| = Q$ where Q is a differential hypoelliptic operator of degree 2. For v even, Q is obtained by combining the signature operator $d_V d_V^* - d_V^* d_V$ with the transverse Dirac operator, where d_V is the vertical differential. (We use the metaplectic cover $M\ell(n)$ of $GL(n)$ to define the spin structure on M.) The explicit formula for Q uses an affine connection on M. The choice of this connection does not affect the *principal hypoelliptic symbol* of Q and therefore of D which ensures that D is invariant as in (3.8) under diffeomorphisms of M.

Let us give the explicit formula for Q in the case $n = 1$, i.e. for $M = S^1$. We replace P by the suspension $SP = \mathbb{R} \times P$ in order to consider the case where the vertical dimension is even. A point of $SP = \mathbb{R} \times P$ is parametrized by three coordinates $\alpha \in \mathbb{R}$ and $p = (s, \theta)$ where $\theta \in S^1$ and $s \in \mathbb{R}$ defines the metric $e^{2s}(d\theta)^2$ for $\theta \in S^1$.

We endow SP with the measure $\nu = d\alpha\, ds\, d\theta$ and represent the algebra $C_c^\infty(SP)$ by multiplication operators on $\mathcal{H} = L^2(SP, \nu) \otimes \mathbb{C}^2$. Functoriality of the construction above gives the following unitary representation of the group $\mathrm{Diff}^+(S^1)$,

$$(U(\varphi)^{-1}\xi)(\alpha, s, \theta) = \varphi'(\theta)^{1/2}\, \xi(\alpha, s - \log \varphi'(\theta), \varphi(\theta)) . \tag{3.14}$$

The operator Q is given by the formula

$$Q = -2\partial_\alpha \partial_s \sigma_1 + \frac{1}{i}\, e^{-s} \partial_\theta \sigma_2 + \left(\partial_s^2 - \partial_\alpha^2 - \frac{1}{4}\right) \sigma_3 \tag{3.15}$$

where $\sigma_1, \sigma_2, \sigma_3 \in M_2(\mathbb{C})$ are the three Pauli matrices.

In the hypoelliptic calculus the operator ∂_θ has *degree* 2 which shows the hypoellipticity of Q.

A long calculation gives the following result (Connes and Moscovici, 1997):

Theorem 3.2 *Let $\mathcal{A}$ be the crossed product of $C_c^\infty(SP)$ by $\mathrm{Diff}^+(S^1)$.*

(1) *The spectral triple $(\mathcal{A}, \mathcal{H}, D)$ (where $\mathcal{A}$ acts on $\mathcal{H}$ by (3.14) and $D|D| = Q$) satisfies the hypotheses (3.1) and (3.2); its dimension spectrum is $\Sigma = \{0, 1, 2, 3, 4\}$.*

(2) *The only nonzero term of the associated cocycle Theorem 3.1 is φ_3, which is cohomologous to 2ψ, where ψ is the 3-cyclic cocycle of the transversal fundamental class of the crossed product.*

From Theorem 3.1 follows the *integrality* of 2ψ, i.e. that the pairing $\langle 2\psi, K_1(\mathcal{A})\rangle$ is an integer. The 3-cocycle ψ is given by the following formula (cf. Connes, 1994)

$$
\begin{aligned}
&\psi(f^0U(\varphi_0), f^1U(\varphi_1), f^2U(\varphi_2), f^3U(\varphi_3)) \\
&= \left\{ \begin{array}{ll} 0 & \text{if } \varphi_0\varphi_1\varphi_2\varphi_3 \neq 1 \\ \int h^0\, dh^1 \wedge dh^2 \wedge dh^3 & \text{if } \varphi_0\varphi_1\varphi_2\varphi_3 = 1 \end{array} \right\}
\end{aligned} \tag{3.16}
$$

where

$$
h^0 = f^0,\ h^1 = (f^1)^{\varphi_0},\ h^2 = (f^2)^{\varphi_0\varphi_1},\ \ h^3 = (f^3)^{\varphi_0\varphi_1\varphi_2}.
$$

The homology between φ_3 and 2ψ involves the action of the Hopf algebra generated by the following linear transformations of the algebra $\mathcal{A}$ (for the relation of δ_3 to the Godbillon–Vey invariant see Connes, 1994)

$$
\begin{aligned}
&\delta_1(fU(\varphi)) = (\partial_\alpha f)\, U(\varphi)\,, \quad \delta_2(fU(\varphi)) = (\partial_s f)\, U(\varphi)\,, \\
&\delta_3(fU(\varphi)) = f\, e^{-s}\, \partial_\theta \log(\varphi^{-1})'\, U(\varphi)\,, \quad X(fU(\varphi)) = e^{-s}(\partial_\theta\, f)\, U(\varphi)\,.
\end{aligned} \tag{3.17}
$$

The compatibility with the multiplication in $\mathcal{A}$ is given by the coproduct rules

$$
\Delta\delta_j = \delta_j \otimes 1 + 1 \otimes \delta_j \qquad j = 1,2,3 \tag{3.18}
$$

(i.e. δ_j are derivations in $\mathcal{A}$)

$$
\Delta X = X \otimes 1 + 1 \otimes X - \delta_3 \otimes \delta_2\,. \tag{3.19}
$$

The last equation shows that X is of degree 2 not only from its degree as a pseudodifferential operator in the hypoelliptic calculus but also in an algebraic manner since ΔX involves a tensor product of two derivations.

4 The notion of manifold and the axioms of geometry

Let us first characterize the spectral triples corresponding to ordinary Riemannian geometry (Theorem 4.1 below). Let the dimension n be given and $(\mathcal{A}, \mathcal{H}, D)$ be a spectral triple with $\mathbb{Z}/2$-grading γ when n is even.

The axioms for *commutative* geometry are the following:

(1) (Dimension) $ds = D^{-1}$ is an infinitesimal of order $\frac{1}{n}$.

(2) (Order one) $[[D, f], g] = 0 \quad \forall\, f, g \in \mathcal{A}$.

(3) (Smoothness) For all $f \in \mathcal{A}$, both f and $[D, f]$ belong to $\bigcap\limits_k$ Domain δ^k, where δ is the derivation $\delta(T) = [|D|, T]$.

(4) (Orientability) There exists a Hochschild cocycle $c \in Z_n(\mathcal{A}, \mathcal{A})$ such that $\pi(c) = 1$ (n odd) or $\pi(c) = \gamma$ (n even), where π: $\mathcal{A}^{\otimes(n+1)} \to \mathcal{L}(\mathcal{H})$ is the unique linear map such that

$$
\pi(a^0 \otimes a^1 \otimes \cdots \otimes a^n) = a^0[D, a^1] \ldots [D, a^n] \quad \forall\, a^j \in \mathcal{A}.
$$

(5) (Finiteness) The $\mathcal{A}$-module $\mathcal{E} = \bigcap\limits_k$ Domain D^k is finite projective. The following identity defines a Hermitian structure on $\mathcal{E}$,

$$
\langle a\xi, \eta\rangle = \int\!\!\!\!\!\!- \; a(\xi,\eta)\, ds^n \qquad \forall\, \xi, \eta \in \mathcal{E}\ ,\ a \in \mathcal{A}\,.
$$

(6) (Poincaré duality) The intersection form $K_*(\mathcal{A}) \times K_*(\mathcal{A}) \to \mathbb{Z}$ given by the composition of the Fredholm index of D with the diagonal,

$$m_* : K_*(\mathcal{A}) \times K_*(\mathcal{A}) \to K_*(\mathcal{A} \otimes \mathcal{A}) \to K_*(\mathcal{A}),$$

is *invertible.*

(7) (Reality) There exists an antilinear isometry J on $\mathcal{H}$ such that

$$Ja^*J^{-1} = a \quad \forall a \in \mathcal{A} \ \text{ and } \ J^2 = \varepsilon, \quad JD = \varepsilon' DJ, \quad J\gamma = \varepsilon'' \gamma J$$

where the values of $\varepsilon, \varepsilon', \varepsilon'' \in \{-1, 1\}$ are given by the table (1.16) as functions of n modulo 8.

Axioms (2) and (4) describe the presentation of the abstract algebra denoted by $(\mathcal{A}, ds)$ generated by $\mathcal{A}$ and $ds = D^{-1}$.

Theorem 4.1

Let $\mathcal{A} = C^\infty(M)$, where M is a compact smooth manifold.

(1) *Let π be a unitary representation of $(\mathcal{A}, ds)$ satisfying Axioms (1) to (7). Then there exists a unique Riemannian structure g on M such that the geodesic distance is given by*

$$d(x,y) = \operatorname{Sup}\{|a(x) - a(y)|\,;\ a \in \mathcal{A}\ ,\ \|[D,a]\| \le 1\}\,.$$

(2) *The metric $g = g(\pi)$ depends only on the unitary equivalence class of π. Fibres of the map {unitary equivalence classes} $\to g(\pi)$ are a finite union of affine spaces $\mathcal{A}_\sigma$ parametrized by the spin structures σ on M.*

(3) *The functional $⨍ ds^{n-2}$ is a positive quadratic form on each $\mathcal{A}_\sigma$ with a unique minimum π_σ.*

(4) *π_σ is the representation of $(\mathcal{A}, ds)$ in $L^2(M, S_\sigma)$ given by multiplication operators and the Dirac operator associated to the Levi–Civita connection of the metric g.*

(5) *The value of $⨍ ds^{n-2}$ in π_σ is the Hilbert–Einstein action of the metric g,*

$$⨍ ds^{n-2} = -c_n \int r\sqrt{g}\; d^n x\ ,\ c_n = \tfrac{n-2}{12}\,(4\pi)^{-n/2}\,2^{[n/2]}\,\Gamma\left(\tfrac{n}{2}+1\right)^{-1}.$$

To understand the meaning of this theorem consider the simplest example, namely, the verification that the geometry of the circle S^1 of length 2π is completely specified by the presentation:

$$U^{-1}[D,U] = 1\ ,\ \text{where } UU^* = U^*U = 1\,. \tag{4.1}$$

Then the algebra $\mathcal{A}$ is the algebra of smooth functions of the single element U. One has $S^1 = \text{Spectrum}\ (\mathcal{A})$ and the equality (4.1) is the simplest case of Axiom (4).

Remarks

(a) We should not have to assume in the statement of Theorem 4.1 that the algebra $\mathcal{A}$ is equal to the algebra of smooth functions on a manifold. For a commutative $\mathcal{A}$ it should in fact follow from the axioms that the spectrum of $\mathcal{A}$ is a smooth manifold M and that $\mathcal{A} = C^\infty(M)$. From Axioms (3) and (5) it follows that if $\mathcal{A}''$ is the von Neumann algebra generated by $\mathcal{A}$ ($\mathcal{A}''$ is the weak closure and the bicommutant of $\mathcal{A}$ in $\mathcal{H}$) one has

$$\mathcal{A} = \left\{ T \in \mathcal{A}'' \,;\; T \in \bigcap_{k>0} \operatorname{Dom} \delta^k \right\} \tag{4.2}$$

$\mathcal{A}$ is uniquely specified inside $\mathcal{A}''$ by fixing D (i.e. the geometry $(\mathcal{A}, \mathcal{H}, D)$ is determined by $(\mathcal{A}'', \mathcal{H}, D)$). This also implies that $\mathcal{A}$ is stable under the smooth functional calculus in its norm closure $\bar{\mathcal{A}} = A$ and in particular

$$\operatorname{Spectrum} \mathcal{A} = \operatorname{Spectrum} A\,. \tag{4.3}$$

Let $X = \operatorname{Spectrum} A$. It is a compact space. One should deduce from the axioms that the map from X to $\mathbb{R}^N$ given by $a_i^j \in \mathcal{A}$ (the components of the Hochschild cocycle c given by Axiom (4)) is an embedding of X as a smooth submanifold of $\mathbb{R}^N$ (cf. Proposition 15, p. 312, Connes, 1994).

(b) Let us recall that a Hochschild cycle $c \in Z_n(\mathcal{A}, \mathcal{A})$ is an element of $\mathcal{A}^{\otimes(n+1)}$, $c = \sum a_i^0 \otimes a_i^1 \ldots \otimes a_i^n$ such that $bc = 0$, where b is the linear map $b : \mathcal{A}^{\otimes n+1} \to \mathcal{A}^{\otimes n}$ (cf. (3.5)). The class of the Hochschild cycle c determines the *volume form.*

(c) We use the convention that the scalar curvature r is positive for the sphere S^n, in particular, the sign of the action $⨍ ds^{n-2}$ is the correct one for the Euclidean formulation of gravity. For example for $n = 4$ the Hilbert–Einstein action

$$-\frac{1}{16\pi G} \int r \sqrt{g}\, d^4x$$

coincides with the area $\frac{1}{\ell_p^2} ⨍ ds^2$ in Planck units.

(d) When M is a spin manifold the map $\pi \to g(\pi)$ of Theorem 4.1 is surjective and if one fixes the cycle $c \in Z_n(\mathcal{A}, \mathcal{A})$ its image is the set of metrics whose volume form (Remark (b)) is given by the class of c.

(e) If one omits Axiom (7) , one gets a result analogous to Theorem 4.1 replacing spin structures by spin^c-structures (Lawson and Michelson, 1989), but then there will no longer be uniqueness in Theorem 4.1 (3) because of the choice of spin connection.

(f) It follows from Axiom (1) (see Theorem 8, p. 309, Connes, 1994) that the operators $a\, ds^n$, $a \in \mathcal{A}$, are automatically measurable so that $⨍$ is well defined in Axiom (5).

Now let us consider the general noncommutative case. Given an involutive algebra of operators $\mathcal{A}$ on the Hilbert space $\mathcal{H}$, Tomita's theory associates to all vectors $\xi \in \mathcal{H}$, cyclic for $\mathcal{A}$ and for its commutant $\mathcal{A}'$

$$\overline{\mathcal{A}\xi} = \mathcal{H}\,,\ \overline{\mathcal{A}'\xi} = \mathcal{H} \tag{4.4}$$

an antilinear isometric involution $J : \mathcal{H} \to \mathcal{H}$ obtained from the polar decomposition of the operator

$$S\, a\xi = a^*\xi \qquad \forall\, a \in \mathcal{A}\,. \tag{4.5}$$

It satisfies the following commutation relation:

$$J\mathcal{A}'' J^{-1} = \mathcal{A}'\,. \tag{4.6}$$

In particular $[a, b^0] = 0 \quad \forall\, a, b \in \mathcal{A}$ where

$$b^0 = Jb^*J^{-1} \qquad \forall b \in \mathcal{A} \tag{4.7}$$

so $\mathcal{H}$ becomes an $\mathcal{A}$-bimodule using the representation of the opposite algebra $\mathcal{A}^0$ given by (4.7). There is no difference between module and bimodule structures in the commutative case because one has $a^0 = a \quad \forall\, a \in \mathcal{A}$.

Tomita's theorem is the key ingredient which guarantees the substance of the axioms in the general case. The axioms (1), (3), and (5) are left untouched, but in the axiom of reality (7) the equality $Ja^*J^{-1} = a \quad \forall\, a \in \mathcal{A}$ is replaced by

(7′) $\qquad [a, b^0] = 0 \qquad \forall\, a, b \in \mathcal{A}$ where $b^0 = Jb^*J^{-1}$

also Axiom (2) (order one) becomes

(2′) $\qquad [[D, a], b^0] = 0 \qquad \forall\, a, b \in \mathcal{A}\,.$

(Notice that since a and b^0 commute (2′) is equivalent to $[[D, a^0], b] = 0 \quad \forall\, a, b \in \mathcal{A}$.)

The Hilbert space $\mathcal{H}$ becomes an $\mathcal{A}$-bimodule by Axiom (7′) and gives a class μ of KR^n-homology for the algebra $\mathcal{A} \otimes \mathcal{A}^0$ equipped with the antilinear automorphism τ,

$$\tau(x \otimes y^0) = y^* \otimes x^{*0}\,.$$

The Kasparov intersection product (Kasparov, 1980) allows one to formulate Poincaré duality in terms of the invertibility of μ,

(6′) $\exists\, \beta \in KR_n(\mathcal{A}^0 \otimes \mathcal{A})\,, \quad \beta \otimes_{\mathcal{A}} \mu = \mathrm{id}_{\mathcal{A}^0}\,, \quad \mu \otimes_{\mathcal{A}^0} \beta = \mathrm{id}_{\mathcal{A}}\,.$

It implies the isomorphism $K_*(\mathcal{A}) \xrightarrow{\cap\mu} K^*(\mathcal{A})$. The intersection form

$$K_*(\mathcal{A}) \times K_*(\mathcal{A}) \to \mathbb{Z}$$

is obtained from the Fredholm index of D with coefficients in $K_*(\mathcal{A} \otimes \mathcal{A}^0)$. Note that it is defined without using the diagonal map $m : \mathcal{A} \otimes \mathcal{A} \to \mathcal{A}$, which is not a homomorphism in the noncommutative case. This form is quadratic or symplectic according to the value of n modulo 8.

The Hochschild homology with coefficients in a bimodule makes perfect sense in the general case and Axiom (4) takes the following form:

(4′) There exists a Hochschild cycle $c \in Z_n(\mathcal{A}, \mathcal{A} \otimes \mathcal{A}^0)$ such that $\pi(c) = 1$ (n odd), or $\pi(c) = \gamma$ (n even)

(where $\mathcal{A} \otimes \mathcal{A}^0$ is the $\mathcal{A}$-bimodule obtained by restriction of the structure of the $\mathcal{A} \otimes \mathcal{A}^0$-bimodule of $\mathcal{A} \otimes \mathcal{A}^0$ to the subalgebra $\mathcal{A} \otimes 1 \subset \mathcal{A} \otimes \mathcal{A}^0$, i.e.

$$a(b \otimes c^0)d = abd \otimes c^0 \qquad \forall\, a, b, c, d \in \mathcal{A}\,).$$

Axioms (1), (3) and (5) are unchanged in the noncommutative setting. The proof showing that the operators $a(ds)^n$, $(a \in \mathcal{A})$, are measurable stays valid in general.

We adopt Axioms (1), (2′), (3), (4′), (5), (6′) and (7′) in the general case as a definition of a *spectral manifold* of dimension n. After fixing the algebra $\mathcal{A}$ one can talk about the spectral geometry of $\mathcal{A}$ as in (1.20) and (1.21). One can show that the von Neumann algebra $\mathcal{A}''$ generated by $\mathcal{A}$ in $\mathcal{H}$ is automatically finite and hyperfinite and there is a complete list of such algebras up to isomorphism (Connes, 1994). The algebra $\mathcal{A}$ is stable under smooth functional calculus in its norm closure $A = \bar{\mathcal{A}}$ so that $K_j(\mathcal{A}) \simeq K_j(A)$, i.e. $K_j(\mathcal{A})$ depends only on the underlying topology (defined by the C^* algebra A). The integer $\chi = \langle \mu, \beta \rangle \in \mathbb{Z}$ gives the Euler characteristic in the form

$$\chi = \operatorname{Rank} K_0(\mathcal{A}) - \operatorname{Rank} K_1(\mathcal{A})$$

and Theorem 3.1 gives a local formula for it.

The group of automorphisms of the involutive algebra $\mathcal{A}$, $\operatorname{Aut}(\mathcal{A})$, in general plays the role of the group of diffeomorphisms, $\operatorname{Diff}(M)$, of a manifold M. (There is a canonical isomorphism $\operatorname{Diff}(M) \xrightarrow{\alpha} \operatorname{Aut}(C^\infty(M))$ given by

$$\alpha_\varphi(f) = f \circ \varphi^{-1} \qquad \forall\, f \in C^\infty(M)\ ,\ \varphi \in \operatorname{Diff}(M)\,.)$$

In the general noncommutative case, parallel to the normal subgroup $\operatorname{Int}\mathcal{A} \subset \operatorname{Aut}\mathcal{A}$ of inner automorphisms of $\mathcal{A}$,

$$\alpha(f) = ufu^* \qquad \forall\, f \in \mathcal{A} \tag{4.8}$$

where u is a unitary element of $\mathcal{A}$ (i.e. $uu^* = u^*u = 1$), there exists a natural foliation of the space of spectral geometries on $\mathcal{A}$ by equivalence classes of *inner deformations* of a given geometry. Such a deformation is obtained by the following formula without modifying either the representation of $\mathcal{A}$ in $\mathcal{H}$ or the antilinear isometry J

$$D \to D + A + JAJ^{-1} \tag{4.9}$$

where $A = A^*$ is an arbitrary self-adjoint operator of the form

$$A = \Sigma\, a_i[D, b_i]\ ,\ a_i, b_i \in \mathcal{A}\,. \tag{4.10}$$

The newly obtained spectral triple also satisfies Axioms (1) through (7′).

The action of the group $\mathrm{Int}(\mathcal{A})$ on the spectral geometries (cf. (1.21)) is simply the following gauge transformation of A

$$\gamma_u(A) = u[D, u^*] + uAu^* . \tag{4.11}$$

The required unitary equivalence is implemented by the following representation of the unitary group of $\mathcal{A}$ in $\mathcal{H}$,

$$u \to uJuJ^{-1} = u(u^*)^0 . \tag{4.12}$$

The transformation (4.9) is the identity operator for the usual Riemannian case. To get a nontrivial example it suffices to consider the product of a Riemannian triple by the unique spectral geometry on the finite-dimensional algebra $\mathcal{A}_F = M_N(\mathbb{C})$ of $N \times N$ matrices on $\mathbb{C}$, $N \geq 2$. One then has $\mathcal{A} = C^\infty(M) \otimes \mathcal{A}_F$, $\mathrm{Int}(\mathcal{A}) = C^\infty(M, PSU(N))$ and inner deformations of the geometry are parametrized by the gauge potentials for the gauge theory of the group $SU(N)$. The space of pure states of the algebra $\mathcal{A}$, $P(\mathcal{A})$, is the product $P = M \times P_{N-1}(\mathbb{C})$ and the metric on $P(\mathcal{A})$ determined by the formula (1.10) depends on the gauge potential A. It coincides with the Carnot metric (Gromov, 1994) on P defined by the horizontal distribution given by the connection associated to A, cf. Connes (1997). The group $\mathrm{Aut}(\mathcal{A})$ of automorphisms of $\mathcal{A}$ is the following semi-direct product

$$\mathrm{Aut}(\mathcal{A}) = \mathcal{U} \rtimes \mathrm{Diff}(M) \tag{4.13}$$

of the local gauge transformation group $\mathrm{Int}(\mathcal{A})$ by the group of diffeomorphisms. In dimension $n = 4$, the Hilbert–Einstein action functional for the Riemannian metric and the Yang–Mills action for the vector potential A appear in the asymptotic expansion in $\frac{1}{\Lambda}$ of the number $N(\Lambda)$ of eigenvalues of D which are $\leq \Lambda$. One regularizes this expression by replacing it by

$$\mathrm{Trace}\ \varpi\left(\frac{D}{\Lambda}\right) \tag{4.14}$$

where $\varpi \in C_c^\infty(\mathbb{R})$ is an even function which is 1 on $[-1, 1]$, cf. Chamseddine and Connes (1997). Other nonzero terms in the asymptotic expansion are cosmological, Weyl gravity and topological terms.

A more sophisticated example of a spectral manifold is provided by the noncommutative torus $\mathbb{T}^2_\theta$. The parameter $\theta \in \mathbb{R}/\mathbb{Z}$ defines the following deformation of the algebra of smooth functions on the torus $\mathbb{T}^2$, with generators U, V. The relations

$$VU = \exp 2\pi i\theta\ UV \quad \text{and} \quad UU^* = U^*U = 1\ ,\ VV^* = V^*V = 1 \tag{4.15}$$

define the presentation of the involutive algebra $\mathcal{A}_\theta = \{\Sigma\, a_{n,m} U^n V^n\, ;\ a = (a_{n,m}) \in \mathcal{S}(\mathbb{Z}^2)\}$ where $S(\mathbb{Z}^2)$ is the Schwartz space of sequences with rapid decay. Geometries on $\mathbb{T}^2_\theta$ are parametrized by complex numbers τ with positive imaginary part as in the case of elliptic curves. Up to isometry the geometry depends only on the orbit of τ under the action of $PSL(2,\mathbb{Z})$ (Connes, 1994). However, a new phenomenon appears in the noncommutative case, namely, the *Morita equivalence* which relates the algebras $\mathcal{A}_{\theta_1}$ and $\mathcal{A}_{\theta_2}$ if θ_1 and θ_2 are in the same orbit of the $PSL(2,\mathbb{Z})$ action on $\mathbb{R}$ (Rieffel, 1974; 1981).

Given a spectral manifold $(\mathcal{A},\mathcal{H},D)$ and the Morita equivalence between $\mathcal{A}$ and an algebra $\mathcal{B}$ where

$$\mathcal{B} = \mathrm{End}_{\mathcal{A}}(\mathcal{E}) \tag{4.16}$$

where $\mathcal{E}$ is a finite, projective, Hermitian right $\mathcal{A}$-module, one gets a spectral geometry on $\mathcal{B}$ by the choice of a *Hermitian connection* on $\mathcal{E}$. Such a connection ∇ is a linear map $\nabla : \mathcal{E} \to \mathcal{E} \otimes_{\mathcal{A}} \Omega^1_D$ satisfying the rules (Connes, 1994)

$$\nabla(\xi a) = (\nabla\xi)a + \xi \otimes da \qquad \forall\, \xi \in \mathcal{E}\,,\ a \in \mathcal{A} \tag{4.17}$$

$$(\xi, \nabla\eta) - (\nabla\xi, \eta) = d(\xi,\eta) \qquad \forall\, \xi, \eta \in \mathcal{E} \tag{4.18}$$

where $da = [D,a]$ and where $\Omega^1_D \subset \mathcal{L}(\mathcal{H})$ is the $\mathcal{A}$-bimodule generated by operators of the form (4.10).

Any algebra $\mathcal{A}$ is Morita equivalent to itself (with $\mathcal{E} = \mathcal{A}$) and when one applies the above construction one gets exactly the inner deformations of the spectral geometry.

5 The spectral geometry of space–time

The experimental and theoretical data which one has about the structure of space–time is summarized in the following action functional:

$$\mathcal{L} = \mathcal{L}_E + \mathcal{L}_G + \mathcal{L}_{G\varphi} + \mathcal{L}_\varphi + \mathcal{L}_{\varphi f} + \mathcal{L}_f \tag{5.1}$$

where

$$\mathcal{L}_E = -\frac{1}{16\pi G}\int r\,\sqrt{g}\,d^4x$$

is the Hilbert–Einstein action while the five other terms constitute the standard model of particle physics with minimal coupling to gravity. Besides the metric $g_{\mu\nu}$ this Lagrangian involves several bosonic and fermionic fields. Spin 1 bosons are the photon γ, the intermediate bosons $W^\pm$ and Z, and the eight gluons. The zero spin bosons are the Higgs fields φ which are introduced in order to provide masses to various particles through the mechanism of spontaneous symmetry breaking without contradicting the renormalizability of nonabelian gauge fields. All fermions have spin $\frac{1}{2}$ and form three families of quarks and leptons.

The fields involved in the standard model have a priori a completely different status from the one of the metric $g_{\mu\nu}$. The symmetry group of these fields, namely, the group of local gauge transformations:

$$\mathcal{U} = C^\infty(M, U(1) \times SU(2) \times SU(3)) \tag{5.2}$$

is a priori quite different from the group $\mathrm{Diff}(M)$ of symmetries of the total Lagrangian $\mathcal{L}_F$. The natural group of symmetrics of $\mathcal{L}$ is the semi-direct product $\mathcal{U} \rtimes \mathrm{Diff}(M) = G$ of $\mathcal{U}$ by the natural action of $\mathrm{Diff}(M)$ by automorphisms of $\mathcal{U}$. The first requirement if one wants to obtain a pure geometrical understanding of $\mathcal{L}$ unifying gauge theory with gravity is to find a geometric space X such that $G = \mathrm{Diff}(X)$. This determines the algebra $\mathcal{A}$ [1]

$$\mathcal{A} = C^\infty(M) \otimes \mathcal{A}_F\,,\ \mathcal{A}_F = \mathbb{C} \oplus \mathbb{H} \oplus M_3(\mathbb{C}) \tag{5.3}$$

where the involutive algebra $\mathcal{A}_F$ is the direct sum of the algebras $\mathbb{C}$, $\mathbb{H}$ (the quaternions), and $M_3(\mathbb{C})$ of 3×3 complex matrices.

The algebra $\mathcal{A}_F$ corresponds to a *finite* space where the standard model fermions and the Yukawa parameters (masses of fermions and mixing matrix of Kobayashi Maskawa) determine the spectral geometry in the following manner. The Hilbert space is finite–dimensional and admits the set of elementary fermions as a basis. For example for the first generation of leptons this set is

$$e_L,\ e_R,\ \nu_L,\ \bar{e}_L,\ \bar{e}_R,\ \bar{\nu}_L\,. \tag{5.4}$$

The algebra $\mathcal{A}_F$ admits a natural representation in $\mathcal{H}_F$ (see Connes, 1997). Denote by J_F the unique antilinear involution which exchanges f and $\bar{f}$ for all vectors in the base. One checks the commutation relation

$$[a, Jb^* J^{-1}] = 0 \qquad \forall\, a, b \in \mathcal{A}_F\,, \tag{5.5}$$

which shows that Axiom (2) holds. The operator D_F is given simply by the matrix $\begin{bmatrix} Y & 0 \\ 0 & \bar{Y} \end{bmatrix}$ where Y is the Yukawa coupling matrix. The detailed structure of Y (and in particular the fact that colour is not broken) allows one to check the following relation

$$[[D_F, a], b^0] = 0 \qquad \forall\, a, b \in \mathcal{A}_F\,. \tag{5.6}$$

The natural $\mathbb{Z}/2$-grading of $\mathcal{H}_F$ gives 1 for left–handed fermions $(e_L, \nu_L \ldots)$ and -1 for right–handed fermions; one has

$$\gamma_F = \varepsilon\,\varepsilon^0 \text{ where } \varepsilon = (1, -1, 1) \in \mathcal{A}_F\,. \tag{5.7}$$

We refer to Connes (1997) for the verification of the axioms (1) through (7′). The only drawback of this construction is that the number of families introduces a multiplicity in the intersection form, $K_0(\mathcal{A}) \times K_0(\mathcal{A}) \to \mathbb{Z}$, given by an integer multiple of the matrix

$$\begin{bmatrix} -1 & 1 & -1 \\ 1 & 0 & 1 \\ -1 & 1 & 0 \end{bmatrix}. \tag{5.8}$$

---

[1] taking into account the lifting of diffeomorphisms to the spinors

We will dwell on the significance of the specific spectral geometry $(\mathcal{A}_F, \mathcal{H}_F, D_F)$ at the end of this article.

The next step consists of the computation of internal deformations (formula (4.9)) of the product geometry $M \times F$ where M is a 4-dimensional Riemannian spin manifold. The computation gives the standard model gauge bosons γ, $W^{\pm}$, Z, the eight gluons and the Higgs fields φ with accurate quantum numbers. It also shows that

$$\mathcal{L}_{\varphi f} + \mathcal{L}_f = \langle \psi, D\psi \rangle \tag{5.9}$$

where $D = D_0 + A + JAJ^{-1}$ is the inner deformation of the product geometry (given by the operator $D_0 = \not{\partial} \otimes 1 + \gamma_5 \otimes D_F$).

The product structure of $M \times F$ gives a bigrading of Ω_D^1 and a decomposition $A = A^{(1,0)} + A^{(0,1)}$ of A which corresponds to the decomposition (5.9). The term $A^{(1,0)}$ is the sum of the vector potentials of all spin 1 bosons, the term $A^{(0,1)}$ is the Higgs boson which appears from the finite space $(\mathcal{A}_F, \mathcal{H}_F, D_F)$ as finite difference terms. This bigrading exists also on Ω_D^2, the analogue of 2-forms (Connes, 1994), and decomposes the curvature $\theta = dA + A^2$ in three terms $\theta = \theta^{(2,0)} + \theta^{(1,1)} + \theta^{(0,2)}$, which are pairwise orthogonal with respect to the scalar product

$$\langle \omega_1, \omega_2 \rangle = \int\!\!\!\!\!- \, \omega_1 \, \omega_2^* \, ds^4 \, . \tag{5.10}$$

Hence, the Yang–Mills action, $\langle \theta, \theta \rangle = \int\!\!\!\!\!- \, \theta^2 \, ds^4$, also decomposes into the sum of three terms. One can show that these terms are $\mathcal{L}_G$, $\mathcal{L}_{G\varphi}$ and $\mathcal{L}_\varphi$ respectively (Connes, 1994).

The Yang–Mills action $\int\!\!\!\!\!- \, \theta^2 \, ds^4$ uses the decomposition $D = D_0 + A + JAJ^{-1}$ and hence depends on more than just the geometry fixed by D. Since we want to unify matter with gravity by an action which is purely geometric we need a better formula that only involves D alone. In the simplest case, as shown in formula (4.14), the sum $\mathcal{L}_E + \mathcal{L}_G$ appears directly in the asymptotic expansion of the number of eigenvalues of D which are smaller than Λ. The same principle (cf. Chamseddine and Connes, 1997) applies to the standard model and it is governed by the following functional

$$\operatorname{Trace}\left(\varpi \left(\frac{D}{\Lambda} \right) \right) + \langle \psi, D\psi \rangle \tag{5.11}$$

whose asymptotic expansion (Chamseddine and Connes, 1997) gives our original Lagrangian $\mathcal{L}$ (5.1) together with a Weyl gravity term and a term in $r\varphi^2$, the only term which could be added to $\mathcal{L}$ without changing the standard model. For the physical interpretation of this result see Chamseddine and Connes (1997).

The finite geometry $(\mathcal{A}_F, \mathcal{H}_F, D_F)$ above was dictated by the experimental results, encapsulated in the details of the standard model. One still has to understand its conceptual significance using as a tool the analogue of Lie groups in noncommutative geometry, i.e. the quantum groups. A simple fact (cf. Manin, 1988) is that the spin cover Spin(4) of $SO(4)$ is not the maximal cover in the quantum group category. One has Spin(4) $= SU(2) \times SU(2)$ and the group $SU(2)$ admits according to Lusztig a finite cover of the form (Frobenius at ∞):

$$1 \to H \to SU(2)_q \to SU(2) \to 1 \tag{5.12}$$

where q is a root of unity, $q^m = 1$, m odd. The simplest case is $m = 3$, $q = \exp\left(\frac{2\pi i}{3}\right)$. The finite quantum group H has a finite dimensional Hopf algebra very similar to $\mathcal{A}_F$, and the spinor representation of H defines a bimodule whose structure is very similar to the $\mathcal{A}_F$-bimodule $\mathcal{H}_F$. This suggests that we extend spin geometry (Lawson and Michelson, 1989) to quantum covers of the spin group. This requires for the description of principal G bundles the introduction of a minimum of noncommutativity (of the type $C^\infty(M) \otimes \mathcal{A}_F$) in the algebras of functions.

Finally let us mention that we neglected the important difference between Riemannian and Lorentzian signatures in this article.

Bibliography

Atiyah, M. F. (1966). K-theory and reality. *Quart. J. Math. Oxford* (2), **17**, 367–386.

Beals, R. and Greiner, P. (1988). Calculus on Heisenberg manifolds. *Annals of Math. Studies* **119**, Princeton Univ. Press, Princeton, N.J.

Branson, T. P. (1996). An anomaly associated with 4-dimensional quantum gravity. *Comm. Math. Phys.* **178**, no. 2, 301–309.

Branson, T. P. and Ørsted, B. (1991). Explicit functional determinants in four dimensions. *Proc. Amer. Math. Soc.*, **113**, 669–682.

Bernstein, A. R. and Wattenberg, F. (1969). Nonstandard measure theory. In *Applications of model theory to algebra analysis and probability.* Edited by W. A. J. Luxenburg. Holt, Rinehart and Winston.

Chamseddine, A. and Connes, A. (1997). The spectral action principle. To appear.

Connes, A. (1986). Cyclic cohomology and the transverse fundamental class of a foliation. Geometric methods in operator algebras. Kyoto, 1983, 52–144. *Pitman Res. Notes in Math.* **123**. Longman, Harlow.

Connes, A. (1994). *Noncommutative geometry.* Academic Press.

Connes, A. (1997). Gravity coupled with matter and the foundation of non-commutative geometry. Preprint.

Connes, A. and Moscovici, H. (1990) Cyclic cohomology, the Novikov conjecture and hyperbolic groups. *Topology* **29**, 345–388.

Connes, A. and Moscovici, H. (1995). The local index formula in noncommutative geometry. *GAFA*, **5**, 174–243.

Connes, A. and Moscovici, H. (1997). Hypoelliptic Dirac operator, diffeomorphisms and the transverse fundamental class. Preprint.

Connes, A., Sullivan, D., and Teleman, N. (1994). Quasiconformal mappings, operators on Hilbert space, and local formulae for characteristic classes. *Topology*, **33**, no. 4, 663–681.

Damour, T. and Taylor, J. H. (1992). Strong field tests of relativistic gravity and binary pulsars. *Physical Review D*, **45** no. 6, 1840–1868.

Dixmier, J. (1966). Existence de traces non normales. *C.R. Acad. Sci. Paris*, Ser. A-B **262**.

Doplicher, S., Fredenhagen, K., and Roberts, J. E. (1995) Quantum structure of space time at the Planck scale and Quantum fields. *Comm. Math. Phys.* **172**, no. 1, 187–220.

Fröhlich, J. (1994). The noncommutative geometry of two dimensional supersymmetric conformal field theory. Preprint ETH.

Gromov, M. (1994). Carnot–Caratheodory spaces seen from within. Preprint IHES/ M/94/6.

Hilsum, M. and Skandalis, G. (1987). Morphismes K-orientés d'espaces de feuilles et fonctorialité en théorie de Kasparov. *Ann. Sci. Ecole Norm. Sup.* (4) **20**, 325–390.

Kalau, W. and Walze, M. (1995). Gravity, noncommutative geometry and the Wodzicki residue. *J. of Geom. and Phys.* **16**, 327–344.

Kasparov, G. (1980). The operator K-functor and extensions of C^*-algebras. *Izv. Akad. Nauk. SSSR Ser. Mat.*, **44**, 571–636.

Kassel, C. (1989). Le résidu non commutatif. *Séminaire Bourbaki, exposé* 708, *Astérisque* Vol. 175–178.

Kastler, D. (1995). The Dirac operator and gravitation. *Commun. Math. Phys.* **166**, 633–643.

Kirbi, R. C. (1969). Stable homeomorphisms and the annulus conjecture. *Ann. Math.* **89**, 575–582.

Lawson, B., and Michelson, M. L. (1989). *Spin Geometry*, Princeton.

Manin, Y. (1988). Quantum groups and noncommutative geometry. *Centre Recherche Math. Univ. Montréal.*

Milnor, J. and Stasheff, D. (1974). Characteristic classes. *Ann. of Math. Stud.*, **76**. Princeton University Press, Princeton, N.J.

Novikov, S. P. (1965). Topological invariance of rational Pontrjagin classes. *Doklady A.N. SSSR*, **163**, 921–923.

Power, S. (1982). Hankel operators on Hilbert space. *Res. Notes in Math.*, **64** Pitman, Boston, Mass.

Rieffel, M. A. (1974). Morita equivalence for C^*-algebras and W^*-algebras. *J. Pure Appl. Algebra* **5**, 51–96.

Rieffel, M. A. (1981). C^*-algebras associated with irrational rotations. *Pacific J. Math.* **93**, 415–429; MR 83b:46087.

Riemann, B. (1953). *Mathematical Werke.* Dover, New York.

Stein, E. (1970). *Singular integrals and differentiability properties of functions.* Princeton University Press, Princeton, N.J.

Stern, J. (1985). Le problème de la mesure. *Séminaire Bourbaki* **1983/84**, Exp. 632, 325–346. *Astérisque* N. 121/122. *Soc. Math. France*, Paris.

Sullivan, D. (1971). Geometric periodicity and the invariants of manifolds. *Lecture Notes in Math.* **197**, Springer, Berlin.

Sullivan, D. (1977). Hyperbolic geometry and homeomorphisms. In *Geometric Topology, Proceed. Georgia Topology Conf. Athens*, Georgia, 543–555.

Takesaki, M. (1970). Tomita's theory of modular Hilbert algebras and its applications. *Lecture Notes in Math.* **128**, Springer, Berlin.

Weinberg, S. (1972). *Gravitation and Cosmology.* John Wiley and Sons, New York.

Wodzicki, M. (1987). Noncommutative residue, Part I. Fundamentals K-theory, arithmetic and geometry. *Lecture Notes in Math.*, 1289, Springer, Berlin.

5

Einstein's Equation and Conformal Structure

Helmut Friedrich
Albert-Einstein-Institut, Max-Planck-Institut für Gravitationsphysik, Schlaatzweg 1, 14473 Potsdam, Germany

Abstract
The status of Penrose's idea of asymptotic simplicity is reviewed. The underlying relationship between Einstein's equation and the conformal structure of its solutions is discussed.

1 Introduction

Locally, Einstein propagation is governed by the null cone structure or, equivalently, the conformal geometry of the solutions. The "physical" characteristics of the Einstein equation, i.e. those which are independent of gauge conditions or specific representations of the equation, are the null hypersurfaces of its solutions.

Considering now gravitational fields at larger scales, we may expect the relation between Einstein propagation and conformal geometry to get blurred due to the variation of the null cone structure in the course of its evolution. That there might also be a tight relationship between the conformal structure of the solutions and the propagational properties of the Einstein equation in the large was realized by Roger Penrose.

His idea of "asymptotic simplicity" was one of the important results of an intensive discussion about gravitational radiation and asymptotic structure of gravitational fields in the late 1950s and the early 1960s. For an account of this development and the ideas which went into it I refer to Friedrich (1992), which also contains the basic references.

Today it appears obvious that any notion of gravitational radiation in the non-linear theory must be closely tied up with the null geometry of the field and if one wished to analyse the asymptotic behaviour of gravitational waves which are emitted from a compact radiating system, one would have to examine the propagation of the field in null directions. It is not so clear, however, what kind of asymptotic behaviour of the field one could expect and how to characterize it in a precise way.

Penrose conjecture (Penrose 1963) *The asymptotic structure of gravitational fields in null directions can be characterized by the requirement that the conformal structure be smoothly extendible through null infinity.*

The word conjecture is used here for easy reference; perhaps it is not the most suitable one. The beauty and elegance of the idea, which is so much in accordance with the geometric nature of Einstein's theory, should have a persuasive appeal to every geometrically oriented mind.

There remained open questions though and the purpose of this article is to review the status of the conjecture. This will be done with a bias towards my own interests which are concerned as much with the interplay between Einstein's equation and conformal structure in general as with the specific question concerning the asymptotic behaviour of gravitational fields. For a view on the subject from a different angle I refer to the recent article by Bičák (1996).

2 Asymptotic simplicity and conformal Einstein equations

The concise form in which the conjecture has been presented above is not detailed enough for further discussions. Therefore I shall recall the precise formulation of the idea as given in Penrose (1963, 1965) and later literature.

Suppose that $(\tilde{M}, \tilde{g})$ is a smooth, connected, 4-dimensional Lorentz space. It is called "asymptotically simple" if it admits a smooth "conformal extension" (M, g, Ω) with the following properties.

(i) (M, g) is a smooth, orientable, time orientable Lorentz space, possibly with boundary, which satisfies the causality condition. Ω is a smooth real function on M with non-empty set of zeros.

(ii) There is a smooth embedding $\Phi : \tilde{M} \to M$ with $\Phi(\tilde{M})$ a component of the set $\{\Omega > 0\}$ such that its boundary in M is contained in $\{\Omega = 0\}$. We use Φ to identify $\tilde{M}$ with $\Phi(\tilde{M})$ and set $\mathcal{J} = \{p \in \partial\tilde{M} : \Omega(p) = 0,\ d\Omega(p) \neq 0\}$. This set is called the "conformal boundary" of $(\tilde{M}, \tilde{g})$ and also referred to as *"scri"*.

(iii) The embedding is conformal such that

$$g = \Omega^2\, \tilde{g} \quad \text{on} \quad \tilde{M}. \tag{2.1}$$

(iv) Any null geodesic of $(\tilde{M}, \tilde{g})$ acquires on $\mathcal{J}$ an endpoint in the past and an endpoint in the future.

(v) The space $(\tilde{M}, \tilde{g})$ satisfies in a neighbourhood of $\mathcal{J}$ Einstein's equation

$$Ric(\tilde{g}) = \lambda\, \tilde{g} \tag{2.2}$$

with cosmological constant λ. If $\lambda = 0$, we call $(\tilde{M}, \tilde{g})$ also "asymptotically flat".

The conditions imply that all null geodesics are complete. The conformal boundary forms a smooth hypersurface in M generated by the endpoints of null geodesics in the infinite past and future respectively. The conditions above are not violated, if the function Ω is replaced by $\Theta\,\Omega$ with some function $\Theta > 0$. Therefore the definition involves in fact only the conformal structure of $(\tilde{M}, \tilde{g})$. Of course, the region in M beyond the closure of $\tilde{M}$ is not determined by the conformal structure of $(\tilde{M}, \tilde{g})$.

In many circumstances it is useful to relax the requirements. As an example, the completeness condition (iv) may be weakened to allow for the occurrence of black holes and singularities. Furthermore, we may require lower smoothness for the space $(\tilde{M}, \tilde{g})$ and, separately, for the conformal extension. What exactly should be stipulated here remains largely a matter of taste and convenience as long as field equations are not involved.

Most important for us is that the existence of a smooth conformal extension entails a certain fall-off behaviour of the "physical" metric $\tilde{g}$ near *scri*. In a sense, conditions (i) to (iv) characterize the fall-off condition in a most precise way. No freedom is left for further specification. Apart from its precision and elegance, this geometric way of describing the asymptotic behaviour has many technical advantages. This has already been discussed in Penrose (1963, 1965).

On the other hand, we require in (v) Einstein's equation (2.2) to hold near *scri*. It impresses its own specific fall-off condition on the field. That this should match exactly with the geometric conditions (i) to (iv) is the gist of the Penrose conjecture.

Even the Cauchy problem local in time for Einstein's equation was not completely understood at the time when the idea of asymptotic simplicity was put forward. There was no way to derive any result of a global nature. The rigorous analysis of the behaviour of solutions "locally" near some point on the conformal boundary appeared out of reach. Moreover, in some cases the conformal boundary necessarily consists of a component in the infinite past and a component in the infinite future. Specifying the behaviour of the field in these regions with such a degree of precision must have been considered by those working with standard PDE methods as a very bold act indeed. Though there was a certain amount of evidence in favour of it, the conjecture remained controversial.

The situation has changed in recent years. There are now theorems available which guarantee the existence of general classes of asymptotically simple space–times. We also have a clear understanding of the structural basis for the specific asymptotic properties of these solutions. The existence proofs are based on the use of "conformal representations of the Einstein equation" (Friedrich 1981, 1995). In such a representation the Einstein equation (2.2) is replaced by a system of equations for the conformal metric g, the conformal factor Ω, possibly a torsion free conformal connection $\hat{\nabla}$ (see below), and a number of fields derived from them, in particular the rescaled conformal Weyl tensor $d^{\mu}{}_{\nu\lambda\rho} = \Omega^{-1}\, C^{\mu}{}_{\nu\lambda\rho}$. In regions where the conformal factor does not vanish this system is equivalent to Einstein's equation. The system implies, after a suitable choice of gauge, propagation equations which are hyperbolic, irrespective of the sign of the conformal factor. We refer briefly to any such system as the "conformal Einstein equations". The derivation of these representations includes as an important ingredient the covariant transformation law under conformal rescalings of the Bianchi equation for the conformal Weyl tensor. The importance of this conformal covariance has already been emphasized in Penrose (1965).

I shall discuss situations which are close to or modelled after the conformally flat, simply connected, geodesically complete, asymptotically simple standard solutions to (2.2) which are also spaces of highest possible symmetry: de Sitter space, anti-de Sitter covering space (AdS), and Minkowski space.

It has been observed in Penrose (1965) that the causal nature of the conformal boundary is determined by the sign of the cosmological constant. As a consequence the three cases above are not only distinguished by the sign of the cosmological constant but present problems of increasing difficulty. Each of them requires new insights into the "conformal structure of Einstein's equation".

3 De Sitter-type space–times

We consider first de Sitter space–time. It can be characterized as the maximal solution to Einstein's field equations (2.2) with cosmological constant $\lambda < 0$ which is determined by Cauchy data (S, h, χ), where the initial hypersurface S is diffeomorphic to S^3, the metric h is the standard metric on the unit 3-sphere, and the extrinsic curvature χ vanishes. In this case we have a very concise result.

Theorem 3.1. (Friedrich 1986) *Solutions to Einstein's field equation (2.2) arising from Cauchy data sufficiently close to de Sitter data are asymptotically simple.*

General stability results for solutions to hyperbolic equations allow us to show the existence of solutions to the conformal field equations which are close to conformal de Sitter space and extend beyond the set $\mathcal{J} = \{\Omega = 0\}$. The conformal field equations themselves imply then that $d\Omega \neq 0$ on $\mathcal{J}$ and we can conclude that this set represents the conformal boundary of the physical space–time we have constructed. Important for the conciseness of our statement is also the compactness of the initial hypersurface; asymptotic conditions for the initial data need not be considered here.

The theorem shows that asymptotic simplicity is stable under small but finite perturbations of the de Sitter data. I shall not dwell on the generalizations of this result. They can be found in the references.

4 Anti-de Sitter-type space–times

We consider now solutions of (2.2) with positive cosmological constant. In this case the conformal boundary $\mathcal{J}$ is a time-like hypersurface at null and space-like infinity. This implies that such solutions are not globally hyperbolic and it indicates that we need to consider initial boundary value problems if we want to analyse general solutions. The data need to be given on a space-like slice $\tilde{S}$ extending to $\mathcal{J}$ and also on $\mathcal{J}$ itself. This makes the analysis more complicated. In particular, the hypersurface $\mathcal{J}$, which is defined in terms of the solution, has yet to be constructed. To obtain "Anti-de Sitter-type" solutions we consider the following data and assumptions.

(i) Suppose $(\tilde{S}, \tilde{h}, \tilde{\chi})$ is a smooth space-like Cauchy data set for Einstein's equation (2.2) with cosmological constant $\lambda > 0$. Assume that it has a smooth conformal compactification (S, h, χ) which is asymptotically simple in the sense that it induces smooth initial data for the conformal field equations. In particular S is a smooth, orientable 3-manifold with boundary ∂S and interior $\tilde{S}$.

(ii) Suppose $\alpha > 0$ and on $\mathcal{J}_\alpha = \,]-\alpha, \alpha[\, \times \partial S$ is given a smooth conformal structure $\mathcal{C}$ of signature $(+,-,-)$.

(iii) Suppose that the data above are compatible, i.e. they satisfy the "corner conditions" implied by the conformal field equations on $\partial S \simeq \{0\} \times \partial S$.

Theorem 4.1. (Friedrich 1995) *Given data as above, there exists for some $\alpha > 0$ a unique smooth solution $\tilde{g}$ to Einstein's equation (2.2) on $\tilde{M}_\alpha = \,]-\alpha, \alpha[\, \times \tilde{S}$ with the following properties. The metric $\tilde{g}$ induces $\tilde{h}$ as the first and $\tilde{\chi}$ as the second fundamental form on $\tilde{S} \simeq \{0\} \times \tilde{S}$ and for any smooth function Ω on $M_\alpha = \,]-\alpha, \alpha[\, \times S$ with $\Omega > 0$ on $\tilde{M}_\alpha$, $\Omega = 0$, $d\Omega \neq 0$ on $\mathcal{J}_\alpha$, the metric $\Omega^2 \tilde{g}$ on $\tilde{M}_\alpha$ extends to a smooth Lorentz metric g on M_α which induces on $\mathcal{J}_\alpha$ the conformal structure $\mathcal{C}$.*

The result shows the existence of general AdS-type space–times. In fact, all smooth AdS-type space–times are characterized here by initial and boundary data. No smallness assumption has been imposed. The solutions obtained here could be called global in space-like directions. In particular, there exist null geodesics, complete either in the future or in the past, whose endpoints generate the conformal boundary $\mathcal{J}_\alpha$. The completeness condition (iv) is not satisfied. The problem of constructing solutions which are global in conformal time is of a very peculiar nature and has not been studied yet.

The conformal properties of the field equations are again the basis of the result, but the existence proof requires more technicalities than in the previous case. There are subtleties, hidden in the formulation of the theorem, which arise from the non-compactness of the physical initial hypersurface $\tilde{S}$. The conformal constraints are regular. Nevertheless, if one tries to construct solutions to them, one realizes that they "remember" that the physical data extend to infinity and that fall-off conditions have to be assumed. To assess the value of the theorem for our discussion we need to look at this in more detail.

In the articles by Andersson *et al.* (1992) and Andersson and Chruściel (1994) the existence of smooth "hyperboloidal initial data" (see below) for the case $\lambda = 0$ has been discussed. In Kánnár (1996) it has been shown how to derive from these data space-like Cauchy data with $\lambda \neq 0$ as required in the present case. It follows from these articles that there exist smooth space-like Cauchy data sets $(\tilde{S}, \tilde{h}, \tilde{\chi})$ allowing conformal compactifications (S, h, χ) with smooth fields h, χ on S for which the implied data for the conformal field equations, in particular for the rescaled conformal Weyl tensor, are not smooth. The requirement of asymptotic simplicity imposes as an extra condition on the free data that certain fields determined from the free data have to vanish on the boundary ∂S at infinity.

It turns out, however, that in general the rescaled conformal Weyl tensor for such non-smooth data is strongly divergent at ∂S. Most likely, due to the hyperbolic nature of the conformal field equations, this singularity will spread in such a way that the null geodesics in the maximally extended solution will not be complete. In that case there will be no null infinity and we cannot even ask whether it allows a smooth structure.

I consider the statements above as perfect vindications of the Penrose conjecture for the case of non-vanishing cosmological constant. The attempt to remove the smallness assumption in the first theorem will lead to qualitatively different problems, the occurrence of singularities and black holes. The analysis of such situations seems at present beyond the scope of our technical means.

More could be said about the cases $\lambda \neq 0$ and the solutions considered above, but I shall devote the rest of my talk to the case $\lambda = 0$ which has not been resolved yet.

5 Minkowski-type space–times

5.1 Conformal Minkowski space

The standard example of an asymptotically flat solution of Einstein's equation (2.2) is, of course, Minkowski space with manifold $\tilde{M} = \mathbb{R}^4$ and metric $\tilde{g} = dt^2 - (dr^2 + r^2\, d\sigma^2)$, where $t \in \mathbb{R}$, $r \geq 0$ is a radial coordinate, and $d\sigma^2$ is the line element on the standard unit sphere S^2.

A suitable conformal extension is provided by (M, g, Ω) with $M = \mathbb{R} \times S^3$, $g = d\tau^2 - (d\chi^2 + sin^2\chi\, d\sigma^2)$, and $\Omega = cos\,\tau + cos\,\chi$ with $\tau \in \mathbb{R}$ and $0 \leq \chi \leq \pi$ one of the standard polar coordinates on the 3-sphere. An embedding Φ of Minkowski space into the Einstein cosmos (M, g) is defined by $t - r = \tan(\frac{\tau-\chi}{2})$, $t + r = \tan(\frac{\tau+\chi}{2})$. It allows the identification of Minkowski space with the image $\{|\tau + \chi| < \pi,], |\tau - \chi| < \pi\}$ of Φ such that after identification (2.1) holds on this set.

The set $\mathcal{I}$ consists of the two connected components $\mathcal{I}^{\pm} = \{\tau = \pm\,(\pi - \chi),\ 0 < \chi < \pi\}$, called future and past null infinity respectively. The boundary of $\tilde{M}$ in M contains furthermore the points $i^{\pm} = \{\chi = 0, \tau = \pm\pi\}$, future respectively past time-like infinity, as well as the point $i^0 = \{\chi = \pi, \tau = 0\}$ which represents space-like infinity. Any time-like geodesic in Minkowski space approaches i^- in the infinite past and i^+ in the infinite future, while any space-like geodesic approaches i^0 if any of its affine parameters goes to infinity. The sets $\mathcal{I}^{\pm}$ represent (parts of) the future resp. past light cone swept out by the null geodesics through i^0 in the Einstein cosmos; these null geodesics reconverge in the past at i^- and in the future at i^+.

In the following we shall be concerned with the question of whether there exist more general solutions $(\tilde{M}, \tilde{g})$ of Einstein's equation $Ric(\tilde{g}) = 0$ which satisfy conditions (i) to (iv). Such solutions will be called "Minkowski-type space–times" if they arise from standard Cauchy data on some initial hypersurface diffeomorphic to $\mathbb{R}^3$ analogous to the Cauchy hypersurface $\{t = 0\}$ in Minkowski space. These data should be "asymptotically flat at space-like infinity" in the sense in which this is used in the study of the Cauchy problem for Einstein's equations.

We recall that for such solutions the smoothness of *scri* will imply the "Sachs-peeling property". This says that the components Ψ_k of the conformal Weyl tensor in a suitable normalized frame satisfy along outgoing null geodesics the particular fall-off behaviour

$$\Psi_k(r) = O\left(\frac{1}{r^{k-5}}\right) \quad \text{as} \quad r \to \infty, \quad k = 0, 1, \ldots, 4,$$

if r denotes a suitable parameter on the null geodesics.

Finally we remark that analogues of the points $i^{\pm}$ can be expected to occur in smooth conformal extensions of Minkowski-type solutions but we shall not be much concerned with them here. In contrast, for solutions with positive ADM mass no analogue of the point i^0 can exist in smooth conformal extensions and this fact motivates almost all of the following discussions. Nevertheless, we shall follow the usual custom in thinking of space-like infinity as being represented by a point i^0 in some non-smooth conformal extension.

5.2 Some existence results

A result intimating the naturalness of the idea of asymptotic flatness is obtained by analysing the "hyperboloidal initial value problem" for the conformal field equations (Friedrich, 1986). Here data are given which represent the geometry of a space-like hypersurface S in an asymptotically flat solution which extends to future null infinity. It turns out that the smoothness of null infinity is preserved by the evolution off the initial hypersurface and, assuming data sufficiently close to Minkowskian data, a regular point at future time-like infinity is obtained. This indicates that the decision for the smoothness of the structure at null infinity is made in arbitrarily "small" neighbourhoods of space-like infinity. For this reason most of the following discussion will be concerned with the structure of the field near space-like infinity.

The importance of analysing the situation near space-like infinity has in fact already been stressed in Penrose (1965) where we read at the end of the article: "... of great interest would be an examination of the geometry in the neighbourhood of the point i^0. For this provides the most immediate link between *scri*$^-$ and *scri*$^+$. Without such a link, it is difficult to see how the outgoing radiation is to be correlated with the incoming radiation."

Christodoulou and Klainerman demonstrated the existence of geodesically complete solutions for small data (Christodoulou and Klainerman, 1993). Their work reproduces the qualitative overall picture suggested by the concept of asymptotic flatness. However, they do not confirm the Sachs-peeling but instead a weaker fall-off behaviour for the conformal Weyl tensor. Thus they are led to speculate that the requirement of asymptotic simplicity might be too strong to admit non-trivial Minkowski-type space–times. Before any such conclusion can be drawn from their work, two questions need to be answered.

(i) Are their results sharp?

(ii) If this is the case, can the Sachs–peeling property be established for solutions arising from data which satisfy stronger fall-off conditions?

These questions may be related to the observation alluded to above that the construction of smooth hyperboloidal initial data requires free data which not only extend smoothly to the conformal boundary $S \cap \mathcal{J}^+$ but which satisfy there certain extra conditions. Two different explanations may be offered for this.

(i) Hyperboloidal data arising by Einstein evolution from standard Cauchy data satisfy these conditions due to the nature of the evolution process. In this case the estimates of Christodoulou and Klainerman would not be sharp.

(ii) The necessity of extra conditions near $S \cap \mathcal{J}^+$ indicates that we also have to impose extra conditions on the standard Cauchy data near space-like infinity beyond the conventional fall-off conditions.

Space-like infinity being thought of as a point, one might wonder about the nature of such conditions. In any case, I believe that the question of the appropriateness of the *scri*-concept should not be a matter of choosing between a few Sobolev classes. If there were obstructions to the smoothness of the conformal boundary it should be possible to identify them in the evolution process and to trace them back to the structure of the data.

5.3 Difficulties

In the conformal version of the Cauchy problem for Minkowski-type space–times the initial hypersurface S will be a 3-sphere on which to one point $i \in S$ is assigned the meaning of space-like infinity for the interior physical metric $\tilde{h} = \Omega^{-2}\, h$ on the physical initial hypersurface $\tilde{S} = S \setminus \{i\}$. The 3-dimensional notion of space-like infinity should be distinguished conceptually from the 4-dimensional one which is referred to by the symbol i^0.

The basic reason for most of the following discussion is the fact that in the case of non-vanishing ADM mass some of the data on S for the conformal field equations diverge at i. This "structural singularity" cannot be avoided. Thus the conformal field equations cannot be used in a direct way to show the existence of solutions which behave near i^0 like Minkowski-type space–times.

The stability arguments used in the case of de Sitter-type solutions no longer apply. In particular, it is not an easy problem to see how null infinity "connects" to space-like infinity or how we could locate null infinity in the solution to the conformal field equations. Furthermore, gauge conditions become a very delicate matter. An unfortunate choice may produce irregularities along null infinity which possibly hide the smoothness of the underlying structure.

On top of the problem of global existence there is a further difficulty. We will have to find a geometrical setting in which we can investigate the consequences of the field equations near i^0 to such a precision that the behaviour of the fields near null and space-like infinity can be discussed in terms of properties imposed on the initial data. The analysis needs to take care in a very detailed way of the interplay between the constraint equations, the evolution equations, and the geometric nature of the solution.

The constraint equations impose quite weak conditions at space-like infinity. They allow us to "shift some non-smoothness of the free data to infinity". This non-smoothness will not be noticed in finite regions but it may spoil the asymptotic behaviour of the solutions. We need to make a judicious choice of data which distinguishes between such "spurious singularities" and the unavoidable structural singularity of the conformal data at space-like infinity.

5.4 Assumptions on the data

To discard such spurious singularities we choose in the following discussion free data which satisfy the strongest smoothness requirements which are compatible with the geometric situation. Thus we shall assume that the conformal structure on the initial hypersurface is represented by a smooth Riemann metric h on S. In fact, we shall assume for convenience that h is analytic near i.

To simplify the necessarily quite detailed analysis we shall assume further that our data be time–symmetric. Then (S, h) represents the initial data set completely. From other investigations this case is known to catch the essential features of the problems.

The basic data for the conformal field equations are then derived from h and the conformal factor Ω which determine the physical metric $\tilde{h} = \Omega^{-2}\, h$ on $\tilde{S}$. Let x^α be h-normal coordinates centred at i and define the radial function $\rho = |x|$ and the analytic function $\Gamma = |x|^2$ near i which both vanish at i. Then the conformal factor is near i of the form $\Omega = \frac{\Gamma}{(U+\rho\, W)^2}$ with an analytic function U which is calculated locally from h near i and satisfies $U(i) = 1$, and an analytic function W which contains global information on (S, h) since $2\, W(i) = m$ is the ADM mass. It is useful to split the rescaled conformal Weyl tensor, given here in space spinor notation, into two components $\phi_{abcd} = \phi'_{abcd} + \phi^W_{abcd}$. Here

$$\begin{aligned}
\phi'_{abcd} \quad &= \quad \Gamma^{-2}\,\{U^2\, D_{(ab}\, D_{cd)}\,\Gamma - 4\, U\, D_{(ab}\Gamma\, D_{cd)}U \\
&\quad - 2\,\Gamma\, U\, D_{(ab}\, D_{cd)}\, U + 6\,\Gamma\, D_{(ab}U\, D_{cd)}U + \Gamma\, U^2\, s_{abcd}\}
\end{aligned} \tag{5.1}$$

is the part determined by the local geometry near i. The spinor field s_{abcd} is the trace free part of the Ricci tensor of h and covariant derivatives are defined by h. The part

$$\begin{aligned}
\phi^W_{abcd} \quad &= \quad \rho^{-3}\left\{-\frac{3}{2}\,\frac{1}{\Gamma}\, U\, W\, D_{(ab}\,\Gamma\, D_{cd)}\,\Gamma + U\, W\, D_{(ab}\, D_{cd)}\,\Gamma\right. \\
&\quad +2\, W\, D_{(ab}\Gamma\, D_{cd)}U - 6\, U\, D_{(ab}\Gamma\, D_{cd)}W \\
&\quad -2\,\Gamma\,(U\, D_{(ab}\, D_{cd)}\, W + W\, D_{(ab}\, D_{cd)}\, U \\
&\quad \left. - \,6\, D_{(ab}U\, D_{cd)}\, W - U\, W\, s_{abcd})\right\} \\
&\quad -2\, W\, D_{(ab}\, D_{cd)}\, W + 6\, D_{(ab}W\, D_{cd)}\, W + W^2\, s_{abcd},
\end{aligned} \tag{5.2}$$

contains global information. We shall call ϕ'_{abcd} the "massless" and ϕ^W_{abcd} the "massive" part of the rescaled Weyl tensor. The smoothness of the massless part depends on the choice of free data near i, but we have at best

$$\phi^W_{abcd} = O(\rho^{-3}) \quad \text{as} \quad \rho \to 0 \quad \text{unless} \quad m = 0.$$

It will be convenient to generalize our discussion by allowing subsets of $S \setminus \{i\}$ to be removed. Then we may consider data where $\phi^W_{abcd} = 0$ near i without being told by the positive mass theorem that necessarily $\phi_{abcd} \equiv 0$.

5.5 Gauge conditions and conformal field equations

We have seen that the usual conformal representation of Minkowski-type space–times suggests an initial value problem which is local but singular at i. We shall now replace this by a problem which is finite and regular. For this purpose the previous analysis of the conformal field equations, where the possibility of performing arbitrary conformal rescalings has been introduced as an additional gauge freedom, will be completed. We will also admit transitions to arbitrary torsion free connections which respect the conformal class, i.e. for which parallel transport maps g-conformal frames again into such frames. The natural background for this and the chosen representation of the conformal field equations is the theory of normal conformal Cartan connections (Cartan 1923) (cf. Friedrich, 1995, for further references). The resurgence of this theory in 4 dimensions in the context of "local twistors" (Dighton, 1974; Friedrich, 1977) may suggest that twistors could play a role here. Our use of the conformal Cartan connection appears, however, not to be related in an obvious way to twistor ideas.

To define the gauge we make use of "conformal geodesics" (Yano, 1938, 1939). These are space–time curves $x^\mu(\tau)$ with a 1-form $b_\mu(\tau)$ along them which are associated with conformal structures in a similar way as geodesics are associated with metric structures. They are governed by the system of ODE's

$$(\tilde{\nabla}_{\dot{x}}\dot{x})^\mu + S(b)_\nu{}^\mu{}_\rho\, \dot{x}^\nu\, \dot{x}^\rho = 0, \quad (\tilde{\nabla}_{\dot{x}}b)_\nu - \frac{1}{2} b_\mu\, S(b)_\nu{}^\mu{}_\rho\, \dot{x}^\rho - \tilde{L}_{\nu\mu}\, \dot{x}^\mu = 0,$$

where $S(b)_\mu{}^\nu{}_\rho = \delta^\nu{}_\mu\, b_\rho + \delta^\nu{}_\rho\, b_\mu - \tilde{g}_{\mu\rho}\, \tilde{g}^{\nu\lambda}\, b_\lambda$ is of the form of a difference tensor defined by two conformal connections and $\tilde{L}_{\nu\mu} = \frac{1}{2}\,(\tilde{R}_{\nu\mu} - \frac{1}{6}\,\tilde{g}_{\nu\mu}\,\tilde{R})$. Using these curves, we construct "conformal Gauss coordinates" based on S in analogy to standard Gauss coordinates. The 1-form b defines a symmetric connection $\hat{\nabla} = \tilde{\nabla} + S(b)$ along the curves which is conformal but not necessarily metric. We use this connection. We use further a conformal frame c_j which is adapted to S with parallel propagation with the connection $\hat{\nabla}$. All fields are given in this frame. Finally we use the conformal factor Ω satisfying $\Omega^2\, \tilde{g}(c_i, c_k) = \eta_{ik}$ and the Levi–Civita connection ∇ of the metric $g = \Omega^2\, \tilde{g}$ which can be determined in our formalism from $\hat{\nabla}$.

The basic fields entering the field equations are the components $c^\mu{}_k$ of the frame in the conformal Gauss coordinates, the connection coefficients $\hat{\Gamma}_i{}^j{}_k$ in that frame, the tensor $\hat{\Gamma}_{lj} = \frac{1}{4}\,\hat{R}_{[lj]} - \frac{1}{2}\,\hat{R}_{(lj)} + \frac{1}{12}\,\hat{R}_k{}^k\,\eta_{lj}$ determined from the Ricci tensor of $\hat{\nabla}$, and the rescaled conformal Weyl tensor $d^i{}_{jkl} = \Omega^{-1}\, C^i{}_{jkl}$.

The conformal factor Ω and the 1-form $d = \Omega\, b$ will also appear in the conformal equations. In terms of these quantities the Einstein equations have the conformal representation

$$[c_p, c_q] = (\hat{\Gamma}_p{}^l{}_q - \hat{\Gamma}_q{}^l{}_p)c_l, \tag{5.3}$$

$$c_p(\hat{\Gamma}_q{}^i{}_j) - c_q(\hat{\Gamma}_p{}^i{}_j) - \hat{\Gamma}_k{}^i{}_j(\hat{\Gamma}_p{}^k{}_q - \hat{\Gamma}_q{}^k{}_p) + \hat{\Gamma}_p{}^i{}_k\hat{\Gamma}_q{}^k{}_j - \hat{\Gamma}_q{}^i{}_k\hat{\Gamma}_p{}^k{}_j \tag{5.4}$$

$$\equiv \hat{R}^i{}_{jpq} = \delta^i{}_q\,\hat{\Gamma}_{pj} - \delta^i{}_p\,\hat{\Gamma}_{qj} + \delta^i{}_j(\hat{\Gamma}_{pq} - \hat{\Gamma}_{qp}) - \eta^{ik}(\hat{\Gamma}_{pk}\eta_{qj} - \hat{\Gamma}_{qk}\eta_{pj}) + \Omega\, d^i{}_{jpq},$$

$$\hat{\nabla}_p\,\hat{\Gamma}_{qj} - \hat{\nabla}_q\,\hat{\Gamma}_{pj} = -d_i\, d^i{}_{jpq}, \tag{5.5}$$

$$\nabla_i d^i{}_{jkl} = 0. \tag{5.6}$$

5.6 The finite regular initial value problem near space-like infinity

The use of conformal geodesics is important to us for the following reasons.

(i) The gauge is associated directly with the conformal structure, a fact which will be important for the interpretation of our results.

(ii) In the equations above the fields Ω, d_k occur. These quantities, which reflect the conformal gauge freedom, are not determined by field equations. Using conformal geodesics to fix the gauge, we find the following result.

Lemma 5.1. (Friedrich 1995) *If the data determine a solution to Einstein's equation (2.2), the fields Ω, d_k can be expressed explicitly in terms of the conformal Gauss coordinates and the conformal initial data on S. It follows from their expressions that at points where $\Omega = 0$ the conditions $\nabla_k\,\Omega = b_k \neq 0$, $\nabla_k\,\Omega\,\nabla^k\,\Omega = 0$ are satisfied if $\lambda = 0$.*

Thus we have a complete differential system. Most importantly, arranging the initial data in analogy to the situation in conformal Minkowski space we find that Ω has a zero set $\mathcal{J} = \mathcal{J}^- \cup \mathcal{J}^+$ which is near i similar to that found in the Minkowski case. Thus, provided the solutions extend to $\mathcal{J}$, we have in conformal Gauss coordinates complete *a priori* information on the location of null infinity.

(iii) In our gauge we can extract from the equations above a symmetric hyperbolic system of propagation equations which is of maximal simplicity in the sense that the fields $c^\mu{}_k$, $\hat{\Gamma}_i{}^j{}_k$, $\hat{\Gamma}_{jk}$ obey ordinary differential equations along the conformal geodesics. This fact simplifies the analysis of the solution near the singular point i considerably. These equations are coupled to the partial differential equations for the rescaled Weyl tensor extracted from the Bianchi equation. The latter can be chosen such that the complete system is symmetric hyperbolic. A solution to these equations satisfying the conformal constraints defines in fact a solution to Einstein's equation.

(iv) The example of Minkowski space suggests that the conformal geodesics will pass through null infinity undisturbed by the singularity at space-like infinity and that our gauge conditions are not affected by the presence of this singularity.

(v) By adapting the remaining gauge freedom on $\tilde{S}$ in a suitable way to the geometric situation we end up with a problem of the following type. The initial manifold S is replaced by a compact manifold $\bar{S}$ with boundary by blowing up the point i to a 2-sphere I^0. We have $\tilde{S} = \bar{S} \setminus I^0$. The radial function ρ induces a coordinate on $\bar{S}$ near I^0, denoted again by ρ, which vanishes on I^0 and is positive elsewhere. Near I^0 the manifold on which the solution is to be constructed may be given in a suitable gauge in the form $\bar{M} = \{(\tau, q) \in \mathbb{R} \times \bar{S} : |\tau| \leq 1 + \rho(q)\}$. Space-like infinity, thought of as a point i^0 before, is now represented by the cylinder $\bar{I} = I \cup I^+ \cup I^-$, where $I = \{|\tau| < 1, \rho = 0\}$ and $I^\pm = \{\tau = \pm 1, \rho = 0\}$. This cylinder can be thought of as the set of non-time-like (directed) directions at the point i^0. We could in fact extend it formally to include time-like directions at i^0 and in the case of conformal Schwarzschild space–time mentioned below this inclusion is not formal but does occur naturally in the analytic extension beyond $I^\pm$.

If the solution extends far enough in a smooth way off the initial hypersurface $\bar{S} = \{\tau = 0\}$, we know a priori that null infinity will be given near space-like infinity by the hypersurfaces $\mathcal{J}^\pm = \{\tau = \pm(1 + \rho(q)),\ q \in \tilde{S}\}$ which "touch" I at the sets $I^\pm$. In this setting the curves $\{|\tau| \leq 1 + \rho(q)\}$, with $q \in \tilde{S}$ fixed, are conformal geodesics with natural parameter τ.

The conformal Cauchy data on $\tilde{S}$ extend in this setting smoothly (in fact analytically) to I^0. The equations are symmetric hyperbolic equations for an unknown u and are of the form

$$\{A^\tau\,\partial_\tau + A^\rho\,\partial_\rho + A^+\,X_+ + A^-\,X_-\}\,u = C\,u$$

with matrix–valued functions A^τ, A^ρ, $A^\pm$, C which depend on u and the coordinates. Here the $X_\pm$ are essentially operators on the 2-sphere.

5.7 The total characteristic at space-like infinity

From the point of view of the 4-dimensional geometry we have here a singular representation of space-like infinity. From the point of view of the conformal field equations we have a regular situation. The set I is a characteristic of the system. We can call the hypersurface I "totally characteristic", because the symmetric hyperbolic system of propagation equations reduces completely to an interior system on I. This allows us to determine the unknown u on I from the data on I^0. As is to be expected, no boundary values can be prescribed on I. Moreover, by taking formal derivatives of the equations with respect to ρ, we get transport equations for the functions $u^p = \partial_\rho^p\,u|_I$ for $p = 0, 1, 2, \ldots$.

This offers us a way to analyse the fields on the set $\bar{I}$ to any desired degree of precision. The setting should also allow us to relate concepts like Bondi–energy–momentum, angular momentum, NP constants, etc., on null infinity to concepts at space-like infinity. It is the first time that a setting has become available that allows us to analyse in the presence of non-vanishing mass the relation, as mediated by the field equations, between the structure of the initial data and the behaviour of the solutions near space-like and null infinity.

The setting will become useless if the conformal geodesics develop caustics before reaching null infinity. Close to space–like infinity the essential quantity determining the behaviour of the field is the ADM mass. Therefore, if the presence of the mass causes the conformal geodesics to converge, this should be seen already in the case of the Schwarzschild solution. It turns out that for conformal Schwarzschild data the solution of the finite regular initial value problem extends in an analytic way through $\mathcal{J}^{\pm}$ near $\bar{I}$. No caustics occur near $\bar{I}$ before the conformal geodesics enter the region beyond $\mathcal{J}^{\pm}$. Moreover, the solution shows the remarkable behaviour that it approaches conformal Minkowski space near $\bar{I}$ in a smooth and uniform way when we let the mass go to zero.

Calculating now in the general case the unknown u on I, we find, not unexpectedly, that $A^{\tau} = \mathrm{diag}\{1+\tau, 1-\tau, 1, \ldots, 1\}$ on I. Thus A^{τ} degenerates on the sets $I^{\pm}$ where the total characteristic I meets the characteristics $\mathcal{J}^{\pm}$ transversely.

The resulting degeneracy of the propagation equations on the sets $I^{\pm}$ has important consequences. Expanding the unknowns in terms of some function system on the sphere, we find that in a sense the solutions u^p to the transport equations can be described explicitly. The actual expressions are defined recursively and are given as integrals. They are quite complicated and I haven't worked out all their interesting details yet. However, the following statement has been obtained. It is most easily expressed in terms of the symmetric space–spinor b_{abcd} which represents the Cotton tensor $K_{pij} = 2\,D_{[j}\,L_{i]p}$. Here $L_{ij} = R_{ij} - \frac{1}{4}\,R\,h_{ij}$ and R_{jk}, R denote the Ricci tensor and Ricci scalar of h respectively.

Theorem 5.2. (Friedrich 1997) *The solution u of the propagation equations extends smoothly to $I^{\pm}$ only if the condition*

$$D_{(a_q b_q} \ldots D_{a_1 b_1}\, b_{abcd)}(i) = 0 \tag{5.7}$$

is satisfied at all orders q. If it is not satisfied for some order q, the solution develops logarithmic singularities at $I^{\pm}$.

5.8 Comments on our procedure

Due to our choice of gauge conditions we know that the logarithmic singularities occurring here are associated with the conformal structure and not an artefact of an obscure choice of gauge. The condition is conformally invariant and thus identifies obstructions to smoothness explicitly in terms of the free data. It is the first time that implications on the data from requirements of asymptotic smoothness have been derived in the presence of mass terms.

We could still hope that the complete analysis will yield a picture where a singularity is residing on the sets $I^{\pm}$ while the solution extends smoothly through null infinity. However, because of the hyperbolicity of the propagation equations we should rather expect that these singularities "spread along null infinity", thus destroying any hope of finding a smooth structure there. In the case where the massive part of the conformal Weyl tensor vanishes near space-like infinity condition (5.7) is also sufficient for the smoothness of the solution at null infinity. In that case it can be shown that if condition (5.7) is satisfied for $q = 0, 1, \ldots, N$ with sufficiently large integer N, the solution still allows us to define a differentiability structure at null infinity with some finite differentiability depending on N.

To the extent to which I have analysed the transport equations, no problems arose from the massive part of the conformal Weyl tensor. The logarithmic singularities observed above may well be the basic source for the possible failure of the Christodoulou–Klainerman solutions to satisfy the Sachs–peeling property (assuming data which are as clean at space-like infinity as ours).

The condition (5.7) imposes a very weak restriction on the free data in the class considered above. It still allows us to prescribe the metric h arbitrarily on any fixed compact subset of S not containing i. The analysis may appear to have led to a negative result. I rather like to think of it as a constructive step. It identifies and possibly removes a stumbling block on the way to establishing the existence of Minkowski-type solutions. Moreover, it is interesting from a conceptual point of view. This will be discussed in the last section.

Inevitably, some features of the construction indicated above are reminiscent of other work concerned with the asymptotic structure of the fields near space-like infinity, notably of the work by Ashtekar (cf. Ashtekar, 1984) and by Beig and Schmidt (1982) (cf. also Beig, 1984; Ashtekar and Romano, 1992). However, there are basic differences. In the approach of Ashtekar (1984) questions of regularity are controlled by fiat. Thus regularity conditions cannot be derived. The approach of Beig and Schmidt (1982) also supplies transport equations near space-like infinity. It is hard to see, however, how the relationship between space-like and null infinity could be analysed in this setting.

To avoid the usual misunderstandings: The fact that the analysis is still complicated does not indicate that in the end it will not be of practical use. The fact that the situation has been analysed so far for a restricted class of data does not indicate that it cannot be extended to more general situations. The properties of the propagation equations and the gauge conditions used in the discussion above are independent of conditions on the data.

The assumption of analyticity near space-like infinity is made here mainly for convenience. At the expense of higher calculational complexity the result given in the last theorem can be suitably generalized to situations of lower differentiability. However, from the point of view of modelling situations of physical interest such generalizations appear to be quite irrelevant.

6 Concluding remarks

In view of the complicated analysis one may ask whether it is worthwhile to spend time on it. The question, to what extent the Penrose conditions reflect the fall-off behaviour of solutions to Einstein's equation, could be regarded as an interesting but purely mathematical exercise of an advanced nature. There is, however, much more to it.

The investigation gives insight into the structure of the theory. The important role of the conformal structure for global studies of general space–times has been known for a long time. Here we see that it is equally effective in the global analysis of solutions to the Einstein equation. Since important open problems of general relativity are questions about the global conformal structure of the gravitational field, the understanding of the conformal structure of Einstein's equation should prove profitable in their investigation.

The idea of an "isolated system" refers to an idealization or approximation of physical reality which is important for the analysis and interpretation of various observations (cf. also the discussion in Geroch, 1977). How precisely this concept should be formulated is not a simple matter. To see some of the problems let us look at the universe at a scale where the object we want to study, a star or a system of stars, in brief "the system", appears to be far away from other such systems. Since we are interested in the behaviour of our system and not in processes going on at very large distances, we separate it by a cut along a time-like hypersurface T from the rest of the universe. This cut should be far enough from the system to retain everything which appears important to characterize its properties but still close enough such that the gravitational field is essentially determined by our system and possibly by gravitational radiation falling onto it from other parts of the universe.

Having thus isolated the system, we would like to analyse its behaviour. Its characterization in terms of boundary data and field equations would require the knowledge of standard Cauchy data on some space-like slice S' with boundary $\partial S'$ on T and suitable data on T. In an initial boundary value problem one would expect to prescribe on the part T^+ in the future of $\partial S'$ those components of the fields as free data which are related to inward going characteristics. Similarly one would prescribe on the part T^- in the past of $\partial S'$ components related to outward going characteristics. This will put S', or rather the class of space-like hypersurfaces intersecting T at $\partial S'$, into an unjustifiedly prominent position. Also, which data would correspond to the type of data which were induced on T before the system had been isolated from the rest of the universe is not easy to see. Finally, if we were interested in the "radiative components" of the field there would be no obvious way to characterize them in terms of data on T. We would rather want these components to be filtered out by the propagation process itself which is defined by the field equations. Therefore we take recourse in an approximation.

We try to extend the 3-manifold S' and the data on it beyond the boundary $\partial S'$ to obtain "asymptotically flat initial data" on a 3-manifold $S \supset S'$ such that the system is specified completely in terms of standard Cauchy data. However, this not only raises the question of how the data should be chosen beyond $\partial S'$, i.e. near "space-like infinity", and what "asymptotically flat" should mean, but it also poses the difficult if not impossible task of constructing an extension which satisfies the constraint equations on S.

Thus we rather approximate the data on the 3-manifold S' in some suitable sense by asymptotically flat initial data on the manifold S. Assuming S' as being embedded into S such that smooth extensions (not necessarily satisfying the constraints) to S of the data on S' exist, we can choose on S' the "free data" required to construct solutions to the constraints to be as close to the given data on S' as we wish. Since elliptic equations need to be solved to determine the solution of the constraints from the free data, the final data will in general depend on the chosen extension of the free data also on S'. It remains to decide how the extension to $S \setminus S'$ should be chosen. Note that the concept of an isolated system at which we arrive by these considerations would not make much sense if there were not relevant features of the field which remained essentially unaffected by the specific extension of the data near space-like infinity.

The evolution of the fields will be of particular interest in regions far away from the system. Here "observers" will analyse the field to deduce statements on the behaviour of the sources. And with this "far field", or rather by way of idealization with the asymptotic field at null infinity, will be associated concepts which characterize the dynamics of the system in a general way (e.g. the Bondi–energy–momentum). Choosing the extension to $S \setminus S'$, in particular its fall-off behaviour near space-like infinity, in the most general way consistent with the constraints and the evolution equations, will give us the safe feeling that nothing important has been left out. However, this may have the effect that essential features will be missed when we analyse the solution, because relevant information contained somewhere in the field may be drowned by a wealth of unimportant information fed into the field by the extension process. We should rather narrow down the structure of the data near space-like infinity in such a way that the essential features of the system are preserved and characterizing features of the system (e.g. the radiation field) can be read off the asymptotic field in as precise and simple a way as possible. If we should find out later that our assumptions are too narrow for modelling certain systems of physical importance, we will have learned something interesting.

In this context it is particularly remarkable that the definition of asymptotic flatness proposed in Penrose (1963) leads via condition (5.7) to a proposal of how to narrow down the concept of asymptotic flatness for Cauchy data. The restriction we have found is of a very mild nature. Of course, it should be generalized to the case of data with non-trival second fundamental forms. If the conditions so obtained turn out to be the only type of restrictions arising from smoothness requirements at null infinity, there will indeed be a very rich class of data leading to space–times which satisfy Penrose's conditions.

Since we are discussing here an approximation, practical considerations concerning calculability should also play a role. The completion of the analysis which has been outlined above will yield an enormous amount of direct information on the solutions of the field equations near space-like and null infinity. The discussion of the relationship between physical quantities near space-like infinity and corresponding quantities on null infinity will reduce essentially to straightforward calculations.

The conformal field equations proved quite efficient in the numerical construction of space–times from hyperboloidal data (Hübner 1996). However, from such data Minkowskian-type space–times cannot be completely determined because the domain of dependence of a hyperboloidal hypersurface covers only a part of such a space–time. The new setting opens the possibility to construct numerically complete solutions to Einstein's equation in finite grids and to read off the radiation field on the conformal boundary. The completion of our investigation should also have consequences for the design of approximation procedures, which ought to take into account analytical results on the global structure of the solutions.

Bibliography

Andersson, L. and Chruściel, P. T. (1994). On "Hyperboloidal" Cauchy data for vacuum Einstein equations and obstruction to the smoothness of scri. *Commun. Math. Phys.*, **161**, 533–568.

Andersson, L., Chruściel, P. T. and Friedrich, H. (1992). On the regularity of solutions to the Yamabe equation and the existence of smooth hyperboloidal initial data for Einstein's field equations. *Commun. Math. Phys.*, **149**, 587–612.

Ashtekar, A. (1984). Asymptotic properties of isolated systems: Recent developments. In: *General Relativity and Gravitation*, Bertotti, B. *et al.* (eds). Reidel, Dordrecht.

Ashtekar, A. and Romano, J. D. (1992). Spatial infinity as a boundary of space–time. *Class. Quantum Grav.*, **9**, 1069–1100.

Beig, R. (1984). Integration of Einstein's equations near spatial infinity. *Proc. Roy. Soc. A*, **391**, 295–304.

Beig, R. and Schmidt, B. G. (1982). Einstein's equations near spatial infinity. *Commun. Math. Phys.*, **87**, 65–80.

Bičák, J. (1996). Radiative spacetimes: Exact approaches. In: *Astrophysical Sources of Gravitational Radiation*, Marck, J.-A. and Lasota, J.-P. (eds). Cambridge University Press.

Cartan, E. (1923). Les espaces à connexion conforme. *Ann. Soc. Po. Math.*, **2**, 171–221. Reprinted in: Cartan, Élie (1955): *Oeuvres Complètes*, Partie III, Vol. 1, 747–797. Gauthier–Villars, Paris.

Christodoulou, D. and Klainerman, S. (1993). *The Global Nonlinear Stability of the Minkowski Space*. Princeton University Press, Princeton.

Dighton, K. (1974). An introduction to the theory of local twistors. *Int. J. Theor. Phys.*, **11**, 31–43.

Friedrich, H. (1977). Twistor connection and normal conformal Cartan connection. *Gen. Rel. Grav.*, **8**, 303–312.

Friedrich, H. (1981). The asymptotic characteristic initial value problem for Einstein's vacuum field equations as an initial value problem for a first-order quasilinear symmetric hyperbolic system. *Proc. Roy. Soc. A*, **378**, 401–421.

Friedrich, H. (1986). On the existence of n-geodesically complete or future complete solutions of Einstein's field equations with smooth asymptotic structure. *Commun. Math. Phys.*, **107**, 587–609.

Friedrich, H. (1992). Asymptotic structure of space–time. In: *Recent Advances in General Relativity*, Janis, A. I. and Porter, J. R. (eds). Birkhäuser, Basel.

Friedrich, H. (1995). Einstein equations and conformal structure: existence of anti-de Sitter-type space–times. *J. Geom. Phys.*, **17**, 125–184.

Friedrich, H. (1997). Gravitational fields near space-like and null infinity. *J. Geom. Phys.*, in press.

Geroch, R. (1977). Asymptotic structure of space–time. In: *Asymptotic Structure of Space-Time*, Esposito, B. F. P. and Witten, L. (eds). Plenum Press, New York.

Hübner, P. (1996). Method for calculating the global structure of (singular) spacetimes. *Phys. Rev. D*, **53**, 701.

Kánnár, J. (1996). Hyperboloidal initial data for the vacuum Einstein equations with cosmological constant. *Class. Quantum Grav.*, **13**, 3075–3084.

Penrose, R. (1963). Asymptotic properties of fields and space-time. *Phys. Rev. Lett.*, **10**, 66–68.

Penrose, R. (1965). Zero rest-mass fields including gravitation: asymptotic behaviour. *Proc. Roy. Soc. Lond. A*, **284**, 159–203.

Yano, K. (1938). Sur le circonférences généralisées dans les espaces à connexion conforme. *Proc. Imp. Acad. Tokyo*, **14**, 329–332.

Yano, K. (1939). Sur la théorie des espaces à connexion conforme. *J. Fac. Sci. Univ. Tokyo*, Sect. 1, **4**, 1–59.

6
Twistors, Geometry, and Integrable Systems

R. S. Ward
Department of Mathematical Sciences, University of Durham, Durham DH1 3LE

1 Introduction

I am delighted to be able to make this contribution to the celebration of Roger Penrose's 65th birthday. The subject is that of integrable systems and their relation to geometry and twistor theory. When twistors were created, about thirty years ago, their creator did not have integrable systems as his primary motivation: Roger was aiming at a wider class of physical problems. In a sense, integrable systems are a cop-out: they deliberately avoid the messiest interactions and dynamics that occur in the real world. But they do provide a window onto nonlinear, coherent, nonperturbative phenomena—these become accessible, beautiful, and (if you look at them in the right way) geometrical. In the words of Mason and Woodhouse [8], "they combine tractability with nonlinearity, so they make it possible to explore nonlinear phenomena while working with explicit solutions".

I want to describe some of the features of a particular integrable system, in (2+1)-dimensional flat space-time. Almost all of this material was ultimately motivated by things that were introduced by Roger, such as compactifying space-time by attaching a surface $\mathscr{I}$ at infinity [9–11], twistor space [12], using twistor functions to solve linear field equations [13, 14], and using holomorphic twistor structures to solve nonlinear integrable systems [16, 15]. I hope that what I describe might serve as a reminder of some of his contributions.

2 Twistors for 3-dimensional space-time

The twistor correspondence for Euclidean R^3 is well-known in connection with the construction of static Yang-Mills-Higgs monopoles. It first appeared, in effect, as a special case of the correspondence for R^4 (see, for example, [26]). We shall use a more direct description [2], in terms of minitwistor space T. This talk deals with (2+1)-dimensional Minkowski space-time M (rather than Euclidean R^3), which corresponds to using a slightly different reality condition (antiholomorphic involution) on T. In addition, we shall extend the correspondence to compactifications $\overline{M}$ and $\overline{T}$ of the space-time and twistor spaces respectively. In the (Euclidean) monopole case, such compactifications would be inapproriate, since monopoles (unlike the rational-algebraic objects of this talk) involve transcendental data which do not extend to the compactified spaces.

The twistor space T is a two-dimensional complex manifold, which can be visualized in several ways. To begin with, let us think of T as a quadric cone (minus its vertex O) in CP^3. The space of conics on T, coming from planes $\hat{p}$ in CP^3 not passing through O, is isomorphic to C^3—this is the complexification of our space-time M. To get $M \cong R^3$, we restrict to "real" planes $\hat{p}$ (think of the coefficients in the linear form defining $\hat{p}$ as being real-valued rather than complex-valued). So points $p \in M$ correspond to real planes $\hat{p} \in CP^3$ (or real conics on T) avoiding the given point O. This correspondence endows M with a Weyl structure (conformal structure plus affine connection), which in our case is flat. For example, two points p and q in M are null-separated if and only if the corresponding conics $\hat{p}$ and $\hat{q}$ touch at a double-point. And if p and q are not null-separated, so that $\hat{p}$ and $\hat{q}$ intersect at two distinct points Z and Y in T, then the set of all real conics through Z and Y gives the geodesic in M which contains p and q. See [3] for more details of this correspondence and its curved version.

In terms of this picture, M has an obvious compactification: one simply includes the real planes which pass through the vertex O. So $M \cong R^3$ is compactified to $\overline{M} \cong RP^3$. At infinity in space-time, one adds on the sphere S^2 of all directions, and then identifies antipodal points. Let us denote by $\mathcal{I}$ this surface RP^2 at infinity. All geodesics in a given direction end up at the same point on $\mathcal{I}$. In particular, $\mathcal{I}$ contains a circle of null points, and open subsets of timelike and spacelike points.

In the familiar (3+1)-dimensional case [9–11], the conformal geometry is homogeneous on the compactified space—points on $\mathcal{I}$ are on the same footing as points in the interior of the space-time. But that is not the case with our projective compactification; this is clear from the structure of the compactified twistor space $\overline{T}$, which we now examine.

To get $\overline{T}$ from the cone T, we add in its vertex O, and then blow it (the vertex) up into a CP^1. It is convenient at this stage to view T in a different way, namely as a holomorphic line bundle over CP^1 (in fact, the holomorphic tangent bundle $\mathcal{O}(2)$). The conics $\hat{p}$ are exactly the holomorphic sections of this bundle. See Figure 1. To get $\overline{T}$, one attaches an additional section L_∞ (which is the blow-up of O). Each fibre E_λ above $\lambda \in CP^1$ is now itself a copy of CP^1. And each point of $\mathcal{I}$ corresponds to a union $L_\infty \cup E_\lambda \cup E_\mu$ of three CP^1s, rather than being a single CP^1 like the section $\hat{p}$ corresponding to a finite point p. (This structure was first brought to my attention by A. J. Small in 1989; see [1] for more details of its use.)

If λ and μ referred to above are unrestricted, then of course one gets the complexification of $\mathcal{I}$. For real $\mathcal{I}$, one restricts λ and μ to be the roots of a quadratic polynomial with real coefficients (these coefficients are the homogeneous coordinates for $\mathcal{I} \cong RP^3$). More specifically, timelike points on $\mathcal{I}$ correspond to complex conjugate pairs $\lambda = \bar{\mu}$, null points to the double case $\lambda = \mu$ real (including ∞),

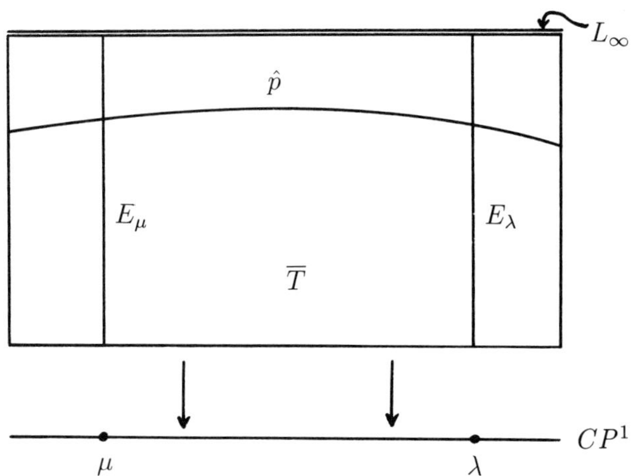

FIG. 1. Compact minitwistor space $\overline{T}$, as a bundle over CP^1.

and spacelike points to distinct real pairs λ, μ (including ∞).

3 An integrable Yang–Mills–Higgs system

The Penrose transform relates geometric/holomorphic objects on twistor space T to solutions of "geometrical" field equations on space-time M. In particular, holomorphic vector bundles on T correspond to solutions of certain gauge-field equations on M. Indeed, there is a one-to-one correspondence between

- holomorphic SU(2) vector bundles V over T, satisfying

$$V\big|_{\hat{p}} \text{ is trivial for all } p \in M; \tag{3.1}$$

- real-analytic solutions of the Yang-Mills-Higgs equations

$$D_\mu \Phi = \tfrac{1}{2}\varepsilon_{\mu\alpha\beta}F^{\alpha\beta} \tag{3.2}$$

  on M, with gauge group SU(2).

Here Φ, the Higgs field, is a function on M with values in the Lie algebra su(2); its covariant derivative is $D_\mu \Phi = \partial_\mu \Phi + [A_\mu, \Phi]$ where $\partial_\mu \Phi$ is the partial derivative of Φ with respect to the standard space-time coordinates; the SU(2) gauge potential is A_μ and the corresponding gauge field is $F_{\mu\nu} = \partial_\mu A_\nu - \partial_\nu A_\mu + [A_\mu, A_\nu]$; space-time indices $\mu, \nu, \ldots$ are raised and lowered with the Minkowski metric (signature $-++$); and $\varepsilon_{\mu\alpha\beta}$ denotes the totally-skew tensor with $\varepsilon_{012} = -1$.

Saying that V is an SU(2) bundle means that V is of rank 2, and that the gauge group is reduced from GL(2,C) to SU(2) by requiring $\det V$ to be trivial and an appropriate antiholomorphic involution on T to lift to V.

These Yang-Mills-Higgs equations form a hyperbolic system on our (2+1)-dimensional Minkowski space-time M. This system can be reformulated in terms of a chiral field, in two distinct ways [7, 22, 23, 24, 19, 20, 25, 5, 1]; the solutions described later were mostly obtained in this chiral-field context. The equations are integrable, in the sense of being the consistency condition for an overdetermined system of linear equations. This system is the pair

$$(\lambda D_x - D_t - D_y + \lambda\Phi)\psi = 0, \tag{3.3}$$

$$(\lambda D_t - \lambda D_y - D_x + \Phi)\psi = 0, \tag{3.4}$$

where ψ is a 2×2 matrix function of the space-time coordinates (t, x, y) and of the complex parameter λ (which corresponds to the λ of the previous section).

The analogous elliptic Yang-Mills-Higgs system on Euclidean R^3 is of course well-known: it consists of the Bogomol'nyi equations for static monopoles. The space of monopoles (in a given topological class) is finite-dimensional, whereas the space of solutions of our hyperbolic system is infinite-dimensional (even if one imposes boundary conditions at spatial infinity). My interest here is in describing finite-dimensional families of solutions, namely those for which the corresponding vector bundles V extend to the compactified twistor space $\overline{T}$. This amounts to imposing some severe global restrictions on the Yang-Mills-Higgs field. The restricted set of fields is then classified topologically, by a positive integer n (the corresponding bundle V will have second Chern class equal to $2n$). And the space of solutions of class n has dimension $8n - 3$.

One way to see how the topology arises is as follows (it was suggested by a remark made to me by C. K. Anand). Choose a spacelike plane, say $t = 0$. The unit vectors tangent to this plane are $v^\mu(\theta) = (0, \cos\theta, \sin\theta)$. Let $\psi(x, y, \theta)$ be an SU(2)-valued function satisfying the ordinary differential equation

$$v^\mu(\theta) D_\mu \psi + \Phi\psi = 0. \tag{3.5}$$

This is actually the spatial part of the linear system described above, namely (3.3) plus λ^{-1} times (3.4), where $\lambda = -\cot\frac{1}{2}\theta$. Gauge transformations act by $\psi \mapsto \Lambda(x, y)\psi$; choose a gauge such that

$$\psi(x, y, 0) = I, \tag{3.6}$$

where I is the identity. There is some additional freedom in ψ, namely $\psi \mapsto \psi K$, where K is an SU(2)-valued function of θ and $x\sin\theta - y\cos\theta$ (so K is annihilated by $v^\mu \partial_\mu$), and $K = I$ when $\theta = 0$. We can think of (3.5) as propagating ψ along the line $(x, y) = (\sigma\cos\theta, \sigma\sin\theta)$, from $\sigma = -\infty$ to $\sigma = +\infty$. Let us use the freedom $\psi \mapsto \psi K$ to ensure that $\psi \to I$ as $\sigma \to -\infty$. For this to make sense, we need to assume that the gauge field and Higgs field tend to zero at spatial infinity, and we henceforth make this assumption (a boundary condition). But now we add the assumption that $\psi \to I$ as $\sigma \to +\infty$ as well, in other words that (3.5) gives "zero scattering". Consequently,

$$\psi(x, y, \theta) \to I \quad \text{as } x^2 + y^2 \to \infty. \tag{3.7}$$

This is a global condition, involving the Yang-Mills-Higgs field throughout space, and is much stronger than the boundary condition. But it holds for our restricted class of fields, namely those coming from vector bundles over $\overline{T}$. Note that it corresponds to the scattering matrix S of [7] being trivial.

Because of (3.7), ψ is defined on $S^2 \times S^1$, where the S^2 is the one-point compactification of xy-space. But since we also have (3.6), we may regard ψ as being defined on S^3. In other words, ψ maps S^3 to SU(2)$\cong S^3$. The integer n is the winding number (degree) of this map (choose orientations so that n is positive).

4 SU(2) bundles over $\overline{T}$

Holomorphic vector bundles of rank 2 over the algebraic surface $\overline{T}$ are classified topologically by the two Chern numbers (integers) c_1 and c_2. Imposing condition (3.1) of the previous section implies that $c_1 = 0$; and in our case c_2 has to be even, say $c_2 = 2n$ where n is a positive integer. The moduli space of SU(2) bundles with $c_1 = 0$ and $c_2 = 2n$ has (real) dimension $8n - 3$, and the integer n turns out to be the same as the winding number of the space-time field defined in the previous section.

The bundles V we are interested in have the additional property that $V\big|_{L_\infty}$ is trivial, and $V\big|_{E_\lambda}$ is trivial except for a finite number of non-real values of $\lambda \in CP^1$. This might exclude some exceptional bundles; it is connected with the Yang-Mills-Higgs field being zero at spatial infinity. Having excluded such possibilities, we know that $V\big|_{E_\lambda}$ fails to be trivial for exactly $2n$ values of λ (counted with multiplicity), consisting of n complex-conjugate pairs $\{\lambda_1, \overline{\lambda_1}, \ldots, \lambda_n, \overline{\lambda_n}\}$: these are "jumping lines" [4, 24]. Typically, jumping lines correspond to singularities in the space-time field, so (bearing in mind our description of points on $\mathcal{J}$) we expect that the Yang-Mills-Higgs field will be smooth on $\overline{M}$ except at n timelike points on $\mathcal{J}$ (corresponding to the λ_j). This is indeed what happens.

Bundles can be constructed by specifying the jumping lines (i.e. the λ_j) and giving some data at each one [4, 24]. If E_λ is a jumping line, then $V\big|_{E_\lambda} \cong \mathcal{O}(\alpha) \oplus \mathcal{O}(-\alpha)$ for some positive integer α; we say that such a jumping line is of type $(\alpha, -\alpha)$. The multiplicity of λ is at least α. The generic bundle has jumping lines of multiplicity one, and hence of type $(1, -1)$; the extra data at each jumping line then consists of a meromorphic fractional-linear function defined on it (specified by three complex parameters). Since each λ_j is itself a complex parameter, we then have a total of $4n$ complex $= 8n$ real parameters. (The structure attached to the conjugate lines is determined by the reality condition, so does not get counted.) Finally, removing an overall SU(2) phase leaves $8n - 3$ real parameters, as claimed earlier.

In the next section, we shall see what the corresponding space-time fields look like, in particular for $n = 1$ and $n = 2$. We shall also examine the limiting cases, for $n = 2$, in which λ has multiplicity 2. Here the general situation is for the jump to be of type $(1, -1)$, with type $(2, -2)$ occurring as a further special

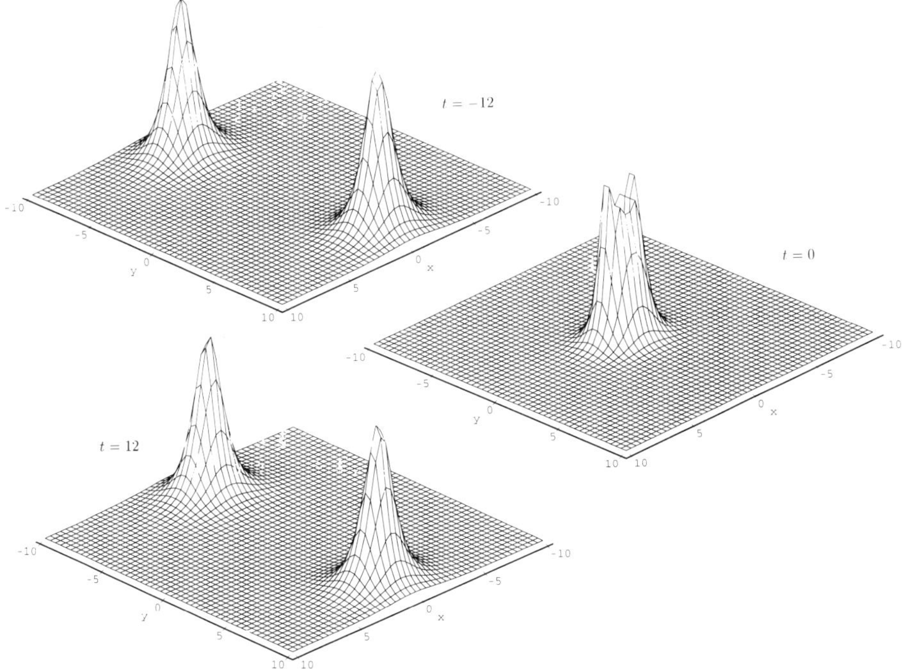

FIG. 2. Example of a 2-soliton solution, with $\lambda_1 \neq \lambda_2$.

case.

5 Soliton solutions

Here we shall see something of the space-time fields that correspond to bundles over the compact minitwistor space $\overline{T}$, for the cases $n = 1$ and $n = 2$. The fields are rational functions of the space-time coordinates, and can be most simply expressed in terms of the matrix $\psi(x^\mu, \lambda)$; see, for example, [7, 22, 25]. For a generic n-soliton solution, ψ has n simple poles in λ (at the points λ_j); in the degenerate cases, higher-order poles can occur.

If $n = 1$, then we have a single-soliton solution, localized in space, and depending on five real parameters. Two of the parameters, corresponding to the complex number λ_1, determine a timelike direction, and hence the velocity of the soliton. A further two parameters specify its location in space (at some given time); and the final parameter determines its size. The field is smooth on $\overline{M} \setminus \{p\}$, where $p \in \mathfrak{J}$ corresponds to λ_1. The solution looks like a localized lump in space, and is radially-symmetric in its rest-frame.

Let us turn briefly to the more general hyperbolic system, namely the partial differential equation (3.2) plus a straightforward boundary condition at spatial infinity. So the solution-space is now infinite-dimensional. One might ask

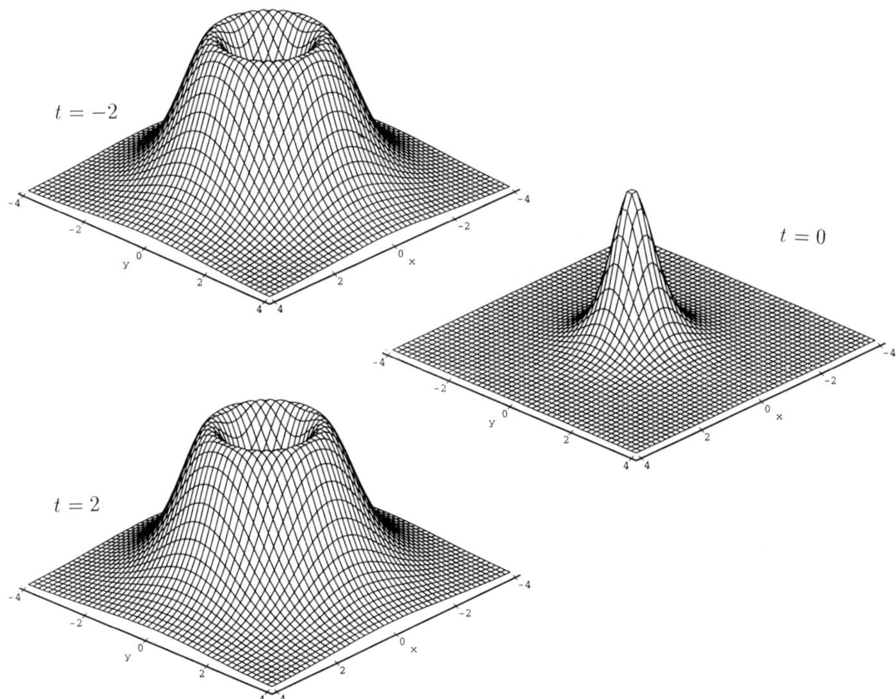

FIG. 3. Example of a $\lambda_1 = \lambda_2$ solution: an imploding-exploding ring.

whether the above soliton solution is stable. There is no topological reason for it to be, but the fact that the system is integrable may well ensure stability. There is strong evidence from numerical experiments that this is indeed the case [20].

We return now to the compact case, and look at some $n = 2$ examples. Here there are 13 parameters, which generically correspond to position+velocity+size for each of two solitons (a total of 10 parameters), plus a relative SU(2) phase (the remaining 3 parameters). The solution is smooth on $\overline{M} \setminus \{p_1, p_2\}$, where p_1 and p_2 are (possibly equal) points on $\mathcal{J}$, determined by λ_1 and λ_2. If the λ_i are distinct, then we have two solitons, each with constant velocity; there is no scattering as they pass each other (not even the "phase shift" that occurs in (1+1)-dimensional examples such as KdV). This corresponds of course to distinct jumping lines of type $(1, -1)$. See [22] for more details. Figure 2 depicts a particular solution: for three sequential values of time t, the gauge-invariant quantity $-\operatorname{tr}(\Phi^2)$ is plotted as a function of x and y. Two solitons, moving in the positive and negative y-directions with nonzero impact parameter, pass by each other.

In the special case $\lambda_1 = \lambda_2$, the generic situation is for the jump to be of type $(1, -1)$. In the rest-frame corresponding to λ_1, one might have expected the solution to be static, but this is not the case. Instead, one gets a picture

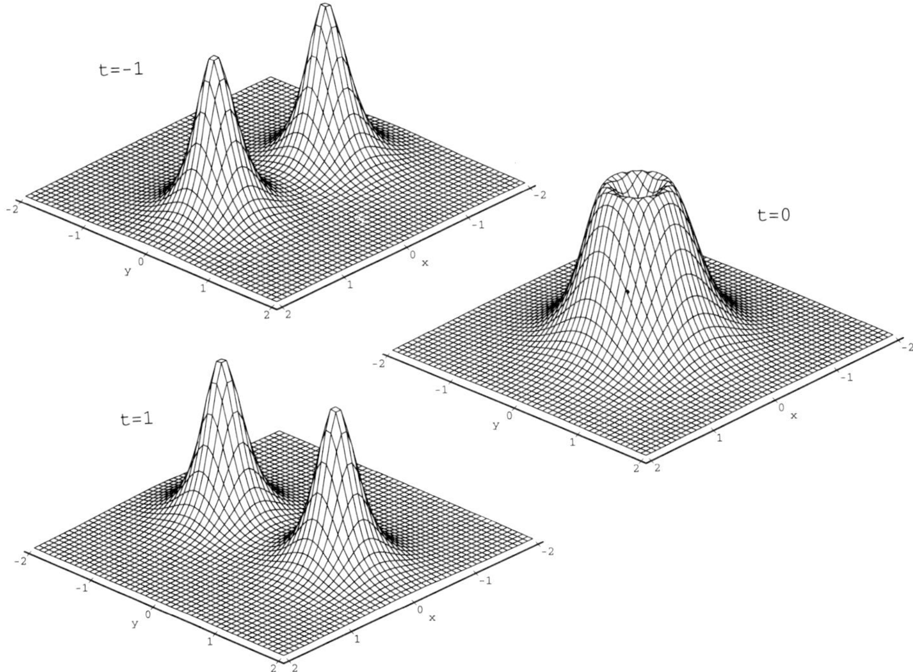

FIG. 4. Another $\lambda_1 = \lambda_2$ solution: a variant of the imploding-exploding ring.

of incoming-outgoing waves. (It may be hard to visualize how these solutions can be limiting cases of the previous nonscattering ones, but they are.) The waves are not "linear"; for example, they move at a speed which goes like $1/\sqrt{t}$ for large $|t|$. In the simplest case, one has radial symmetry, and an imploding-exploding ring [5]; see Figure 3. (In this figure and the next one, the function which is plotted is not $-\operatorname{tr}(\Phi^2)$, but rather the energy density of the chiral-field reformulation of the same system.) Another example, depicted in Figure 4, has two lumps instead of a ring: they approach each other, momentarily form a ring, and then separate at right angles [25]. These lumps are not solitons in the usual sense: in particular, their size and speed are not constant. So the similarity to the right-angle scattering that occurs in non-integrable situations, such as vortex or monopole dynamics, is perhaps spurious.

The final degeneration is where the jump is of type $(2, -2)$. Here we simply get a ring, the shape of which is constant in time.

6 Concluding remarks

There are various different approaches to (and even definitions of) integrability for systems of partial differential equations. Some approaches are geometrical, while others are more analytic. In this talk I have described a system which is very geometrical in flavour. One question is whether all integrable systems are geometrical (at least in some useful sense).

A basic feature of an integrable system is that it is associated with a pair of linear operators of a certain type (the Lax pair) which commute. The nonlinear equations of the system are equivalent to the vanishing of the commutator of these linear operators. In many cases, the linear operators are (or can readily be expressed as) first-order partial differential operators, and hence as vector fields on some manifold. The condition that they commute then has an obvious geometrical meaning: the vector fields are tangent to a family of surfaces. The system I have described is of this type: in fact, the "family of surfaces" corresponds exactly to the vector bundle over twistor space. Most of the well-known integrable systems fall into this category: for a survey, see [8].

But there are several which do not. One example is the KP equation, where the relevant partial differential operators cannot be expressed as vector fields. Another, even simpler, example is that of the infinite Toda chain. The latter case can be viewed as an $N \to \infty$ limit of geometrical/twistor systems involving vector bundles of rank N [21, 8]. It seems likely that some kind of generalized geometry (or generalized twistor correspondence) is needed to deal with such systems, and a couple of suggestions of what this might be have been made by Mason [8] and Strachan [17, 18]; see also [6]. The full significance of these structures has yet to be understood. Such generalizations might point the way beyond integrable systems, and show how twistor theory can be applied in a far wider context, thereby providing yet more support for Roger's original (and continuing) vision.

Bibliography

1. Anand, C. K. (1995). Uniton bundles. *Commun. Anal. Geom.* **3**, 371.

2. Hitchin, N. J. (1982). Monopoles and geodesics. *Commun. Math. Phys.*, **82**, 579–602.

3. Hitchin, N. J. (1983). Complex manifolds and Einstein's equations. In: *Twistor geometry and non-linear systems*, eds. H. D. Doebner and T. D. Palev. *Lecture Notes in Mathematics* **570**, Springer, Berlin, 73–99.

4. Hurtubise, J. (1986). Instantons and jumping lines. *Commun. Math. Phys.* **105**, 107–122.

5. Ioannidou, T. (1996). Soliton solutions and nontrivial scattering in an integrable chiral model in (2+1) dimensions. *J. Math. Phys.* **37**, 3422–3441.

6. Kuperschmidt, B. A. (1985). Discrete Lax equations and differential-difference calculus. *Astérisque* **123**.

7. Manakov, S. V. and Zakharov, V. E. (1981). Three-dimensional model of relativistic-invariant field theory, integrable by the inverse scattering transform. *Lett. Math. Phys.* **5**, 247–253.

8. Mason, L. J. and Woodhouse, N. M. J. (1996). *Integrability, self-duality, and twistor theory.* Oxford University Press.

9. Penrose, R. (1963). Asymptotic properties of fields and space-times. *Phys. Rev. Lett.* **10**, 66–68.

10. Penrose, R. (1964). Conformal treatment of infinity. In: *Relativity, groups and topology*, eds. B. S. DeWitt and C. M. DeWitt, pp. 563–584. Gordon and Breach, New York.

11. Penrose, R. (1965). Zero-rest-mass fields including gravitation: asymptotic behaviour. *Proc. Roy. Soc. Lond. A* **284**, 159–203.

12. Penrose, R. (1967). Twistor algebra. *J. Math. Phys.* **8**, 345–366.

13. Penrose, R. (1968). Twistor quantisation and curved space-time. *Int. J. Theor. Phys.* **1**, 61–99.

14. Penrose, R. (1969). Solutions of the zero-rest-mass equations. *J. Math. Phys.* **10**, 38–39.

15. Penrose, R. (1976). Nonlinear gravitons and curved twistor theory. *Gen. Rel. Grav.* **7**, 31–52.

16. Penrose, R. and MacCallum, M. A. H. (1972). Twistor theory: an approach to the quantisation of fields and space-time. *Phys. Repts. C* **6**, 241–316.

17. Strachan, I. A. B. (1996) The dispersive self-dual Einstein equations and the Toda lattice. *J. Phys. A*, **29**, 6117–6124.

18. Strachan, I. A. B. (1997). A geometry for multidimensional integrable systems. *J. Geom. Phys.*, **21**, 255–278.

19. Sutcliffe, P. M. (1992). Nontrivial soliton scattering in an integrable chiral model in (2+1)-dimensions. *J. Math. Phys.* **33**, 2269–2278.

20. Sutcliffe, P. M. (1993). Yang-Mills-Higgs solitons in 2+1 dimensions. *Phys. Rev. D* **47**, 5470–5476.

21. Ward, R. S. (1987). Multi-dimensional integrable systems. In: *Field theory, quantum gravity and strings II*, eds. H. J. deVega and N. Sanchez. *Lecture Notes in Physics* **280**, Springer, Berlin, 106–116.

22. Ward, R. S. (1988). Soliton solutions in an integrable chiral model in 2+1 dimensions. *J. Math. Phys.* **29**, 386–389.

23. Ward, R. S. (1989). Twistors in 2+1 dimensions. *J. Math. Phys.* **30**, 2246–2251.

24. Ward, R. S. (1990). Classical solutions of the chiral model, unitons, and holomorphic vector bundles. *Commun. Math. Phys.* **128**, 319–332.

25. Ward, R. S. (1995). Nontrivial scattering of localized solitons in a (2+1)-dimensional integrable system. *Phys. Lett. A* **208**, 203–208.

26. Ward, R. S. and Wells, R. O. (1990). *Twistor geometry and field theory.* Cambridge University Press.

7
On Four-Dimensional Einstein Manifolds

Claude LeBrun
The Department of Mathematics, SUNY, Stony Brook, NY 11794, USA

1 Introduction

For the purposes of this lecture, an *Einstein metric* on a smooth manifold M will mean a smooth Riemannian metric g satisfying *Einstein's equations*

$$r = \lambda g; \tag{1.1}$$

here r denotes the Ricci curvature of g and λ is an arbitrary constant. If the manifold M is compact and g is an Einstein metric on M, we will then say that (M, g) is an *Einstein manifold.*

Any Einstein manifold (M, g) of dimension 2 or 3 necessarily has constant sectional curvature, and its universal cover $(\tilde{M}, g)$ is therefore completely determined by the constant λ. In these low dimensions, the study of Einstein manifolds therefore reduces to the study of suitable discrete group actions on the sphere, Euclidean space, and hyperbolic space. On the other hand, all 2-manifolds and "most" 3-manifolds, actually admit such metrics, Besse (1987), Thurston (1982), so it is quite natural to ask if something of the kind is also true in higher dimensions.

In dimensions ≥ 4, equation (1.1) no longer determines the geometry in a local manner, and this vastly increases both the interest and the difficulty of understanding Einstein manifolds in higher dimensions. In dimensions ≥ 5, our knowledge of general Einstein manifolds is quite scant, although an impressive array of methods has been developed for producing and classifying Einstein manifolds with special holonomy or isometry groups, Aubin (1976), Besse (1987), Boyer *et al.* (1994), Joyce (1996), Yau (1977). The borderline case of dimension 4, however, is considerably more tractable, largely because the 4-dimensional rotation group is non-simple:

$$SO(4) = [SU(2) \times SU(2)]/\mathbf{Z}_2.$$

My aim here is therefore to review what we now know about the following fundamental problems:

Supported in part by NSF grant DMS-9505744.

- **Existence:** Which smooth compact 4-manifolds M admit Einstein metrics?
- **Uniqueness:** Modulo diffeomorphisms and rescalings, how many Einstein metrics exist on a given smooth compact 4-manifold M?

Roger Penrose has been one of the true pioneers of the 4-dimensional Einstein equations, and it is largely due to his efforts that spinor techniques, Penrose (1968), Penrose and Rindler (1986), have become standard tools for analysing the local geometry of Einstein manifolds. A principal goal of this lecture will be to explain how, via Seiberg–Witten theory, spinor techniques can also lead to insights into the *global* geometry of 4-dimensional Einstein manifolds.

While I'm going to pretend that Einstein 4-manifolds just constitute a topic in pure mathematics, it is rather curious that all the major ideas we'll use really stem from physics. Now I'm certain that Roger will find this absolutely hilarious. After all, Einstein called equation (1.1) *the biggest blunder of his life*, because the introduction of an unknown constant λ into the gravitational field equations prevented him from predicting the Hubble expansion of the universe (Gamow 1970). And I'll be talking about *Riemannian* ('Euclidean signature') solutions here—on *compact* manifolds, no less! Of course, Stephen Hawking (1979), has argued that such solutions are of fundamental importance to physics, but Roger has never hesitated to tell Stephen that this is the biggest blunder of *his* life! Well, make of this what you may. Happy birthday, Roger—I hope it's entertaining!

2 The curvature of 4-manifolds

The geometric meaning of equation (1.1) on a 4-manifold M^4 can be reformulated in quite elementary terms. If $P \subset T_xM$ is a 2-plane in the tangent space of M at some point $x \in M$, its orthogonal complement $P^\perp$ with respect to a Riemannian metric g is also a 2-plane, and we can consider the sectional curvatures $K(P)$ and $K(P^\perp)$ of these two 2-planes. (Recall that the sectional curvature $K(P)$ is the Gauss curvature at x of the surface obtained by applying the exponential map to $P \subset T_xM$.) Remarkably, the Einstein condition (1.1) holds iff

$$K(P) = K(P^\perp)$$

for every 2-plane P.

For our purposes, this is best seen by noticing that the curvature tensor of a Riemannian 4-manifold naturally decomposes into simpler pieces. Indeed, by raising an index, the curvature tensor of any Riemannian manifold (M, g) may be identified with a linear map $\mathcal{R} : \Lambda^2 \to \Lambda^2$ from 2-forms to 2-forms; this map $\mathcal{R}$ is called the *curvature operator* of g. Now assume that M is an oriented 4-manifold. We then have a Hodge star operator $\star : \Lambda^2 \to \Lambda^2$, corresponding to the replacement of oriented 2-planes P by their oriented orthogonal complements $P^\perp$. Because $\star^2 = 1$, the only eigenvalues of $\star$ are ± 1, and we thus have a decomposition

$$\Lambda^2 = \Lambda^+ \oplus \Lambda^-$$

of the rank-6 bundle of 2-forms as a sum of the two rank-3 eigenbundles of $\star$; elements of these eigenbundles are called *self-dual* or *anti-self-dual* 2-forms, in accordance with the sign of the eigenvalue. Thus the curvature operator $\mathcal{R}$ can be considered to consist of more primitive pieces:

$$\mathcal{R} = \left(\begin{array}{c|c} W_+ + \frac{s}{12} & \overset{\circ}{r} \\ \hline \overset{\circ}{r} & W_- + \frac{s}{12} \end{array}\right), \tag{2.1}$$

where trace $W_\pm = 0$. Here s is the scalar curvature, and $\overset{\circ}{r} = r - \frac{s}{4}g$ is the traceless Ricci curvature, while W_+ and W_- are respectively called the self-dual and anti-self-dual Weyl curvatures. Now Einstein's equations (1.1) are equivalent to the stipulation that $\overset{\circ}{r} = 0$, and we now see that this is equivalent to requiring that $\mathcal{R}$ commute with $\star$. But this is equivalent to saying that

$$\langle \star\varphi, \mathcal{R}(\star\varphi)\rangle = \langle \varphi, \mathcal{R}(\varphi)\rangle$$

for all 2-forms φ, and hence equivalent to requiring that $K(P) = K(P^\perp)$ for all 2-planes.

Now the above discussion was really just a pretext for describing the canonical decomposition of the curvature tensor. Having made the acquaintance of $W_\pm$, $\overset{\circ}{r}$ and s, we are now prepared to explore some relations between topology and curvature.

3 The Hitchin–Thorpe inequality

Two simple but important topological invariants of a smooth compact oriented 4-manifold M are its *Euler characteristic*

$$\chi(M) = 2 - 2b_1(M) + b_+(M) + b_-(M)$$

and *signature*

$$\tau(M) = b_+(M) - b_-(M);$$

here $b_1(M) = \dim H^1(M, \mathbf{R})$ is the first Betti number, while $b_+(M)$ and $b_-(M)$ denote the maximal dimensions of linear subspaces of $H^2(M, \mathbf{R})$ on which the cup product $H^2 \times H^2 \to H^4$ is respectively positive or negative definite.

If g is any Riemannian metric on M, these invariants may be expressed in terms of Gauss–Bonnet-like curvature integrals:

$$\chi(M) = \frac{1}{8\pi^2}\int_M \left[|W_+|^2 + |W_-|^2 + \frac{s^2}{24} - \frac{|\overset{\circ}{r}|^2}{2}\right] d\mu$$

$$\tau(M) = \frac{1}{12\pi^2}\int_M \left[|W_+|^2 - |W_-|^2\right] d\mu$$

where the curvatures, norms $|\cdot|$, and volume form $d\mu$ are those of g. In particular,

$$(2\chi \pm 3\tau)(M) = \frac{1}{4\pi^2}\int_M \left[2|W_\pm|^2 + \frac{s^2}{24} - \frac{|\overset{\circ}{r}|^2}{2}\right] d\mu.$$

Now notice that the integrand is non-negative when g is Einstein. It follows, Hitchin (1974), Thorpe (1969), that not all 4-manifolds admit Einstein metrics!

Theorem 3.1 *If the smooth compact orientable 4-manifold M admits an Einstein metric g, then*

$$2\chi(M) \geq 3\tau|(M)|.$$

If, moreover, equality holds, then, provided we choose the orientation for which $\tau(M) \leq 0$, the Einstein metric g must be Ricci-flat and anti-self-dual; and this can only happen if M is finitely covered by K3 or the 4-torus.

Here K3 denotes the underlying smooth oriented 4-manifold of the complex quartic surface

$$\{[s:t:u:v] \in \mathbf{CP}_3 \mid s^4 + t^4 + u^4 + v^4 = 0\},$$

and is the only compact simply connected 4-manifold which admits metrics for which Λ^+ is flat with respect to the Riemannian connection. A metric is called anti-self-dual if $W_+ = 0$; and a metric is Ricci-flat and anti-self-dual iff Λ^+ is flat. The final clause of the theorem therefore follows from the Cheeger–Gromoll splitting theorem and Bieberbach's theorem; for details, cf. Besse (1987).

Example 3.2 Let $M = k\mathbf{CP}_2\#\ell\overline{\mathbf{CP}}_2$ be the connected sum of k copies of the complex projective plane $\mathbf{CP}_2$ and ℓ copies of the *reverse-oriented* complex projective plane $\overline{\mathbf{CP}}_2$. (The connected sum operation # just connects the given manifolds together with $S^3 \times \mathbf{R}$ "hoses" in a manner compatible with the given orientations.) Then $\chi(M) = 2 + k + \ell$ and $\tau(M) = k - \ell$, so we have $2\chi \geq 3|\tau|$ only if $4 + 5k \geq \ell \geq (k-4)/5$. Moreover, each of these manifolds is non-spin and simply connected, so none of them is a quotient of K3 or the 4-torus. Hence the existence of an Einstein metric on M can definitely be ruled out unless $4 + 5k > \ell > (4-k)/5$.

In addition to predicting that Einstein metrics don't exist on certain manifolds, the Hitchin–Thorpe inequality allows us to completely classify the Einstein metrics on manifolds with $2\chi = 3|\tau|$. Without loss of generality, we may assume that M is the 4-torus or K3, and choose the orientation of M so that $b^+ = 3$. For any Einstein metric on M, there is a 3-dimensional space of parallel harmonic self-dual 2-forms, and these determine a 3-dimensional subspace $\mathcal{H}^+$ of the de Rham cohomology $H^2(M,\mathbf{R})$. An Einstein metric g on M then turns out to be completely determined modulo diffeomorphisms and rescalings by this 3-plane.

The proof, cf. Barth *et al.* (1984), of this *Torelli theorem* relies heavily on the fact, deduced above, that every Einstein metric on M is Kähler with respect to a 2-sphere of complex structures.

4 Recent results

The Hitchin–Thorpe argument shows that any Einstein metric on a 4-torus is simply the obvious flat metric on $\mathbf{R}^4$ modulo some lattice $\cong \mathbf{Z}^4$. One way of generalizing these examples is to replace Euclidean space $\mathbf{R}^4$ with hyperbolic 4-space $\mathcal{H}^4 = SO(4,1)/SO(4)$ or complex-hyperbolic 2-space $\mathbf{C}\mathcal{H}_2 = SU(2,1)/U(2)$. Both of these negative-curvature homogeneous Einstein manifolds $\approx \mathbf{R}^4$ have (Borel, 1963; Mostow, 1985) infinitely many distinct compact quotient manifolds M^4 obtained by dividing by discrete groups of isometries. It has recently been shown (Besson *et al.*, 1995; LeBrun, 1995) that the only Einstein metrics on these spaces are the obvious ones:

Theorem 4.1. (Besson–Courtois—Gallot) *Let M^4 be a smooth compact quotient of hyperbolic 4-space $\mathcal{H}^4 = SO(4,1)/SO(4)$, and let g_0 be its standard metric of constant sectional curvature. Then every Einstein metric g on M is of the form $g = \lambda\varphi^* g_0$, where $\varphi : M \to M$ is a diffeomorphism and $\lambda > 0$ is a constant.*

Theorem 4.2. (LeBrun) *Let M^4 be a smooth compact quotient of complex-hyperbolic 2-space $\mathbf{C}\mathcal{H}_2 = SU(2,1)/U(2)$, and let g_0 be its standard complex-hyperbolic metric. Then every Einstein metric g on M is of the form $g = \lambda\varphi^* g_0$, where $\varphi : M \to M$ is a diffeomorphism and $\lambda > 0$ is a constant.*

Despite the apparent similarities between these results, their proofs rest on entirely different principles. The proof of Theorem 4.1 proceeds (Besson *et al.*, 1995) by studying the entropy of the geodesic flow, and exploits Bishop's comparison theorem for the Ricci curvature. Instead, Theorem 4.2 is proved (LeBrun, 1995) by estimating the scalar curvature of Riemannian metrics by means of the Seiberg–Witten invariants of smooth 4-manifolds.

These same techniques are well-adapted to proving non-existence results (LeBrun, 1996):

Theorem 4.3. (LeBrun) *Infinitely many compact smooth simply connected 4-manifolds with $2\chi > 3|\tau|$ do not admit Einstein metrics.*

In fact, I will describe a sequence of smooth manifolds homeomorphic to $k\mathbf{CP}_2\#\ell\overline{\mathbf{CP}}_2$, where $\ell : k$ is roughly $4:1$, which do not admit Einstein metrics. By contrast, the Hitchin–Thorpe obstruction comes into play only when $\ell : k$ is roughly $5:1$.

While the Seiberg–Witten techniques also give non-existence results for manifolds with large fundamental groups, the entropy estimates of Besson *et al.* (1995) allow one to say rather more (Sambusetti, 1996):

Theorem 4.4. (Sambusetti) *Any integer pair (χ, τ) with $\tau \equiv \chi \bmod 2$ can be realized as the Euler characteristic and signature of a compact smooth 4-*

manifold (with infinite fundamental group) which is not homeomorphic to any Einstein manifold.

In fact, we'll see how to explicitly construct such manifolds as connected sums of hyperbolic manifolds and other pieces. The fundamental groups of the key examples will thus not only be infinite, but actually have exponential growth.

5 Seiberg–Witten techniques

If (X^4, J) is a compact complex surface—i.e. a complex manifold of real dimension 4—then there is a process called *blowing up* which produces a new complex surface by replacing some given point $x \in X$ with a complex projective line $\mathbb{CP}_1$. The resulting surface is diffeomorphic to a connected sum $X\#\overline{\mathbb{CP}}_2$, where $\overline{\mathbb{CP}}_2$ is the complex projective plane with the *non-standard* orientation. This process can then be iterated, and in particular one may blow up any given collection of k distinct points of X so as to produce new complex surfaces diffeomorphic to $X\#k\overline{\mathbb{CP}}_2$ for any positive integer k.

Conversely, any compact complex surface (M, J) can be expressed as $X\#k\overline{\mathbb{CP}}_2$ with $k \geq 0$, an iterated blow-up of some complex surface X which is not itself the blow-up of anything else, cf. Barth *et al.* (1984). One says that X is a *minimal model* for M.

A compact complex surface (M, J) is said to be of *general type* if its minimal model X satisfies

$$(2\chi + 3\tau)(X) > 0$$

and X is neither $\mathbb{CP}_2$ nor a $\mathbb{CP}_1$-bundle over a complex curve. For example, the degree-m hypersurface

$$\{[u : v : w : z] \in \mathbb{CP}_3 \mid u^m + v^m + w^m + z^m = 0\}$$

in complex projective 3-space is of general type if $m > 4$; these examples are all simply connected, and are their own minimal models.

Theorems 4.2 and 4.3 both follow from the following scalar curvature estimate:

Theorem 5.1 *Let (M, J) be a compact complex surface of general type, and let X be its minimal model. Then any Riemannian metric g on M satisfies*

$$\int_M s_g^2 d\mu_g \geq (2\chi + 3\tau)(X),$$

with equality iff $M = X$ and g is Kähler–Einstein with respect to some complex structure on M.

Proof The complex structure J is *a priori* completely unrelated to the metric g under discussion, but its deformation class is enough to allow one to define twisted spinor bundles $V_\pm = \mathbf{S}_\pm \otimes L^{1/2}$, where L is a Hermitian line bundle with $c_1(L) = c_1(M, J)$. Now assume for simplicity that $b^+(M) > 1$. For any g, it then turns out that the *Seiberg–Witten equations*

$$D^\theta \Phi = 0 \tag{5.1}$$
$$F^{\theta +} = i\sigma(\Phi) \tag{5.2}$$

must be satisfied by some smooth connection θ on L and some smooth section Φ of V_+. Here D^θ is the Dirac operator coupled to θ, the purely imaginary 2-form $F^{\theta+}$ is the self-dual part of the curvature of θ, and the real-quadratic map $\sigma : V_+ \to \wedge^2_+$ induced by the isomorphism $\wedge^+ \otimes \mathbf{C} = \odot^2 \mathbf{S}_+$ satisfies $|\sigma(\Phi)|^2 = |\Phi|^4/8$. This can be made more explicit by choosing some Hermitian local trivialization of L, so that the connection θ is represented by a purely imaginary 1-form ϑ; in Penrose's spinorial abstract-index notation (Penrose, 1968; Penrose and Rindler, 1986) the Seiberg–Witten equations then become

$$\begin{aligned}(\nabla_{AA'} + \frac{1}{2}\vartheta_{AA'})\Phi^A &= 0\\ \nabla^{A'}_{(A}\vartheta_{B)A'} &= \frac{1}{2}\Phi_{(A}\overline{\Phi}_{B)}\end{aligned}$$

with the convention that $|\Phi|^2 = \Phi^A\overline{\Phi}_A$. The number of solutions, modulo gauge equivalence and counted with appropriate multiplicities, can be shown to be independent[1] of g; and because the equations can be solved explicitly when the metric happens to be Kähler, it is not difficult to show that this invariant is 1. It follows that there must be at least one solution for every metric g on M.

Now the Seiberg–Witten equations imply the Weitzenböck formula

$$0 = 4\Delta^\theta\Phi + s\Phi + |\Phi|^2\Phi,$$

where $\Delta^\theta = -g^{ab}\nabla^{\overline{\theta}}_a\nabla^\theta_b$ is the Bochner Laplacian of the connection on V_+. Taking the inner product with Φ and integrating by parts, one therefore has

$$0 = \int_M [|2\nabla^\theta\Phi|^2 + s|\Phi|^2 + |\Phi|^4] d\mu$$

so that the Cauchy–Schwartz inequality tells us that

$$\left(\int_M s^2 d\mu\right)^{1/2}\left(\int_M |\Phi|^4 d\mu\right)^{1/2} \geq \int_M (-s)|\Phi|^2 d\mu \geq \int_M |\Phi|^4 d\mu.$$

Let c_1^+ denote the self-dual part of the harmonic representative of c_1 with respect to g. If $c_1^+ \neq 0$, equation (5.2) tells us that $\Phi \not\equiv 0$, so from the last inequality we deduce that

$$\int_M s^2 d\mu \geq \int_M |\Phi|^4 d\mu = 8\int |F^{\theta+}|^2 d\mu \geq 32\pi^2 (c_1^+)^2.$$

---

[1]If $b^+ = 1$, this is not quite true, but the spirit of the proof remains essentially the same. For details, see LeBrun (1996).

It follows that

$$\int_M s^2 d\mu \geq 32\pi^2 (c_1^+)^2$$

for every metric on M. In particular,

$$\int_M s^2 d\mu \geq 32\pi^2 (c_1)^2 = 32\pi^2(2\chi + 3\tau)(M),$$

and it is not difficult to check that equality can occur iff g is Kähler–Einstein.

On the other hand, this inequality can be improved if M is not its own minimal model. Indeed, in this case there are diffeomorphisms of

$$M = X \# \overline{\mathbb{CP}}_2 \# \cdots \# \overline{\mathbb{CP}}_2$$

which act trivially on the homology of X and act as -1 on H_2 of any $\overline{\mathbb{CP}}_2$ we choose. Pushing J forward via such diffeomorphisms gives rise to 2^k complex structures on M with

$$c_1 = c_1(X) \pm E_1 \pm \cdots E_k$$

where the E's are Poincaré dual to the generators of H_2 of the $\overline{\mathbb{CP}}_2$'s. But then $(c_1^+)^2 = [c_1(X)^+]^2 + (\pm E_1^+ \pm \cdots \pm E_k^+)^2 + 2c_1(X)^+ \cdot (\pm E_1^+ \pm \cdots \pm E_k^+)$ and we can make this $\geq [c_1(X)^+]^2$ by choosing our signs so that the last term is non-negative. Hence our previous argument tells us that

$$\int_M s^2 d\mu \geq 32\pi^2 c_1(X)^2 = 32\pi^2(2\chi + 3\tau)(X).$$

A little extra work (LeBrun, 1996) now allows one to rule out the possibility of equality when $k > 0$. □

To prove Theorem 4.2, first notice that a complex hyperbolic 4-manifold $M = \mathbf{C}\mathcal{H}_2/\Gamma$ automatically satisfies $(2\chi+3\tau) = 3(2\chi-3\tau)$ because the standard metric is Einstein with $W_- = 0$ and $|W_+|^2 = s^2/24$. One the other hand, Theorem 5.1, applied to an arbitrary Einstein metric g on M now yields

$$\begin{aligned} 3(2\chi - 3\tau)(M) &= \frac{3}{4\pi^2}\int_M \left(|2W_-|^2 + \frac{s^2}{24}\right) d\mu \\ &\geq \frac{1}{32\pi^2}\int_M s^2 d\mu \\ &\geq (2\chi + 3\tau)(M) \end{aligned}$$

so the metric must be Kähler–Einstein with $W_- = 0$. Moreover, we must have $s < 0$ because $\int s^2 d\mu > 0$ and there are solutions of the Seiberg–Witten equations with respect to g. The universal cover of (M, g) is therefore a rescaled version of $\mathbf{C}\mathcal{H}_2$, and Theorem 1.1 therefore follows from Mostow rigidity (Mostow, 1985).

Theorem 5.1 also implies the following result:

Theorem 5.2 *Let X be a minimal complex algebraic surface of general type, and let $M = X\#k\overline{\mathbf{CP}}_2$ be obtained from X by blowing up $k > 0$ points. If $k \geq \frac{2}{3}(2\chi + 3\tau)(X)$, then M does not admit Einstein metrics.*

Proof Suppose g is an Einstein metric on M. Using our Seiberg–Witten scalar curvature estimate and the appropriate Gauss–Bonnet formula, we have

$$\begin{aligned}(2\chi + 3\tau)(X) - k &= (2\chi + 3\tau)(M)\\ &= \frac{1}{4\pi^2}\int_M \left(2|W_+|^2 + \frac{s^2}{24}\right) d\mu\\ &> \frac{1}{24\cdot 4\pi^2}[32\pi^2(2\chi + 3\tau)(X)]\\ &= \frac{1}{3}(2\chi + 3\tau)(X),\end{aligned}$$

so that

$$\frac{2}{3}(2\chi + 3\tau)(X) > k.$$

Hence M cannot admit an Einstein metric if $k \geq \frac{2}{3}(2\chi + 3\tau)(X)$. □

But this in turn immediately implies Theorem 4.3. Indeed, if X is any minimal complex surface of general type with $2\chi + 3\tau \geq 3$, there is then at least one integer k satisfying $(2\chi + 3\tau)(X) > k \geq \frac{2}{3}(2\chi + 3\tau)(X)$. The complex surface $M = X\#k\overline{\mathbf{CP}}_2$ then satisfies $2\chi > 3|\tau|$, but does not admit Einstein metrics by the above result. Theorem 4.3 therefore follows by considering the sequence of X's given by the surfaces of degree > 4 in $\mathbb{CP}_3$.

6 Entropy inequalities

In the previous section I explained how the Seiberg–Witten equations give rise to certain scalar-curvature estimates in dimension 4. These techniques belong to the realm of *smooth topology*, in the sense that one typically gets different estimates for different differentiable structures on the same topological manifold. And while these techniques apply equally to simply connected manifolds and manifolds with enormous fundamental groups, they tell us absolutely nothing about dimensions other than 4.

Seiberg–Witten theory is the study of a system of partial differential equations which first arose in quantum field theory (Witten, 1994). A much more classical area of physics, namely thermodynamics, is the ultimate inspiration of another important invariant. Let (M, g) be a compact Riemannian manifold, and let $(\tilde{M}, g)$ be its universal cover. Let $x \in \tilde{M}$, and let $B_\rho(x) \subset \tilde{M}$ denote the set of points of distance $< \rho$ from x; let $\mathrm{Vol}(B_\rho(x))$ denote the Riemannian volume of this distance-ball. Then the *entropy* of (M, g) is defined to be

$$\mathcal{E}(M, g) = \lim_{\rho\to\infty} \frac{\log \mathrm{Vol}(B_\rho(x))}{\rho}.$$

This is independent of the base-point x, and can be non-zero only if the fundamental group of M is infinite. The reader may wish to check that any compact n-manifold of constant sectional curvature $K \leq 0$ has entropy $(n-1)\sqrt{|K|}$.

The last example brings out the fact that the entropy is not scale-invariant; if (M, g) has entropy $\mathcal{E}$, then, for any positive constant λ, $(M, \lambda^2 g)$ has entropy $\mathcal{E}/\lambda$. We can remedy this by instead considering the *normalized entropy*

$$\hat{\mathcal{E}}(M, g) = (\mathrm{Vol}(M))^{1/n}\mathcal{E}(M, g).$$

Theorem 6.1. (Besson–Courtois–Gallot) *Let M be any compact quotient of a real, complex, quaternionic, or octonionic hyperbolic space, and let g_0 be the standard metric on M. Then any other metric g on M satisfies*

$$\hat{\mathcal{E}}(M, g) \geq \hat{\mathcal{E}}(M, g_0),$$

with equality iff g is locally symmetric.

The proof involves embedding the universal cover of M in L^2 of its sphere at infinity, and comparing the entropy with the integral of a certain closed differential form over a fundamental domain; for details see Besson *et al.* (1995).

An important application is the following:

Corollary 6.2 *Let (M, g_0) be a compact quotient of hyperbolic space $\mathcal{H}^n$, and let g be any metric on M with Ricci curvature $r \geq -(n-1)g$. Then* $\mathrm{Vol}(M, g) \geq \mathrm{Vol}(M, g_0)$, *with equality iff g has constant sectional curvature -1.*

Notice that this is still true for all n, but that M is now assumed to be *real* hyperbolic. The reason is that *Bishop's inequality* (Besse, 1987; Bishop, 1963) says that a ball of radius ρ in $(\tilde{M}, g)$ must be no bigger than that of the corresponding ball in $\mathcal{H}^n$, since we've assumed that the Ricci curvature of $(\tilde{M}, g)$ is no smaller than that of $\mathcal{H}^n$. We therefore have

$$\mathcal{E}(M, g) \leq (n-1) = \mathcal{E}(M, g_0).$$

But we also know that $(\mathrm{Vol}(M, g))^{1/n}\mathcal{E}(M, g) \geq (\mathrm{Vol}(M, g_0))^{1/n}\mathcal{E}(M, g_0)$, so it follows that $\mathrm{Vol}(M, g) \geq \mathrm{Vol}(M, g_0)$, and that equality only occurs if g has constant curvature -1.

If we now further assume that $n = 4$ and that g is Einstein, with $r = -3g$, then this argument tells us that

$$\frac{1}{8\pi^2}\int_M \frac{s_g^2}{24} d\mu_g = \frac{3}{4\pi^2}\mathrm{Vol}(M, g) \geq \frac{3}{4\pi^2}\mathrm{Vol}(M, g_0) = \frac{1}{8\pi^2}\int_M \frac{s_{g_0}^2}{24} d\mu_{g_0},$$

with equality iff g has constant sectional curvature -1. But the 4-dimensional Gauss–Bonnet formula tells us that

$$\frac{1}{8\pi^2}\int_M \frac{s_{g_0}^2}{24} d\mu_{g_0} = \chi(M) = \frac{1}{8\pi^2}\int_M \left[|W_g|^2 + \frac{s_g^2}{24}\right] d\mu_g \geq \frac{1}{8\pi^2}\int_M \frac{s_g^2}{24} d\mu_g,$$

so equality holds, and g must have constant sectional curvature -1. The result therefore follows by Mostow rigidity.

The same reasoning that leads to Theorem 6.1 also proves the following:

Theorem 6.3. (Besson–Courtois–Gallot) *Let (M, g_0) be as in Theorem 6.1, and let N be a compact manifold of the same dimension. If there is a smooth map $f : N \to M$ of degree $j > 0$, then any metric g on N satisfies*

$$\hat{\mathcal{E}}(N, g) \geq j\hat{\mathcal{E}}(M, g_0).$$

If $\dim M > 2$, *moreover, equality is only achieved if f is homotopic to a covering map which is a local isometry.*

Theorem 4.4 is then a consequence (Sambusetti, 1996). Indeed, let (M, g_0) be a compact 4-dimensional hyperbolic manifold, and let N to be a connected sum of a number of copies of M, $S^1 \times S^3$'s, and $\mathbb{CP}_2$'s with either orientation:

$$N = jM\#k(S^1 \times S^3)\#\ell\mathbb{CP}_2\#m\overline{\mathbb{CP}}_2.$$

Thus

$$\begin{aligned} \chi(N) &= 2 + j(\chi(M) - 2) - 2k + \ell + m \\ \tau(N) &= \ell - m. \end{aligned}$$

Without an extra constraint, then, for each positive integer j, we could sweep out all (χ, τ) with $\chi \equiv \tau \bmod 2$ by varying the non-negative integers k, ℓ, and m. However, we'll now only consider those choices for which $\chi(N) < j\chi(M)$; that is, we also stipulate that

$$2 - 2k + \ell + m < 2j.$$

Now if N admitted an Einstein metric with $r = -3g$, we would have $\mathcal{E}(N, g) \leq 3 = \mathcal{E}(M, g_0)$ by Bishop's inequality. But there is a nearly tautological map $f : N \to M$ of degree j, so

$$(\mathrm{Vol}(N, g))^{1/n}\mathcal{E}(N, g) \geq j(\mathrm{Vol}(M, g_0))^{1/n}\mathcal{E}(M, g_0)$$

by the above result, and hence $\mathrm{Vol}(N, g) \geq j\mathrm{Vol}(M, g_0)$. Hence

$$\chi(N) \geq \frac{1}{8\pi^2}\int_N \frac{s_g^2}{24} d\mu_g \geq \frac{j}{8\pi^2}\int_M \frac{s_{g_0}^2}{24} d\mu_{g_0} = j\chi(M)$$

which is of course a contradiction because we have arranged that $\chi(N) < j\chi(M)$. Since $\chi(M) > 0$, varying j, k, ℓ, and m therefore allows us to realize every integer pair (χ, τ) with $\chi \equiv \tau \bmod 2$ as the Euler characteristic and signature of a 4-manifold which does not admit Einstein metrics.

Since the above argument only depends on the existence of a suitable homotopy class of maps $f : N \to M$, changing the differentiable structure of these examples would change absolutely nothing; these N's are not even *homeomorphic* to Einstein manifolds.

Once again, notice how these entropy arguments involve the gigantic size of the fundamental group in an essential way. These beautiful results can therefore shed no light at all on manifolds with small fundamental group.

7 Concluding remarks

In this article, I have tried to explain two new sets of techniques which yield results concerning 4-dimensional Einstein manifolds. Seiberg–Witten theory gives us differential-topological invariants which allow one to estimate the scalar curvature of a metric in relation to its volume. The entropy method instead allows one to deduce Ricci-curvature estimates from homotopy-theoretic assumptions. The mystery is that while these techniques sometimes lead to analogous results, they seem profoundly unrelated to one another. I am therefore convinced that there must be a deeper explanation of these results, involving principles which remain to be discovered.

The most tantalizing parallel between results obtained by these different means must surely be the analogy between Theorem 4.1 and Theorem 4.2. But let me now highlight an important technical distinction. The proof of Theorem 4.1 actually shows that any Einstein 4-manifold which is homotopy equivalent to a hyperbolic manifold must itself be hyperbolic. But the proof of Theorem 4.2 does not lead to a similar conclusion in the complex-hyperbolic case. Does this merely illustrate a limitation of the method of proof, or does it capture a factual difference between the real- and complex-hyperbolic cases?

In the same vein, it is interesting to consider what the two techniques tell us about blow-ups of complex-hyperbolic manifolds. The Seiberg–Witten argument tells us that blowing up such a space at $\frac{2}{3}(2\chi + 3\tau)$ points will result in a smooth manifold without Einstein metrics. The entropy argument is slightly less efficient, but it does reach the same conclusion provided $\frac{17}{25}(2\chi + 3\tau)$ points are blown up. However, the entropy argument reaches a stronger conclusion: the non-existence assertion applies equally to all smooth structures on the manifold. Presumably neither of these results is sharp, though. Could it be that blowing up at just *one* point is enough to obstruct the existence of an Einstein metric?

Bibliography

Aubin, T. (1976). Equations du Type Monge-Ampère sur les Variétés Kählériennes Compactes. *C. R. Acad. Sci. Paris 283A*, 119–121.

Barth, W. Peters, C., and Van de Ven, A. (1984). *Compact Complex Surfaces.* Springer-Verlag, Berlin.

Besse, A. (1987). *Einstein Manifolds.* Springer-Verlag, Berlin.

Besson, G., Courtois, G., and Gallot, S. (1995). Entropies et Rigidités des Espaces Localement Symétriques de Courbure Strictement Négative. *Geom. and Func. An.*, **5**, 731–799.

Bishop, R.L. (1963). A Relationship Between Volume, Mean Curvature, and Diameter, *Notices Am. Math. Soc.*, **10**, 364.

Borel, A. (1963). Compact Clifford-Klein Forms of Symmetric Spaces. *Topology*, **2**, 111–222.

Boyer, C.P., Galicki, K., and Mann, B.M. (1994). The Geometry and Topology of 3-Sasakian Manifolds. *J. Reine Angew. Math.*, **455**, 183–220.

Gamow, G. (1970). *My World Line.* Viking Press, New York, p. 44.

Hawking, S.W. (1979). The Path-Integral Approach to Quantum Gravity. In *General Relativity: An Einstein Centenary Survey*, eds. S.W. Hawking and W. Israel, Cambridge University Press, pp. 746–789.

Hitchin, N.J. (1974). On Compact Four-Dimensional Einstein Manifolds. *J. Diff. Geom.*, **9**, 435–442.

Joyce, D. (1996). Compact 8-Manifolds with Holonomy Spin(7). *Inv. Math.*, **123**, 507–552.

LeBrun, C. (1995). Einstein Metrics and Mostow Rigidity. *Math. Res. Lett.*, **2**, 1–8.

LeBrun, C. (1996). Four-Manifolds without Einstein Metrics. *Math. Res. Lett.*, **3**, 133–147.

Mostow, G.D. (1985). Discrete Subgroups of Lie Groups. In *The Mathematical Heritage of Élie Cartan (Lyon, 1984). Astérisque, Hors Séries*, 289–309.

Penrose, R. (1968). Structure of Space-Time. In *Battelle Rencontres: 1967 Lectures in Mathematics and Physics*, eds. C. DeWitt and J.A. Wheeler, Benjamin, New York.

Penrose, R. and Rindler, W. (1986). *Spinors and Space-Time, Volume 2.* Cambridge University Press.

Sambusetti, A. (1996). An Obstruction to the Existence of Einstein Metrics on 4-Manifolds. *C. R. Acad. Sci. Paris I*, **322**, 1213–1218.

Thorpe, J.A. (1969). Some Remarks on the Gauss-Bonnet Formula. *J. Math. Mech.*, **18**, 779–786.

Thurston, W., (1982). Three Dimensional Manifolds, Kleinian Groups and Hyperbolic Geometry. *Bull. Am. Math. Soc.*, **6**, 357–381.

Witten, E. (1994). Monopoles and Four-Manifolds. *Math. Res. Lett.*, **1**, 809–822.

Yau, S.-T. (1977). Calabi's Conjecture and Some New Results in Algebraic Geometry. *Proc. Nat. Acad. USA*, **74**, 1789–1799.

8
Loss of Information in Black Holes

Stephen Hawking
Department of Applied Mathematics and Theoretical Physics
University of Cambridge
Silver Street, Cambridge CB3 9EW

1 Personal and historical remarks

It is a pleasure to be here to celebrate Roger's birthday. Like many people, I find it difficult to believe he is about to be 65. I just hope I am in such good form when I reach that age.

When I began research in the early 1960s, general relativity was at a low ebb. Many of you will have read Feynman's comments, on a conference on general relativity that he attended. I believe it was at Chapel Hill. At that time, there was no contact with observations. Most people in the field spent their time doing messy calculations of Christoffel symbols. They were so pleased if they found a solution that they did not care what physical significance it had, if any.

Roger stood out, head and shoulders, above other people in the field. His background in pure mathematics meant that he understood that the global structure of space–time might need more than a single coordinate patch to cover it. That may seem obvious now, but in those days, many people thought that the surface, $r = 2m$, in the Schwarzschild solution was a singularity, and one could not go inside it. I remember a relativist, whose lectures at Kings College I attended in my first year as a research student, saying that someone had found a coordinate transformation that removed half the singularity. But clearly he did not think it was physically significant.

Roger took a more global view of space–time. While Bondi, Newman, and their collaborators, studied gravitational radiation in terms of messy asymptotic expansions, Roger had the brilliant idea of applying a conformal transformation, and making infinity into a finite null surface, *scri*. This meant that the behaviour of gravitational radiation at infinity could be studied in local terms. This work on conformal properties brought Roger to consider the global causal structure of space–time, and led to his next great step forward, the Penrose singularity theorem of 1965. It opened a new dimension. Previous work on gravitational collapse had been in particular coordinate systems, and had made assumptions of symmetry, or hand waving approximations. But Roger showed that once gravitational collapse had gone beyond a certain point, a singularity of space–

time was inevitable. This theorem held for any reasonable form of matter, and it applied to fully general space–times, without any symmetries. It was in direct contradiction to the Russians, Lifshitz and Khalatnikov, who claimed that a general collapse would not lead to singularities. But it is hard to argue with a mathematical theorem, so in the end they had to recant, though it took them another five years.

Roger's theorem showed that gravitational collapse had to end in a singularity, at which general relativity broke down. We estimate that gravitational collapse occurs at least once a year in our galaxy, and at a similar rate in other galaxies. So how is it that we can predict anything? The answer is that general relativity seems to obey what Roger called the Cosmic Censorship Conjecture. This says that any singularities formed in collapse from non-singular data will be hidden behind an event horizon, another term Roger introduced. The Cosmic Censorship Conjecture is fundamental to all work on black holes. I therefore had a bet with Kip Thorne and John Preskill that it was true. Unfortunately, I was not careful enough about the wording of the bet, and the similarity solutions that represent the threshold of black hole formation were a counterexample. But all the evidence suggests that the conjecture is true for generic initial data. The world is safe from naked singularities, at least in classical general relativity.

The next major development in the field came with the unexpected discovery by Werner Israel, that a static black hole had to be spherically symmetric. My use of the term 'black hole' here is an anachronism, because it was not introduced until later, but it is a convenient shorthand. At first, many people, including Israel himself, thought this result implied that black holes would form only in collapses that were exactly spherical, which would be a set of measure zero amongst all collapses. However, Roger and John Wheeler put forward a different interpretation. This was that in gravitational collapse, an asymmetric body would lose all multi–pole moments, except mass, angular momentum, and electric charge, which were protected by coupling to gauge fields. If this interpretation was correct, black holes could have no hair. In other words, time independent black holes would be determined completely by their gauge charges. This was proved for Einstein–Maxwell theory, by the combined work of Israel, Carter, Robinson, and myself. For gravity coupled to Yang–Mills, Higgs, or dilation fields, there are additional complications, but the result remains essentially true: stable non-extreme black hole solutions are completely determined by their gauge charges.

2 Information loss

The no hair results indicated that a lot of information was lost in gravitational collapse. One ended up with a black hole that was determined by its mass, angular momentum, and electric charge, but which was otherwise independent of the nature of the body that collapsed. In a purely classical theory, there would be no information loss, because one would never actually lose sight of the collapsing body. It would appear to slow down, and hover just outside the horizon, getting rapidly more red shifted, but never actually disappearing. So in classical theory, someone at infinity could in principle always observe the collapsing body, and there would be no loss of information. But in quantum theory, the number of photons emitted by the collapsing body before the event horizon forms will be finite. There will be far too few of them to carry all the information about the collapsing body. Thus an outside observer will lose information about what collapsed to form the black hole.

This loss of information did not matter too much, when it was thought the black holes went on forever. One could still say that the information was inside the black hole. It was just that it was not accessible to an observer at infinity. However, the situation changed when it was discovered that black holes radiate with a thermal spectrum. The radiation would be in a mixed quantum state, and would not carry away information about what was inside the black holes. However, it would carry away energy, and hence, mass. Unless they were stabilized by something like a magnetic charge, it seemed black holes would radiate down to zero mass, and disappear. What then would happen to the information they contained?

Physicists seem to have a strong emotional attachment to information. I think this comes from a desire for a feeling of permanence. They have accepted that they will die, and even that the baryons which make up their bodies will eventually decay. But they feel that information, at least, should be eternal. There have therefore been three main suggestions, for preserving information in black holes.

The first is that the black hole does not evaporate completely, but leaves a Planck sized remnant containing the information. The second is that all the information in the black hole comes out in the final burst of radiation when the black hole gets down to the Planck size. That is where the semi-classical calculation of the radiation will break down, so information might emerge. The third is that the radiation coming out of the black hole contains subtle correlations between different modes, which are not seen in the semi-classical calculation, and which encode the information.

I think one can dismiss the first two suggestions. The remnant hypothesis would not be CPT invariant, if black holes could form, but never disappear completely. And one might expect more than the cosmological density of remnants, left over from the evaporation of black holes in the very early universe. The final burst hypothesis is in trouble because it takes energy to carry information, and there is very little energy left in the final stages. The only possibility is for the information to trickle out very slowly. So this possibility is similar to the remnant one.

In my opinion, the only way of preserving information that has not been ruled out is if there are correlations in the outgoing radiation that carry the information. Such correlations would occur if black holes can be replaced by collections of strings, attached to D-branes, with the same gauge charges. These systems of D-branes and strings have a number of internal states, or amount of information, that is the same function of the gauge charges, as e to the quarter area of the horizon of the black hole ($e^{A/4}$). However, the D-brane calculations are valid only for weak coupling, at which string loops can be neglected. But at these weak couplings, the D-branes are definitely not black holes: there are no horizons, and the topology of space–time is that of flat space. One can foliate such a space–time with a family of non-intersecting surfaces of constant time. One can then evolve forward in time with the Hamiltonian, and get a unitary transformation from the initial state to the final state. A unitary transformation would be a one to one mapping from the initial Hilbert space to the final Hilbert space. This would imply that there was no loss of information, or quantum coherence.

To get something that corresponds to a black hole, one has to increase the string coupling constant, until it becomes strong. This means that string loops can no longer be neglected. However, it is argued that for gauge charges that correspond to extreme, or near extreme, black holes, the number of internal states will be protected by non-renormalization theorems, and will remain the same. On the other hand, as the coupling is increased, and event horizons appear, the Euclidean topology of space–time will change discontinuously. The change in topology will mean that any vector field that agrees with time translations at infinity will necessarily have zeroes in the interior of the space–time. In turn, this will mean that one cannot foliate space–time with a family of time surfaces. If one tries, the surfaces will intersect at the zeroes of the vector field. One therefore cannot use the Hamiltonian to get a unitary evolution from initial state to final. But if the evolution is not unitary, there will be loss of information. An initial state that is a pure quantum state can evolve to a quantum state that is mixed. That is, it is described by a density matrix, which is a two index tensor on Hilbert space, rather than by a state vector. This process of evolution from a pure state to a mixed state is called loss of quantum coherence. It is what those who are attached to unitarity object to so violently.

One cannot just ignore topology, and pretend one is in flat space, which in effect is what those that want to preserve unitarity are doing. The recent progress in duality is based on non-trivial topology. One considers small perturbations about different vacua of the product form, flat space cross internal space, and shows that one gets equivalent Kaluza–Klein theories. But if one can have small perturbations about product metrics, one should also consider larger fluctuations that change the topology from the product form. Indeed, such non-product topologies are necessary to describe pair creation or annihilation of solitons, like black holes, or D-branes.

It is often claimed that supergravity is just a low energy approximation to the fundamental theory, which is string theory. However, I think the recent work on duality is telling us that string theory, D-branes, and supergravity, are all on a similar footing. None of them is the whole picture, but each are valid in different, but overlapping regions. There may be some fundamental theory, from which they can all be derived as different approximations. Or it may be that theoretical physics is like a manifold, that cannot be covered by a single coordinate patch. Instead, we may have to use a collection of apparently different theories, that are valid in different regions, but which agree on the overlaps. After all, we know from Goedel's theorem that even arithmetic cannot be formulated in terms of a single set of axioms. Why should theoretical physics be different?

Even if there is a single formulation of the underlying fundamental theory, we do not have it yet. What is called string theory has a good loop expansion. But it is only perturbation theory about some background, generally flat space. So it will break down when the fluctuations become large enough to change the topology. Supergravity, on the other hand, is better at dealing with topological fluctuations, but it will probably diverge at some high number of loops. Such divergences do not mean that supergravity predicts infinite answers. It is just that it cannot predict beyond a certain degree of accuracy. But in that, it is no different from string theory. The string loop perturbation series almost certainly does not converge, but is only an asymptotic expansion. Thus at finite coupling, it will only have limited accuracy.

I shall take what I have just said as justification for discussing information loss in terms of general relativity or supergravity, rather than D-branes and strings. If one accepts that this is the right arena, it follows that information loss must occur, independent of what might occur in the final stages of evaporation, when Planck scale physics is involved. Consider, for example, an extreme magnetically charged black hole. This will have zero temperature, and so will not radiate anything. Suppose one now sends in a certain number of particles in a pure quantum state. This will increase the information content of the black hole. It will also increase the mass, and the black hole will radiate thermally, until it gets back to the extreme zero temperature state. The net effect is that the information about the particles sent in has been lost. All one gets back is thermal radiation in a mixed quantum state. Yet the curvature outside the horizon has been small at all times. So the semi-classical calculation of the radiation should be valid, and in particular, the conclusion that the radiation is in a mixed state, with no correlations between different modes.

For years, I have felt that information loss would occur, not only for macroscopic black holes, but on a microscopic scale as well, because of virtual black holes that would appear and disappear in the vacuum state. Particles could fall into these virtual black holes, which would then radiate other particles in a mixed quantum state. However, I did not know how to describe such processes, because I could not see how a black hole could appear and disappear, in a manner that was non-singular, at least in the Euclidean regime. I now realize that my mistake was to try to picture a single virtual black hole appearing and disappearing. Instead, black holes appear and disappear in pairs, like other virtual particles. Equivalently, one can think of the virtual pair as a single black hole moving on a closed loop.

In d dimensions, a single black hole has a Euclidean section with topology $S^{d-2} \times R^2$. A real or virtual loop of black holes has Euclidean topology $S^{d-2} \times S^2$, minus a point that has been sent to infinity by a conformal transformation. For simplicity, I shall consider $d = 4$, but the treatment for higher d will be similar.

On the manifold $S^2 \times S^2$, minus a point, I shall consider Euclidean metrics that are asymptotic to flat space at infinity. Such metrics can be interpreted as closed loops of virtual black holes. Because they are off-shell, they need not satisfy any field equations. But they will contribute to the path integral, just as off-shell loops of particles contribute to the path integral, and produce measurable effects. The effect that I will be concerned with for virtual black holes is loss of quantum coherence. This is a distinctive feature of such topological fluctuations that distinguishes them from ordinary unitary scattering, which is produced by fluctuations that do not change the topology.

One can calculate scattering in an asymptotically Euclidean metric on $S^2 \times S^2$, minus a point. One then weights with e to the minus the action of the metric, and integrates over all asymptotically Euclidean metrics. This would give the full scattering, with all quantum corrections. However, we can neither calculate the scattering in a general metric, nor integrate over all metrics. Instead, what I shall do is point out some qualitative features of the scattering in general metrics, that indicate that quantum coherence is lost. I shall then report on some scattering calculations that Simon Ross and I are doing in a specific metric, the c metric. It is sufficient to show that quantum coherence is lost in some metrics in the path integral, because the integral over other metrics cannot restore the quantum coherence lost in our examples.

In general, an asymptotically Euclidean metric cannot be analytically continued to a region where it is real, and Lorentzian. This does not matter for scattering calculations, because the metric can be analytically continued to Minkowski space at infinity, and ingoing and outgoing states defined. These can be propagated to each other, by the analytically continued Euclidean Green functions. In scattering calculations, one measures only states at infinity, and not in the interior of the space–time. One should therefore integrate over all metrics for the interior. The idea is that a path integral over all Euclidean metrics is equivalent in a contour integral sense to a path integral over all Lorentzian metrics.

Despite twenty years of advocating the Euclidean approach to quantum gravity, I do not have much intuition for Euclidean scattering. However, if the Euclidean metric has a hypersurface orthogonal Killing vector, it can be analytically continued to a real Lorentzian metric, in which it is much easier to see what is happening. I shall therefore consider scattering in such metrics. Metrics on $S^2 \times S^2$ minus a point can at most admit one other Killing vector, but they cannot be spherically symmetric. Thus they do not behave like the two dimensional models of black holes.

The Lorentzian section will contain a pair of black holes that accelerate to infinity. One might think that this is not very physical, but it is no different from a closed loop of a particle like an electron. Closed particle loops are really defined in Euclidean space. If you analytically continue them to Minkowski space, you get an electron–positron pair accelerating away from each other. This kind of behaviour is typical. Any topologically non-trivial asymptotically Euclidean metric will appear to have solitons accelerating to infinity, in the Lorentzian section. But this does not mean that there are actual black holes at infinity, any more than there are runaway electrons and positrons, with a virtual electron loop. One can regard the use of the Lorentzian metric, with its black holes accelerating to infinity, as just a mathematical trick to evaluate a Euclidean path integral.

The idea is to consider quantum fields propagating on an accelerating black hole background, rather like my original calculation of black hole radiation. I remember when I first presented my results at a conference near Oxford, people were rather scornful. What I was doing was not real quantum gravity. That was a matter of Feynman diagrams. That attitude still persists, though the Feynman diagrams have now been replaced by string diagrams. But as I said earlier, string theory has only a limited range of validity. It is good for calculating perturbative quantum corrections, but not well adapted to dealing with topological fluctuations.

To understand the structure of these accelerating black hole metrics, it is helpful to draw Penrose diagrams, or should I say, Penrose–Carter diagrams. In interpreting these diagrams, one has to remember that these spaces are not spherically symmetric, like Schwarzschild or Reissner–Nordstrom. A single two dimensional section therefore does not express the whole structure. Start with the Penrose diagram for Rindler space, with the two acceleration horizons, and $scri^+$, and $scri^-$. A uniformly accelerated particle moves on a world line, that goes out to $scri^-$ and $scri^+$, at the points where they intersect the acceleration horizons. We now want to replace the accelerating particle, and the similar accelerating particle on the other side, with black holes. So we replace the regions of Rindler space, to the right and left of the accelerating world lines, with intersecting black hole horizons. It turns out that the two accelerating black holes are just the two sides of the same three–dimensional wormhole, so one has to identify the two sides of the Penrose diagram. At first sight, it looks as if one has lost half of *scri*. But this is because this Penrose diagram applies only exactly in the direction of the acceleration. Generators of $scri^+$ that are in different directions do not intersect the acceleration horizons, and lie entirely to their future. One can see this by drawing $scri^+$ as a cylinder. The acceleration horizons are then the past light cones of points on the two generators, in the direction of the acceleration. I shall ignore data on these two generators, as being of measure zero on *scri*.

The idea now is to do quantum field theory in curved space, on the background of the accelerating black hole metric. One calculates the probability to go from initial to final state, in each background, and integrates over all background metrics to get the full answer. This procedure will incorporate the back reaction of the particles on the metric, automatically.

The analytically continued Euclidean Green function will define a vacuum state, 0_h, which is the analogy of the so called Hartle–Hawking state, for a static black hole. I find it a bit awkward to use that term, so I shall call it the Euclidean quantum state, but it is equivalent. The Euclidean quantum state can be characterized by saying that positive frequency means positive frequency with respect to the affine parameters on the horizons. In the accelerating black hole metrics, there are two kinds of horizons, black hole, and acceleration. Each kind of horizon consists of two intersecting null hypersurfaces, which I will refer to as left and right. To get a Cauchy surface for the space–time, one has to break the symmetry between left and right, and choose say the left acceleration horizon, and the right black hole horizon. The quantum state defined by positive frequency, with respect to the affine parameters on these horizons, is the same as the quantum state defined by the other choice of horizons.

Another Cauchy surface in the future (modulo a set of measure zero) is formed by *scri*$^+$, and the future parts of the black hole horizon, as shown in the diagram. There is a natural notion of positive frequency on *scri*. This is defined by the affine parameter on the generators, in the conformal metric in which *scri* is a cylinder. On the black hole horizons, the concept of positive frequency is less well defined. One could use Rindler time, but what one observes on *scri* is independent of the choice of positive frequency on the black hole horizons.

Because *scri*$^+$ and the future black hole horizons are a Cauchy surface, the quantum state of a field ϕ will be determined by data on these surfaces. This means that the Hilbert space of the space–time, $\mathcal{H}$, will be isomorphic to the tensor product, $\mathcal{F}_{\mathcal{J}^+} \times \mathcal{F}_{b^+}$, the Fock spaces on *scri*$^+$ and the future black hole horizons respectively. The Euclidean vacuum state, 0_h, can be represented as a state in this tensor product. But because of frequency mixing, it will not be the vacuum state on *scri*, cross the vacuum state on the black hole horizon. Rather it will be a state containing pairs of particles. Both members of the pair may go out to *scri*$^+$, or both may fall into the hole, or one go out to *scri*$^+$, and one fall in. It is this last category that has the interesting effects. An observer on *scri*$^+$ cannot measure the part of the quantum state on the black hole horizon. He or she would therefore have to trace out over all possibilities on the future black hole horizons, and describe the observations by a density matrix. Thus there will be loss of quantum coherence.

The calculation of the density matrix in the accelerating black hole metric is very similar to that for a static black hole, though a lot harder to do quantitatively. One takes positive frequency data on *scri*$^+$, and propagates it backward, with nothing coming out of the future black hole horizons. Part of the waves will enter the black holes, and register on the past horizon. However, the geodesic continuation of the past horizon of one black hole is the future horizon of the other. But the data has been set to zero on the future black hole horizons. So the wave is non-zero on only half the horizon. This means it is not purely positive frequency with respect to the affine parameter. Thus there is outgoing radiation at *scri*$^+$, contrary to what has been claimed by Yi, who said that accelerating black holes would not radiate. His argument was basically that an observer moving with the black hole would think that the black hole was in equilibrium with the acceleration radiation. But an observer at infinity in Rindler space will not see the acceleration radiation. So Yi reasoned that an observer at infinity should not see radiation from the black holes either. But the frequency mixing argument I have given shows that this expectation was wrong, just as the frequency mixing argument refuted the earlier expectation that static black holes should not radiate.

The quantum state defined by the analytically continued Euclidean Green functions will contain both incoming and outgoing radiation. Unlike the Euclidean state for static black holes, there will not be radiation to infinity at a steady rate for an infinite time. Instead, the radiation will be peaked around the points on *scri*, where the acceleration horizons intersect *scri*. The radiation will die off at early and late times on *scri*, and the total energy radiated in any given direction will be finite.

How should we interpret this? In the case of a static black hole, one usually imposes the boundary condition that there is no incoming radiation on $scri^-$. This means that one has to subtract the incoming radiation from the Euclidean state, to give what is called the Unruh state. This is singular on the past horizon, but that does not matter, as one normally replaces this region of the metric with the metric of a collapsing body. The energy for the steady rate of outgoing radiation comes from a slow decrease of the mass of the black hole formed by the collapse. However, in the case of a virtual black hole loop, there is no collapse process to remove the singularities on the past horizons of the black holes, or supply the energy of the outgoing radiation. My view, therefore, is that integrating over virtual black hole metrics will cause the amplitude to be zero, unless the energy of the outgoing particle or particles is matched by particles with the same energy falling in. One might object that one would never have exactly the combination of incoming particles that corresponded to the quantum state obtained from the Euclidean Green functions. But I do not think that matters. All the metric cares about is the energy–momentum tensor. It does not matter what species of particles produces it. I think all that is important is that there should be an energy–momentum balance, between the particles that fall into the accelerating black holes, and those that come out. The black holes do not emit a definite collection of particles. Instead, the outgoing radiation is in a mixed quantum state, with different probabilities for sending out different collections of particles. All that one does, is compare the relative probabilities for those collections of particles with the same energy as the particles that fall in. Thus one can consider processes in which one or two particles fall into a virtual black hole loop, and one or two come out. These processes look like Feynman diagrams, but the difference is that the outgoing particles will be in a mixed state. There will be loss of quantum coherence.

These virtual black holes will presumably be of Planck size. The cross–section for a low energy particle to fall into a Planck size static black hole is very low unless the particle is spin zero. This suggests the effects of virtual black holes will be small except for scalar particles. To check this, Simon and I wanted to do a scattering calculation in an accelerating black hole metric. We chose the c metric. This does not really qualify as a virtual black hole metric, because it has string singularities pulling the black holes apart. But it has the same structure as a virtual black hole, and it has the great advantage that the wave equation separates.

The c metric can be written in the form

$$ds^2 = \frac{1}{A^2(x-y)^2}\left\{G(y)dt^2 - \frac{dy^2}{G(Y)} + \frac{dx^2}{G(x)} + G(x)d\phi^2\right\}.$$

Here G is a quartic with four real roots, ξ_1 to ξ_4. The coordinate x is an angular variable, like the angle θ, and runs from ξ_3 to ξ_4. The coordinate y can be thought of as a radial coordinate. The black hole horizon is at $y = \xi_2$. The acceleration horizon is at $y = \xi_3$. And infinity is at $y = x$, between ξ_3 and ξ_4. The coordinate t is like Rindler time, rather than Minkowski time, or retarded time on *scri*. The Euclidean extension of the c metric is periodic in it, with period 2π. Thus someone moving on the orbits of the Killing vector would see the black holes in thermal equilibrium with the acceleration horizons, at a temperature of $\frac{1}{2\pi}$.

The minimally coupled scalar wave equation, $\Box\phi = 0$, is the same as the conformally coupled wave equation, because the scalar curvature, R, of the c metric, is zero. One can therefore make a conformal transformation with factor $\Omega = A(x-y)$, and obtain a metric in which the wave equation separates.

The equation in the angular variable, x, has solutions that are regular on both axes, only for discrete values of the separation constant, D. In the limit in which the accelerating black holes are small, the angular eigenfunctions will be spherical harmonics. One can then use these values of D in the radial y equation. The black hole horizons will radiate particles according to the Bose–Einstein factor, where ω is the frequency with respect to Rindler time. Some of these particles will be reflected back into the black holes, and some will propagate to the acceleration horizon, and then on to *scri*$^+$. On *scri*$^+$, one can take Fourier transforms of the mode with respect to the affine parameter on the generators to obtain the particle content.

Simon and I are still working on this programme. However, we have shown that the transmission factor between the black hole and acceleration horizons is proportional to ω^2, for small ω. This will cancel the ω^{-1} divergence in the Bose–Einstein factor, and give a finite emission in each angular mode. We also find that the dominant contribution comes from the $l = 0$ mode, and that higher modes are suppressed. This suggests that the main effect of virtual black holes will be on scalar fields, and that the interaction with higher spin fields will be suppressed.

To summarize, the no hair theorems and the quantum nature of light, mean that one cannot measure the information inside black holes. This information then gets lost, when the black hole evaporates and disappears. I have taken it that general relativity or supergravity, rather than string theory or D-branes, are the right arena to discuss horizons and non-trivial topology. On this basis, it is clear that information is lost, and the evolution is non-unitary. One can see this in the Lorentzian regime, because *scri*$^+$ on its own is not a Cauchy surface for space–time. One has to add the future black hole horizons. Because one cannot measure the data on the black hole horizons, one loses information and quantum coherence. In the Euclidean regime, the non-trivial topology shows that one cannot foliate with a family of time surfaces, and get a unitary Hamiltonian evolution. Finally I presented calculations on virtual black holes. These indicate that quantum coherence can be lost on a microscopic scale, and that the main effect would be on scalar fields. Could this be why we have not seen the Higgs?

Roger has suggested that quantum gravity causes what he calls OR, or objective reduction of the wave function. This is a form of loss of quantum coherence. So I have provided a mechanism for Roger's proposal. But I very much doubt whether it has anything to do with Schroedinger's cat, or the operation of the brain. Decoherence through interaction with the environment will be more important in both cases.

I have gone on long enough to lose quantum coherence. Thank you for listening.

9

Funda-mental Geometry: the Penrose–Hameroff 'Orch OR' Model of Consciousness

Stuart Hameroff
Health Sciences Center, University of Arizona,
Tucson, Arizona 85724-5114

1 Introduction: on the trail of an enigma

In *The Emperor's New Mind* and *Shadows of the Mind*, Roger Penrose followed a delicate thread to begin to unravel the problem of consciousness. He concluded that conscious thought is more than computation—our minds utilize some type of non-algorithmic processing unavailable in computers. Regarding the identity of the proposed processing, the intrepid sleuth also searched for a 'subtle and largely unknown physical principle' capable of performing the needed non-computational actions. That principle was deduced to be 'objective reduction' (OR) of quantum coherent superposition, Penrose's quantum gravity solution for the problem of wave function collapse in quantum mechanics. The next piece of the puzzle then became: where and how could quantum coherent superposition and OR occur in the brain? To that end, Penrose began collaboration with this author, who suggested that the logical and most likely source of macroscopic quantum effects in the brain lies in neuronal microtubules in the cell cytoskeleton.

Critics referred to the juxtaposition of consciousness, quantum gravity and brain microtubules as a mere 'minimization of mysteries', and a 'marriage of convenience', and questioned how quantum coherence could possibly occur and environmental decoherence be avoided in the 'noisy', thermal brain. Addressing these and other issues, Penrose and Hameroff (1995; Hameroff and Penrose 1996a, 1996b) developed a specific tripartite (philosophy, physics, biology) model of consciousness: orchestrated objective reduction (Orch OR) in neuronal microtubules.

In this paper, relevant points of philosophy, physics and biology are surveyed, the Orch OR model is summarized, and a list of testable predictions is presented. In conclusion, Penrose's 'Platonic world' is considered in the context of new ideas in physics and biology.

2 Philosophy: a panexperiential 'funda-mentality'

In addition to non-computable processing, certain features of consciousness are difficult to explain by conventional neural connectionist theories. For example, can the nature of conscious experience (Chalmers' 'hard problem') be fully explained solely by dynamical firing patterns of the brain's neural networks? Reductionist neuroscientists and philosophers (e.g. Dennett, 1991) claim that, yes, consciousness can be fully explained in terms of neural firing activity. However others caution that even if the activity of every neuron, channel and molecule in an entire brain were known and precisely correlated with a particular mental state, that state's subjective 'qualia', or experience (the smell of a rose, the sound of an oboe,...) would remain enigmatic (Chalmers, 1996a, 1996b). Other features of consciousness for which reductionist explanations appear to fail include 'binding' (e.g. in vision, of unitary 'self'), pre-conscious to conscious transitions, and non-deterministic free will.

In addition to reductionism, philosophical approaches to consciousness include 'dualism' (consciousness is separate and distinct from the brain), 'idealism' (consciousness and reality are one) and various forms of 'panpsychism' or 'panexperientialism' (conscious experience is somehow intrinsic to physical reality). The latter may hold the most promise—the raw precursor to conscious experience being somehow embedded in the fundamental physics of the universe.

In the historic panpsychism of Spinoza (1677), every atom and material object was considered to have a rudimentary aspect of proto-consciousness. Many philosophers refined this position, for example Leibniz (e.g. 1768) saw the universe as an infinite number of fundamental units ('monads')—each having a primitive psychological being. Whitehead (1929, 1933) described dynamic monads with greater spontaneity and creativity, interpreting them as mind-like entities of limited duration ('occasions of experience'—each bearing a quality akin to 'feeling'). More recently, Wheeler (e.g. 1990) described a 'pre-geometry' of fundamental reality comprised of information. Chalmers (1996a, 1996b) contends that fundamental information includes 'experiential aspects' leading to consciousness.

The word 'panpsychism' is often used to describe these positions, however Griffin (1988) suggests that the term 'panexperientialism' is preferable: 'psyche' implies the complex conscious mind, whereas experience proposed to be embedded at the micro-scale is raw, undifferentiated proto-consciousness (Farleigh 1998).

The panexperiential view most consistent with modern physics is that of Alfred North Whitehead: 1) consciousness is a process of events occurring in a wider, basic field of raw proto-conscious experience, 2) Whitehead's events (discrete occasions of experience) are comparable to quantum state reductions (Shimony, 1993).

This suggests that consciousness could involve a self-organizing process of quantum events occurring in a 'basic field' of raw experience. If experience *is* fundamental, the basic field must exist at the most fundamental level of reality.

The universe is generally described in terms of spacetime geometry. Modern physics holds that reality is rooted in 3-dimensional space and a 1-dimensional time, combined together into a 4-dimensional spacetime. This spacetime is

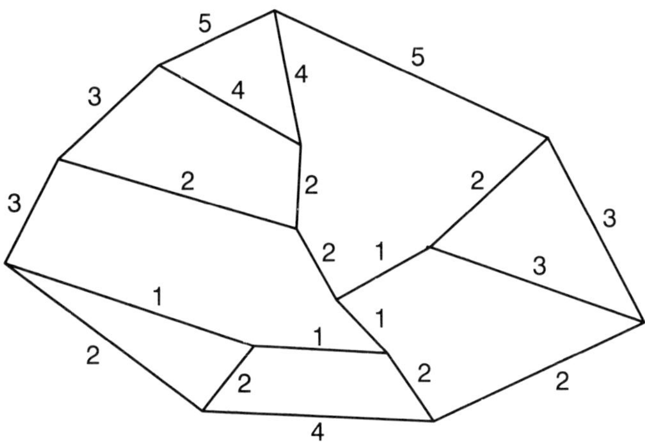

FIG. 1. A spin network. Introduced by Roger Penrose (1971) as a quantum mechanical description of the geometry of space, spin networks describe spectra of discrete Planck scale volumes and configurations (with permission, Rovelli and Smolin, 1995a). Scale bar (lower right): Planck length (10^{-33}cm; 10^{-25}nm).

slightly curved, in accordance with Einstein's general theory of relativity which precisely describes distributions of mass density in terms of spacetime curvature. Gravitational fields of all mass objects are encoded as curvatures in spacetime geometry; large objects have large curvatures, small objects have small curvatures. (Another cause of spacetime curvature is the Weyl tensor, which Penrose likens to 'a kind of gravitational analogue of the electromagnetic field quantity'.) But what exactly *is* the fundamental nature of the spacetime being curved?

The Planck scale (10^{-33}cm) is the length at which spacetime is no longer smooth. To provide a quantum mechanical description of the geometry of space at this scale, Penrose (1971) introduced 'spin networks' (e.g. Figure 1, Rovelli and Smolin, 1995a, 1995b) in which spectra of discrete Planck scale volumes and configurations are obtained. It is interesting to compare these fundamental spacetime volumes and configurations with the philosophical concept of (quantum) monads.

In a panexperiential philosophical view consistent with modern physics, quantum spin networks defining Planck-scale spacetime geometry are a possible site for proto-conscious 'funda-mental' experience. In this view, various configurations of quantum spin geometries represent different varieties of raw experience. A self-organizing quantum process operating in funda-mental spacetime could then select a particular geometry of conscious experience. What type of quantum process could do this?

3 Physics: objective reduction (OR)

Quantum theory depends upon quantum superposition, whereby individual particles, or even large collections of them (quantum coherence) can coexist in separate locations or states at once, sometimes spread over macroscopic distances. Yet quantum superpositions seem to perish when systems get 'too large', and mysteriously 'collapse' (or 'reduce') to definite, particular states and locations. As simple isolated quantum systems do not, by themselves, become 'unsuperposed', conventional theory contends that systems remain in superposition until consciously observed. Accordingly, the mythical 'Schrödinger's cat' remains both dead and alive in its closed box until externally examined.

However many physicists now believe that intermediate between tiny quantum-scale systems and 'large' cat-size systems some objective factor disturbs the superposition causing collapse, or reduction. This putative process is called objective reduction (OR).

One increaingly popular OR viewpoint (initiated by Károlyházy in 1966—Károlyházy *et al.*, 1986) suggests this 'largeness' is to be gauged in terms of gravitational effects—and in Einstein's general relativity, gravity *is* spacetime curvature. According to Penrose (1989, 1994, 1996), quantum superposition—actual separation (displacement) of mass from itself—causes underlying spacetime (?spin networks) to also separate (simultaneous curvatures in opposite directions, e.g. Figure 2). Such separations are unstable and a critical degree of separation (related to quantum gravity) results in spontaneous self-collapse (OR).

Particular configurations in spacetime geometry and corresponding mass distributions to which superpositions self-collapse are chosen in the OR process 'non-computably' (i.e. by non-algorithmic processing, Penrose, 1989, 1994, 1996). These states are only chosen non-computably if the quantum collapse occurs by the OR process, rather than by environmental interaction causing decoherence (in which case the states are chosen randomly). Thus a quantum system requires isolation from its environment to reach threshold for non-computable OR. A self-organizing OR process could account for non-computability and—by accessing funda-mental spacetime—the nature of conscious experience (as well as binding and other enigmatic features of consciousness). Could such a process occur in the brain?

In the gravitational OR scheme described by Penrose (1989, 1994; 1996), the degree of superposed mass/spacetime separation is inversely related to the coherence time—the time for which superposition must be sustained to induce collapse—according to the indeterminacy principle $E = \hbar/T$. E is the gravitational self-energy of the superposed, displaced mass (separated from itself), $\hbar$ is Planck's constant over 2π, and T is the time until self-collapse. For superpositions with complete separation (i.e. greater than the mass radius), E may be calculated from the gravitational constant G, superposed mass m, and it's displacement separation a by $E = Gm^2/a$. (Separations less than the mass radius are slightly more complicated. See e.g. Hameroff and Penrose, 1996a.)

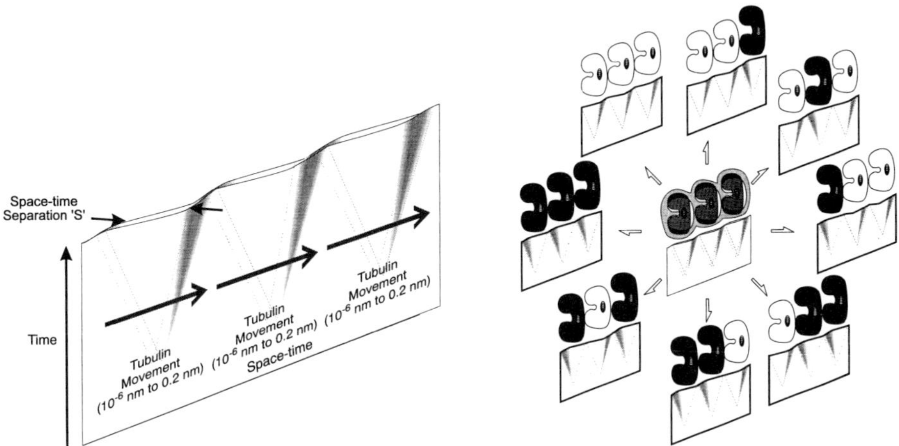

FIG. 2. Spacetime representations of tubulin superpositions. (Left): Schematic spacetime separation illustration of three superposed tubulins. The Planck-scale spacetime separations S are very tiny in ordinary terms, but relatively large mass movements (e.g. hundreds of tubulin conformations, each moving from 10^{-6} nm to 0.2 nm) indeed have precisely such very tiny effects on the spacetime curvature. A critical degree of separation causes an abrupt selection of one curvature, or the other. (Right): Center—Three superposed tubulins with corresponding schematic spacetime separation illustrations. Surrounding the superposed tubulins are the eight possible post-reduction 'eigenstates' for tubulin conformation, and corresponding spacetime geometry. The post-reduction state is chosen non-computably in Orch OR events.

Using these equations (in absolute units in which G, $\hbar$ and the speed of light c are equal to one—Penrose, 1994, page 338), the times T at which isolated, quantum superposed masses would self-collapse (OR) may be calculated:

nucleon (proton or neutron)	$T \sim 10^7$	years
beryllium atom (Monroe *et al.*, 1996)	$T \sim 10^6$	years
protein ($\sim 10^5$ nucleons)	$T \sim 2$	years
water droplet (radius $= 10^{-5}$ cm)	$T \sim 2$	hours
water droplet (radius $= 10^{-4}$ cm)	$T \sim .05$	sec
water droplet (radius $= 10^{-3}$ cm)	$T \sim 10^{-3}$	sec
Schrödinger's mythical cat ($m = 1\,\text{kg}, a = 100\,\text{cm}$)	$T \sim 10^{-36}$	sec

A brain OR process capable of non-computable selection of fundamental experience would require coherence times T in the range of known neurophysiological processes (tens to hundreds of msec) implying total superposed mass in the range of nanograms. What brain structures involved in known neural activities could support and isolate large scale quantum superposition?

Ideally, such structures should be highly prevalent in the brain, functionally important (e.g. regulating synapses), sensitive to quantum-level events and have crystal-like long range order. Most importantly they should be capable of information processing, and able to be isolated from external interaction to avoid decoherence. Membranes, membrane proteins, synapses, cell water, DNA, clathrins, myelin, centrioles and other brain structures have been suggested, but in our view cytoskeletal microtubules are best-suited.

4 Biology: quantum coherence in microtubules?

4.1 Microtubules

Interiors of neurons and other living cells ('cytoplasm') are spatially and dynamically organized by self-assembling protein networks: the cytoskeleton (Figure 3). The three major cytoskeletal components are microtubules, actin and intermediate filaments. The versatile and prominent microtubules ('MTs') are hollow cylindrical polymers of individual proteins known as tubulin (Figure 4). In dividing cells, tubulins assemble into MTs to form the mitotic spindles which (guided by MT-based centrioles) separate chromosomes, then disassemble and reassemble to establish 'daughter cell' architecture. In stable non-dividing cells such as neurons, MTs are interconnected by linking proteins (microtubule-associated proteins: 'MAPs') to other MTs and cell structures to form cytoskeletal networks which maintain neuronal form and regulate synaptic connections.

MTs are hollow cylinders 25 nanometers (nm) in diameter. Their lengths vary and may be quite long within some nerve processes. MT cylinder walls are hexagonal lattices of tubulin subunit proteins—polar, 8 nm dimers which consist of two slightly different 4 nm monomers (alpha and beta tubulin). Tubulin dimers are dipoles, with surplus negative charges localized toward alpha monomers (DeBrabander, 1982); MTs are thus predicted to have ('electret') ferroelectric and piezoelectric behaviours (Tuszynski *et al.*, 1995).

Traditionally viewed as the cell's 'bone-like' scaffolding, MTs and other cytoskeletal structures also appear to fill communicative and information processing roles. Numerous types of studies link the cytoskeleton to cognitive processes (for reviews, cf. Dayhoff *et al.*, 1994; Hameroff and Penrose, 1996a). Theoretical models and simulations suggest conformational states of tubulins within MT lattices are influenced by quantum events, and can interact with neighbouring tubulins to represent, propagate and process information as in molecular-level 'cellular automata', or 'spin-glass' type computing systems (e.g. Hameroff and Watt, 1982; Rasmussen *et al.*, 1990; Tuszynski *et al.*, 1995). There is some suggestion that quantum coherence could be involved in MT computation (Jibu *et al.*, 1994; Hameroff, 1994). In the Orch OR model, quantum coherent computing occurs in MTs in a pre-conscious mode, and (with isolation) continues until an OR threshold for self-collapse is reached (Figure 5). The self-collapse which

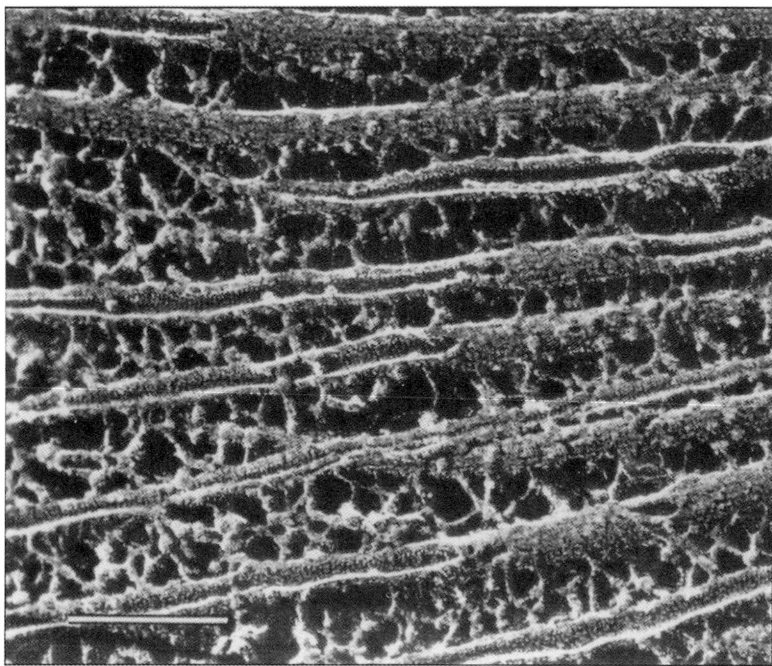

FIG. 3. Electron micrograph of neuronal microtubules (MTs) interconnected by microtubule-associated-proteins (MAPs). Several MTs have been opened during preparation, showing hollow inner cores. Scale bar (lower left): 100 nanometers (nm). With permission, from Hirokawa (1991).

reconfigures funda-mental spacetime meets philosophical criteria for a conscious event (Whitehead's 'occasion of experience'). Can quantum coherence occur in MTs?

4.2 Fröhlich's biological coherence

Herbert Fröhlich, an early contributor to the understanding of superconductivity, also predicted quantum coherence in living cells (based on earlier work by Oliver Penrose and Lars Onsager, 1956). Fröhlich theorized that sets of protein dipoles in a common electromagnetic field (e.g. proteins within a polarized membrane, subunits within an electret polymer like microtubules) undergo coherent conformational excitations when sufficient energy is supplied. Fröhlich postulated that biochemical and thermal energy from the surrounding cytoplasmic 'heat bath' provides such energy. Cooperative, organized processes leading to coherent excitations emerged, according to Fröhlich, because of structural coherence of dipoles in a common voltage gradient.

Coherent excitation frequencies on the order of 10^9 to 10^{11} Hz (identical to the time domain for functional protein conformational changes, and in the microwave or gigaHz spectral region) were deduced by Fröhlich who termed them acousto-conformational transitions, or coherent (pumped) phonons. Such

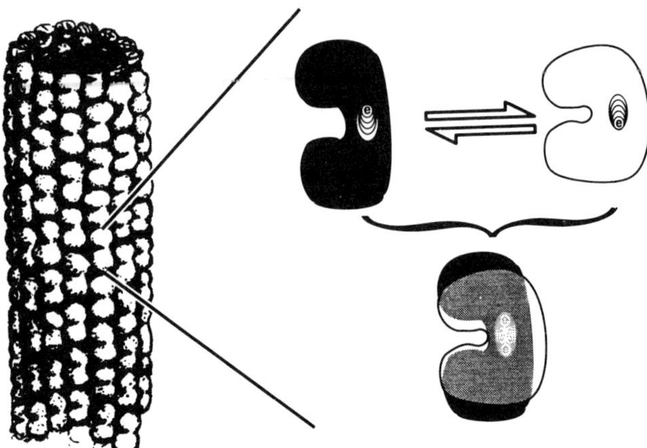

FIG. 4. Left: Microtubule (MT) structure: a hollow tube of 25 nanometers diameter, consisting of 13 columns of tubulin dimers arranged in hexagonal lattice (Penrose, 1994). Right (top): Each tubulin molecule can switch between two (or more) conformations, coupled to a quantum event such as electron location in tubulin hydrophobic pocket. Right (bottom): each tubulin can also exist in quantum superposition of both conformational states (Hameroff and Penrose, 1996).

coherent states are termed Bose-Einstein condensates in quantum physics and have been suggested by Marshall (1989) to provide macroscopic quantum states which support the unitary binding of consciousness.

Experimental evidence for Fröhlich-like coherent excitations in biological systems includes observation of gigaHz-range phonons in proteins (Genberg *et al.*, 1991), sharp-resonant non-thermal effects of microwave irradiation on living cells (Grundler and Keilman, 1983), gigaHz induced activation of microtubule pinocytosis in rat brain (Neubauer *et al.*, 1990), and laser Raman spectroscopy detection of Fröhlich frequency energy (Genzel *et al.*, 1983; Vos *et al.*, 1992). Can coherent MT excitations result in macroscopic quantum states?

4.3 Quantum isolation—avoiding environmental interaction and decoherence

Certain properties suggest that quantum effects are important in individual proteins (e.g. Roitberg *et al.*, 1995; Tejada *et al.*, 1996). For quantum coherence to occur in and among MTs, quantum states must be at least transiently isolated from decoherence by the brain's apparently noisy thermal environment. Feasibility of quantum coherence in the internal cell environment is supported by the observation that quantum spins from biochemical radical pairs which become separated retain their correlation in cytoplasm (Walleczek, 1995).

MTs are embedded in cytoplasm, which exists in alternating phases of 'sol' (solution, or liquid), and 'gel' (gelatinous phase of various sorts). Transition between sol and gel phases depends on actin polymerization. Actin co-polymer-

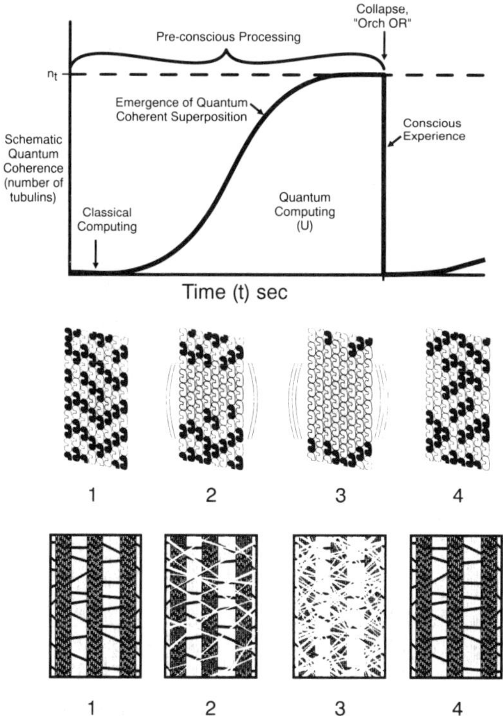

FIG. 5. Steps in an Orch OR event. Top: Schematic graph of proposed quantum coherence (number of tubulins) emerging vs time in microtubules. Area under curve connects mass-energy differences with collapse time in accordance with gravitational OR ($E = \hbar/T$). n_t is the number of tubulins whose mass separation (and separation of underlying spacetime) for time T will self-collapse. For example, for time $T = 25$ msec (e.g. 40 Hz oscillations), $T = 2 \times 10^{10}$ tubulins. Middle: Microtubule simulation in which classical computing (step 1) leads to emergence of quantum coherent superposition (and quantum computing—steps 2–3) in certain (grey) tubulins. Step 3 (in coherence with other microtubule tubulins) meets critical threshold related to quantum gravity for self-collapse (Orch OR). A conscious event (Orch OR) occurs in the step 3 to 4 transition. Tubulin states in step 4 are non-computably chosen in the collapse, and evolve by classical computing to regulate neural function. Bottom: Schematic sequence of phases of actin gelation (quantum isolation) and solution (environmental communication) around MTs.

izes with different types of 'actin cross-linking proteins' to form microfilaments and various types of gels (Figures 6 and 7). Specific actin gel characteristics are determined by particular actin cross-linkers. Of the various types of gels, some are viscoelastic, but others (e.g. those induced by the actin cross-linker

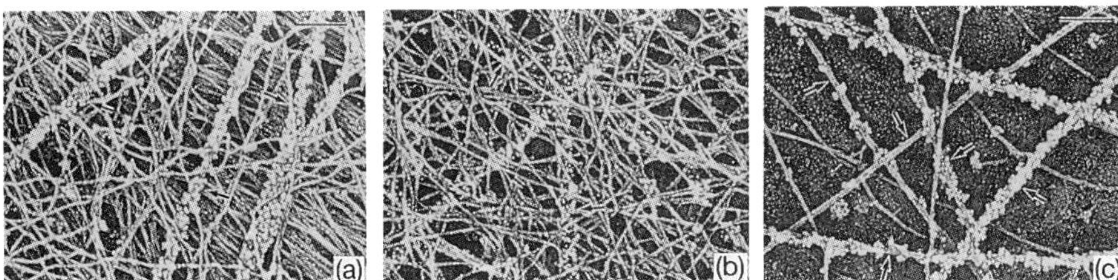

FIG. 6. Microtubules (arrows) immunogold labelled with antibody to detyrosinated tubulin. In (a) and (b), actin filaments surround microtubules. In (c), gelsolin has dissolved actin and MTs are more clearly revealed. MTs are labelled to different levels corresponding to the extent of tubulin detyrosination (a 'post-translational modification'). Scale bars: 200 nm. With permission from Svitkina *et al.* (1995).

avidin) are solid and can be deformed by an applied force without any response (Wachsstock *et al.*, 1994). Quantum states in MTs surrounded by this type of gel may be isolated from environmental decoherence.

Gels depolymerize back to liquid phase by calcium ions acting on the enzyme gelsolin which severs actin. Nuclear magnetic resonance studies have shown that actin gelation binds and orders cell water (e.g. up to 55%—Pauser *et al.*, 1995). Cycles of actin gelation/solution are essential to cytoplasmic rearrangements such as amoeboid movements, mitosis and cleavage, neurite growth and synaptic formation. Some rearrangements can be quite rapid: Miyamoto (1995) and Muallem *et al.* (1995) have shown that cycles of actin gelation/solution correlate with each release of neurotransmitter vesicles from pre-synaptic axon terminals. Cycles of actin gelation/solution in dendrites, for example, could provide alternating phases of a) gelation (environmental isolation/MT quantum coherence/OR) with b) solution (environmental communication of post-OR outcome states). If quantum coherence can occur in MTs within individual neurons, how might MTs in many neurons be involved in macroscopic quantum coherence?

4.4 Macroscopic quantum coherence and gap junctions

Consciousness must surely involve neural activity in many neurons throughout large volumes of the brain. If actin gelation can isolate MT quantum coherence within individual neurons (and glia), how could macroscopic quantum coherence occur among MTs in different and distant neurons (and glia)? How could active synapses and membranes be traversed without environmental interaction and

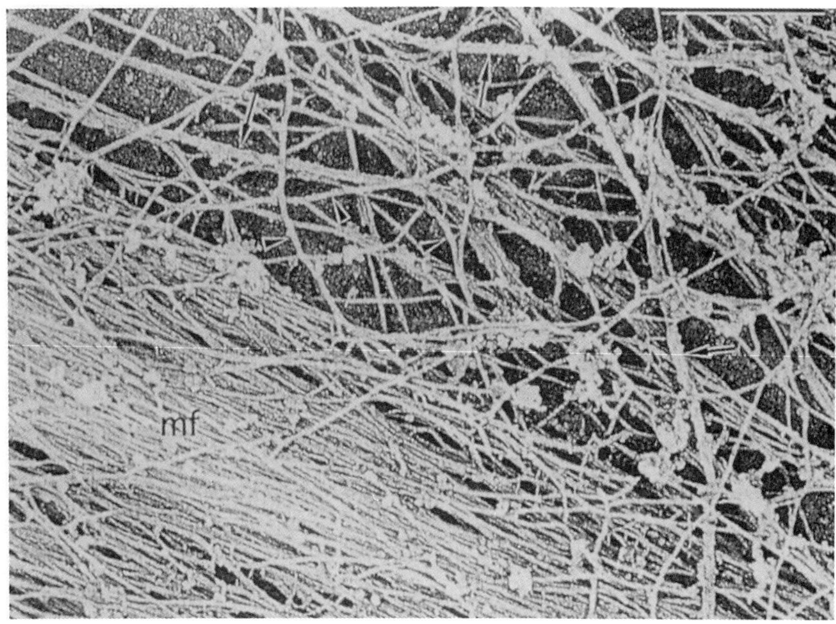

FIG. 7. Electron micrograph of rat embryo fibroblast showing all cytoskeletal elements: actin microfilaments (mf), microtubules (arrows), intermediate filaments (arrowheads). Dense actin microfilament gel (lower left) completely obscures microtubules. Scale bar (upper right): 500 nm. With permission from Svitkina *et al.* (1995).

decoherence?

There are two types of brain synapses: 1) chemical (neurotransmitter), and 2) electrical, or electrotonic synapses. Electrical synapses (about 15% of all synapses) are now recognized to be largely 'gap junctions', membrane pores which connect with similar pores in an adjacent neuron's membrane. Cell cytoplasm is continuous between the two gap junction-connected neurons so that ions, cytoplasm, injected dye and current flows freely between the cells. A group of neurons interconnected by gap junctions forms a network which 'fires synchronously, behaving like one giant neuron' (Kandel *et al.*, 1991). Such 'electrotonic' networks have been demonstrated in mammalian cortex, hippocampus, hypothalamus and other brain areas (MacVicar and Dudek, 1981; Jaslove and Brink, 1987). Unlike chemical synapses which separate neural processes by 30-50 nanometres, gap junction separations are 3.5 nanometres, within range for quantum tunneling. Jibu (1990) proposed that gap junctions/electrical synapses account for synchronized 40 Hz neural activity. Such activity has been suggested to provide 'binding' in classical consciousness models (e.g. Crick and Koch, 1990). Networks of gap junction-connected neurons with actin gelation/solution phases

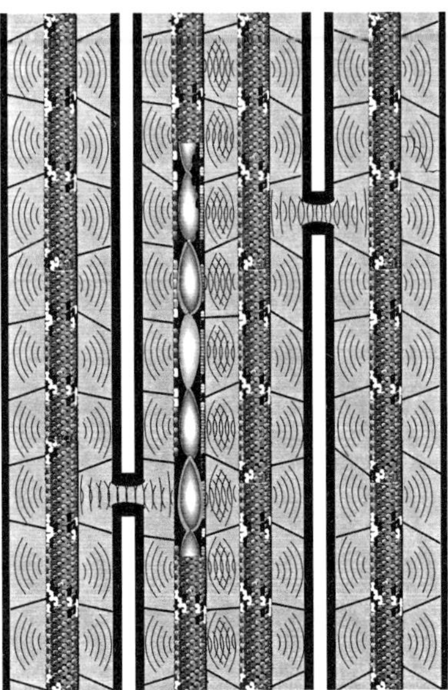

FIG. 8. Schematic quantum coherence in microtubules in three synchronized dendrites interconnected by tunnelling through gap junctions. Within each neuronal dendrite, MAP (microtubule-associated-protein) attachments breach isolation and prevent quantum coherence; MAP attachment sites thus act as 'nodes' which tune and orchestrate quantum oscillations and set possibilities and probabilities for collapse outcomes ('orchestrated objective reduction: Orch OR'). Gap junctions enable quantum tunnelling among dendrites in macroscopic quantum coherent state.

coupled to synchronized 40 Hz activity could isolate MTs across large brain volumes and provide cycles of macroscopic quantum coherence (Figure 8).

4.5 Evolution, Orch OR and the Cambrian explosion

Did consciousness emerge recently in evolution, for example in humans, or in primates with language, tool making, or other purposeful behaviour? Or did consciousness appear earlier, in simpler life forms? Perhaps primitive consciousness is a property of the living state—accompanying even unicellular organisms (e.g. Margulis and Sagan, 1995)? Whenever it appeared, would the emergence of consciousness have altered the course of evolution?

According to fossil records, life on earth originated about 4 billion years ago. For its first 3.5 billion years or so (pre-Cambrian period), life on earth seems to have evolved slowly, producing only single cells and a few simple multicellular organisms. During this pre-Cambrian period about 1.5 billion years ago, eukaryotic cells appeared—apparently as a symbiotic merger of organelles (mitochondria, plastids, microtubules) with bacterial cells (MTs and the dynamically functional cytoskeleton apparently originated as motile spirochetes—Margulis, 1975; Margulis and Sagan, 1995). Pre-Cambrian eukaryotic cells continued to evolve slowly for another billion or so years until a relatively brief 10 million years beginning about 540 million years ago (the Cambrian period). A dramatic acceleration in the rate of evolution apparently then occurred—the Cambrian explosion. There emerged a vast array of diversified life—all the phyla from which today's animals are derived (e.g. Gould, 1989).[1]

Did a critical level of biological complexity precipitate the Cambrian explosion (e.g. Kauffman, 1995)? If so, can a particular critical factor be identified? Cytoskeletal dynamics greatly enhanced eukaryotic cell adaptability by providing movement, internal organization, separation of chromosomes and numerous other functions. As cells became more specialized with MT-based extensions like cilia, axonemes and eventually neural processes, increasingly larger cytoskeletal arrays provided sensory perception, locomotion and intelligent behaviour. Critical cytoskeletal complexity and cooperativity could have accelerated evolution and precipitated the Cambrian explosion—even without consciousness.

However, as Margulis and Sagan (1995) observe (echoing similar, earlier thoughts by Erwin Schrödinger): 'If we grant our ancestors even a tiny fraction of the free will, consciousness, and culture we humans experience, the increase in [life's] complexity on Earth over the last several thousand million years becomes easier to explain: life is the product not only of blind physical forces but also of selection in the sense that organisms choose ...' (Scott, 1996). By itself, the ability to make choices is insufficient evidence for consciousness (e.g. computers can choose intelligently). However conscious experience (of surprise, pleasure, fear—Margulis and Sagan 1995) and non-computable, seemingly random conscious choices with an element of unpredictability may have been particularly advantageous for survival (e.g. Barinaga, 1996).

Did consciousness catalyze the Cambrian explosion? Conditions for conscious events according to the Orch OR model (Section 5) apparently were present in early Cambrian animals. Quantum coherence may have ensued in MT assemblies via the Fröhlich mechanism as a by-product of coordinated dynamics and biochemical energy. Development of actin gel isolation and larger geometric cytoskeletal arrangements within cells, and gap junction quantum tunnelling connections among cells would have, at some point, reached sufficient quantum coherence to elicit OR. Rudimentary 'conscious' events would have then occurred, presumably resulting in experience, choice, and perhaps enhanced survivability resulting in the Cambrian evolutionary explosion.

Here three scenarios (consistent with the Orch OR model) for early Cambrian emergence of consciousness are considered: 1) sufficiently complex gap junction-

[1]The Cambrian explosion theory has been questioned (e.g. Wray *et al.*, 1996) by fossil nucleotide substitution analysis suggesting a more linear process, with animals appearing about one billion years ago. But the more gradual, linear case assumes a constant rate of nucleotide substitution. It seems more likely (Vermeij, 1996) that the nucleotide substitution rate also increases during increased rates of evolution, and the Cambrian explosion theory still holds.

connected neural networks (hundreds of neurons, 10^9 tubulins—e.g. small worms such as *C elegans*). 2) primitive vision (ciliated ectoderm eye cup, 'cephalic organ'—e.g. annelid worms) 3) geometrical microtubule arrays (e.g. axonemes in small sea urchins such as *actinosphaerium*)

Many early Cambrian organisms preserved in fossil form are small worms with simple nervous systems. The Orch OR model (unlike other approaches) is able to suggest a critical degree of neural network complexity for emergence of consciousness. In Hameroff and Penrose (1996b) we speculated that among current organisms, the threshold for rudimentary Orch OR conscious events (500 msec pre-conscious time) may be very roughly at the level of 300 neurons 3×10^9 neural tubulin nematode worms such as *C elegans*. This is the same degree of neural network complexity as in early Cambrian worms. With optimal conditions (e.g. actin gel isolation, inter-MT geometry, biochemical energy) longer pre-conscious times could be sustained, and primitive conscious events have occurred in smaller, simpler organisms.

Another candidate for the Cambrian emergence of Orch OR consciousness involves the evolution of visual photoreceptors. Primitive unicellular organisms such as amoeba respond to light by a diffuse sol-gel alteration of their cytoplasm/cytoskeleton (Cronly-Dillon and Gregory, 1991). Other single cell organisms such as *euglena* and *erythrodynium* have localized 'eye spots'— e.g. regions at the root of the microtubule-based flagellum. Cytoplasm may focus light toward the eye spots, and pigment shield material located there suggests a mechanism for directional light detection (e.g. Insinna 1998). These and other single cell organisms respond to light behaviourally, despite the lack of neurons and synapses.

Even mammalian cells (such as ours) respond to light. Albrecht-Buehler (e.g. 1994) showed that single fibroblast cells move toward red/infra-red light by utilizing their MT-based centrioles for directional detection and guidance. (Albrecht-Buehler points out that centrioles—mega-cylinders of nine MT doublets— appear to be well designed light detectors and waveguides.) Jibu and Yasue (1995; c.f. Jibu *et al.*, 1997) have suggested quantum optical waveguide effects in MT cores, and that efficient photon propagation through cytoplasm requires a quantum state of ordered water. Hagan (1995) has proposed that quantum effects/cellular vision provided an evolutionary advantage for cytoskeletal arrays capable of quantum coherence.

Photoreceptors such as the rods and cones in our retinas are similar to *euglena* and fibroblasts—all relying on MTs in cilia and centrioles. Evolutionary development of vision began in simple multicellular organisms, where some ciliated cells became specialized (like centrioles and flagella, cilia are MT mega-cylinders). Light-sensitive ciliated cells formed primitive eyes (the first example is possibly the 'cephalic organ' in nemertine worms). The next stage in complexity is the simple 'eye cup' (up to 100 photoreceptor cells) found in many phyla including flatworms, annelid worms, molluscs, crustacea, echinoderms and chordates (our original evolutionary branch—Cronly–Dillon and Gregory, 1991). The retinas in our eyes today include over 10^8 photoreceptors (10^8 rods, 10^6 cones—Leibovic, 1990) comprised of an inner and outer segment connected by a ciliated stalk. As each cilium is comprised of about 300,000 tubulins, the 10^8 rods and cones in our retinas contain about 3×10^{13} tubulins per eye. Conventional vision science assumes the cilium is purely structural, but perhaps the same centriole/cilium/flagella MT structure—which Albrecht-Buehler has analysed as an ideal directional photoreceptor—detects or guides photons in eye spots of single cells, primitive eye cups in early multicellular organisms, and rods and cones in our retinas? Quantum coherence leading to consciousness could have emerged in sheets of gap junction-connected ciliated cells in eye cups or cephalic organs of early Cambrian worms.

Perhaps consciousness occurred in even simpler organisms? Many Cambrian fossils similar to present day species have particularly interesting MT arrangements. For example *actinosphaerium* is a present-day echinoderm, a tiny sea-urchin heliozoan with about one hundred rigid protruding axonemes about 300 microns in length. Axoneme extensions which sense the environment and provide locomotion (and appear similar to those of Cambrian creatures) are each comprised of several hundred MTs (about 3×10^9 tubulins per axoneme—Roth *et al.*, 1970) interlinked in a double spiral (Figure 9).

Allison and Nunn (1968; c.f. Allison *et al.*, 1970) studied living *actinosphaerium* in the presence of the anesthetic gas halothane. They observed that the axoneme MTs disassembled in the presence of halothane (although at rather high concentrations) and proposed that anesthetic effects on neuronal MTs in the brain may cause loss of consciousness.[2] Axonemes and similar appendages of geometric MT arrays involved in sensing and movement in early Cambrian organisms comparable to present-day *actinosphaerium* are candidates for emergence of quantum coherence, OR and rudimentary conscious events.

Three possible scenarios for the emergence of biological OR in evolution have been described: 1) sufficiently large networks of gap junction-connected neurons containing MT networks (e.g. very roughly the 300 neuron *C elegans*, 2) MT-cilia in gap junction-linked visual photoreceptor cells, 3) geometric MTs arrayed in axonemes of simple organisms such as *actinosphaerium*. All three scenarios are consistent with emergence of conscious events during the Cambrian evolutionary explosion.

The Orch OR model synthesizes new elements from philosophy (funda-mental panexperientialism), physics (Penrose's objective reduction), and biology (quantum coherence in microtubules). The following section summarizes the Orch OR model.

---

[2]Subsequent studies (e.g. Saubermann and Gallagher, 1973) showed that at minimal anesthetic concentrations capable of causing loss of consciousness, neuronal MTs remained structurally intact. However effects of clinically relevant anesthetic concentrations on MT dynamical functions including the proposed quantum coherence remain unknown. Delon and Legendre (1995) and Whatley *et al.* (1994) have shown that the activities of GABAa and glycine receptors—whose functions are known to be altered by relevant anesthetic concentrations—are modulated by neuronal MTs (Franks and Lieb, 1998).

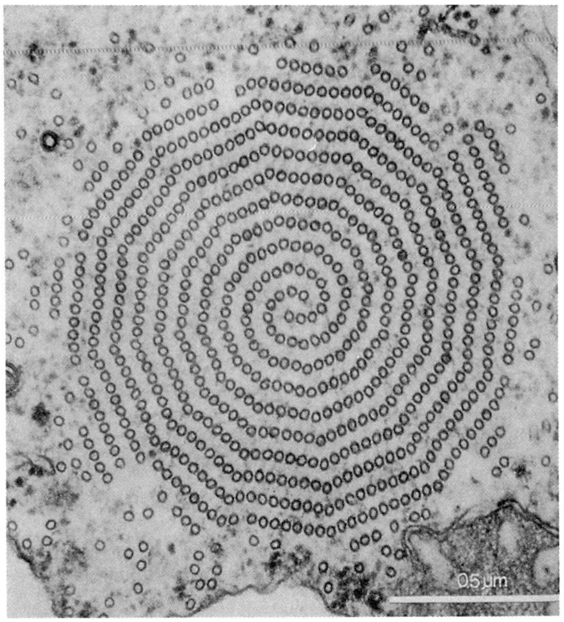

FIG. 9. Cross-section of double spiral array of interconnected MTs in axoneme of *actinosphaerium*. The heliozoan organism has about one hundred of these 300 micron long and rigid protruding axonemes, each comprised of several hundred inter-linked MTs (3×10^9 tubulins per axoneme). The axonemes are disassembled by anesthetics. This organism is similar to those which emerged during the Cambrian evolutionary explosion which began roughly 540 million years ago. Scale bar: 500 nm. (With kind permission from L.E. Roth.)

5 Summary of the 'Orch OR' model of consciousness

Full rationale and details of the Orch OR model are given in Penrose and Hameroff (1995) and Hameroff and Penrose (1996a, 1996b). An eleven point summary is presented here:

(1) Conformational states of tubulins in neuronal MTs are coupled to internal quantum events, and cooperatively interact with other MT tubulins in both classical and quantum computation (Hameroff *et al.*, 1992; Rasmussen *et al.*, 1990—Figures 4 and 5).

(2) Quantum coherent superposition occurs among MT tubulins, pumped by thermal and biochemical energy in a manner described by Fröhlich (1968, 1975). Actin gel states surround and isolate MT quantum coherence (Figures 5, 6 and 7).

(3) Quantum coherence in MTs in individual neuronal and glial cells are linked by gap junction mediated synchronization (e.g. 40 Hz—Jibu, 1990) providing macroscopic quantum coherence in networks of many gap junction-

connected cells (neurons and glia) throughout large brain volumes (Figure 8).

(4) Quantum coherent superposition/computation in neural (and glial) MTs corresponds to pre-conscious processing, which continues until the mass-distribution differences among the separated tubulins reaches a threshold related to quantum gravity $E = \hbar/T$. OR self-collapse then occurs (Figure 5). As an example, for $T = 25$ msec (i.e. 40 Hz), E is roughly the superposition/separation of 2×10^{10} tubulins.

(5) Each OR self-collapse selects (non-computably) particular outcome states of MT tubulins which implement neural functions including synaptic regulation. Particular experience in funda-mental spacetime geometry is also selected (Figure 2).

(6) Probabilities and possibilities for post-OR states are influenced by factors including attachments of MT-associated proteins ('MAPs'), which tune and orchestrate quantum oscillations. We thus term the self-tuning OR process in MTs 'orchestrated' objective reduction— Orch OR (Figure 8).

(7) Orch OR conscious events may be of variable intensity and duration. Calculating from $E = \hbar/T$, for a pre-conscious processing time of, for example, T=500 msec (e.g. shown by Libet *et al.*, 1979, to be one possible pre-conscious processing time), E is equivalent to 10^9 tubulins. Thus 10^9 tubulins in isolated quantum coherence for 500 msec will self-collapse (Orch OR) and constitute a conscious event (Whitehead 'occasion of experience'). As previously mentioned, for pre-conscious times T of, say, 25 msec (e.g. 40 Hz neural excitations) 2×10^{10} quantum superposed tubulins would be required. A more intense 5 msec event would involve 10^{11} tubulins.

(8) Each brain neuron is estimated to contain about 10^7 tubulins (Yu and Bass, 1994). If, say, 10% of each neuron's tubulins became coherent, then tubulins within roughly 1,000 neurons would be required for a 500 msec Orch OR, or 20,000 for a 25 msec Orch OR.

(9) Each instantaneous Orch OR event can bind various superpositions which have evolved in separated spatial distributions and over differing time scales, but whose net displacement reaches threshold at a particular moment: information is bound into a conscious now. Cascades of Orch ORs can then represent our familiar stream of consciousness.

(10) According to the arguments for OR put forth in Penrose (1994), superpositioned states each have their own spacetime geometry. When the degree of coherent mass-energy difference leads to sufficient separation of spacetime geometry, the system must choose and decay (reduce, collapse) to a single universe state. Thus Orch OR involves self-selections in fundamental spacetime geometry (Figure 2).

(11) Following a panexperiential view which has evolved from Spinoza and Leibniz to Whitehead, Wheeler and Chalmers, we assume qualia, or experience is fundamental to the universe, and is therefore constitutive of spacetime

geometry. The Penrose spin network model of Planck scale volume quantization (Rovelli and Smolin, 1995a, 1995b) is suggested here to be a possible makeup of physical reality suitable for containing qualia and raw experience ('quantum monads'). Each Orch OR event is a self-configuration of funda-mental geometry.

6 Assumptions and testable predictions of Orch OR

Detailed arguments and supportive evidence for the major assumptions in Orch OR are given in Penrose (1989, 1994, 1996), Penrose and Hameroff (1995), Dayhoff *et al.* (1993), Hameroff (1987, 1994), Hameroff and Penrose (1996a, 1996b). Here major assumptions (**bold**) and corresponding testable predictions of the model are listed.

Neuronal MTs are directly necessary for consciousness

Testable predictions:

(1) Synaptic sensitivity and plasticity correlate with cytoskeletal architecture/activities in both pre-synaptic and post-synaptic neuronal cytoplasm.

(2) Actions of psychoactive drugs including antidepressants involve neuronal MTs.

(3) Neuronal MT-stabilizing/protecting drugs may prove useful in Alzheimer's disease, ischemia, and other conditions.

MTs communicate by cooperative dynamics of tubulin subunits

Testable predictions:

(4) Laser spectroscopy (e.g. Vos *et al.*, 1993) will demonstrate coherent gigaHz Fröhlich excitations in MTs.

(5) Dynamic vibrational states in MT networks correlate with cellular activity.

(6) Stable patterns of MT-cytoskeletal networks (including neurofilaments) and intra-MT diversity of tubulin states correlate with memory and neural behaviour.

(7) Cortical dendrites contain largely 'A-lattice' MTs (compared to 'B-lattice' MTs, A-lattice MTs are preferable for information processing—Mandelkow and Mandelkow, 1994; Tuszynski *et al.*, 1995).

Quantum coherence occurs in microtubules

Testable predictions:

(8) Studies similar to the famous 'Aspect experiment' in physics (which verified non-local quantum correlations— Aspect *et al.*, 1982) will demonstrate quantum correlations between spatially separated MT subunit states a) on the same MT, b) on different MTs in the same neuron, c) different neurons connected by gap junctions (see below).

(9) Experiments with SQUIDs (Superconducting Quantum Interference Device) such as those suggested by Leggett (1984) will detect phases of quantum coherence in MTs.

MT quantum coherence requires isolation by cycles of surrounding actin gelation

Testable predictions:

(10) Neuronal microtubules in cortical dendrites and other brain areas are intermittently surrounded by tightly cross-linked actin gels.

(11) Cycles of gelation and dissolution in neuronal cytoplasm occur concomitantly with membrane electrical activity (e.g. synchronized 40 Hz activities in dendrites).

(12) The sol-gel cycles surrounding MTs are regulated by calcium ions released and reabsorbed by calmodulin associated with MTs.

Macroscopic quantum coherence occurs among MTs in hundreds/ thousands of distributed neurons and glia linked by gap junctions

Testable predictions:

(13) Electrotonic gap junctions link synchronously firing networks of cortical neurons.

(14) Quantum tunnelling occurs across gap junctions.

(15) Quantum correlation occurs between microtubule subunit states in different neurons connected by gap junctions (the MT 'Aspect experiment' in different neurons).

The amount of neural tissue involved in a conscious event is inversely proportional to the event time ($E = \hbar/T$)

Testable prediction:

(16) The amount of neural mass involved in a particular cognitive task or conscious event (as measurable by near-future advances in brain imaging techniques) is inversely proportional to the pre-conscious time (e.g. visual perception).

An isolated, unperturbed quantum system self-collapses according to ($E = \hbar/T$)

Testable prediction:

(17) Technological quantum superpositions (e.g. Bose-Einstein condensates and quantum computers) will self-collapse according to ($E = \hbar/T$).

MT-based cilia/centriole structures are quantum optical devices

Testable prediction:

(18) MT-based cilia in rods and cones directly detect visual photons and connect with retinal glial cell MTs via gap junctions.

A critical degree of cytoskeletal assembly (coinciding with the onset of rudimentary consciousness) had significant impact on the rate of evolution

Testable prediction:

(19) Fossil records and comparison with present-day biology will show that organisms which emerged during the early Cambrian period with onset roughly 540 million years ago had critical degrees of MT-cytoskeletal size, complexity and capability for quantum isolation (e.g. tight actin gels, gap junctions).

7 Conclusion: Penrose's Platonic world

In *Shadows of the Mind*, Roger Penrose described three worlds: the physical world, the mental world and the Platonic world. The physical world and the mental world are familiar and agreed upon as actual realities—clearly, the physical world exists and thoughts exist. However the Platonic world of mathematical truths, laws and relationships appears purely abstract. As Penrose suggests, the Platonic world may also describe aesthetics and ethics—our senses of beauty and morality. If mathematical truths, aesthetics and ethics are actual, perhaps (like experience) they are encoded at the Planck scale? Penrose has suggested the Weyl tensor component in spacetime curvature may provide a fine grain at Planck-scale spin networks influencing non-computable outcome states in objective reductions including Orch OR conscious events.

The trail of non-computability in human thought has led Roger Penrose to a collaborative model which portrays consciousness as a self-organizing quantum brain process reconfiguring Planck-scale spacetime geometry. It requires innovation in the following areas:

Philosophy A 'funda-mental' panexperiential view in which primitive experience exists at the Planck scale of spacetime geometry (e.g. spin networks).

Physics Penrose's 'objective reduction' in which quantum gravity mediates wave function self-collapse at a critical degree of spacetime separation.

Biology Quantum coherence and self-organized ('orchestrated') objective reduction (Orch OR) among brain microtubules (MTs) are isolated by actin gel states and connected to other brain neurons and glia by gap junctions to provide macroscopic coherence.

The Orch OR model is consistent with known neural and cognitive activities, and attempts to link brain functions with self-configurations in 'funda-mental' spacetime geometry. Orch OR addresses enigmatic features of consciousness (e.g. nature of experience, binding, pre-conscious to conscious transition, flow of time, free will and non-computability) and suggests an intimate link between consciousness and the universe.

Acknowledgements I thank the organizers for the opportunity to participate in the conference and book honouring Roger Penrose. I also thank Roger for the privilege of our collaboration (and for just being Roger). Although some of the newer ideas expressed here (actin gel isolation, gap junctions, spin networks, Weyl tensor, vision, evolution, etc.) come from my recollection of our discussions, they may not necessarily reflect Roger's view (especially if incorrect). For discussions regarding spin networks, thanks also to Lee Smolin who doesn't necessarily endorse their linkage to 'funda-mental' experience.

Bibliography

Albrecht-Buehler, G. (1994). Cellular infra-red detector appears to be contained in the centrosome. *Cell Motility and the Cytoskeleton*, **27**(3), 262–71.

Allison, A. C. and Nunn, J. F. (1968). Effects of general anesthetics on microtubules. A possible mechanism of anesthesia. *Lancet*, **2**, 1326–9.

Allison, A. C., Hulands, G. H., Nunn, J. F., Kitching, J. A., and MacDonald A. C. (1970). The effects of inhalational anaesthetics on the microtubular system in Actinosphaerium nucleofilum. *J. Cell Science*, **7**, 483–99.

Aspect, A., Grangier, P., and Roger, G. (1982). Experimental realization of Einstein-Podolsky-Rosen-Bohm Gedankenexperiment: a new violation of Bell's inequalities. *Physical Review Letters*, **48**, 91–4.

Barinaga, M. (1996). Neurons put the uncertainty into reaction times. *Science*, **274**, 344.

Chalmers, D. J. (1996a). *The Conscious Mind—In Search of a Fundamental Theory*. Oxford University Press, New York.

Chalmers, D. J. (1996b). Facing up to the problem of consciousness In: *Toward a Science of Consciousness—The First Tucson Discussions and Debates*, pp 5–28. Eds S. R. Hameroff, A. Kaszniak and A. C. Scott. MIT Press, Cambridge, MA.

Crick, F. and Koch, C. (1990). Towards a neurobiological theory of consciousness. *Seminars in the Neurosciences*, **2**, 263–75.

Cronly-Dillon J. R. and Gregory, R. L. (1991). *Evolution of the Eye and Visual System (Vision and Visual Dysfunction*, Vol 2*)*. CRC Press, Boca Raton.

Dayhoff, J. E., Hameroff, S., Lahoz-Beltra, R. and Swenberg, C. E. (1994). Cytoskeletal involvement in neuronal learning: a review. *Eur. Biophys. J.*, **23**, 79–93.

De Brabander, M. (1982). A model for the microtubule organizing activity of the centrosomes and kinetochores in mammalian cells. *Cell Biol. Intern. Rep.*, **6**, 901–15.

Delon, J. and Legendre, P. (1995). Effects of nocodazole and taxol on glycine evoked currents on rat spinal-cord neurons in culture. *Neuroreport* **6**, 1932–6.

Dennett, D. C. (1991). *Consciousness Explained.* Little Brown, Boston.

Farleigh, P. (1998). Whitehead's even more dangerous idea. In: *Toward a Science of Consciousness II—The Second Tucson Discussions and Debates.* Eds S. Hameroff, A. Kaszniak, and A. Scott. MIT Press, Cambridge, MA (in press).

Franks, N. and Lieb, W.R. (1998). On the molecular mechanism of general anesthesia. In: *Toward a Science of Consciousness II—The Second Tucson Discussions and Debates*. Eds S. Hameroff, A. Kaszniak, and A. Scott. MIT Press, Cambridge, MA (in press).

Fröhlich, H. (1968). Long-range coherence and energy storage in biological systems. *Int. J. Quantum Chem.*, **2**, 641–9.

Fröhlich, H. (1970). Long range coherence and the actions of enzymes. *Nature*, **228**, 1093.

Fröhlich, H. (1975). The extraordinary dielectric properties of biological materials and the action of enzymes. *Proc. Natl. Acad. Sci.*, **72**, 4211–15.

Genberg, L., Richard, L., McLendon, G., and Dwayne-Miller, R. J. (1991). Direct observation of global protein motion in hemoglobin and myoglobin on picosecond time scales. *Science*, **251**, 1051–4.

Genzel, L., Kremer, F., Poglitsch, A., and Bechtold, G. (1983). Relaxation processes on a picosecond time scale in hemoglobin and polylysine observed by millimeter-wave spectroscopy. *Biopolymers*, **22**, 1715–29.

Gould, S. J. (1989). *Wonderful Life—The Burgess Shale and the Nature of History*. W.W. Norton, New York.

Griffin, D. R. (1988). *Of Minds and Molecules, in the Reenchantment of Science*. Ed. D. R. Griffin. SUNY Press, Albany New York.

Grundler, W. and Keilmann, F. (1983). Sharp resonances in yeast growth prove nonthermal sensitivity to microwaves. *Physical Reveiw Letters*, **51**, 1214–16.

Hagan, S. (1995). Personal communication.

Hameroff, S. R. (1987). *Ultimate computing: Biomolecular Consciousness and Nanotechnology*. Elsevier-North Holland, Amsterdam.

Hameroff, S. R. (1994). Quantum coherence microtubules: a neural basis for emerging consciousness. *Journal of Consciousness Studies* **1**(1) 91–118.

Hameroff, S. R. and Penrose, R. (1996a). Orchestrated reduction of quantum coherence in brain microtubules: A model for consciousness. In: *Toward a Science of Consciousness—The First Tucson Discussions and Debates*. Eds S. R. Hameroff, A. Kaszniak and A. C. Scott. MIT Press, Cambridge, MA. Also published in *Mathematics and Computers in Simulation* **40**, 453-480.

Hameroff, S. R. and Penrose, R. (1996b). Conscious events as orchestrated spacetime selections. *Journal of Consciousness Studies*, **3**(1), 36–53.

Hameroff S. R. and Watt R. C. (1982). Information processing in microtubules. *Journal of Theoretical Biology*, **98**, 549–61.

Hameroff S. R., Dayhoff, V. E., Lahoz-Beltra, R., Samsonovich, A., and Rasmusssen, S. (1992). Conformational automata in the sytoskeleton: models for molecular computation. IEEE Computer (October Special Issue on Molecular Comp.) pp 30–39.

Hirokawa, N. (1991). Molecular architecture and dynamics of the neuronal cytoskeleton. In: *The Neuronal Cytoskeleton*, pp 5–74. Ed. R. D. Burgoyne. Wiley-Liss, New York.

Insinna, E. (1998). Co-evolution of vision and consciousness. In: *Toward a Science of Consciousness II—The Second Tucson Discussions and Debates.* Eds S. Hameroff, A. Kaszniak, and A. Scott. MIT Press, Cambridge, MA (in press).

Jaslove, S. W. and Brink, P. R. (1987). Electrotonic coupling in the nervous system. In: *Cell-to-Cell Communication.* Ed. W. C. De Mello. Plenum Press, New York.

Jibu, M. (1990). On a heuristic model of the coherent mechanism of the global reaction process of a group of cells. *Bussei Kenkyuu* (Material Physics Research), **53**(4), 431–6 (in Japanese).

Jibu, M. and Yasue, K. (1995). *Quantum Brain Dynamics: An Introduction.* John Benjamins, Amsterdam.

Jibu, M., Hagan, S., Hameroff, S. R., Pribram, K. H., and Yasue, K. (1994). Quantum optical coherence in cytoskeletal microtubules: implications for brain function. *BioSystems*, **32**, 195–209.

Jibu, M., Yasue, K., and Hagan, S. (1997). *Water Laser Effect in Cellular 'Vision'* (submitted).

Károlyházy, F. (1966). Gravitation and quantum mechanics of macroscopic bodies. *Nuo. Cim. A*, **42**, 390–402.

Károlyházy, F., Frenkel, A., and Lukacs, B. (1986). On the possible role of gravity on the reduction of the wave function. In: *Quantum Concepts in Space and Time.* Eds R. Penrose and C. J. Isham. Oxford University Press, Oxford.

Kandel, E. R., Siegelbaum S. A., and Schwartz, J. H. (1991). Synaptic transmission. In: *Principles of Neural Science* (3rd edn), pp 121–34. Eds E. R. Kandel, J. H. Schwartz, T. M. Jessell. Elsevier, New York.

Kauffman, S. (1995). *At Home in the Universe.* Oxford Press, New York.

Leggett, A. J. (1984). Schrödinger's cat and her laboratory cousins. *Contemp. Physics*, **25**(6), 583.

Leibniz, G. W. (1768). *Opera Omnia*, 6 volumes. Ed. Louis Dutens. Geneva.

Leibovic, K. N. (1990). *Science of Vision.* Springer-Verlag, New York.

Libet, B., Wright, E.W., Jr., Feinstein, B., and Pearl, D. K. (1979). Subjective referral of the timing for a conscious sensory experience. *Brain*, **102**, 193–224.

MacVicar, B. A. and Dudek, F. E. (1981). Electrotonic coupling between pyramidal cells: a direct demonstration in rat hippocampal slices. *Science*, **213**, 782–5.

Mandelkow, E. and Mandelkow, E.-M. (1994). Microtubule structure. *Curr. Opinions Structural Biology*, **4**, 171–9.

Margulis, L. (1975). *Origin of Eukaryotic Cells.* Yale University Press, New Haven.

Margulis, L. and Sagan, D. (1995). *What is Life?* Simon and Schuster, New York.

Marshal, I. N. (1989) Consciousness and Bose–Einstein condensates. *New Ideas in Psychology* **7**, 73–83.

Miyamoto, S. (1995). Changes in mobility of synaptic vesicles with assembly and disassembly of actin network. *Biochimica et Biophysica Acta*, **1244**, 85–91.

Monroe, C., Meekhoff, D. M., King, B. E., and Wineland, D. J. (1996). A 'Schrödinger cat' superposition of an atom. *Science*, **272**, 1131–5.

Muallem, S., Kwiatkowska, K., Xu, X., and Yin, H. L. (1995). Actin filament disassembly is a sufficient final trigger for exocytosis in nonexcitable cells. *Journal of Cell Biology*, **128**, 589–98.

Neubauer, C., Phelan, A. M., Keus, H., and Lange, D. G. (1990). Microwave irradiation of rats at 2.45 GHz activates pinocytotic-like uptake of tracer by capillary endothelial cells of cerebral cortex. *Bioelectromagnetics*, **11**, 261–8.

Pauser, S., Zschunke, A., Khuen, A., and Keller, K. (1995). Estimation of water content and water mobility in the nucleus and cytoplasm of Xenopus lacvis vortex by NMR microscopy Magentic Resonance Imaging **13**(2) 269–276.

Penrose, R. (1971). In: *Quantum Theory and Beyond.* Ed. E. A. Bastin. Cambridge University Press, Cambridge.

Penrose, R. (1989). *The Emperor's New Mind.* Oxford University Press, Oxford.

Penrose, R. (1994). *Shadows of the Mind.* Oxford Univeristy Press, Oxford.

Penrose, R. (1996). On gravity's role in quantum state reduction. *General Relativity and Gravitation*, **28**(5), 581–600.

Penrose, R. and Hameroff, S. R. (1995). What gaps? Reply to Grush and Churchland. *Journal of Consciousness Studies*, **2**(2), 99–112.

Penrose, O. and Onsager, L. (1956). Bose-Einstein condensation and liquid helium. *Physical Review*, **104**, 576–84.

Rasmussen, S., Karampurwala, H., Vaidyanath, R., Jensen, K. S., and Hameroff, S. (1990). Computational connectionism within neurons: A model of cytoskeletal automata subserving neural networks. *Physica D*, **42**, 428–49.

Roitberg, A., Gerber, R. B., Elber, R., and Ratner, M. A. (1995). Anharmonic wave functions of proteins: quantum self-consistent field calculations of BPTI. *Science*, **268** (5315), 1319–22.

Roth, L. E., Pihlaja, D. J., and Shigenaka, Y. (1970). Microtubules in the heliozoan axopodium. I. The gradion hypothesis of allosterism in structural proteins. *J. Ultrastr. Res.*, **30**, 7–37.

Rovelli, C. and Smolin, L. (1995a). Discreteness of area and volume in quantum gravity. *Nuclear Physics B*, **442**, 593–619.

Rovelli, C. and Smolin, L. (1995b). Spin networks in quantum gravity. Physical *Review D*, **52**(10), 5743–59.

Saubermann, A. J. and Gallagher, M. L. (1973). Mechanisms of general anesthesia: Failure of pentobarbital and halothane to depolymerize microtubules in mouse optic nerve. *Anesthesiology*, **38**, 25–9.

Scott, A. C. (1996). Book review: Lynn Margulis and Dorion Sagan 'What is life?'. *Journal of Consciousness Studies*, **3**(3), 286–7.

Shimony, A. (1993). *Search for a Naturalistic World View*, Vol. II. *Natural Science and Metaphysics.* Cambridge University Press, Cambridge.

Spinoza, B. (1677). *Ethica in Opera quotque reperta sunt* (3rd edn). Eds J. van Vloten and J. P. N. Land. The Hague.

Svitkina, T. M., Verkhovsky, A.B., and Borisy, G. B. (1995). Improved procedures for electron microscopic visualization of the cytoskeleton of cultured cells. *Journal of Structural Biology*, **115**, 290–303.

Tejada, J., Garg, A., Gider, S., Awschalom, D.D., DiVincenzo, D.P., and Loss, D. (1996). Does macroscopic quantum coherence occur in ferritin? *Science*, **272**, 424–6.

Tuszynski, J., Hameroff, S., Sataric, M. V., Trpisova, B., and Nip, M. L. A. (1995). Ferroelectric behavior in microtubule dipole lattices; implications for information processing, signaling and assembly/disassembly. *J. Theor. Biol.*, **174**, 371–80.

Vermeij, G. J. (1996). Animal origins. *Science*, **274**, 525–6.

Vos, M. H., Rappaport, J., Lambry, J. Ch., Breton, J., and Martin, J. L. (1993). Visualization of coherent nuclear motion in a membrane protein by femtosecond laser spectroscopy. *Nature*, **363**, 320–5.

Wachsstock, D. H., Schwarz, W. H., Pollard, T. D. (1994). Cross-linker dynamics determine the mechanical properties of actin gels. *Biophysical Journal*, **66**(3 Pt 1), 801–9.

Walleczek, J. (1995). Magnetokinetic effects on radical pairs: a possible paradigm for understanding sub-kT magnetic field interactions with biological systems. In: *Electromagnetic Fields: Biological Interactions and Mechanisms.* Ed. M. Blank. Advances in Chemistry, No 250. American Chemical Society Books, Washington DC.

Whatley, V. J., Mihic, S. J., Allan, A. M., McQuikin, S. J., and Harris, R. A. (1994). Gamma-aminobutyric acida receptor function is inhibited by microtubule depolymerization. *J. Biol. Chem.*, **269**, 19546–52.

Wheeler, J. A. (1957). Assessment of Everett's 'relative state' formulation of quantum theory. *Rev. Mod. Phys.*, **29**, 463–5.

Wheeler, J. A. (1990). Information, physics, quantum: The search for links. In: *Complexity, Entropy, and the Physics of Information.* Ed. W. Zurek. Addison-Wesley.

Whitehead, A. N. (1926). *Science and the Modern World* (2nd edn). Macmillan, New York.

Whitehead, A. N. (1929). *Science and the Modern World.* Macmillan, New York.

Whitehead, A. N. (1933). *Process and Reality.* Macmillan, New York.

Whitehead, A. N. (1978). *Process and Reality* (Corrected—1929). Macmillan, New York.

Wray, G. A., Levinton, J. S., and Shapiro, L. H. (1996). Molecular evidence for deep precambrian divergences among metazoan phyla. *Science*, **274**, 568-73.

Yu, W. and Bass, P. W. (1994). Changes in microtubule number and length during axon differentiation. *J. Neuroscience*, **14**(5), 2818–29.

10

Implications of Transience for Spacetime Structure

Abner Shimony
Professor Emeritus of Philosophy and Physics
Boston University, Boston, Massachusetts 02215

Abstract

Answers are outlined to arguments of McTaggart and Barbour that temporal transience is illusory or derivative. Eight propositions concerning transience are distilled from Whitehead's process philosophy. The question is posed: what theories of spacetime structure can implement these propositions? Some conjectures are made concerning this question, especially stochastic modifications of the Schroedinger equation. Agreement is expressed with the thesis of process philosophers that a world in which there is transience and mentality is a locus of value.

In the early part of the century two eminent Oxonians made penetrating, and diametrically opposite comments on transience. One was the philosopher J.M.E. McTaggart, who reasons (McTaggart 1927, vol. 2, p. 20) as follows:

> "Past, present, and future are incompatible determinations. Every event must be one or the other, but no event can be more than one. If I say that any event is past, that implies that it is neither present nor future, and so with the others. And this exclusiveness is essential to change, and therefore to time. For the only change we can get is from future to present, and from present to past.
> The characteristics, therefore, are incompatible. But every event has them all. If M is past, it has been present and future. If it is future, it will be present and past. If it is present, it has been future and will be past. Thus all the three characteristics belong to each event. How is this consistent with their being incompatible?"

This argument focuses upon the inconsistency, hence unreality, of transience, but it purports to establish yet more—namely, the unreality of change and time—in view of the brief remark that transience from futurity to presentness to pastness is essential to change and therefore to time.

The second commentator on transience was Max Beerbohm, though not in his own voice but in that of a character, Savonarola, in a play by the fictional playwright Ladbroke Brown. In Act III Savonarola is arrested by Pope Julius

II, and in the following Act he broods in a monologue over the humiliation of imprisonment:

> "What life, when I come out,
> Awaits me? Why, the very Novices
> And callow Postulants will draw aside
> As I pass by, and say 'That man hath done
> Time!' And yet shall I wince? The worst of Time
> Is not in having done it, but in doing't" (Beerbohm 1919, p. 209).

What Beerbohm is saying, through the medium of Savonarola, is that time is real, painfully real, precisely because it exhibits transience—in the case of a period of imprisonment the excruciatingly slow passage of present moments into the past.

My first aim in this lecture is to present arguments showing why McTaggart was wrong and Beerbohm was right. The core of the arguments have been presented elsewhere (Shimony 1993), but some modifications will be made here in order to address the ingenious deconstruction of time by Julian Barbour. The second aim is to explore the possibility of finding a theory of space and time which can accommodate transiency in a natural manner. I am much less confident about accomplishing the second aim than the first, and it is my hope to elicit assistance in this project from the experts gathered here.

The essential flaw in McTaggart's argument is his treatment of pastness, presentness, and futurity as "determinations" of an event, that is to say, as properties, in the way that "having electric field strength E" is a property of the event. According to classical electromagnetic theory (setting aside quantum uncertainties) the property of having a specified electric field strength can either be correctly ascribed to an event or cannot be, and whichever is the case is so for all time. The subject-predicate form shared by the two sentences "Event M has electric field strength E" and "Event M is present" is deceptive, however, and a careful semantical analysis shows a profound difference between them. The classical exposition of the deep difference between these two sentences was made by Kant (1781, A592–A602) in the course of his refutation of Anselm's "ontological" argument for the existence of God. That argument maintains that the very concept of a perfect being ensures the existence of such a being, because existence is a component of the complex property of perfection. Disregarding difficulties inherent in the ill-defined property of perfection, Kant gets to the heart of the matter by pointing out that existence is not a property at all, and hence could not be a component of the property of perfection, however that property is clarified. Kant points out, in a homely way, that the existence of a hundred thalers makes a difference to his financial condition but does not add anything to the concept of a hundred thalers. Kant's refutation of Anselm is readily adapted to McTaggart's argument (Favrholdt 1984; Shimony 1993). Whatever properties are correctly attributable to an event, or similarly to a spacetime point, apply timelessly; e.g., that the electric field strength at the point p is E is true always or is false always, and the fact that the point p is present (momentarily) in no way modifies the electric field strength or any other property. The futurity of p (when it is future) is different from the presentness of p (when it is present) and both are different from the pastness of p (when it is past), but these differences are "existential" and in no way "qualitative." Kant taught us, long before the flowering of linguistic philosophy, to beware of the superficial grammatical similarity of the sentences "Event E is blue" and "Event E is present."

The refutation of McTaggart in a Kantian manner does not, however, diminish the force of various objections to the thesis that transience is an objective physical feature of the world. These objections grant that the subjective feeling of transience is deeply embedded in our psychology and that locutions of transience are deeply embedded in tensed grammar, but they propose to derive or construct the appearance of transience from ingredients that are intrinsically free from transience. I previously (Shimony 1993, pp. 276–9) analysed a derivation of this genre by Grünbaum (1971), but now shall consider some proposals by Julian Barbour (1994), which are especially interesting at this gathering because they are part of a serious program of rethinking general relativity theory.

Barbour interprets both classical physics and quantum physics timelessly. I shall explicitly discuss only the former, which is conceptually simpler, but my objections against his thesis of timelessness apply to both. What a classical physical treatment of the entire world offers, in principle, is a curve in relative configuration space. For Machian reasons the center of mass coordinates of the entire world are meaningless. More important, classical physics provides only a curve, not a time-parametrized trajectory, since there is no clock external to the entire world which would bestow physical meaning upon the time parameter. Subsystems of the world can indeed be described by time-parametrized trajectories, because other subsystems external to the subsystem of interest can serve as clocks. In no case is time a basic concept, because its role can always be supplied by conceptual ingredients available in a timeless world.

> "I first introduce the heap hypothesis. There are two heaps: the heap of all possible configurations, the heap of possibilities, and the heap of realized configurations, the heap of actualities. ... what are our most basic theories ... actually telling us? The heap hypothesis is that these theories are simply rules to establish which configurations from the heap of possibilities go into the heap of actualities. ... $F = ma$ ... says there will be one and only one such curve of actualities. ... how are we to recover that most powerful appearance, the appearance that we move forward in time in one definite direction along such a curve?" (Barbour 1994, pp.406–9).

Barbour's answer to this question employs his concept of a "time capsule." This is

> "a single configuration (either of the entire universe or part of it)

> that seems to be the outcome of a dynamical process of evolution through time in accordance with definite laws. It appears to contain records of the past, and these records are mutually consistent. ... My suggestion is that the belief in time and its passage is solely a consequence of the fact that, at any instant, we find ourselves within a time capsule. ... it is a fact that what we experience psychologically is always a time capsule; for our memory is like a progress book, with snapshots taken every day and faithfully pasted in, one next to another, ... the ones in the supposed past getting fainter and fainter" (Barbour 1994, p. 408).

Of course the emphasis in this proposed answer is upon "seems", "appears", and "supposed".

My approach to Barbour's "derivation" of time from timelessness resembles the trouble-shooting of an experiment that has yielded a bizarre result. A systematic error has probably occurred, one of the commonest of these being the inadvertent (and sometimes advertent) introduction of impurities in the observed specimen of material. The insertion of impurities into the material of a thought experiment is a rather unusual occurrence, but that is precisely what Barbour has done. The strata in a time capsule are nothing more than a timeless collection of configurations. But the psyche of the thought experimenter surveys the strata progressively, and the progressive survey introduces an impurity of temporality. Thought would correspond more accurately to fact, in a timeless world, if the psyche simply compared the strata statically, like the photographs pasted on a page in an album. (It may be the case that static contemplation itself depends on a progressive sweep at a subconscious level, but if so there is all the more reason for admitting temporality as something fundamental.) Furthermore, not only is progressive sweep less natural than static contemplation in a timeless world, but it hardly seems possible at all, unless the psychological domain differs from the physical domain by possessing temporality. That indeed may be a way in which one could accept Barbour's thesis of the timelessness of the physical world and yet recover the appearance of time psychologically. But the price for saving Barbour's thesis in this way is very high indeed. Part of the price is the abandonment of any pretense at making the psychological domain derivative from the physical (and this part of the price Barbour may be willing to pay, for in the works by him that I have read I have never found an explicit commitment to the reduction in principle of psychology to physics). The other part of the price is that if the psyche is ontologically as basic as the entities of physics, and the psychological domain is temporal, then the universe as a whole is fundamentally temporal even though it has a timeless physical domain.

The objection that I have just given to Barbour's proposal of a timeless world is an application of an important philosophical principle that I have called "the phenomenological principle," for lack of a more descriptive name. The principle asserts that a minimal condition on ontology is the existence of a set of realities sufficient to account for appearances (Shimony 1993, vol. I, p. 278). It is a principle that must be deployed with caution, for the abuse of it could blur the distinction between appearance and reality and endow the former with a higher status in the ontological hierarchy than it deserves. I submit, however, that in my presentation of an objection against Barbour's timeless world, I have indeed been cautious in deploying the principle. I have argued only that there is temporality at a fundamental level in the world, and that an uneliminable aspect of this temporality is transience. I have not tried, on the basis of a meagre set of considerations, to disentangle objective and subjective components in the rich phenomenology of time in everyday life or to ascertain how transience fits in the structure of physical spacetime.

It would be irresponsible not to be cautious when we attempt to relate the phenomenology of everyday life to fundamental aspects of the world, in view of the immense complexity lying beneath the surface of the simplest mental presentation. How, for example, are psychological and neuronal time-scales related? And if we refrain from neurophysiology and restrict attention to strictly psychological experimentation, we still cannot fail to be impressed that the "specious present" endures much longer than do the typical stimuli controlled by the experimenter. Consequently, even if the specious present is a transient thing, there is no reason to believe that its mode of transience is shared by more elementary events. When I introspect, I seem to find a continuous slipping away of the specious present, with its tail end merging smoothly into the unequivocal past, and the unequivocal future crowding into its forward end. But I see no reason to postulate, at least at this stage of investigation, that transience at a fundamental level is continuous rather than discrete and that it involves merging with the recent past and with the impending future in the way that phenomenology describes. Rather, I suspect that the psychological mechanism of immediate presentation meshes intricately with the mechanisms of memories and anticipations. Phenomenology is a many-event domain, as macrophysics is a many-body domain; and just as in the latter one finds astonishing emergences of properties (an acknowledgment that in no way denies the "reducibility in principle" of macrophysics to microphysics), one should expect emergences in the former.

Even more problematic is the role of transience in physical theory. Classical mechanics, special relativity, and general relativity differ profoundly in their assumptions about spacetime structure, but in all three the structure is characterized without any reference to the slipping away of the present moment into the past. Furthermore, even if a direction of time is introduced, fundamentally or derivatively, in order to deal with irreversible processes, no assumption of transience is required. Generally, if we restrict our attention to pre-quantum physics, the only reason for suspecting that transience has anything to do with physical spacetime structure is the pervasiveness of transience in everyday life, and presumably the physical structure of spacetime is relevant to everyday life, whether or not the mental is fully reducible to the physical. Quantum mechanics, however, may radically change the situation. If the quantum mechanical state of a physical system objectively characterizes it, and the actualization of quantum mechanical potentialities is a real physical process, then transience will be an objective feature of physical processes. This interpretation of quantum mechanics will be considered seriously later in the lecture.

In sum, the conclusions that I draw from the arguments so far presented for a fundamental status of transiency in the world are minimalist. As Howard Stein once remarked, "A little minimalism goes a long way," but some amplification is still needed. In the twentieth century there have been many formulations of "process philosophy," in all of which the idea of transience is central (e.g., Bergson 1907; Alexander 1920; Whitehead 1929, 1933; Čapek 1991). Of these I have found Whitehead's philosophy the most inspiring, although I have serious reservations about many of his propositions (Shimony 1965). Far from being a minimalist, Whitehead aimed at a comprehensive world view that would explain causality, the mind-matter relation, the evolutionary emergence of natural law, the place of value in a world of fact, and much else. I shall summarize his main theses briefly, but then try to extract a core from them that will serve my limited aim of exploring transience as an essential aspect of time.

For this audience, in which an acquaintance with Leibniz's philosophy is widespread, it may be helpful to characterize Whitehead's philosophy of organism as Leibniz's monadology (1714) transformed by process. According to both Leibniz and Whitehead the elementary concrete entities of the universe are mental in character (or, more carefully, proto-mental, since the "petites perceptions" of those monads that constitute ordinary physical bodies and the "feelings" that Whitehead ascribes to low order occasions must be regarded as extremely remote extrapolations from the experiences of human beings). With this caveat, Leibniz and Whitehead can be characterized as "reductionists" in their treatments of the mind-body problem, but they propose to "derive" the properties of ordinary matter from a fundamental proto-mental reality rather than (as is much more common in contemporary philosophy) "derive" mentality from sufficiently complex aggregations of material systems. (Indeed, both Leibniz and Whitehead can be read as endorsing and applying what I earlier called "the Phenomenological Principle" in order to justify their preference regarding the direction of reduction.) The fundamental difference between Leibniz and Whitehead is that the monads of the former are eternal, whereas the actual occasions of the latter are of brief duration, probably much shorter than the duration of the psychological specious present, though that is a point on which Whitehead is somewhat obscure. The two philosophers also differ with regard to the question of interaction among fundamental entities. According to Leibniz "the monads are windowless." When a monad is created by God, its subsequent internal mental or proto-mental life is completely programmed at the outset, and no other monad affects its career. The apparent phenomena of interaction, communication, and cooperation among concrete entities are ascribed by Leibniz entirely to "preestablished harmony," that is the harmonization of the internal programs of autonomous monads. According to Whitehead, by contrast, an actual occasion is richly endowed with windows, open to those occasions which have completed their internal processes before the new occasion launches its process. The occasions of the past supply proto-mental raw material which the new occasion "prehends" or "re-enacts" or "objectifies" or uses as an "ingredient". Whitehead uses these different suggestive terms in order to convey the sense of the radical extrapolation which he is proposing. At some risk of oversimplifying Whitehead's thought I suggest that an occasion of the past is "ingredient" in an occasion currently in process in roughly the way that a memory of a past experience enters into the present state of mind of a human being.

I propose now to extract from Whitehead's philosophy of organism some propositions on the temporal structure of actual occasions, with only a little further attention to their proto-mentality. This procedure is appropriate partly because the present conference is devoted to geometric issues, and partly because I have recently discussed Whitehead's proto-mentalism in my comments on Roger Penrose's Tanner Lectures of February, 1995 (Penrose 1997).

(1) The fundamental concrete entities of the universe are transient. In Locke's vivid phrase, which Whitehead (1929, pp. 43, 94, 196, 222, 320, 516, 517, 527) quotes with approbation, time is "a perpetual perishing."

(2) Transience is primarily local, manifested in the first instance in single actual occasions. Global time, to the extent that there is such, is derivative from local time. To use a non-Whiteheadian expression, global time is a collective phenomenon.

(3) There is a kind of quantization of local time, in that an actual occasion viewed from the outside is temporally extended but not temporally divisible. It is prehended as an integral process.

(4) There is, however, internal structure in the quantum of time, with distinguishable phases that Whitehead refers to as the "genetic division" of the occasion (ibid., pp. 325–428).

(5) In the internal development of an occasion there is continuity, despite the quantization of time as seen from outside the occasion.

(6) Temporal irreversibility is exhibited at the microscopic level, since the initial and final phases of an occasion are functionally different, the final phase consisting of foreclosures among possibilities that are open in the initial phase.

(7) The actualization of potentialities—in a sense that is not derived from quantum mechanics but dimly anticipates it—is essential to irreversibility in the individual occasion.

(8) This process of actualization is constrained but not entirely deterministic.

How can these rough-hewn philosophical ideas be implemented in a physical theory that is mathematically precise, internally coherent, and capable of experimental confirmation? That, of course, is a very ambitious program, the realization of which would be a scientific revolution as profound and radical as any we have yet witnessed. I am diffident about making any suggestions for such a program, but I can point to certain speculations, already in the forum of physical discussion, that are relevant to the foregoing Whiteheadian propositions.

(A) A number of investigators (e.g., Károlyházy 1974; Károlyházy *et al.*

1986; Gisin 1984; Ghirardi *et al.* 1986; Penrose 1986; Pearle 1986; Diósi 1988; Percival 1994; Golden 1994) have proposed stochastic modifications of the time-dependent Schrödinger equation. Foremost among their motivations is to provide a physical explanation for the occurrence of definite results in measurement processes, without any reliance upon the psyche of the observer. But their ideas are fertile, and I suggest that some such stochastic equation may be appropriate for modelling the internal process of a Whiteheadian occasion. The wave function governed by the stochastic equation would not describe a macroscopic measuring apparatus but rather a microscopic system, viz., an occasion, which is the most microscopic entity in nature according to Whitehead. The potentialities in the initial phase of the occasion are those properties which are not assigned definite values by the initial wave function, and the actualization of potentialities (postulated in proposition (7) above) is governed by the stochastic equation. In typical stochastic modifications of the Schrödinger equation there is a preferred basis, specified by contingencies of the environment in some theories and by fundamental law in others, and this basis determines the array of possible actualizations. The choice from this array is a matter of chance, thus satisfying proposition (8) above. One feature of many of the stochastic modifications of the Schrödinger equation, which is usually regarded as an embarrassment, is the fact that actualization of potentialities is achieved only asymptotically, as the time t goes to infinity. (How could such a theory account for definite results of experiments in a finite time interval?) This feature may turn out to be desirable, however, in the program of modelling Whiteheadian occasions. Propositions (3) and (4) above distinguish between the internal process of the occasion and the views of the occasion from outside. The time t of the stochastic equation may be assumed to refer to the internal process. If there is a well defined clock time variable for the environment, then the relation between t and that variable would have to be worked out, and it is conceivable that the infinite limit of t would correspond to a finite value of the clock time. The primary virtue of the model is that it captures transiency, as postulated in proposition (1), for the internal process goes from a genuine multiplicity of potentialities to an actual choice among them.

(B) Nothing has been said so far about the space of states in the model for a Whiteheadian occasion. One possibility is to take seriously the elementarity of the occasion and look for the simplest non-trivial mathematical representation which can allow for properties with indefinite values. The candidate offered by quantum mechanics is the spin one-half system, associated with a two-dimensional Hilbert space. Since quantum mechanics has a remarkable principle of composition unknown in classical physics, namely the formation of many-particle states by non-factorizable vectors in a product Hilbert space (called "entanglement" by Schrödinger), the resources offered by an indefinite number of spin one-half systems are immense. Penrose (1972), in fact, suggested that a network of such systems may be the deep structure underlying a space-time that looks continuous when it is not probed too finely. If a stochastic modification of the Schrödinger equation were used to treat the dynamics of a spin one-half system, then most of the propositions (1) through (8), which were distilled from Whitehead's ontology, may perhaps be implemented. Nevertheless, one may wonder whether such a simple system is sufficient to model the structure of the proto-mentality that Whitehead ascribes to an actual occasion. At a minimum, one may suppose, a topologically richer model may be needed, such as Kauffman's knots (Kauffman 1987) or the loops of Rovelli and Smolin (1988). After all, at a human level we know too well that feelings are tangled things!

(C) In order to relate the internal time of an occasion to the clock time of its environment it may be useful to adapt the well-known procedure of theoretical physics of separating the fast from the slow coordinates in a complex system. In the Born-Oppenheimer treatment of molecules one proceeds in stages, first fixing the values of the slow coordinates and solving the reduced Schrödinger equation for the fast coordinates, and then using that solution to obtain an effective potential for finding the wave function of the slow coordinates. The Born-Oppenheimer procedure, however, is understood as an approximation that in principle can be improved in higher order, whereas the proposed adaptation, in which the internal process of an occasion is treated as fast and the reaction of the environment as slow, would be a modification of fundamental physics.

I shall abstain from further speculation, however, because I do not want my good questions concerning transience to be spoiled by bad answers. Before concluding, however, I wish to mention a few investigators who have taken transience seriously and have attempted to embed the idea in physical theory: von Weizsaecker (1958), Bohm (1980), Stapp (1993), Golden (1994), and Bialynicki-Birula (1994). In the last reference I find (on p.278) the following passage, which is programmatic but, to my mind, inspiring:

> "When the state of a system at a given instant becomes ill-defined, time itself becomes diffused. ... One may say that the idea of a diffused time offers the ultimate resolution of the Zeno paradox. The paradox disappears since there is no instantaneous, 'static' state; at all times Zeno's arrow carries with it the notion of flight—the internal notion of change. The notion of state is not static; the static state gives way to a dynamic state of motion. ... The concept of diffused time has also consequences for time's arrow. Once we accept the idea that the notion of a state should be replaced by the concept of a dynamic state of motion, we can include the sense of direction of that motion in time."

I conclude by approving an idea frequently expressed by process philosophers—that a world in which there is transience and mentality is a world that has a locus for value. There is a chilling passage at the end of one of Steven Weinberg's books (Weinberg 1977, p.154): "The more the universe seems comprehensible the more it also seems pointless." Underlying this remark seems to be a tacit assumption that the "point" is a plan or an aim, so that the absence of evidence for a plan is tantamount to pointlessness. But must the primary source of value be a plan?

Might it not rather be the zest of the present moment? If so, then the widespread physicists' conception of time without transience not only does injustice to the phenomenology of temporal experience, but loses the prime candidate for the locus of value in the world. It is the transience of the present moment that makes it precious and savory. Of course there is a terrible price to be paid for value via process—that the moment which is enjoyed for its own sake passes away, and hence value is inseparable from loss. At the higher levels of experience there are psychical mechanisms for compensating for the inevitable loss of the treasured present moment—memory, hope, long range plans, aesthetic and intellectual achievement, vicarious participation in the experience of family and friends, and social concern. But none of these mechanisms could supply any satisfaction whatever were it not for the vivacity of the "holy ground", which is Whitehead's characterization of the present moment. My thoughts on the matter are summed up in the following poem, entitled "The Present":

Because the present never halts
Adorn its fleeting with rare skills
Of warbles, melismata, trills,
Cadenzas, canons, and roulades,
Of entrechats, jetés, glissades,
And tumbling triple somersaults.

Bibliography

Alexander, S. (1920). *Space, Time, and Deity.* Macmillan, London.

Barbour, J. (1994). The emergence of time and its arrow from timelessness, in Halliwell, J.J., Perez-Mercader, J., and Zurek, W.H. (eds) *Physical Origins of Time Asymmetry.* Cambridge University Press, 405–14.

Beerbohm, M. (1919). *Seven Men.* William Heinemann Ltd., London.

Bergson, H. (1907). *L'evolution creatrice.* F. Alcan, Paris. (English translation by Mitchell, A. (1911). *Creative Evolution.* Henry Holt, New York.)

Bialynicki-Birula, I. (1994). Is time sharp or diffused?, in Halliwell, J.J., Perez-Mercader, J., and Zurek, W.H. (eds) *Physical Origins of Time Asymmetry.* Cambridge University Press, 274–9.

Bohm, D. (1980). *Wholeness and the Implicate Order.* Routledge and Kegan Paul, London.

Čapek, M. (1991). *The New Aspects of Time.* Kluwer, Dordrecht.

Diósi, L. (1988). Quantum stochastic processes as models for state vector reduction. *Journal of Physics A,* **21**, 2885–98.

Favrholdt, D. (1984). On the concept of time. *Danish Yearbook of Philosophy,* **21**, 85–96. (Acknowledgement is given to two students, Jesper Gorm Madsen and Flemming Raben Anker, for applying Kant's argument to McTaggart.)

Ghirardi, G.C., Rimini, A., and Weber, T. (1986). Unified dynamics of microscopic and macroscopic systems. *Physical Review D,* **34**, 470–91.

Gisin, N. (1984). Quantum measurements and stochastic processes. *Physical Review Letters,* **52**, 1657–60.

Golden, S. (1994). Quantization of time. *Physica A,* **208**, 65-90.

Grünbaum, A. (1971). The meaning of time, in Freeman, E. and Sellars, W. (eds) *Basic Issues in the Philosophy of Time.* Open Court, La Salle, IL, 211.

Kant, I. (1781). *Kritik der reinen Vernunft* Riga. (English translation by Kemp Smith, N. (1929). *Critique of Pure Reason.* Macmillan, London.)

Károlyházy, F. (1974). Gravitation and quantum mechanics of macrscopic bodies. *Magyar Fizikai Folyóirat,* **12**, 24.

Károlyházy, F., Frenkel, A., and Lukács, B. (1986). On the possible role of gravity in the reduction of the wave function, in Penrose, R. and Isham, C. (eds) *Quantum Concepts in Space and Time.* Clarendon Press, Oxford, 109–28.

Kauffman, L. (1987). On knots. *Annals of Mathematics Studies,* **115** Princeton University Press.

McTaggart, J.M.E. (1927). *The Nature of Existence.* Cambridge University Press.

Pearle, P. (1986). Models for reduction, in Penrose, R. and Isham, C. (eds) *Quantum Concepts in Space and Time.* Clarendon Press, Oxford, 84–108.

Penrose, R. (1972). On the nature of quantum gravity, in Klauder, J. (ed) *Magic Without Magic.* Freeman, San Francisco, 333–54.

Penrose, R. (1986). Gravity and state vector reduction, in Penrose, R. and Isham, C. (eds) *Quantum Concepts in Space and Time.* Clarendon Press, Oxford, 129–146.

Penrose, R. (1997). *The Large, the Small and the Human Mind* (with comments by Abner Shimony, Nancy Cartwright, and Stephen Hawking). Cambridge University Press.

Percival, I. (1994). Primary state diffusion. *Proc. Roy. Soc. A,* **447**, 189–209.

Rovelli, C. and Smolin, L. (1988). Knot theory and quantum gravity. *Phys. Rev. Letters,* **61**, 1155–8.

Shimony, A. (1965). Quantum physics and the philosophy of Whitehead, in Black, M. (ed) *Philosophy in America.* George Allen and Unwin, London, 240-261. (Reprinted in Shimony, A. (1993). *Search for a Naturalistic World View.* Cambridge University Press, vol. 2, 291–309.)

Shimony, A. (1993). The transient now, in Shimony, A. *Search for a Naturalistic World View.* Cambridge University Press, vol. 2, 271-287. Also in Hahn, L.E. (ed.) (1995). *The Philosophy of Paul Weiss.* Open Court, Chicago, 331–48.

Stapp, H.P. (1993). *Mind, Matter, and Quantum Mechanics.* Springer, Berlin.

Weinberg, S. (1977). *The First Three Minutes.* Basic Books, New York.

Weizsaecker, C. F. v. (1958). Die Quantentheorie der einfachen Alternative.

Zeitschrift für Naturforschung A, **13**, 245–53.

Whitehead, A.N. (1929). *Process and Reality*. Macmillan, London.

Whitehead, A.N. (1933). *Adventures of Ideas*. Macmillan, London.

Zeilicovici, D. (1986). A (Dis)Solution of McTaggart's Paradox. *Ratio*, **28**, 175–195.

11
Geometric Issues in Quantum Gravity

Abhay Ashtekar
Center for Gravitational Physics and Geometry
Physics Department, Penn State University, University Park, PA 16802-6300

1 Introduction

I was a post-doc in Roger Penrose's group in Oxford in the mid-seventies and I learned a great deal during those two years with Roger; indeed, more than in any comparable period in my subsequent academic career. This, of course, included much technical material in general relativity and geometry. However, I learned something even more valuable, something intangible and deeper, just watching Roger select scientific problems and arrive at solutions with his unique geometric insights, almost effortlessly, with a touch of magic. Since then, his way of approaching science has continued to have a deep influence on all my work. It is therefore an honour and a pleasure for me to contribute to this volume.

1.1 Setting the stage

During my post-doc period in Oxford, Roger introduced the celebrated Penrose transform (Penrose, 1976; Mason and Hughston 1990) to construct, by deforming twistor space, the general self-dual solution of Einstein's equation, or, as he called it, 'the non-linear graviton.' In the opening paragraph of his paper (Penrose 1976), he spelled out his view on quantum gravity as follows:

... *if we remove life from Einstein's beautiful theory by steam-rollering it first to flatness and linearity, then we shall learn nothing from attempting to wave the magic wand of quantum theory over the resulting corpse.*

Coming from Roger, these words made a deep impression especially on the younger relativists who were attracted to quantum gravity.

Let me explain this issue in some detail. In general relativity, the space-time metric plays a dual role. On the one hand, it represents the gravitational potential and is thus a dynamical variable. On the other hand, it determines space-time geometry. Field theoretic approaches to quantum gravity—including the current formulation of string theory—split this role. Typically, one introduces a flat, kinematic metric and regards the difference between the physical metric and this flat background as a perturbation which is then subjected to quantization. Roger's remarks emphasized that the dual role of the metric should be taken seriously and not compromised just because the standard machinery of quantum field theory is inapplicable if we do not have a background space-time at our disposal. Rather, the suggestion was that this machinery itself should undergo suitable modifications so as to be applicable to the problem at hand; we should learn to do quantum field theory in the absence of a background metric.

Later in his paper on the non-linear graviton, Roger illustrated why he thought that non-perturbative effects should be important. In the perturbative treatment, one begins with Minkowski space and describes the gravitational interaction through spin-2 quanta that Roger referred to as 'linear gravitons.' As Roger put it:

If one such 'graviton' is added to the vacuum (Minkowski space) state the space remains flat. The null cones do not shift. If a second such 'graviton' is added, and a third and a fourth, the space still remains flat, with null cones still locked in their original Minkowski position. With such a perturbative viewpoint, it is only after an infinite number of gravitons have been added that the space can become curved. The situation may be compared to a power series expansion. For example, with any finite number of terms, the function $\frac{1}{z} + \frac{1}{z^2} + \ldots + \frac{1}{z^n}$ *has a pole stuck at* $z = 0$. *But the sum to infinity* $\frac{1}{z} + \frac{1}{z^2} + \frac{1}{z^3} + \ldots$ *has its pole shifted to* $z = 1$.

At the time, this emphasis on non-perturbative methods received little attention from the field theoretic community probably because it had yet to be backed by concrete calculations. By now, however, detailed arguments have emerged through exactly soluble models. For example, in 3-dimensional general relativity coupled to matter, the exact Hamiltonian H of the model is non-polynomially related to the naive perturbative Hamiltonian H_0 (Ashtekar and Varadarajan 1994, Varadarajan 1995):

$$H = \frac{1}{4G}(1 - e^{-4GH_0}) \equiv H_0 - 2GH_0^2 + \ldots.$$

The two agree in the weak field limit. But while H_0 (and the truncated Hamiltonian, obtained by terminating the series to any finite order) is unbounded above, H is in fact bounded. Furthermore, the resulting non-perturbative quantum metric exhibits several unforeseen features (Ashtekar 1996) which would have been impossible to notice to any finite order in a perturbative treatment.

Once one accepts the viewpoint that the dual role of the metric should not be sacrificed in the passage to the quantum theory, the nature of quantum geometry becomes a central issue of quantum gravity. For, in a non-perturbative scenario which avoids splitting of the dual role of the metric, quantum gravity must lead to quantum geometry. Is the familiar continuum picture then only an approximation? If so, what are the 'atoms' of geometry? What are its fundamental excitations? Is there a discrete underlying structure? If so, how does the continuum picture arise from this fundamental discreteness? These questions are very different from those at the forefront of field theoretical approaches. There, the emphasis is on such issues as S-matrices, ghosts that are necessary to ensure perturbative unitarity and radiative corrections to graviton-graviton scattering.

The issues of quantum geometry are, by contrast, *local* issues which lie beyond S-matrix calculations.

1.2 Quantum geometry

The non-perturbative perspective advocated by Roger has led to several different avenues to explore the quantum nature of geometry. First of course is twistor theory, developed by Roger and his numerous colleagues (Bailey and Baston, 1990; Mason and Hughston 1990, 1995; Penrose 1997 and references contained therein). A related approach is that of '$\mathcal{H}$-spaces' developed by Newman and his collaborators (Ko *et al.* 1981). Most recently, this approach has led to an exciting development in which quantization of the gravitational radiative degrees of freedom at null infinity leads to a 'fuzzing' of interior space-time points already in the linearized approximation (Fritelli *et al.* 1996). It would be extremely interesting to extend this construction beyond the linear theory. Yet another avenue was initiated by Hawking and his colleagues in terms of Euclidean path integrals (Hawking 1979). For technical reasons, concrete work within this approach has been limited to mini-superspaces. However, recent mathematical developments along the lines indicated in the main body of this article may well enable one to extend these ideas to the full theory in the spirit of the Osterwalder-Schrader approach to familiar quantum field theories. A program in a quite different direction was initiated by Sorkin (Bombelli *et al.* 1987). Here, one puts the causal structure of space-time at the forefront, begins by postulating that the fundamental underlying structure is just a point set with a causal relation—called a poset—and then attempts to recover the continuum picture by a suitable coarse graining.

In this contribution, I will focus on yet another approach which is based on canonical quantization. This approach is 'constructive' in the sense that one does not begin by postulating what the fundamental underlying geometrical structure should be, but allows it to emerge from the theory itself. The starting point is the principles that underlie general relativity (most notably the absence of a background space-time metric or other geometrical structures) and quantum mechanics (as spelled out in the canonical approach à la Dirac). Although in the technical work emphasis is often rather different from that in Roger's work, nonetheless some of Roger's cherished ideas feature prominently. In particular, at the classical level, the program has received much inspiration from the theme 'self-duality simplifies' that has emerged from twistor theory. Similarly, in the quantum part, the spin-networks that Roger introduced twenty-five years ago (Penrose 1971) play a key role. They provide us with 'typical' quantum states, non-perturbative excitations of quantum geometry.

This approach has been pursued by a rather large number of co-workers and there is a fair amount of diversity in viewpoints, areas explored and results obtained to date. This contribution is *not* intended to be a comprehensive survey of all this work. Rather, I will restrict myself to only one main issue: the nature of quantum geometry in non-perturbative quantum gravity. General relativity is based on differential geometry. The question is: What should replace it in a non-perturbative quantum treatment? Recall that, in the classical theory, differential geometry offers us a general kinematic framework which is then used to formulate the actual field equations. I will indicate how the corresponding framework is constructed for quantum gravity and highlight some of the striking predictions it leads to. However, I will not be concerned here with the issues of how to formulate quantum dynamics using this framework. This area is a focal point of much of the current research. (See, e.g., Rovelli and Smolin, 1994; Ashtekar 1995; Thiemann 1996a, b, c). However, due to space limitation, I will forego these developments and, in the spirit of the theme of the conference, only address geometrical issues. Because of this focus and because of the research orientation of most participants, I will primarily address this article to mathematical physicists and mathematicians.

Thus, the key question I wish to consider is: Is geometry quantized? To probe this issue, I will follow the most direct procedure used already in non-relativistic quantum mechanics. What do we mean by 'quantization' there? We mean that there exist physical quantities which can take on continuous values classically but are such that the corresponding quantum operators have a discrete spectrum. For example, this is the sense in which the energy and angular momentum of the hydrogen atom are quantized. The question therefore is whether there exist geometrical quantities for which a similar quantization occurs. In differential geometry, lengths of curves, areas of surfaces and volumes of regions can take on continuous values; they vary continuously as we move in the space of metrics. The question then is: Can one construct corresponding self-adjoint operators in the quantum theory and, if so, do they have discrete spectra? If so, we will say that geometry is quantized. We can then explore physical consequences of this quantization.

Note how simple it is to formulate the basic questions. Indeed, they could have been formulated immediately after the advent of quantum mechanics. Why then have these issues then remained unexplored until recently? The main reason is that the only concrete calculational tools we have had until recently have been perturbative. As we saw already, in these treatments, one begins with a flat background metric and tries to incorporate quantum effects perturbatively. At any finite order in perturbation theory, then, it is hard to see any discreteness in the spectra of geometrical quantities. Thus, the reason why this rather basic issue could not be dealt with satisfactorily is that the necessary non-perturbative methods were simply unavailable.

The article is organized as follows. In section 2, I will introduce the space of quantum states, the arena for doing quantum physics without a background metric. The resulting quantum geometry is discussed in section 3. This geometry is to quantum physics what differential geometry is to classical. I will conclude in section 4 with a brief discussion.

2 Quantum States

2.1 Phase space

In the canonical approach, the starting point is the Hamiltonian formulation of the classical theory under consideration. For definiteness, I will take this to be 4-dimensional, Lorentzian general relativity (although this is not essential.) The first step is to construct the phase space—i.e., the infinite-dimensional symplectic manifold—of this field theory. It will consist of fields on an orientable 3-manifold Σ which, in the classical theory, plays the role of 'an instant of time'. (For later convenience we will assume that Σ has an analytic structure.) The phase space has a natural cotangent bundle structure. To specify the configuration (or base) space, let us fix an $SU(2)$ principal bundle B over Σ. The configuration space $\mathcal{A}$ is then the space of all smooth $SU(2)$ connections A on B. Since all principal $SU(2)$ bundles on an orientable 3-manifold are trivial, in concrete calculations, we can fix a trivialization of B and represent A as a Lie algebra-valued 1-form A_a^i on Σ. (Here a is the 1-form index and i is the $su(2)$-Lie algebra index.) Momenta, i.e. cotangent vectors of $\mathcal{A}$, are represented by 2-forms e_{abi} which take values in the dual of the $su(2)$-Lie algebra. By dualization on Σ, we can represent them as vector densities E_i^a of weight one. In the terminology which is standard in physics, A_a^i and E_i^a are said to be canonically conjugate, i.e. to satisfy the Poisson-bracket relations

$$\{A_a^i(x), E_j^b(y)\} = G\delta_j^i\delta_a^b\delta^{(3)}(x, y) \tag{2.1}$$

where x and y are any two points in Σ and G is Newton's constant.

Thus, the phase space of general relativity can be identified with that of the $SU(2)$ Yang-Mills theory (Ashtekar, 1987a, b). In the passage to quantum theory, this is a significant advantage over the older 'metric formulation' because mathematical techniques are much better developed to handle theories of connections. The dynamics of general relativity is of course entirely different from that of Yang-Mills theory. However, it turns out that this 'connection formulation' of general relativity also simplifies the Einstein dynamics, making it more amenable to a full non-perturbative treatment.

The geometrical interpretation of these variables is as follows. The connection A enables one to parallel-transport $SU(2)$-spinors (or, 'chiral fermions' in the physics terminology) along curves. The 'momenta' E represent triads (with density weight one); $Q^{ab} := E_i^a E^{bi}$ is a contravariant metric on Σ (with density weight two. Here the Lie-algebra indices have been contracted using the Cartan-Killing metric on $su(2)$.) Thus, the connection A enables us to define $SU(2)$-valued holonomies while familiar quantities from Riemannian geometry such as lengths, areas, and volumes can be constructed from E. For example, if S is a 2-dimensional submanifold of Σ defined in local coordinates by $x_3 =$ const, its area is given by:

$$A_S := \int_S d^2x\, [E_i^3 E^{3i}]^{\frac{1}{2}}. \tag{2.2}$$

Similarly, the volume of a 3-dimensional region R is given by

$$V_R := \int_R d^3x \, |\epsilon_{ijk} E_i^a E_j^b E_k^c \eta_{abc}|^{\frac{1}{2}} \tag{2.3}$$

where η_{abc} is the natural Levi-Civita density on Σ.

Points of this phase space represent kinematic states of the classical theory and functions such as A_S and V_R on the phase space are kinematic observables. Our task in the remainder of this article is to construct the quantum analogues of such structures.

2.2 Quantum configuration space

For systems with only a finite number of degrees of freedom, the classical configuration space also serves as the domain space of quantum wave functions, i.e., as the quantum configuration space. For systems with an infinite number of degrees of freedom, on the other hand, this is not true: generically, the quantum configuration space is an enlargement of the classical. In free field theory in Minkowski space (as well as exactly solvable models in low space-time dimensions), for example, while the classical configuration space can be built from suitably smooth fields, the quantum configuration space includes all tempered distributions. This enlargement is important because, for typical measures one encounters in quantum field theory, the classical configuration spaces are contained in sets of zero measure. The overall situation is the same in general relativity. The quantum configuration space $\overline{\mathcal{A}}$ is a certain completion of $\mathcal{A}$ (Ashtekar and Isham, 1992; Ashtekar and Lewandowski, 1994).

To see the nature of the extension involved, let us fix a trivialization of the bundle B. Then, each smooth connection defines a holonomy along analytic paths in Σ: $h_p(A) := \mathcal{P}\exp - \int_p A$, where $\mathcal{P}$ stands for 'path ordered'.[1] Generalized connections capture this notion. That is, each $\bar{A}$ in $\overline{\mathcal{A}}$ can be defined as a map which assigns to each oriented path p in Σ an element $\bar{A}(p)$ of $SU(2)$ such that: i) $\bar{A}(p^{-1}) = (\bar{A}(p))^{-1}$; and, ii) $\bar{A}(p_2 \circ p_1) = \bar{A}(p_2) \cdot \bar{A}(p_1)$, where p^{-1} is obtained from p by simply reversing the orientation, $p_2 \circ p_1$ denotes the composition of the two paths (obtained by connecting the end of p_1 with the beginning of p_2) and $\bar{A}(p_2) \cdot \bar{A}(p_1)$ is the composition in $SU(2)$ (Baez, 1994a, b; Ashtekar and Lewandowski 1995a).

As in Yang-Mills theory, physically interesting quantities—and in particular the geometrical observables discussed above—should be invariant under the action of gauge transformations (i.e., vertical automorphisms of B.) Therefore, we need to specify what the appropriate transformations are. A generalized gauge transformation is a map g which assigns to each point x of Σ an $SU(2)$ element $g(x)$ (in an arbitrary, possibly discontinuous fashion). It acts on $\bar{A}$ in the expected manner, at the end points of paths: $\bar{A}(p) \to g(v_+)^{-1} \cdot \bar{A}(p) \cdot g(v_-)$, where

[1] For technical simplicity, we will assume that all paths are analytic. An extension of the framework to allow for smooth paths is carried out by Baez and Sawin (1995).

v_- and v_+ are respectively the beginning and the end point of p. If $\bar{A}$ happens to be a smooth connection, say A, we have $\bar{A}(p) = h_p(A)$. However, in general, $\bar{A}(p)$ cannot be expressed as a path ordered exponential of a smooth 1-form with values in the Lie algebra of $SU(2)$. Similarly, in general, a generalized gauge transformation cannot be represented by a smooth group valued function on Σ. The quotient $\overline{\mathcal{A}}/\overline{\mathcal{G}} =: \overline{\mathcal{A}/\mathcal{G}}$ is the gauge invariant quantum configuration space.

An interesting property of the space $\overline{\mathcal{A}}$ (as well as of $\overline{\mathcal{A}/\mathcal{G}}$) is that it does *not* depend on the initial choice of the bundle B over Σ: $\overline{\mathcal{A}}$ contains every (smooth) connection on every $SU(2)$ principal bundle over Σ. In addition, it contains genuinely generalized connections, e.g., ones which are 'distributional' in character in the sense that their support is lower dimensional. Therefore, at first sight the spaces $\overline{\mathcal{A}}$, $\overline{\mathcal{G}}$ and $\overline{\mathcal{A}/\mathcal{G}}$ seem too large to be mathematically controllable. However, it turns out that $\overline{\mathcal{A}}$ and $\overline{\mathcal{G}}$ can be regarded as projective limits of a family of compact, Hausdorff, smooth manifolds (labelled by graphs in Σ).[2] This enables one to introduce on them differential geometry (Ashtekar and Lewandowski, 1995a) as well as measure theory (Ashtekar and Lewandowski 1994, 1995b; Marolf and Mourão 1995).

We will conclude this subsection by summarizing the projective limit construction. Let us begin with some definitions. An *edge* is an oriented, 1-dimensional submanifold of Σ with two boundary points, called *vertices*, which is analytic everywhere, including the vertices. A *graph* in Σ is a collection of edges such that if two distinct edges meet, they do so only at vertices. Using standard ideas from lattice gauge theory, we can associate a 'configuration space' with the graph γ. Consider the space $\mathcal{A}_\gamma$, each element A_γ of which assigns to every edge in γ an element of $SU(2)$ and the space $\mathcal{G}_\gamma$ each element g_γ of which assigns to each vertex in γ an element of $SU(2)$. (Thus, if N is the number of edges in γ and V the number of vertices, $\mathcal{A}_\gamma$ is isomorphic with $[SU(2)]^N$ and $\mathcal{G}_\gamma$ with $[SU(2)]^V$). $\mathcal{G}_\gamma$ has the obvious action on $\mathcal{A}_\gamma$: $A_\gamma(e) \to g(v_+)^{-1} \cdot A_\gamma(e) \cdot g(v_-)$. The (gauge invariant) configuration space associated with γ is just $\mathcal{A}_\gamma/\mathcal{G}_\gamma$. It is easy to verify that A_γ, $\mathcal{G}_\gamma$ and $\mathcal{A}_\gamma/\mathcal{G}_\gamma$ provide us with three projective families, labelled by graphs γ. The projective limits of these spaces yield precisely the spaces $\overline{\mathcal{A}}$, $\overline{\mathcal{G}}$ and $\overline{\mathcal{A}/\mathcal{G}}$ discussed above. (Note incidentally that this limit is *not* the usual 'continuum limit' of a lattice gauge theory in which one lets the edge length go to zero. Here, we are already in the continuum and just expressing the quantum configuration space of the continuum theory as a suitable limit of finite dimensional configuration spaces associated with graphs.)

To summarize, the quantum configuration space $\mathcal{A}/\mathcal{G}$ (or, $\overline{\mathcal{A}/\mathcal{G}}$) is a specific 'completion' of the classical configuration space $\mathcal{A}$ (resp. $\mathcal{A}/\mathcal{G}$). Being gauge

[2]The space $\overline{\mathcal{A}/\mathcal{G}}$ admits three characterizations: one as the projective limit of a family of compact Hausdorff topological manifolds, one as the Gel'fand spectrum of an Abelian $C^\star$ algebra, and one as the space of homomorphisms from the so-called 'hoop group' associated with Σ to $SU(2)$ (Ashtekar and Isham, 1992; Ashtekar and Lewandowski, 1994, 1995a). Each serves to bring out certain structures and properties and together they provide a great deal of control on $\overline{\mathcal{A}/\mathcal{G}}$.

invariant, quantum states will be complex-valued, square-integrable functions on $\overline{\mathcal{A}/\mathcal{G}}$, or, equivalently, $\overline{\mathcal{G}}$-invariant square-integrable functions on $\overline{\mathcal{A}}$ (with respect to suitable measures.) As in Minkowskian field theories, while $\mathcal{A}$ is dense in $\overline{\mathcal{A}}$ topologically, measure theoretically it is generally sparse; typically, $\mathcal{A}$ is contained in a subset set of zero measure of $\overline{\mathcal{A}/\mathcal{G}}$ (Marolf and Mourão 1995). Consequently, what matters is the value of wave functions on 'genuinely' generalized connections. In contrast with the usual Minkowskian situation, however, $\overline{\mathcal{A}}$, $\overline{\mathcal{G}}$ and $\overline{\mathcal{A}/\mathcal{G}}$ are all *compact* spaces in their natural (Gel'fand) topologies. This fact simplifies a number of technical issues.

2.3 Kinematical Hilbert space

Since $\overline{\mathcal{A}/\mathcal{G}}$ is compact, it admits regular (Borel, normalized) measures and for every such measure we can construct a Hilbert space of L^2-functions. Thus, to construct the Hilbert space of quantum states, we need to select a specific measure on $\overline{\mathcal{A}/\mathcal{G}}$.

It turns out that $\overline{\mathcal{A}}$ admits a measure μ^o that is preferred by both mathematical and physical considerations (Ashtekar and Lewandowski, 1994, 1995b; Baez 1995a, 1995b). Mathematically, the measure μ^o is natural because its definition does not involve introduction of any additional structure: it is induced on $\overline{\mathcal{A}}$ by the Haar measure on $SU(2)$. More precisely, since $\mathcal{A}_\gamma$ is isomorphic to $[SU(2)]^N$, the Haar measure on $SU(2)$ induces on it a measure μ^o_γ in the obvious fashion. As we vary γ, we obtain a family of measures which turn out to be compatible in the appropriate sense and therefore induce a measure μ^o on the projective limit $\overline{\mathcal{A}}$. This measure has the following attractive properties: i) it is faithful; i.e., for any continuous, non-negative function f on $\overline{\mathcal{A}}$, $\int d\mu^o f \geq 0$, equality holding if and only if f is identically zero; and, ii) it is invariant under the (induced) action of Diff$[\Sigma]$, the diffeomorphism group of Σ. Finally, μ^o induces a natural measure $\tilde{\mu}^o$ on $\overline{\mathcal{A}/\mathcal{G}}$: $\tilde{\mu}^o$ is simply the push-forward of μ^o under the projection map that sends $\overline{\mathcal{A}}$ to $\overline{\mathcal{A}/\mathcal{G}}$. Physically, the measure $\tilde{\mu}^o$ is selected by the so-called 'reality conditions'. More precisely, the classical phase space admits an (over)complete set of naturally defined configuration and momentum variables which are real, and the requirement that the corresponding operators on the quantum Hilbert space be self-adjoint selects for us the measure $\tilde{\mu}^o$ (Ashtekar *et al.* 1995).

Thus, it is natural to use $\mathcal{H}^o := L^2(\overline{\mathcal{A}}, d\mu^o)$ as our Hilbert space. Elements of $\mathcal{H}^o$ are 'kinematic' quantum states; quantum Einstein equations have not been imposed. Thus, $\mathcal{H}^o$ is the quantum analogue of the phase-space of section 2.1. Gauge invariant quantum states are elements of $\tilde{\mathcal{H}}^o = L^2(\overline{\mathcal{A}/\mathcal{G}}, d\tilde{\mu}^o)$; they are the $\overline{\mathcal{G}}$-invariant functions on $\overline{\mathcal{A}}$. In fact, since the spaces $\overline{\mathcal{A}}$ and $\overline{\mathcal{G}}$ are compact and measures normalized, we can regard $\tilde{\mathcal{H}}^o$ as the gauge invariant *subspace* of the Hilbert space $\mathcal{H}^o$. *In what follows, we we will often do so.*

What do 'typical' quantum states look like? To provide an intuitive picture, we can proceed as follows. Fix a graph γ with N edges and consider functions Ψ_γ of generalized connections of the form $\Psi_\gamma(\bar{A}) = \psi(\bar{A}(e_1), ..., \bar{A}(e_N))$ for *some* smooth function ψ on $[SU(2)]^N$, where $e_1, ..., e_N$ are the edges of the graph

γ. Thus, the functions Ψ_γ 'know only about' what the generalized connections do to those paths which constitute the edges of the graph γ. This space of states, although infinite dimensional, is quite 'small' in the sense that it is the L^2-space associated with a *finite* dimensional manifold. However, if we vary γ through all possible graphs, the collection of all states that results is very large. Indeed, one can show that it is *dense* in the Hilbert space $\mathcal{H}^o$. (If we restrict ourselves to Ψ_γ which are $\overline{\mathcal{G}}$-invariant, we obtain a dense subspace in $\tilde{\mathcal{H}}^o$.) Since each of these states depends only on a finite number of variables, borrowing the terminology from the quantum theory of free fields in Minkowski space, they are called *cylindrical functions* and denoted by Cyl. Gauge invariant cylindrical functions represent the 'typical' kinematic states. In many ways, Cyl is analogous to the space $C_o^\infty(R^3)$ of smooth functions of compact support on R^3 which is dense in the Hilbert space $L^2(R^3, d^3x)$ of quantum mechanics. Just as one often defines quantum operators—e.g., the position, the momentum and the Laplacian—on C_o^∞ first and then extends them to an appropriately larger domain in the Hilbert space $L^2(R^3, d^3x)$, we will define our operators first on Cyl and then extend them appropriately.

Cylindrical functions provide considerable intuition about the nature of quantum states we are led to consider. One can visualize them as '1-dimensional polymer-like excitations' of geometry/gravity. They are very different from the graviton states of perturbative treatments which are '3-dimensional wavy undulations' on flat space. Just as a polymer, although intrinsically 1-dimensional, exhibits 3-dimensional properties in sufficiently complex and densely packed configurations, the fundamental 1-dimensional excitations of geometry can be packed appropriately to provide a geometry which, when coarse-grained on scales much larger than the Planck length, resemble continuum geometries (Ashtekar *et al.* 1992). In this description, gravitons can arise only as approximate notions in the low energy regime (Iwasaki and Rovelli, 1993, 1994). At the basic level, states in $\tilde{\mathcal{H}}^o$ are fundamentally different from the Fock states of Minkowskian quantum field theories. The main reason is the underlying diffeomorphism invariance: In the absence of a background geometry, it is not possible to introduce the familiar Gaussian measures and associated Fock spaces.

Finally, it is interesting to note that the obvious generalizations of the spin-network states introduced by Roger also for quantum gravity but in a somewhat different context (Penrose 1971) provide a natural decomposition of $\tilde{\mathcal{H}}^o$ (and $\mathcal{H}^o$) into finite dimensional subspaces. Each of these subspaces can be identified with the space of states of a spin-system. This fact simplifies various constructions and calculations enormously. This is a very interesting development. However, I will not discuss it here because there were two other talks at the conference (by Kauffman and Smolin) devoted to it.

3 Quantum geometry

3.1 Preliminaries

We now wish to introduce the operators corresponding to geometrical quantities. As we saw in section 2.1, in the classical theory Riemannian geometry on Σ is determined by the triads E_i^a. Since these fields are 'canonically conjugate' to the connection A_a^i, heuristically, they are to be represented by $G\hbar\delta/\delta A_a^i$, so that the commutator between the fundamental operators is $i\hbar$ times the Poisson bracket (2.1). The main task is to make this idea precise and show that the resulting triad operators are well-defined on the kinematical Hilbert space introduced in section 2.3. We will see that this is indeed the case. One can then construct other geometric operators from these quantum triads. I will illustrate the procedure for the area operators which have had the most interesting applications so far. The length and the volume operators can be defined similarly (Thiemann 1996d; Rovelli and Smolin 1995; Loll 1995, 1996; Lewandowski 1996, Ashtekar and Lewandowski 1996a, b).

The detailed regularization procedure shows that in this approach there is remarkable 'harmony' between differential geometry and functional analysis: n-form valued operators turn out to be well-defined when smeared in n-dimensions. For example, we already saw that the holonomies of connections lead to well-defined operators on $\mathcal{H}^o$; the connection 1-form is 'smeared' in 1-dimension. The conjugate momentum E_i^a is dual to a 2-form e_{abi} (with values in the dual of the $su(2)$-Lie algebra) and geometrically it is natural to 'smear' it in 2-dimensions. It again turns out that these 2-dimensionally smeared fields lead to well-defined operators on $\mathcal{H}^o$. Indeed, if one attempts to smear connections and triads in 3-dimensions, not only is the procedure geometrically awkward but the resulting operators typically fail to be densely-defined on $\mathcal{H}^o$! This situation is in striking contrast with what happens in the Fock representation in free field theories in Minkowski space. In the Maxwell Fock space, for example, the 1-form valued connection as well as the 2-form valued electric fields are 3-dimensional distributions and fail to be well-defined if smeared in lower dimensions. The natural compatibility between geometry and analysis in the present approach can be traced back to the underlying diffeomorphism invariance, i.e., to the fact that we did not use any background metric to construct either the quantum configuration space $\overline{\mathcal{A}}$ nor the the measure μ^o thereon.

Fix any 2-surface S in Σ and let $f_\epsilon(x,y)$ be a 1-parameter family of fields on S which tend to the $\delta^{(2)}(x,y)$ as ϵ tends to zero; i.e., such that

$$\lim_{\epsilon\to 0}\int_S d^2y\, f_\epsilon(x^1,x^2;y^1,y^2)g(y^1,y^2) = g(x^1,x^2)\,, \tag{3.1}$$

for all smooth densities g of weight 1 and of compact support on S. (Thus, $f_\epsilon(x,y)$ is a density of weight 1 in x and a function in y.) The smeared version of $E_i^a(x)$ will be defined to be:

$$\begin{aligned} [E_i]_f(x) &:= \int_S dy^a \wedge dy^b\, f_\epsilon(x,y)\eta_{abc}E_i^c(y) \\ &= \int_S dy^a \wedge dy^b f_\epsilon(x,y)e_{abi}(y) \end{aligned} \tag{3.2}$$

so that, if the surface S is given in local coordinates by $x^3 = \text{const}$, as ϵ tends to zero $[E_i]_f$ tends to $E_i^3(x)$. The 'point-splitting' strategy now provides a 'regularized expression' of area (see Eq. (2.2)):

$$[A_S]_f := \int_S d^2x\, \left[[E_i]_f(x)[E^i]_f(x)\right]^{\frac{1}{2}}, \tag{3.3}$$

which will serve as the point of departure for our construction of the area operator. To simplify technicalities, we will assume that the smearing field $f_\epsilon(x,y)$ has the following additional properties for sufficiently small $\epsilon > 0$: i) for any given y, $f_\epsilon(x,y)$ has compact support in x which shrinks uniformly to y as ϵ tends to zero; and, ii) $f_\epsilon(x,y)$ is non-negative. These conditions are very mild and we are thus left with a large class of regulators[3]. Our task is to find the operators corresponding to these smeared triads and areas.

3.2 Triad operators

Let us fix a graph γ and consider a cylindrical function Ψ_γ on $\overline{\mathcal{A}}$,

$$\Psi_\gamma(\bar{A}) = \psi(\bar{A}(e_1),..,\bar{A}(e_N)), \tag{3.4}$$

where, as before, N is the total number of edges of γ and where ψ is a smooth function on $[SU(2)]^N$. A careful regularization (Ashtekar and Lewandowski 1996a) shows that the smeared operator $[\hat{E}_i]_f$ has the following action on Ψ_γ:

$$([\hat{E}_i]_f(x) \cdot \Psi_\gamma)(\bar{A}) = (\frac{i\ell_P^2}{2}\Big[\sum_{I=1}^{N} \kappa_I\, f_\epsilon(x,v_I)\; X_I^i\Big] \cdot \psi)(\bar{A}(e_1),...,\bar{A}(e_N)). \tag{3.5}$$

Here $\ell_P = \sqrt{G\hbar}$ is the Planck length, I labels the edges of the graph γ, v_I the point at which the I-th edge intersects S and κ_I is a constant associated with the edge e_I via

$$\kappa_I = \begin{cases} 0, & \text{if } e_I \text{ is tangential to } S \text{ or does not intersect } S, \\ +1, & \text{if } e_I \text{ has an isolated intersection with } S \text{ and lies above } S, \\ -1, & \text{if } e_I \text{ has an isolated intersection with } S \text{ and lies below } S. \end{cases} \tag{3.6}$$

[3]For example, $f_\epsilon(x,y)$ can be constructed as follows. Take *any* non-negative function f of compact support on S such that $\int d^2x f(x) = 1$ and set $f_\epsilon(x,y) = (1/\epsilon^2)f((x-y)/\epsilon)$. Here, we have implicitly used the given chart to give $f_\epsilon(x,y)$ a density weight in x.

The non-trivial part of the operator (3.5) is contained in X^i_I which are right/left invariant vector fields on $SU(2)$. More precisely, X^i_I is an operator assigned to the vertex v_I and and the oriented edge e_I by the following formula

$$(X^i_I \cdot \psi)(\bar{A}(e_1),...,\bar{A}(e_N)) = \begin{cases} (\bar{A}(e_I)\tau^i)^A_B \frac{\partial \psi}{\partial(\bar{A}(e_I))^A_B}, & \text{when } e_I \text{ is outgoing} \\ -(\tau^i\bar{A}(e_I))^A_B \frac{\partial \psi}{\partial(\bar{A}(e_I))^A_B}, & \text{when } e_I \text{ is incoming.} \end{cases} \tag{3.7}$$

Thus, if none of the edges of the graph γ intersect S, or if they all lie within S, the operator $[\hat{E}_i]_f(x)$ simply annihilates the state. The action is non-trivial only if some of the edges intersect S transversely. If they do, at the vertex of intersection each edge contributes via the Lie derivative along the i-th right or left invariant vector field on the copy of $SU(2)$ associated with that edge (depending on whether the edge is oriented to be outgoing at the vertex or incoming). Thus, the operator is geometrical and rather simple.

Denote by $\mathcal{H}^o_\gamma$ the Hilbert space $L^2(\mathcal{A}_\gamma, d\mu^o_\gamma)$ of square integrable cylindrical functions associated with a fixed graph γ. Now, μ^o_γ is the induced Haar measure on $\mathcal{A}_\gamma$ and the operator is just a sum of right/left invariant vector fields. Hence, standard results in analysis imply that, with domain Cyl^1_γ of all C^1 cylindrical functions based on γ, the operator (3.5) is essentially self-adjoint on $\mathcal{H}^o_\gamma$. Now, it is straightforward to verify that the operators on $\mathcal{H}^o_\gamma$ obtained by varying γ are all compatible in the appropriate sense.[4] Hence, it follows from the general results on projective limits (Ashtekar and Lewandowski 1995a) that $[\hat{E}_i]_f(x)$, with domain Cyl^1 (the space of all C^1 cylindrical functions), is an essentially self-adjoint operator on $\mathcal{H}^o$. For notational simplicity, we will denote its self-adjoint extension also by $[\hat{E}_i]_f(x)$. (The context should make it clear whether we are referring to the essentially self-adjoint operator or its extension.) These results will be important in the next subsection in the definition of the area operator.

Finally, it is straightforward to remove the regulator, i.e., take the limit $\epsilon \to 0$. Let us, as before, introduce a chart and denote the surface by $x^3 = \text{const}$. The resulting triad operator-valued distribution is then given by:

$$([\hat{E}^3_i](x) \cdot \Psi_\gamma)(\bar{A}) = (\frac{i\ell^2_P}{2} \sum_v \delta^{(2)}(x,v) \sum_{I_v} \kappa_{I_v}\, X^i_{I_v} \cdot \psi)(\bar{A}(e_1),...,\bar{A}(e_N)), \tag{3.8}$$

where v ranges over vertices of γ and I_v ranges over the edges of γ which meet at v. Alternatively, given any 2-surface S in Σ, and a test field λ^i with values in

[4] Given two graphs, γ and γ', we say that $\gamma \geq \gamma'$ if and only if every edge of γ' can be written as a composition of edges of γ. Given two such graphs, there is a projection map from $\mathcal{A}_\gamma$ to $\mathcal{A}_{\gamma'}$, which, via pull-back, provides an unitary embedding $U_{\gamma,\gamma'}$ of $\mathcal{H}^o_{\gamma'}$ into $\mathcal{H}^o_\gamma$. A family of operators $\mathcal{O}_\gamma$ on the Hilbert spaces $\mathcal{H}^o_\gamma$ is said to be compatible if $U_{\gamma,\gamma'}\mathcal{O}_{\gamma'} = \mathcal{O}_\gamma U_{\gamma,\gamma'}$ and $U_{\gamma,\gamma'}D_{\gamma'} \subset D_\gamma$ for all $\gamma \geq \gamma'$, where D_γ and $D_{\gamma'}$ are the domains of $\mathcal{O}_\gamma$ and $\mathcal{O}_{\gamma'}$.

the $su(2)$-Lie algebra,

$$\begin{aligned}\hat{e}[\lambda] \cdot \Psi_\gamma &:= \int_S dx^a \wedge dx^b \hat{e}_{abi}(x)\lambda^i(x) \cdot \Psi_\gamma \\ &= (\frac{i\ell_P^2}{2} \sum_v \sum_{I_v} \kappa_{I_v}\, X^i_{I_v} \cdot \psi)(\bar{A}(e_1), ..., \bar{A}(e_N)) \,,\end{aligned} \tag{3.9}$$

is an essentially self-adjoint operator with domain Cyl^1, where v ranges over points of S. (The sum is well-defined because, for any cylindrical function Ψ_γ, v ranges over only a finite number of points.)

3.3 Area operators

Let us now turn to the expression (3.3) of the smeared area functional. Since we already have the expression (3.5) of the smeared triad operators, our task is quite straightforward. Let us denote the determinant of the intrinsic metric on S by g_S, and its smeared version by $[g_S]_f$. Then, (3.3) and (3.5) imply:

$$\begin{aligned}[\hat{g}_S]_f(x) \cdot \Psi_\gamma &:= [\hat{E}_i]_f(x)[\hat{E}^i]_f(x) \cdot \Psi_\gamma \\ &= -\frac{\ell_P^4}{4}[\sum_{I,J} \kappa(I,J) f_\epsilon(x, v_I) f_\epsilon(x, v_J)\, X^i_I X^i_J] \cdot \Psi_\gamma \,,\end{aligned} \tag{3.10}$$

where the summation extends over all the oriented pairs (I, J); v_I and v_J are the vertices at which edges e_I and e_J intersect S; $\kappa(I, J) = \kappa_I \kappa_J$ equals 0 if either of the two edges e_I and e_J fails to intersect S or lies entirely in S, $+1$ if they lie on the same side of S, and -1 if they lie on the opposite sides. (For notational simplicity, from now on we shall not keep track of the position of the internal indices i; as noted in section 2.1, they are contracted using the invariant metric on the Lie algebra $su(2)$.) The next step is to consider vertices v at which γ intersects S and, as in (3.8), simply rewrite the above sum by re-grouping terms by vertices. The result simplifies if we choose ϵ sufficiently small so that $f_\epsilon(x, v_I) f_\epsilon(x, v_J)$ is zero unless $v_I = v_J$. We then have:

$$[\hat{g}_S]_f(x) \cdot \Psi_\gamma = -\frac{\ell_P^4}{4}\, [\sum_v (f_\epsilon(x, v))^2 \sum_{I_v, J_v} \kappa(I_v, J_v) X^i_{I_v} X^i_{J_v}] \cdot \Psi_\gamma \,, \tag{3.11}$$

where, as before, the index v labels the vertices on S and I_v and J_v label the edges at the vertex v.

The next step is to take the square-root of this expression. The same reasoning that established the self-adjointness of $[\hat{E}_i]_f(x)$ now implies that $[\hat{g}_S]_f(x)$ is a non-negative self-adjoint operator and hence has a well-defined square-root which is also a positive definite self-adjoint operator. Since we have chosen ϵ to be sufficiently small, for any given point x in S, $f_\epsilon(x, v)$ is non-zero for at most one vertex v. We can therefore take the sum over v outside the square-root. One then obtains

$$([\hat{g}_S]_f)^{\frac{1}{2}}(x) \cdot \Psi_\gamma = \frac{\ell_P^2}{2} \sum_v f_\epsilon(x,v) [\sum_{I_v,J_v} \kappa(I_v,J_v) X^i_{I_v} X^i_{J_v}]^{\frac{1}{2}} \cdot \Psi_\gamma. \tag{3.12}$$

Note that the operator is neatly split; the x-dependence all resides in f_ϵ and the operator within the square-root is 'internal' in the sense that it acts only on copies of $SU(2)$.

Finally, we can remove the regulator, i.e., take the limit as ϵ tends to zero. By integrating both sides against test functions on S and then taking the limit, we conclude that the following equality holds in the distributional sense:

$$\widehat{\sqrt{g_S}}(x) \cdot \Psi_\gamma = \frac{\ell_P^2}{2} \sum_v \delta^{(2)}(x,v) [\sum_{I_v,J_v} \kappa(I_v,J_v) X^i_{I_v} X^i_{J_v}]^{\frac{1}{2}} \cdot \Psi_\gamma. \tag{3.13}$$

Thus, the area element operator is a well-defined 2-dimensional distribution on the surface S (defined by x_3 = const). It has the same geometrical structure (i.e. density character) as its classical analogue. It is somewhat surprising that a well-defined operator-valued distribution could be extracted in spite of the fact that, due to the square-root, $\sqrt{g_S}(x)$ has a non-polynomial dependence on the triad $E^a_i(x)$.

It now follows immediately that the regularized area operator is given by:

$$\hat{A}_S \cdot \Psi_\gamma = \frac{\ell_P^2}{2} \sum_v [\sum_{I_v,J_v} \kappa(I_v,J_v) X^i_{I_v} X^i_{J_v}]^{\frac{1}{2}} \cdot \Psi_\gamma. \tag{3.14}$$

(Here, as before, v labels the vertices at which γ intersects S and I_v labels the edges of γ at the vertex v.) With Cyl^2 as its domain, $\hat{A}_S$ is essentially self-adjoint on the Hilbert space $\mathcal{H}^o$.

So far, for simplicity, we assumed that the surface Σ is covered by a single chart of adapted coordinates. One can remove this assumption using the same arguments as in classical differential geometry. Thus, if such a global chart does not exist, we can cover Σ with a family $\mathcal{U}$ of neighbourhoods such that for each $U \in \mathcal{U}$ there exists a local coordinates system (x^a) adapted to Σ. Let $(\varphi_U)_{U\in\mathcal{U}}$ be a partition of unity associated to $\mathcal{U}$. We just repeat the above regularization for a slightly modified classical surface area functional, namely for

$$A_{S,U} := \int_S dx^1 \wedge dx^2 \, \varphi_U [E^3_i E^{3i}]^{\frac{1}{2}} \tag{3.15}$$

which has support within a domain U of an adapted chart. Thus, we obtain the operator $\hat{A_{SU}}$. Then we just define

$$\hat{A}_S = \sum_{U\in\mathcal{U}} \hat{A}_{S,U}. \tag{3.16}$$

The result is given again by formula (3.14). The reason why the functions φ_U disappear from the result is that the operator obtained for a single domain of an adapted chart is insensitive to changes of this chart. This concludes our technical discussion.

The classical expression A_S of (2.2) is rather complicated. It is therefore somewhat surprising that the corresponding quantum operators can be constructed rigorously and have quite manageable expressions. The essential reason is the underlying diffeomorphism invariance which severely restricts the possible operators. Given a surface and a graph, the only diffeomorphism invariant entities are the intersection vertices. Thus, a diffeomorphism covariant operator can only involve structure at these vertices. In our case, it just acts on the copies of $SU(2)$ associated with various edges at these vertices.

I have presented this derivation in considerable detail to spell out all the assumptions, to bring out the generality of the procedure and to illustrate how regularization can be carried out in a fully non-perturbative treatment. While one is free to introduce auxiliary structures such as preferred charts or background fields in the intermediate steps, the final result must respect the underlying diffeomorphism invariance of the theory. Construction of other geometric operators follows this general procedure.

3.4 Properties of area operators

I will now discuss a few illustrative properties of area operators. (For a more complete discussion of the spectrum and other properties see Ashtekar and Lewandowski 1996a).

1. *Vertex operators:* Most properties of the area operators can be derived by noting that they are constructed from certain 'vertex operators'. To see this, recall first that, in the final expression (3.12) of the area element operator, there is a clean separation between the 'x-dependent' and the 'internal' parts. Given a graph γ, the internal part is a sum of square-roots of the operators

$$\triangle_{S,v} := \sum_{I_v, J_v} \kappa(I_v, J_v) X^i_{I_v} X^i_{J_v} \tag{3.17}$$

associated with the surface S and the vertex v on it. It is straightforward to check that operators corresponding to different vertices commute. Therefore, to analyse the properties of area operators, we can focus just on one vertex operator at a time.

Furthermore, given the surface S and a point v on it, we can define an operator $\triangle_{S,v}$ on the dense subspace Cyl^2 on $\mathcal{H}^o$ as follows:

$$\triangle_{S,v} \cdot \Psi_\gamma := \begin{cases} \sum_{I_v, J_v} \kappa(I_v, J_v) X^i_{I_v} X^i_{J_v} \cdot \Psi_\gamma, & \text{if } \gamma \text{ intersects } S \text{ in } v, \\ 0, & \text{otherwise} \end{cases} \tag{3.18}$$

where, as before, I_v and J_v label the edges of γ which have v as a vertex. (Recall that every cylindrical function is associated with *some* graph γ. If γ intersects

S at v but v is not a vertex of γ, one can simply 'extend' γ by adding a new vertex v and orienting the edges at v to outgoing.) It is straightforward to verify that this definition is unambiguous: if a cylindrical function can be represented in two different ways, say as Ψ_γ and $\Psi_{\gamma'}$, then $\triangle_{S,v} \cdot \Psi_\gamma$ and $\triangle_{S,v} \cdot \Psi_{\gamma'}$ are two representations of the same function on $\overline{\mathcal{A}}$. Finally, there is a precise sense in which $\triangle_{S,v}$ can be regarded as a Laplacian operator on $\mathcal{H}^o$ (Ashtekar and Lewandowski 1995a). The area operator is a sum over all the points v of S of square-roots of Laplacians,

$$\hat{A}_S = \frac{\ell_P^2}{2} \sum_{v \in S} \sqrt{-\triangle_{S,v}}\,. \tag{3.19}$$

The sum is well defined because, for any cylindrical function, it contains only a finite number of non-zero terms corresponding to the isolated intersection points of the associated graph with S.

2. *Discreteness of the spectrum:* By inspection, it follows that the total area operator $\hat{A}_S$ leaves the subspace of Cyl^2_γ which is associated with any one graph γ invariant and is a self-adjoint operator on the subspace $\mathcal{H}^o_\gamma$ of $\mathcal{H}^o$ corresponding to γ. Next, recall that $\mathcal{H}^o_\gamma = L^2(\mathcal{A}_\gamma, d\mu^o)$, where $\mathcal{A}_\gamma$ is a compact manifold, isomorphic with $(SU(2))^N$ where N is the total number of edges in γ. As, explained above, the restriction of $\hat{A}_S$ to $\mathcal{H}^o_\gamma$ is given by certain commuting elliptic differential operators on this compact manifold. Therefore, all its eigenvalues are discrete. Now suppose that the complete spectrum of $\hat{A}_S$ on $\mathcal{H}^o$ has a continuous part. Denote by P_c the associated projector. Then, given any Ψ in $\mathcal{H}^o$, $P_c \cdot \Psi$ is orthogonal to $\mathcal{H}^o_\gamma$ for any graph γ, and hence to the space Cyl of cylindrical functions. Now, since Cyl^2 is dense in $\mathcal{H}^o$, $P_c \cdot \Psi$ must vanish for all Ψ in $\mathcal{H}^o$. Hence, the spectrum of $\hat{A}_S$ has no continuous part.

Note that this method is rather general: It can be used to show that *any* self-adjoint operator on $\mathcal{H}^o$ which maps (the intersection of its domain with) $\mathcal{H}^o_\gamma$ to $\mathcal{H}^o_\gamma$, and whose action on $\mathcal{H}^o_\gamma$ is given by elliptic differential operators, has a purely discrete spectrum on $\mathcal{H}^o$. Geometrical operators, constructed purely from the triad field tend to satisfy these properties.

3. *Spectrum:* In the case of the area operators, one can do much more: the *complete* spectrum can be calculated. If we assume that the surface is open and its closure is contained in Σ, the eigenvalues are given by:

$$a_S = \frac{\ell_P^2}{2} \sum_v \left[2j_v^{(d)}(j_v^{(d)}+1) + 2j_v^{(u)}(j_v^{(u)}+1) - j_v^{(d+u)}(j_v^{(d+u)}+1)\right]^{\frac{1}{2}} \tag{3.20}$$

where v runs over non-negative integers (which label 'vertices' in S) and $j^{(d)}, j^{(u)}$ and $j^{(d+u)}$ are non-negative half-integers assigned to each v and subject to the usual inequality

$$|j^{(d)} - j^{(u)}| \le j^{(d+u)} \le j^{(d)} + j^{(u)}. \tag{3.21}$$

These eigenvalues are realized not only on the full Hilbert space $\mathcal{H}^o$ but also on the gauge invariant subspace $\tilde{\mathcal{H}}^o$ thereof. If the topology is non-trivial, however, not all of these eigenvalues are admitted in the gauge invariant subspace. I will indicate some of these restrictions below.

4. *Area gap:* The minimum eigenvalue is of course zero. Since the spectrum is discrete, there is an 'area gap'. Interestingly, the gap contains information about the topology of the surface. Consider first the case when the surface S is open with its closure contained in Σ. An example is a disk $z = 0, x^2 + y^2 < 1$ in $\mathcal{R}^3$. Then the first non-zero eigenvalue is obtained when (there is a single 'vertex' and, e.g.,) $j^{(d)} = 0$ and $j^{(u)} = j^{(d+u)} = 1/2$ in (3.20). The value is $a_S^o = (\sqrt{3}/4)\ell_P^2$. A second possibility is to consider a surface S which is closed ($\partial S = \emptyset$) and divides Σ into disjoint open sets. An example is given by $S = \mathcal{S}^2$ and $\Sigma = \mathcal{R}^3$. Then, the area gap is higher; $a_S^o = (2\sqrt{2}/4)\ell_P^2$. Finally, consider the case when S is closed but not of the second type above. An example is given by $S = \mathcal{T}^2$, a 2-torus in $\Sigma = \mathcal{T}^3$. In this case the area gap is in between. It is given by $a_S^o = (2/4)\ell_P^2$.

5. *Asymptotic behaviour of eigenvalues:* Although the spectrum is discrete, the level spacing is not equal; the eigenvalues tend to crowd for large areas. This decrease in level spacing occurs very rapidly. In the case of trivial topology, for instance, there is only one non-zero eigenvalue with $a_S < 0.5\ell_P^2$, seven with $a_S < \ell_P^2$ and 98 eigenvalues with $a_S < 2\ell_P^2$ (Fairhurst 1996). A quick estimate shows that, in the large eigenvalue limit, one can bound the spacing between consecutive eigenvalues via:

$$\delta a_S \leq (\frac{\ell_P}{2\sqrt{a_S}})\ell_P^2 + \mathcal{O}(\frac{\ell_P^2}{a_S})\ell_P^2. \tag{3.22}$$

(This is a 'quick' bound, far from being the best.) Thus, although the spectrum *is* purely discrete, for large areas, it approaches the continuum rather rapidly.

This concludes the discussion of quantum geometry. My aim in this subsection was to illustrate how various properties of geometric operators can be analysed in detail within this framework. These properties bring out the striking difference between the Riemannian geometry that underlies the classical theory and 'polymer-like geometry' that emerges from quantum states and operators. It is particularly pleasing to see how the topology of the 2-surface S is coded in the spectrum of the area operator A_S. The fact that the level-spacing in the spectrum of the area operator goes rapidly to zero makes it easy to visualize why the continuum picture is such an excellent approximation even on the smallest laboratory scales (10^{-18} cm $\approx 10^{15}\ell_P$) probed in high energy physics. Furthermore, as we will see in the next section, this asymptotic behaviour of the spectrum in the large area limit has interesting consequences to the issue of black-hole evaporation (Bekenstein and Mukhanov 1995).

4 Discussion

Riemannian geometry provides the mathematical framework to formulate general relativity (as well as other modern theories of gravity). If we follow the line of thought advocated by Roger Penrose and face the problem of quantum gravity non-perturbatively, we are naturally led to the conclusion that a quantum theory of gravity should, in particular, provide us with a quantum theory of geometry. Riemannian geometry can then emerge only as an approximation on a large scale, upon a suitable coarse graining of the semiclassical states of the full-fledged quantum theory. What then is to substitute Riemannian geometry in the small? What is the mathematical framework for describing the 'atoms' of geometry, its 'elementary excitations', its 'basic quanta'? In this contribution, I have presented one possible candidate. The resulting quantum geometry is strikingly different from the familiar continuum descriptions. In particular, the triad operators—as well as area operators associated with intersecting surfaces—fail to commute giving rise to certain intrinsic uncertainties in the simultaneous measurements of geometric quantities (Ashtekar *et al.* 1996). The basic excitations are 1-dimensional, rather like polymers. And geometry is 'quantized' in the old-fashioned sense of the word: geometrical quantities such as lengths, areas and volumes, which take continuous values classically, are represented in the quantum theory by operators with *purely* discrete spectra.

The basic field in the classical (Hamiltonian) theory as well as in the quantum description is the (density weighted) triad E^a_i which can be naturally thought of as a triplet of 2-forms e_{abi}. We saw that, in the passage to quantum theory, differential geometry and functional analysis fuse in a coherent fashion. Classical triads can be most naturally integrated over 2-surfaces. Their quantum analogues turned out naturally to be *2-dimensional* operator-valued distributions; they have to be smeared also along 2-surfaces. In the classical theory, using the Levi-Civita density on the 3-manifold Σ one can also construct 1-forms, e^i_a, the co-triads, which play an important role, e.g., in the definition of lengths of curves (Thiemann, 1996d). It turns out that their quantum analogues are 1-dimensional operator-valued distributions; they have to be smeared along 1-dimensional curves. As noted in section 3.1, this harmonious blending of geometry and analysis is absent in Minkowskian quantum field theories which depend critically on the flat background metric. It is an essentially non-perturbative feature.

A detailed picture of quantum geometry is now emerging. It has several striking features, perhaps the most intriguing of which is the 1-dimensional nature of fundamental excitations. How did this arise? What inputs have gone in this construction? There is in essence only one fundamental assumption: the requirement that traces of holonomies of the connection A around closed loops α—or, in the physics terminology, the Wilson loop variables—$T_\alpha = \mathrm{Tr}\,\mathcal{P} - \exp \int_\alpha A$, be well-defined operators in the quantum theory. This assumption leads one directly to the quantum configuration space $\overline{\mathcal{A}/\mathcal{G}}$ (Ashtekar and Isham, 1992) and hence to the cylindrical functions based on graphs which in turn provide the 'polymer picture' of the basic excitations.

Since there is no background metric, the expressions of geometric operators are severely constrained. The area operator A_S, for example, can have a non-trivial action on a state Ψ_γ only if the surface S intersects the graph γ.

Note however, that the detailed expressions of the operator and its eigenvalues are quite involved; they could not have been guessed from general considerations. The same is true for other geometric operators. This is an important feature and arises because our approach is 'constructive'; we did not pre-suppose the expressions of these operators but arrived at them systematically starting from differential geometry and canonical quantization. By contrast, iconoclastic approaches tend to begin by postulating the nature of the micro-structure of space-time—e.g., by declaring it to be a causal set—and the expressions of geometric quantities are also postulated. Typically, such definitions capture the gross features of our expressions but miss out the subtleties.

The details can have important consequences. I will conclude with an interesting illustration. The simplest eigenvectors Ψ_γ of A_S occur if at each intersection between γ and S there is precisely one incoming edge (lying below S) and one outgoing edge (lying above S). Suppose there are N vertices on S and $2N$ such edges. Then Ψ_γ is an eigenvector of A_S with eigenvalue $a_S = N\,(\sqrt{3}/2)\ell_P^2$. These eigenvalues have 'equal spacing', like the energy levels of the harmonic oscillator. Not surprisingly, these eigenvalues were the first to be discovered. Suppose for a moment (as would be natural in any approach that *begins* by postulating discreteness) that this was the *entire* spectrum of A_S. Then, as Bekenstein and Mukhanov (1995) have argued, the spectrum emitted by an evaporating black-hole would be quite different from that given by the semi-classical calculations of Hawking's (1975).

To see this, recall first that the area of the horizon A goes as M^2, where M is the mass of the black-hole. As the black-hole radiates, the area shrinks. From the perspective of full quantum gravity, the area makes quantum jumps from one eigenvalue to another. In natural units $c = \hbar = G = 1$, the energy emitted in such a transition is given by $\delta M \sim \delta a_S/M$ since $A_S \sim M^2$, where S is now the horizon. For the simple eigenvalues a_S given above, the energy emitted in any one transition is thus N/M, if the transition involves N steps of area eigenvalues. Now, in the Hawking spectrum, the peak of the black-body curve occurs at the frequency $\omega_o \sim 1/M$. Hence, if the area spectrum were the naive one, one would not see any radiation at frequencies below the peak and one would only see lines at frequencies $\omega_o, 2\omega_o, ..., N\omega_o, ...$; the spectral lines would not approximate the continuum back body spectrum at all! Thus, if the spectrum of the area operator were the naive one, quantum gravity would predict major deviations from the black-body spectrum predicted by Hawking.

What is the situation if one uses the correct, full spectrum of the area operator? Now, for large black-holes, the eigenvalues crowd; the level spacing is not equal but tends to zero as a_S increases. As we saw in section 3.4, a (crude) bound is given by $\delta a_S \leq 1/\sqrt{a_S} \sim 1/M$. Hence the energy emitted in a transition goes as $\delta M \sim \omega_o/M$. Thus, there are many, many spectral lines and these, together, can easily approximate the continuous black-body spectrum. There is no a priori conflict with Hawking's semi-classical calculation.[5]

Acknowledgements

I would like to thank John Baez, Chris Isham, Don Marolf, Jose Mourao, Carlo Rovelli, Thomas Thiemann and especially Jerzy Lewandowski for numerous discussions and collaboration. This work was supported in part by the NSF Grants PHY93-96246 and PHY95-14240, by the Erwin Schrödinger International Institute for Mathematical Sciences, and by the Eberly fund of the Pennsylvania State University.

Bibliography

Ashtekar, A. (1987a). *Physical Review*, **D9**, 1433.

Ashtekar, A. (1987b). In *Mathematics and general relativity*, J. Isenberg (ed) American Mathematical Society, Providence.

Ashtekar, A. (1995). Polymer geometry at Planck scale and quantum Einstein's equations, CGPG-95/11-5 pre-print, hep-th/9601054. To appear in the *Proceedings of the 14th International Conference on General Relativity and Gravitation*, M. Francaviglia (ed) World Scientific, Singapore.

Ashtekar, A. (1996). Large quantum gravity effects: Unforeseen limitations of the classical theory, gr-qc/9610008, *Physical Review Letters* in press.

Ashtekar, A. and Isham, C. J. (1992). *Classical and Quantum Gravity*, **9**, 1433.

Ashtekar, A. and Lewandowski, J. (1994). In *Knots and quantum gravity*, J. Baez (ed), Oxford University Press, Oxford

Ashtekar, A. and Lewandowski, J. (1995a). *Journal of Geometry and Physics*, **17**, 191.

Ashtekar, A. and Lewandowski, J. (1995b). *Journal of Mathematical Physics*, **36**, 2170.

Ashtekar, A. and Lewandowski, J. (1996a). Quantum theory of geometry I, gr-qc/9602046, *Classical and Quantum Gravity* in press.

Ashtekar, A. and Lewandowski, J. (1996b). Quantum theory of geometry III, CGPG-preprint.

Ashtekar, A. and Varadarajan, M. (1994). *Physical Review*, **D50**, 4944.

Ashtekar, A., Rovelli, C., and Smolin, L. (1992). *Physical Review Letters*, **69**, 237.

Ashtekar, A., Lewandowski, J., Marolf, D., Mourão, J., and Thiemann, T. (1995). *Journal of Mathematical Physics*, **36**, 6456.

[5]Of course, to be really sure that there is no conflict, one has to calculate all the selection rules that will arise from quantum dynamics and verify that these still leave 'enough' permissible transitions to approximate the continuous black-body spectrum. It is only at the kinematic level implicit in all the discussions so far that the full spectrum removes the apparent problem.

Ashtekar A., Corichi, A., Lewandowski, J., and Zapata J.-A. (1996). Quantum theory of geometry II, CGPG pre-print.

Baez, J. (1994a). *Letters in Mathematical Physics*, **31**, 213.

Baez, J. (1994b). In *The Proceedings of the conference on quantum topology.* D. Yetter (ed) World Scientific, Singapore. hep-th/9305045.

Baez, J. and Sawin, S. (1995). Functional integration on spaces of connections, q-alg/9507023.

Baily, T. N. and Baston, R. J. (1990). *Twistors in mathematics and physics.* Cambridge University Press, Cambridge.

Bekenstein, J. and Mukhanov, S. (1995). *Physics Letters* **B360**, 7.

Bombelli, L., Lee, J., Meyer, D., and Sorkin, R.D. (1987). *Physical Review Letters*, **59**, 521.

Fairhurst, S. (1996). Private communication.

Frittelli, S., Kozameh, C., Newman, E. T., Rovelli, C., and Tate, R. (1996). Fuzzy space-time from a null-surface version of GR, gr-qc/9603061.

Hawking, S. W. (1975). *Communications in Mathematical Physics* **43**, 199.

Hawking, S. W. (1979). In *General relativity: An Einstein centenary survey.* S.W. Hawking and W. Israel (eds), Cambridge University Press, Cambridge.

Iwasaki, J. and Rovelli, C. (1993). *International Journal of Modern Physics*, **D1**, 533.

Iwasaki, J. and Rovelli, C. (1994). *Classical and Quantum Gravity*, **11**, 2899.

Ko, M., Ludvigsen, M., Newman, E. T., and Tod, K.P. (1981). *Physics Reports*, **71**, 51.

Lewandowski, J. (1996). Volume and quantizations, gr-qc/ 9602035.

Loll, R. (1995). *Physical Review Letters*, **75**, 3084.

Loll, R. (1996). Spectrum of the volume operator in quantum gravity, *Nuclear Physics*, **B**, in press.

Marolf, D. and Mourão, J. (1995) *Communications in Mathematical Physics* **170**, 583.

Mason, L. J. and Hughston, L. P. (1990). *Further advances in twistor theory: Volume I: Applications of the Penrose transform.* Longman Scientific and Technical, London.

Mason, L. J., Hughston, L. P., and Kobak, P. Z. (1995). *Further advances in twistor theory: Volume II: Integrable systems, conformal geometry and gravitation.* Longman Scientific and Technical, London.

Penrose, R. (1971). In *Quantum Theory and Beyond*, T. Bastin (ed), Cambridge University Press, Cambridge.

Penrose, R. (1976). In *General Relativity and Gravitation*, **7**, 31.

Penrose, R. (1997). Article in this volume.

Rovelli, C. and Smolin, L. (1994). *Physical Review Letters*, **72**, 446.

Rovelli, C. and Smolin, L. (1995). *Nuclear Physics* **B442**, 593.

Thicmann, T. (1996a). Anomaly-free formulation of non-perturbative, four-dimensional Lorentzian quantum gravity, gr-qc/9606088, *Physics Letters*, **B**, in press.

Thiemann, T. (1996b). Quantum spin dynamics (QSD), gr-qc/9606089.

Thiemann, T. (1996c). Quantum spin dynamics (QSD) II, gr-qc/9606090.

Thiemann, T. (1996d). A length operator for canonical quantum gravity, gr-qc/9606092.

Varadarajan, M. (1995). *Physical Review Letters* **D52**, 2020.

12
From Quantum Code-making to Quantum Code-breaking

Artur Ekert
Clarendon Laboratory, Parks Road, Oxford OX1 3PU.

1 What is wrong with classical cryptography?

Human desire to communicate secretly is at least as old as writing itself and goes back to the beginnings of our civilisation. Methods of secret communication were developed by many ancient societies, including those of Mesopotamia, Egypt, India, and China, but details regarding the origins of cryptology[1] remain unknown (Kahn 1967).

We know that it was the Spartans, the most warlike of the Greeks, who pioneered military cryptography in Europe. Around 400 BC they employed a device known as a *scytale*. The device, used for communication between military commanders, consisted of a tapered baton around which was wrapped a spiral strip of parchment or leather containing the message. Words were then written lengthwise along the baton, one letter on each revolution of the strip. When unwrapped, the letters of the message appeared scrambled and the parchment was sent on its way. The receiver wrapped the parchment around another baton of the same shape and the original message reappeared.

Julius Caesar allegedly used, in his correspondence, a simple letter substitution method. Each letter of Caesar's message was replaced by the letter that followed it alphabetically by three places. The letter A was replaced by D, the letter B by E, and so on. For example, the English word COLD after the Caesar substitution appears as FROG. This method is still called the Caesar cipher, regardless of the size of the shift used for the substitution.

These two simple examples already contain the two basic methods of encryption which are still employed by cryptographers today, namely *transposition* and *substitution*. In transposition (*e.g. scytale*) the letters of the *plaintext*, the technical term for the message to be transmitted, are rearranged by a special permutation. In substitution (*e.g.* Caesar's cipher) the letters of the plaintext are replaced by other letters, numbers or arbitrary symbols. In general the two

[1]The science of secure communication is called cryptology from Greek *kryptos* hidden and *logos* word. Cryptology embodies cryptography, the art of code-making, and cryptanalysis, the art of code-breaking.

techniques can be combined (for an introduction to modern cryptology see, for example, Menezes *et al.* 1996; Schneier 1994; Welsh 1988).

Originally the security of a cryptotext depended on the secrecy of the entire encrypting and decrypting procedures; however, today we use ciphers for which the algorithm for encrypting and decrypting could be revealed to anybody without compromising the security of a particular cryptogram. In such ciphers a set of specific parameters, called a *key*, is supplied together with the plaintext as an input to the encrypting algorithm, and together with the cryptogram as an input to the decrypting algorithm. This can be written as

$$\hat{E}_k(P) = C, \text{ and conversely, } \hat{D}_k(C) = P, \tag{1.1}$$

where P stands for plaintext, C for cryptotext or cryptogram, k for cryptographic key, and $\hat{E}$ and $\hat{D}$ denote an encryption and a decryption operation respectively.

The encrypting and decrypting algorithms are publicly known; the security of the cryptogram depends entirely on the secrecy of the key, and this key must consist of a *randomly chosen*, sufficiently long string of bits. Probably the best way to explain this procedure is to have a quick look at the Vernam cipher, also known as the one-time pad.

If we choose a very simple digital alphabet in which we use only capital letters and some punctuation marks such as

A	B	C	D	E	...	...	X	Y	Z		?	,	.
01	02	03	04	05	...	...	24	25	26	27	28	29	30

we can illustrate the secret-key encrypting procedure by the following simple example:

H	E	L	L	O		R	O	G	E	R	
08	05	12	12	15	27	18	15	07	05	18	30
24	14	26	25	29	17	28	12	01	18	27	03
02	19	08	07	14	14	16	07	08	23	15	03

In order to obtain the cryptogram (sequence of digits in the bottom row) we add the plaintext numbers (the top row of digits) to the key numbers (the middle row), which are randomly selected from between 1 and 30, and take the remainder after division of the sum by 30, that is we perform addition modulo 30. For example, the first letter of the message "H" becomes a number "08"in the plaintext, then we add $08 + 24 = 32$; $32 = 1 \times 30 + 2$, therefore we get 02 in the cryptogram. The encryption and decryption can be written as $P + k \pmod{30} = C$ and $C - k \pmod{30} = P$ respectively.

The cipher was invented by Major Joseph Mauborgne and AT&T's Gilbert Vernam in 1917 and we know that if the key is secure, the same length as the message, truly random, and never reused, this cipher is really unbreakable! So what is wrong with classical cryptography?

There is a snag. It is called *key distribution.* Once the key is established, subsequent communication involves sending cryptograms over a channel, even one which is vulnerable to total passive eavesdropping (e.g. public announcement in mass-media). However in order to establish the key, two users, who share no secret information initially, must at a certain stage of communication use a reliable and a very secure channel. Since the interception is a set of measurements performed by the eavesdropper on this channel, however difficult this might be from a technological point of view, *in principle* any classical key distribution can always be passively monitored, without the legitimate users being aware that any eavesdropping has taken place.

Cryptologists have tried hard to solve the key distribution problem. The 1970s, for example, brought a clever mathematical discovery in the shape of "public key" systems (Diffie and Hellman 1976). In these systems users do not need to agree on a secret key before they send the message. They work on the principle of a safe with two keys, one public key to lock it, and another private one to open it. Everyone has a key to lock the safe but only one person has a key that will open it again, so anyone can put a message in the safe but only one person can take it out. These systems exploit the fact that certain mathematical operations are easier to do in one direction than the other. The systems avoid the key distribution problem but unfortunately their security depends on unproven mathematical assumptions, such as the difficulty of factoring large integers.[2] This means that if and when mathematicians or computer scientists come up with fast and clever procedures for factoring large integers the whole privacy and discretion of public–key cryptosystems could vanish overnight. Indeed, recent work in quantum computation shows that quantum computers can, at least in principle, factor much faster than classical computers (Shor 1994)!

In the following I will describe how quantum entanglement, singled out by Erwin Schrödinger (1935) as the most remarkable feature of quantum theory, became an important resource in the new field of quantum data processing. After a brief outline of entanglement's key role in philosophical debates about the meaning of quantum mechanics I will describe its current impact on both cryptography and cryptanalysis. Thus this is a story about quantum code-making and quantum code-breaking.

2 Is the Bell theorem of any practical use?

Probably the best way to agitate a group of jaded but philosophically-inclined physicists is to buy them a bottle of wine and mention *interpretations of quantum mechanics.* It is like opening Pandora's box. It seems that everybody agrees with the formalism of quantum mechanics, but no one agrees on its meaning. This is despite the fact that, as far as lip-service goes, one particular orthodoxy established by Niels Bohr over 50 years ago and known as the "Copenhagen in-

[2]RSA—a very popular public key cryptosystem named after the three inventors, Ron Rivest, Adi Shamir, and Leonard Adleman (1979)—gets its security from the difficulty of factoring large numbers.

terpretation" still effectively holds sway. It has never been clear to me how so many physicists can seriously endorse a view according to which the equations of quantum theory (e.g. the Schrödinger equation) apply only to unobserved physical phenomena, while at the moment of observation a completely different and mysterious process takes over. Quantum theory, according to this view, provides merely a calculational procedure and does not attempt to describe *objective physical reality.* A very defeatist view indeed.

One of the first who found the pragmatic instrumentalism of Bohr unacceptable was Albert Einstein who, in 1927 during the fifth Solvay Conference in Brussels, directly challenged Bohr over the meaning of quantum theory. The intellectual atmosphere and the philosophy of science at the time were dominated by positivism which gave Bohr the edge, but Einstein stuck to his guns and the Bohr-Einstein debate lasted almost three decades (after all they could not use e-mail). In 1935 Einstein together with Boris Podolsky and Nathan Rosen (EPR) published a paper in which they outlined what a 'proper' fundamental theory of nature should look like (Einstein *et al.* 1935). The EPR programme required completeness ("In a complete theory there is an element corresponding to each element of reality"), locality ("The real factual situation of the system A is independent of what is done with the system B, which is spatially separated from the former"), and defined the element of physical reality as "If, without in any way disturbing a system, we can predict with certainty the value of a physical quantity, then there exists an element of physical reality corresponding to this physical quantity". EPR then considered a thought experiment on two entangled particles which showed that quantum states cannot in all situations be complete descriptions of physical reality. The EPR argument, as subsequently modified by David Bohm (1951), goes as follows. Imagine the singlet-spin state of two spin $\frac{1}{2}$ particles

$$|\Psi\rangle = \frac{1}{\sqrt{2}}\left(|\uparrow\rangle|\downarrow\rangle - |\downarrow\rangle|\uparrow\rangle\right), \tag{2.1}$$

where the single particle kets $|\uparrow\rangle$ and $|\downarrow\rangle$ denote spin up and spin down with respect to some chosen direction. This state is spherically symmetric and the choice of the direction does not matter. The two particles, which we label A and B, are emitted from a source and fly apart. After they are sufficiently separated so that they do not interact with each other we can predict with certainty the x component of spin of particle A by measuring the x component of spin of particle B. This is because the total spin of the two particles is zero and the spin components of the two particles must have opposite values. The measurement performed on particle B does not disturb particle A (by locality) therefore the x component of spin is an element of reality according to the EPR criterion. By the same argument and by the spherical symmetry of state $|\Psi\rangle$ the y, z, or any other spin components are also elements of reality. However, since there is no quantum state of a spin $\frac{1}{2}$ particle in which all components of spin have definite values the quantum description of reality is not complete.

The EPR programme asked for a different description of quantum reality but until John Bell's (1964) theorem it was not clear whether such a description was possible and if so whether it would lead to different experimental predictions. Bell showed that the EPR propositions about locality, reality, and completeness are incompatible with some quantum mechanical predictions involving entangled particles. The contradiction is revealed by deriving from the EPR programme an experimentally testable inequality which is violated by certain quantum mechanical predictions. Extension of Bell's original theorem by John Clauser and Michael Horne (1974) made experimental tests of the EPR programme feasible and quite a few of them have been performed. The experiments have supported quantum mechanical predictions. Does this prove Bohr right? Not at all! The refutation of the EPR programme does not give any credit to the Copenhagen interpretation and simply shows that there is much more to 'reality', 'locality' and 'completeness' than the EPR envisaged (for a contemporary realist's approach see, for example, (Penrose 1989, 1994)).

What does it all have to do with data security? Surprisingly, a lot! It turns out that the very trick used by Bell to test the conceptual foundations of quantum theory can protect data transmission from eavesdroppers! Perhaps it sounds less surprising when one recalls again the EPR definition of an element of reality: "If, without in any way disturbing a system, we can predict with certainty the value of a physical quantity, then there exists an element of physical reality corresponding to this physical quantity". If this particular physical quantity is used to encode binary values of a cryptographic key then all an eavesdropper wants is an element of reality corresponding to the encoding observable (well, at least this was my way of thinking about it back in 1990). Since then several experiments have confirmed the 'practical' aspect of Bell's theorem making it quite clear that the border between blue sky and down-to-earth research is quite blurred.

3 Quantum key distribution

The quantum key distribution which I am going to discuss here is based on distribution of entangled particles (Ekert 1991). Before I describe how the system works let me mention that quantum cryptography does not have to be based on quantum entanglement. In fact a quite different approach based on partial indistinguishibility of non-orthogonal state vectors, pioneered by Stephen Wiesner (1983), and subsequently by Charles Bennett and Gilles Brassard (1984), preceded the entanglement-based quantum cryptography. Entanglement, however, offers quite a broad repertoire of additional tricks such as, for example, 'quantum privacy amplification' (Deutsch *et al.* 1996) which makes the entanglement-based key distribution secure and operable even in the presence of environmental noise.

The key distribution is performed via a quantum channel which consists of a source that emits pairs of spin $\frac{1}{2}$ particles in the singlet state as in eqn (2.1). The particles fly apart along the z-axis towards the two legitimate users of the channel, Alice and Bob, who, after the particles have separated, perform measurements and register spin components along one of three directions, given by unit vectors $\vec{a}_i$ and $\vec{b}_j$ $(i, j = 1, 2, 3)$, respectively for Alice and Bob. For simplicity both $\vec{a}_i$ and $\vec{b}_j$ vectors lie in the x–y plane, perpendicular to the trajectory of the particles, and are characterized by azimuthal angles: $\phi_1^a = 0$, $\phi_2^a = \frac{1}{4}\pi$, $\phi_3^a = \frac{1}{2}\pi$, and $\phi_1^b = \frac{1}{4}\pi$, $\phi_2^b = \frac{1}{2}\pi$, $\phi_3^b = \frac{3}{4}\pi$. Superscripts "a" and "b" refer to Alice's and Bob's analysers respectively, and the angle is measured from the vertical x-axis. The users choose the orientation of the analysers randomly and independently for each pair of the incoming particles. Each measurement, in $\frac{1}{2}\hbar$ units, can yield two results, $+1$ (spin up) and -1 (spin down), and can potentially reveal one bit of information.

The quantity

$$E(\vec{a}_i, \vec{b}_j) = P_{++}(\vec{a}_i, \vec{b}_j) + P_{--}(\vec{a}_i, \vec{b}_j) - P_{+-}(\vec{a}_i, \vec{b}_j) - P_{-+}(\vec{a}_i, \vec{b}_j) \tag{3.1}$$

is the correlation coefficient of the measurements performed by Alice along $\vec{a}_i$ and by Bob along $\vec{b}_j$. Here $P_{\pm\pm}(\vec{a}_i, \vec{b}_j)$ denotes the probability that result ± 1 has been obtained along $\vec{a}_i$ and ± 1 along $\vec{b}_j$. According to the quantum rules

$$E(\vec{a}_i, \vec{b}_j) = -\vec{a}_i \cdot \vec{b}_j. \tag{3.2}$$

For the two pairs of analysers of the same orientation ($\vec{a}_2$, $\vec{b}_1$ and $\vec{a}_3$, $\vec{b}_2$) quantum mechanics predicts total anticorrelation of the results obtained by Alice and Bob: $E(\vec{a}_2, \vec{b}_1) = E(\vec{a}_3, \vec{b}_2) = -1$.

One can define a quantity S composed of the correlation coefficients for which Alice and Bob used analysers of different orientation

$$S = E(\vec{a}_1, \vec{b}_1) - E(\vec{a}_1, \vec{b}_3) + E(\vec{a}_3, \vec{b}_1) + E(\vec{a}_3, \vec{b}_3). \tag{3.3}$$

This is the same S as in the generalised Bell theorem proposed by Clauser, Horne, Shimony, and Holt (1969) and known as the CHSH inequality. Quantum mechanics requires

$$S = -2\sqrt{2}. \tag{3.4}$$

After the transmission has taken place, Alice and Bob can announce in public the orientations of the analysers they have chosen for each particular measurement and divide the measurements into two separate groups: a first group for which they used different orientation of the analysers, and a second group for which they used the same orientation of the analysers. They discard all measurements in which either or both of them failed to register a particle at all. Subsequently Alice and Bob can reveal publicly the results they obtained but within the first group of measurements only. This allows them to establish the value of S, which if the particles were not directly or indirectly "disturbed" should reproduce the result of eqn (3.4). This assures the legitimate users that the results they obtained within the second group of measurements are anticorrelated and can be converted into a secret string of bits—the key.

An eavesdropper, Eve, cannot elicit any information from the particles while in transit from the source to the legitimate users, simply because there is no information encoded there! The information "comes into being" only after the legitimate users perform measurements and communicate in public afterwards. Eve may try to substitute her own prepared data for Alice and Bob to misguide them, but as she does not know which orientation of the analysers will be chosen for a given pair of particles there is no good strategy to escape being detected. In this case her intervention will be equivalent to introducing elements of *physical reality* to the spin components and will lower S below its 'quantum' value. Thus the Bell theorem can indeed expose eavesdroppers.

4 Quantum eavesdropping

The key distribution procedure described above is somewhat idealised. The problem is that there is in principle no way of distinguishing entanglement with an eavesdropper (caused by her measurements) from entanglement with the environment caused by innocent *noise*, some of which is presumably always present. This implies that all existing protocols which do not address this problem are, strictly speaking, inoperable in the presence of noise, since they require the transmission of messages to be suspended whenever an eavesdropper (or, therefore, noise) is detected. Conversely, if we want a protocol that is secure in the presence of noise, we must find one that allows secure transmission to continue even in the presence of eavesdroppers. To this end, one might consider modifying the existing protocols by reducing the statistical confidence level at which Alice and Bob accept a batch of qubits. Instead of the extremely high level envisaged in the idealised protocol, they would set the level so that they would accept most batches that had encountered a given level of noise. They would then have to assume that some of the information in the batch was known to an eavesdropper. It seems reasonable that classical privacy amplification (Bennett *et al.* 1995) could then be used to distill, from large numbers of such qubits, a key in whose security one could have any desired level of confidence. However, no such scheme has yet been proved to be secure. Existing proofs of the security of classical privacy amplification apply only to classical communication channels and classical eavesdroppers. They do not cover the new eavesdropping strategies that become possible in the quantum case: for instance, causing a quantum ancilla to interact with the encrypted message, storing the ancilla and later performing a measurement on it that is chosen according to the data that Alice and Bob exchange publicly. The security criteria for this type of eavesdropping have only recently been analysed (Gisin and Huttner 1996, Fuchs *et al.* 1997, Cirac and Gisin 1997).

The best way to analyse eavesdropping in the system is to adopt the scenario that is most favourable for eavesdropping, namely where Eve herself is allowed to prepare all the pairs that Alice and Bob will subsequently use to establish a key. This way we take the most conservative view which attributes all disturbance in the channel to eavesdropping even though most of it (if not all) may be due to innocent environmental noise.

Let us start our analysis of eavesdropping in the spirit of the Bell theorem and consider a simple case in which Eve knows precisely which particle is in which state. Following Ekert (1991) let us assume that Eve prepares each particle in the EPR pairs separately so that each individual particle in the pair has a well defined spin in some direction. These directions may vary from pair to pair so we can say that she prepares with probability $p(\vec{n}_a, \vec{n}_b)$ Alice's particle in state $|\,\vec{n}_a\rangle$ and Bob's particle in state $|\,\vec{n}_b\rangle$, where $\vec{n}_a$ and $\vec{n}_b$ are two unit vectors describing the spin orientations. This kind of preparation gives Eve total control over the state of *individual* particles. This is the case where Eve will always have the edge and Alice and Bob should abandon establishing the key; they will learn about it by estimating $|S|$ which in this case will always be smaller than $\sqrt{2}$. To see this let us write the density operator for each pair as

$$\rho = \int p(\vec{n}_a, \vec{n}_b)\,|\,\vec{n}_a\rangle\,\langle\vec{n}_a\,| \otimes |\,\vec{n}_b\rangle\,\langle\vec{n}_b\,|\; d\vec{n}_a d\vec{n}_b. \tag{4.1}$$

Equation (3.3) with appropriately modified correlation coefficients reads

$$\begin{aligned} S = \int p(\vec{n}_a, \vec{n}_b) d\vec{n}_a d\vec{n}_b \quad & [(\vec{a}_1 \cdot \vec{n}_a)(\vec{b}_1 \cdot \vec{n}_b) - (\vec{a}_1 \cdot \vec{n}_a)(\vec{b}_3 \cdot \vec{n}_b) \\ & + (\vec{a}_3 \cdot \vec{n}_a)(\vec{b}_1 \cdot \vec{n}_b) + (\vec{a}_3 \cdot \vec{n}_a)(\vec{b}_3 \cdot \vec{n}_b)], \end{aligned} \tag{4.2}$$

and leads to

$$S = \int p(\vec{n}_a, \vec{n}_b) d\vec{n}_a d\vec{n}_b [\sqrt{2}\vec{n}_a \cdot \vec{n}_b] \tag{4.3}$$

which implies

$$-\sqrt{2} \leq S \leq \sqrt{2}, \tag{4.4}$$

for any state preparation described by the probability distribution $p(\vec{n}_a, \vec{n}_b)$.

Clearly Eve can give up her perfect control of quantum states of individual particles in the pairs and entangle at least some of them. If she prepared all the pairs in perfectly entangled singlet states she would lose all her control and knowledge about Alice's and Bob's data who can then easily establish a secret key. This case is unrealistic because in practice Alice and Bob will never register $|S| = 2\sqrt{2}$. However, if Eve prepares only partially entangled pairs then it is still possible for Alice and Bob to establish the key with an absolute security provided they use a *Quantum Privacy Amplification* algorithm (QPA) (Deutsch *et al.* 1996). The case of partially entangled pairs, $\sqrt{2} \leq |S| \leq 2\sqrt{2}$, is the most important one and in order to claim that we have an operable key distribution scheme we have to prove that the key can be established in this particular case. Skipping technical details I will present only the main idea behind the QPA, details can be found in Deutsch *et al.* (1996).

Firstly, note that any two particles that are jointly in a pure state cannot be entangled with any third physical object. Therefore any procedure that delivers EPR-pairs in pure states must also have eliminated the entanglement between any of those pairs and any other system. The QPA scheme is based on an iterative quantum algorithm which, if performed with perfect accuracy, starting with a collection of EPR-pairs in mixed states, would discard some of them and leave the remaining ones in states converging to the pure singlet state. If (as must be the case realistically) the algorithm is performed imperfectly, the density operator of the pairs remaining after each iteration will not converge on the singlet but on a state close to it, however, the degree of entanglement with any eavesdropper will nevertheless continue to fall, and can be brought to an arbitrary low value. The QPA can be performed by Alice and Bob at distant locations by a sequence of local unitary operations and measurements which are agreed upon by communication over a public channel and could be implemented using technology that is currently being developed (c.f. Turchette *et al.* 1995).

The essential element of the QPA procedure is the 'entanglement purification' scheme, the idea originally proposed by Charles Bennett, Gilles Brassard, Sandu Popescu, Benjamin Schumacher, John Smolin, and Bill Wootters (1996a). It has been shown recently that any partially entangled states of two-state particles can be purified (Horodecki *et al.* 1997). Thus as long as the density operator cannot be written as a mixture of product states, i.e. is not of the form (4.1), then Alice and Bob may outsmart Eve!

Finally let me mention that quantum cryptography today is more than a theoretical curiosity. Experimental work in Switzerland (Muller *et al.* 1995), in the UK (Townsend *et al.* 1996), and in the USA (Hughes *et al.* 1996, Franson and Jacobs 1995) shows that quantum data security should be taken seriously!

5 Public key cryptosystems

In the late 1970s Whitfield Diffie and Martin Hellman (1976) proposed an interesting solution to the key distribution problem. It involved two keys, one public key e for encryption and one private key d for decryption:

$$\hat{E}_e(P) = C, \text{ and } \hat{D}_d(C) = P. \tag{5.1}$$

As I have already mentioned, in these systems users do not need to agree on any key before they start sending messages to each other. Every user has his own two keys; the public key is publicly announced and the private key is kept secret. Several public–key cryptosystems have been proposed since 1976; here we concentrate our attention on the most popular one, which was already mentioned in Section 1, namely the RSA (Rivest *et al.* 1979).

If Alice wants to send a secret message to Bob using the RSA system the first thing she does is look up Bob's personal public key in a some sort of yellow pages or an RSA public key directory. This consists of a pair of positive integers (e, n). The integer e may be relatively small, but n will be gigantic, say a couple of hundred digits long. Alice then writes her message as a sequence of numbers using, for example, our digital alphabet from Section 1. This string of numbers is subsequently divided into blocks such that each block when viewed as a number P satisfies $P \leq n$. Alice encrypts each P as

$$\hat{E}(P) = C = P^e \bmod n. \tag{5.2}$$

and sends the resulting cryptogram to Bob who can decrypt it by calculating

$$\hat{D}(C) = P = C^d \bmod n. \tag{5.3}$$

Of course, for this system to work Bob has to follow a special procedure to generate both his private and public key:

- He begins with choosing two large (100 or more digits long) prime numbers p and q, and a number e which is relatively prime to both $p-1$ and $q-1$.
- He then calculates $n = pq$ and finds d such that $ed = 1 \bmod (p-1)(q-1)$. This equation can be easily solved, for example, using the extended Euclidean algorithm for the greatest common divisor.[3]
- He releases to the public n and e and keeps p, q, and d secret.

The mathematics behind the RSA is a lovely piece of number theory which goes back to the 17th century when a French lawyer Pierre de Fermat discovered that if a prime p and a positive integer a are relatively prime, then

$$a^{p-1} = 1 \bmod p. \tag{5.4}$$

A century later, Leonhard Euler found the more general relation

$$a^{\phi(n)} = 1 \bmod n, \tag{5.5}$$

for relatively prime integers a and n. Here $\phi(n)$ is Euler's ϕ function which counts the number of positive integers smaller than n and coprime to n. Clearly for any prime integer such as p or q we have $\phi(p) = p-1$ and $\phi(q) = q-1$; for $n = pq$ we obtain $\phi(n) = (p-1)(q-1)$. Thus the cryptogram $C = P^e \bmod n$ can indeed be decrypted by $C^d \bmod n = P^{ed} \bmod n$ because $ed = 1 \bmod \phi(n)$; hence for some integer k

$$P^{ed} \bmod n = P^{k\phi(n)+1} \bmod n = P. \tag{5.6}$$

For example, let us suppose that Roger's public key is $(e, n) = (179, 571247)$.[4] He generated it following the prescription above choosing $p = 773$, $q = 739$ and

[3] Fortunately an easy and very efficient algorithm to compute the greatest common divisor has been known since 300 BC. This truly 'classical' algorithm is described in Euclid's *Elements*, the oldest Greek treatise in mathematics to reach us in its entirety. Knuth (1981) provides an extensive discussion of various versions of Euclid's algorithm.

[4] Needless to say, the number n in this example is too small to guarantee security, do not try this public key with Roger.

$e = 179$. The private key d was obtained by solving $179\,d = 1 \bmod 772 \times 738$ using the extended Euclidean algorithm which yields $d = 515627$. Now if we want to send Roger encrypted "HELLO ROGER." we first use our digital alphabet from Section 1 to obtain the plaintext which can be written as the following sequence of six digit numbers

$$080512 \qquad 121527 \qquad 181507 \qquad 051830. \tag{5.7}$$

Then we encipher each block P_i by computing $C_i = P_i^e \bmod n$; e.g. the first block $P_1 = 080512$ will be enciphered as

$$P_1^e \bmod n = 080512^{179} \bmod 571247 = 458467 = C_1, \tag{5.8}$$

and the whole message is enciphered as:

$$458467 \qquad 180137 \qquad 323954 \qquad 025252. \tag{5.9}$$

The cryptogram C composed of blocks C_i can be send over to Roger. He can then decrypt each block using his private key $d = 515627$, e.g. the first block is decrypted as

$$458467^{515627} \bmod 571247 = 080512 = P_1. \tag{5.10}$$

In order to recover plaintext P from cryptogram C, an outsider, who knows C, n, and e, would have to solve the congruence

$$P^e \bmod n = C, \tag{5.11}$$

for example, in our case,

$$P_1^{179} \bmod 571247 = 458467, \tag{5.12}$$

which is hard, that is it is not known how to compute the solution efficiently when n is a large integer (say 200 decimal digits long or more). However, if we know the prime decomposition of n it is a piece of cake to figure out the private key d; we simply follow the key generation procedure and solve the congruence $ed = 1 \bmod (p-1)(q-1)$. This can be done efficiently even when p and q are very large. Thus, in principle, anybody who knows n can find d by factoring n, but factoring big n is a hard problem. What does "hard" mean ?

6 Fast and slow algorithms

Difficulty of factoring grows rapidly with the size, i.e. number of digits, of a number we want to factor. To see this take a number n with l decimal digits ($n \approx 10^l$) and try to factor it by dividing it by $2, 3, \ldots \sqrt{n}$ and checking the remainder. In the worst case you may need approximately $\sqrt{n} = 10^{l/2}$ divisions to solve the problem—an exponential increase as a function of l. Now imagine a computer capable of performing 10^{10} divisions per second. The computer can then factor any number n, using the trial division method, in about $\sqrt{n}/10^{10}$

seconds. Take a 100-digit number n, so that $n \approx 10^{100}$. The computer will factor this number in about 10^{40} seconds, much longer than 10^{17} seconds—the estimated age of the Universe!

Skipping details of computational complexity I only mention that there is a rigorous way of defining what makes an algorithm fast (and efficient) or slow (and impractical) (see, for example Welsh 1988). For an algorithm to be considered fast, the time it takes to execute the algorithm must increase no faster than a polynomial function of the size of the input. Informally think about the input size as the total number of bits needed to specify the input to the problem, for example, the number of bits needed to encode the number we want to factorise. If the best algorithm we know for a particular problem has execution time (viewed as a function of the size of the input) bounded by a polynomial then we say that the problem belongs to class P. Problems outside class P are known as hard problems. Thus we say, for example, that multiplication is in P whereas factorisation is apparently not in P and that is why it is a hard problem. We can also design non-deterministic algorithms which may sometimes produce incorrect solutions but have the property that the probability of error can be made arbitrarily small. For example, the algorithm may produce a candidate factor p of the input n followed by a trial division to check whether p really is a factor or not. If the probability of error in this algorithm is ϵ and is independent of the size of n then by repeating the algorithm k times, we get an algorithm which will be successful with probability $1-\epsilon^k$ (i.e. having at least one success). This can be made arbitrarily close to 1 by choosing a fixed k sufficiently large.

There is no known efficient classical algorithm for factoring even if we allow it to be probabilistic in the above senses. The fastest algorithms run in time roughly of order $\exp((\log n)^{1/3})$ and would need a couple of billion years to factor a 200-digit number. It is not known whether a fast classical algorithm for factorisation exists or not—none has yet been found.

It seems that factoring big numbers will remain beyond the capabilities of any realistic computing devices and unless we come up with an efficient factoring algorithm the public–key cryptosystems will remain secure. Or will they? As it turns out we know that this is not the case; the classical, purely mathematical, theory of computation is not complete simply because it does not describe all physically possible computations. In particular it does not describe computations which can be performed by quantum devices. Indeed, recent work in quantum computation shows that a quantum computer, at least in principle, can efficiently factor large integers (Shor 1994).

7 Quantum computers

Quantum computers can compute faster because they can accept as the input not a single number but a coherent superposition of many different numbers and subsequently perform a computation (a sequence of unitary operations) on all of these numbers simultaneously. This can be viewed as a massive parallel computation, but instead of having many processors working in parallel we have only one quantum processor performing a computation that affects all numbers in a superposition, i.e. all components of the input state vector.

The exponential speed-up of quantum computers takes place at the very beginning of their computation. Qubits, i.e. physical systems which can be prepared in one of the two orthogonal states labelled as $|\,0\rangle$ and $|\,1\rangle$ or in a superposition of the two, can store superpositions of many 'classical' inputs. For example, the equally weighted superposition of $|\,0\rangle$ and $|\,1\rangle$ can be prepared by taking a qubit initially in state $|\,0\rangle$ and applying to it transformation $\mathbf{H}$ (also known as the Hadamard transform) which maps

$$|\,0\rangle \longrightarrow \frac{1}{\sqrt{2}}(|\,0\rangle + |\,1\rangle), \tag{7.1}$$

$$|\,1\rangle \longrightarrow \frac{1}{\sqrt{2}}(|\,0\rangle - |\,1\rangle). \tag{7.2}$$

If this transformation is applied to each qubit in a register composed of two qubits it will generate the superposition of four numbers

$$|\,0\rangle\,|\,0\rangle \longrightarrow \frac{1}{\sqrt{2}}(|\,0\rangle + |\,1\rangle)\,\frac{1}{\sqrt{2}}(|\,0\rangle + |\,1\rangle) \tag{7.3}$$

$$= \tfrac{1}{2}(|\,00\rangle + |\,01\rangle + |\,10\rangle + |\,11\rangle) \tag{7.4}$$

where 00 can be viewed as binary for 0, 01 binary for 1, 10 binary for 2 and finally 11 as binary for 3. In general a quantum register composed of l qubits can be prepared in a superposition of 2^l different numbers (inputs) with only l elementary operations. This can be written, in decimal rather than in binary notation, as

$$|\,0\rangle \longrightarrow 2^{-l/2} \sum_{x=0}^{2^l-1} |\,x\rangle\,. \tag{7.5}$$

Thus l elementary operations generate exponentially many, that is 2^l different inputs!

The next task is to process all the inputs in parallel within the superposition by a sequence of unitary operations. Let us describe now how quantum computers compute functions. For this we will need two quantum registers of length l and k. Consider a function

$$f:\{0,1,\dots 2^l-1\} \longrightarrow \{0,1,\dots 2^k-1\}. \tag{7.6}$$

A classical computer computes f by evolving each labelled input, $0,1,\dots,2^l-1$ into a respective labelled output, $f(0), f(1),\dots, f(2^l-1)$. Quantum computers, due to the unitary (and therefore reversible) nature of their evolution, compute functions in a slightly different way. In order to compute functions which are not one-to-one and to preserve the reversibility of computation, quantum computers have to keep the record of the input. Here is how it is done. The first register is loaded with value x, i.e. it is prepared in state $|\,x\rangle$, the second register may initially contain an arbitrary number y. The function evaluation is then determined by an appropriate unitary evolution of the two registers,

$$|\,x\rangle\,|\,y\rangle \xrightarrow{U_f} |\,x\rangle\,|\,y + f(x)\rangle\,. \tag{7.7}$$

Here $y + f(x)$ means addition modulo the maximum number of configurations of the second register, i.e. 2^k in our case.

The computation we are considering here is not only reversible but also quantum and we can do much more than computing values of $f(x)$ one by one. We can prepare a superposition of all input values as a single state and by running the computation U_f *only once*, we can compute *all* of the 2^l values $f(0), \ldots, f(2^l-1)$, (here and in the following we ignore the normalisation constants),

$$\sum_{x=0}^{2^l-1} |\,x\rangle\,|\,y\rangle \xrightarrow{U_f} |\,y + f\rangle = \sum_{x=0}^{2^l-1} |\,x\rangle\,|\,y + f(x)\rangle\,. \tag{7.8}$$

It looks too good to be true so where is the catch? How much information about f does the state

$$|\,f\rangle = |\,0\rangle\,|\,f(0)\rangle + |\,1\rangle\,|\,f(1)\rangle + \ldots + \left|\,2^l - 1\right\rangle \left|\,f(2^l-1)\right\rangle \tag{7.9}$$

really contain?

Unfortunately no quantum measurement can extract all of the 2^l values $f(0), f(1), \ldots, f(2^l-1)$ from $|\,f\rangle$. If we measure the two registers after the computation U_f we register one output $|\,x\rangle\,|\,y+f(x)\rangle$ for some value x. However, there are measurements that provide us with information about joint properties of all the output values $f(x)$, such as, for example, periodicity, without providing any information about particular values of $f(x)$. Let us illustrate this with a simple example.

Consider a Boolean function f which maps $\{0,1\} \to \{0,1\}$. There are exactly four functions of this type: two constant functions ($f(0) = f(1) = 0$ and $f(0) = f(1) = 1$) and two balanced functions ($f(0) = 0, f(1) = 1$ and $f(0) = 1, f(1) = 0$). Is it possible to compute function f *only once* and to find out whether it is constant or balanced, i.e. whether the binary numbers $f(0)$ and $f(1)$ are the same or different? N.B. we are not asking for particular values $f(0)$ and $f(0)$ but for a global property of f.

Classical intuition tells us that we have to evaluate both $f(0)$ and $f(1)$, that is to compute f twice. This is not so. Quantum mechanics allows us to perform the trick with a single function evaluation. We simply take two qubits, each qubit serves as a single qubit register, prepare the first qubit in state $|\,0\rangle$ and the second in state $|\,1\rangle$ and compute

$$\begin{aligned} |\,0\rangle\,|\,1\rangle &\longrightarrow (|\,0\rangle + |\,1\rangle)(|\,0\rangle - |\,1\rangle) \longrightarrow \\ &\longrightarrow |\,0\rangle\,(|\,f(0)\rangle - |\,1 + f(0)\rangle) + |\,1\rangle\,(|\,f(1)\rangle - |\,1 + f(1)\rangle). \end{aligned} \tag{7.10}$$

We start with transformation **H** applied both to the first and the second qubit, followed by the function evaluation. Here $1+f(0)$ denotes addition modulo 2 and simply means taking the negation of $f(0)$. At this stage, depending on values $f(0)$ and $f(1)$, we have one of the four possible states of the two qubits. We apply **H** again to the first and the second qubit and evolve the four states as follows

$$(|\,0\rangle + |\,1\rangle)(|\,0\rangle - |\,1\rangle) \longrightarrow +|\,0\rangle\,|\,1\rangle\,, \tag{7.11}$$
$$(|\,0\rangle + |\,1\rangle)(|\,1\rangle - |\,0\rangle) \longrightarrow -|\,0\rangle\,|\,1\rangle\,, \tag{7.12}$$
$$|\,0\rangle\,(|\,0\rangle - |\,1\rangle) + |\,1\rangle\,(|\,1\rangle - |\,0\rangle) \longrightarrow +|\,1\rangle\,|\,1\rangle\,, \tag{7.13}$$
$$|\,0\rangle\,(|\,1\rangle - |\,0\rangle) + |\,1\rangle\,(|\,0\rangle - |\,1\rangle) \longrightarrow -|\,1\rangle\,|\,1\rangle\,. \tag{7.14}$$

The second qubit returns to its initial state $|\,1\rangle$ but the first qubit contains the relevant information. We measure its bit value—if we register "0" the function is constant if we register "1" the function is balanced!

This example, due to Richard Cleve, Artur Ekert and Chiara Macchiavello (1996), is an improved version of the first quantum algorithm proposed by David Deutsch (1985) and communicated to the Royal Society by Roger Penrose. (The original Deutsch algorithm provides the correct answer with probability 50%.) Deutsch's paper laid the foundation for the new field of quantum computation. Since then quantum algorithms have been steadily improved and in 1994 Peter Shor came up with the efficient quantum factoring algorithm which, at least in theory, leads us directly to quantum cryptanalysis.

8 Quantum code-breaking

Shor's quantum factoring of an integer n is based on calculating the period of the function $F_n(x) = a^x \bmod n$ for a randomly selected integer a between 0 and n. It turns out that for increasing powers of a, the remainders form a repeating sequence with a period which we denote r. Once r is known the factors of n are obtained by calculating the greatest common divisor of n and $a^{r/2} \pm 1$.

Suppose we want to factor 15 using this method. Let $a = 11$. For increasing x the function $11^x \bmod 15$ forms a repeating sequence $1, 11, 1, 11, 1, 11, \ldots$. The period is $r = 2$, and $a^{r/2} \bmod 15 = 11$. Then we take the greatest common divisor of 10 and 15, and of 12 and 15 which gives us respectively 5 and 3, the two factors of 15. Classically calculating r is at least as difficult as trying to factor n; the execution time of calculations grows exponentially with the number of digits in n. Quantum computers can find r in time which grows only as a cubic function of the number of digits in n.

To estimate the period r we prepare two quantum registers; the first register, with l qubits, in the equally weighted superposition of all numbers it can contain, and the second register in state zero. Then we perform an arithmetical operation that takes advantage of quantum parallelism by computing the function $F_n(x)$ for each number x in the superposition. The values of $F_n(x)$ are placed in the second register so that after the computation the two registers become entangled:

$$\sum_x |\, x\rangle \,|\, 0\rangle \longrightarrow \sum_x |\, x\rangle \,|\, F_n(x)\rangle \,. \tag{8.1}$$

Now we perform a measurement on the second register. We measure each qubit and obtain either "0" or "1". This measurement yields the value $F_n(k)$ (in binary notation) for some randomly selected k. The state of the first register right after the measurement, due to the periodicity of $F_n(x)$, is a coherent superposition of all states $|\, x\rangle$ such that $x = k,\ k+r,\ k+2r, \ldots$, i.e. all x for which $F_n(x) = F_n(k)$. The periodicity in the probability amplitudes in the first register cannot be simply measured because the offset, i.e. the value k, is randomly selected by the measurement. However the state of the first register can be subsequently transformed via a unitary operation which effectively removes the offset and modifies the period in the probability amplitudes from r to a multiple of $2^l/r$. This operation is known as the quantum Fourier transform (QFT) and can be written as

$$\mathrm{QFT}_s : |\, x\rangle \longmapsto 2^{-l/2} \sum_{y=0}^{2^l-1} \exp(2\pi i a c/2^l)\ |\, y\rangle \,. \tag{8.2}$$

After QFT the first register is ready for the final measurement which yields with high probability an integer which is the best whole approximation of a multiple of $2^l/r$, i.e. $x = k2^l/r$ for some integer k. We know the measured value x and the size of the register l hence if k and r are coprime we can determine r by cancelling $x/2^l$ down to an irreducible fraction and taking its denominator. Since the probability that k and r are coprime is sufficiently large (greater than $1/\log r$ for large r) this gives an efficient randomized algorithm for determination of r. More detailed description of Shor's algorithm can be found in (Shor 1994) and in (Ekert and Jozsa 1996).

Let me mention in passing that there exists a direct quantum attack on RSA which does not require factoring, but employs Shor's algorithm to determine the order of cryptogram modulo n (Mosca and Ekert 1997).

An open question has been whether it would ever be practical to build physical devices to perform such computations, or whether they would forever remain theoretical curiosities. Quantum computers require a coherent, controlled evolution for a period of time which is necessary to complete the computation. Many view this requirement as an insurmountable experimental problem, however, technological progress may prove them wrong (see review papers e.g. (DiVincenzo 1995; Ekert and Jozsa 1996; Lloyd 1993, 1995)). When the first quantum factoring devices are built the security of classical public–key cryptosystems will vanish. But, as was pointed out by Roger Penrose, by the time we acquire desk–top quantum computers we will probably be able to construct quantum public–key cryptosystems with security based on quantum rather than classical computational complexity. Meanwhile thumbs up for quantum cryptography.

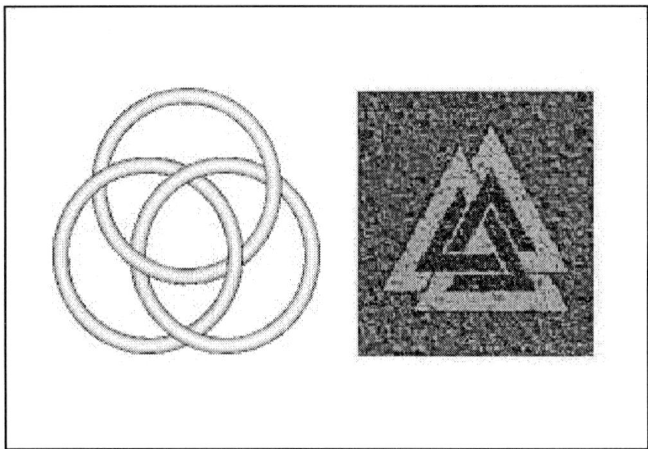

FIG. 1. The Borromean rings consist of three 'entangled' rings with the property that if any one of them is removed, then all three separate. The name Borromean comes from the Italian Borromeo family who used them for their coat of arms. Odin's triangle was found in picture-stones on Gotland, an island in the Baltic sea off the southeast coast of Sweden. These are dated around the ninth century and are thought to tell tales from the Norse myths. To the Norse people of Scandinavia the interlocked triangles are known as "Odin's triangle" of the "Walknot" (or "valknut"—the knot of the slain). The symbol was also carved on the bedposts used in their burials at sea (see, for example, http://sue.csc.uvic.ca/ cos/venn/borromean.html).

9 Concluding remarks

In the last decade quantum entanglement became a sought-after physical resource which allows us to perform qualitatively new types of data processing. Here I have described the role of entanglement in connection with data security and have skipped many other fascinating applications such as quantum teleportation (Bennett *et al.* 1993), quantum dense coding (Bennett and Wiesner 1992), entanglement swapping (Zukowski *et al.* 1993), quantum error correction (Shor 1995; Steane 1996; Ekert and Macchiavello 1996; Calderbank and Shor 1996; Bennett *et al.* 1996b), fault tolerant quantum computing (Shor 1996; DiVincenzo and Shor 1996) and only mentioned in passing entanglement purification (Bennett *et al.* 1996a) and quantum privacy amplification (Deutsch *et al.* 1996). There is much more to say about some peculiar features of two- and many-particle entanglement and there is even an interesting geometry behind it such as, for example, the 'magic dedocahedra' (Penrose 1994). Even the 'simplest' three-particle entangled states, the celebrated GHZ states of three qubits (Greenberger *et al.* 1989) such as $|\,000\rangle \pm |\,111\rangle$, have interesting properties; the three particles are all together entangled but none of the two qubits in the triplet are entangled. That is, the pure GHZ state of three particles is entangled as a whole but the reduced density operator of any of the pairs is separable. This is very reminiscent of some geometric constructions such as "Odin's triangle" or "Borromean rings" (Aravind 1997) shown in Fig. 1. This and many other interesting properties of multi-particle entanglement have been recently studied by Sandu Popescu (1997) and (in a disguised form as quantum error correction) by Rob Calderbank, Eric Rains, Peter Shor, and Neil Sloane (1996). Despite remarkable progress in this field it seems to me that we still know very little about the nature of quantum entanglement; we can hardly agree on how to classify, quantify and measure it (but c.f. Horodecki 1997; Vedral *et al.* 1997). Clearly we should be prepared for even more surprises both in understanding and in utilisation of this precious quantum resource.

In this quest to understand the quantum theory better and better Roger Penrose has played a very prominent role, defending both common sense and the realist's view of the quantum world. I would like to thank him for this and for many fascinating discussions about the quantum–mechanical Z and X-mysteries which so often became 'entangled' with the nature of computation and helped me to understand things better.

Acknowledgements I am greatly indebted to David DiVincenzo, Chiara Macchiavello and Michele Mosca for their comments and help in preparation of this manuscript. The author is supported by the Royal Society, London. This work was supported in part by the European TMR Research Network ERP–4061PL95-1412, Hewlett–Packard and Elsag–Bailey.

Bibliography

Aravind, P. K. (1997). *Borromean entanglement of the GHZ state* in *Potentiality, entanglement and passion-at-a-distance*, edited by Cohen, R., Horne, M. and Stachel, J. Kluwer Academic Publishers.

Barenco, A., Deutsch, D., Ekert, A. and Jozsa, R. (1995). *Phys. Rev. Lett.* **74**, 4083.

Bell, J. S. (1964). *Physics* **1**, 195.

Bennett, C. H. (1996). *Phys. Today* **48**, 27.

Bennett, C. H. and Brassard, G. (1984). in "Proc. IEEE Int. Conference on Computers, Systems and Signal Processing", IEEE, New York.

Bennett, C. H. and Wiesner, S. J. (1992). *Phys. Rev. Lett.* **69**, 2881.

Bennett, C. H., Brassard, G., Crépeau, C., Jozsa, R., Peres, A. and Wootters, W.K. (1993). *Phys. Rev. Lett.* **70**, 1895.

Bennett, C. H., Brassard, G., Crépeau, C. and Maurer, U. M. (1995). *IEEE Trans. Inf. Th.* **IT–41**, 1915.

Bennett, C. H., Brassard, G., Popescu, S., Schumacher, B., Smolin, J. and Wootters, W. K. (1996a). *Phys. Rev. Lett.* **76**, 722.

Bennett, C. H., DiVincenzo, D. P., Smolin, J. A., and Wootters, W. K.

(1996b). *Phys. Rev.* A **54**, 3824.

Bohm, D. (1951). *Quantum Theory*, Prentice–Hall.

Calderbank, A. R. and Shor, P. W. (1996). *Phys. Rev. A* **54**, 1098.

Calderbank, A. R., Rains, E. M., Shor, P. W. and Sloane, N. J. A. (1966). *Quantum Error Correction via Codes over GF(4)*. quant-ph/9608006.

Cirac, J. I. and Gisin N. (1997). *Coherent eavesdropping strategies for the 4 state quantum cryptography protocol.* quant-ph/9702002.

Clauser, J. F. and Horne, M. A. (1974). *Phys. Rev. D* **10**, 526.

Clauser, J. F. and Horne, M. A., Shimony, A. and Holt, R. A. (1969). *Phys. Rev. Lett.* **23**, 880.

Cleve, R., Ekert, A. and Macchiavello, C. (1996). During long afternoon discussions on quantum algorithms and Californian wine at the Santa Barbara Workshop on Quantum Computation and Decoherence. More general analysis of quantum algorithms will be provided in Cleve, R., Ekert, A., Macchiavello, C. and Mosca, M. *Quantum Algorithms Revisited* (in preparation).

Deutsch, D. (1985). *Proc. R. Soc. London A* **400**, 97.

Deutsch, D., Ekert, A., Jozsa, R., Macchiavello, C., Popescu, S. and Sanpera, A. (1996). *Phys. Rev. Lett.* **77**, 2818.

Diffie, W. and Hellman, M. E. (1976). *IEEE Trans. Inf. Theory* **IT–22**, 644.

DiVincenzo, D. P. (1995). *Science* **270**, 255.

DiVincenzo, D. P. and Shor, P. W. (1996). *Phys. Rev. Lett.* **77**, 3260.

Einstein, A., Podolsky, B. and Rosen N. (1935). *Phys. Rev.* **47**, 777.

Ekert, A. (1991). *Phys. Rev. Lett.* **67**, 661.

Ekert, A. and Jozsa, R. (1996). *Rev. Mod. Phys.* **68**, 733.

Ekert, A. and Macchiavello, C. (1996). *Phys. Rev. Lett.* **77**, 2585.

Franson, J. D. and Jacobs, B. C. (1995). *Electron. Lett.* **31**, 232.

Fuchs, C. A., Gisin, N., Griffiths, R. B., Niu, C-S. and Peres, A. (1997). *Optimal eavesdropping in quantum cryptography I.* quant-ph/9701039.

Gisin, N. and Huttner, B. (1996). *Quantum cloning, eavesdropping, and Bell's inequality.* quant-ph/9611041.

Greenberger, D. M., Horne, M. and Zeilinger, A. (1989). Going beyond Bell's theorem, in *Bell's Theorem, Quantum Theory, and Conceptions of the Universe*, ed. by M. Kafatos, Kluwer, Dordrecht, 69–72.

Horodecki, M., Horodecki, P. and Horodecki, R. (1997). *Phys. Rev. Lett.* **78**, 574.

Hughes, R. J., Luther, G. G., Morgan, G. L., Peterson, C. G. and Simmons, C. (1996). *Quantum Cryptography over Underground Optical Fibres*, Advances in Cryptology — Proceedings of Crypto'96, Springer–Verlag.

Kahn, D. (1967). *The Codebreakers: The Story of Secret Writing*, Macmillan.

Knuth, D. E. (1981). *The Art of Computer Programming, Volume 2/ Seminu-*

merical Algorithms. Addison–Wesley.

Lloyd, S. (1993). *Science* **261**, 1569.

Lloyd, S. (1995), *Scient. Am.* **273**, 44.

Menezes, A. J., van Oorschot, P. C. and Vanstone S. A. (1996). *Handbook of Applied Cryptography.* CRC Press.

Mosca, M. and Ekert, A. (1997). *A Note on Quantum Attack on RSA.* Unpublished, available from the authors.

Muller, A., Zbinden, H. and Gisin, N. (1995). *Nature,* **378**, 449.

Penrose, R. (1989). *The Emperor's New Mind.* Oxford University Press.

Penrose, R. (1994). *Shadows of the Mind.* Oxford University Press.

Popescu, S. (1997). Private communication.

Rivest, R., Shamir, A. and Adleman, L. *On Digital Signatures and Public–Key Cryptosystems,* MIT Laboratory for Computer Science, Technical Report, MIT/LCS/TR–212 (January 1979).

Schneier, B. (1994). *Applied Cryptography: Protocols, Algorithms, and Source Code in C.* John Wiley & Sons.

Schrödinger, E. (1935). Die gegenwärtige Situation in der Quantenmechanik, *Naturwissenschaften,* **23**, 807–812; 823–828; 844–849. English translation, The Present Situation in Quantum Mechanics, *Proc. Amer. Phil. Soc.* **124**, 323–338 (1980); reprinted in *Quantum Theory and Measurement* edited by Wheeler, J. A. and Zurek, W. H. (Princeton, 1983), 152–167.

Shor, P. W. (1994). In *Proceedings of the 35th Annual Symposium on the Foundations of Computer Science,* edited by S. Goldwasser (IEEE Computer Society Press, Los Alamitos, CA), 124; Expanded version of this paper is available at LANL quant-ph archive.

Shor, P. W. (1995). *Phys. Rev. A,* **52**, 2493.

Shor, P. W. (1996). *Fault-tolerant quantum computation.* quant-ph/9605011.

Steane, A. (1996). *Phys. Rev. Lett.* **77**, 793.

Townsend, P. D., Marand, C., Phoenix, S. J. D., Blow, K. J. and Barnett, S. M. (1996). *Phil. Trans. Roy. Soc. London A* **354**, 805.

Turchette, Q. A., Hood, C. J., Lange, W., Mabuchi, H. and Kimble, H. J. (1995). *Phys. Rev. Lett.* **75**, 4710.

Vedral, V., Plenio, M. P., Rippin, M. A., and Knight, P. L. (1997). *Quantifying Entanglement.* quant-ph/9702027 and *Phys. Rev. Lett.* (to appear).

Welsh, D. (1988). *Codes and Cryptography.* Clarendon Press, Oxford.

Wiesner, S. (1983). *SIGACT News,* **15**, 78. Original manuscript written *circa* 1970.

Zukowski, M., Zeilinger, A., Horne, M. and Ekert, A. (1993). *Phys. Rev. Lett.* **71**, 4287.

13

Penrose Tilings and Quasicrystals Revisited

Paul J. Steinhardt
Department of Physics and Astronomy
University of Pennsylvania
Philadelphia, PA 19104 USA

Abstract

Roger Penrose's ingenious invention of five-fold symmetric, non-periodic tilings has inspired the proposal of a new form of solid, known as a quasicrystal. Quasicrystals are solids with quasiperiodic atomic structures and symmetries forbidden to ordinary, periodic crystals, e.g., three-dimensional icosahedral symmetry. Here we discuss some new properties of Penrose tilings uncovered in recent months which suggest a new, simple picture of the structure of quasicrystals and shed new light on why they form.

1 Introduction

For the most part, this volume celebrates theoretical issues which Roger Penrose has actively pursued and led for many years. This paper, however, represents an example of his 'inadvertent influence,' a byproduct of one of his creative avocations — a true sign of the breadth of his genius. In 1974, Penrose discovered a set of tiles plus matching rules which constrain how they may match edge-to-edge such that the only space-filling tiling is non-periodic (Penrose 1974). Ten years later, the Penrose tiling inspired the proposal of a new phase of solid matter, known as the quasicrystal (Levine and Steinhardt 1984). As it turns out, the occasion is a timely moment to revisit the subject of Penrose tilings and quasicrystals. A burst of progress is underway that is forcing us to reconsider our intuition about Penrose tilings and that may have a lasting impact on the study of quasicrystals.

Quasicrystals gained world-wide attention when, for the first time, a material was found whose atomic structure has icosahedral symmetry (Shechtman *et al.* 1984), a symmetry strictly forbidden by the mathematical laws of crystallography established over 150 years ago. It is possible to violate the conventional laws of crystallography if the atoms or atomic clusters are arranged in a three-dimensional, quasiperiodic pattern analogous to a Penrose tiling (Levine and Steinhardt 1984; Levine and Steinhardt 1986). Quasiperiodic means that the structure can be expressed in terms of functions which repeat with incommensurate frequencies. For example, the frequency of fat rhombi divided by the frequency of thin rhombi that repeat in a Penrose tiling along any given direction is an irrational number, inexpressible as the ratio of integers. It is precisely because they are quasiperiodic that quasicrystals can violate Bravais' mathematical laws of crystal symmetry developed 150 years ago, which presume periodicity. All previously forbidden symmetries are allowed for quasicrystals. In addition to five-fold symmetry, eight- and twelve-fold symmetry axes have all been observed in different materials since 1984. Some quasicrystals consist of periodically stacked layers in which atoms are packed quasiperiodically within each layer in one of the forbidden symmetry patterns. Others are quasiperiodic in all three dimensions and exhibit five-fold axes arranged with icosahedral symmetry, the symmetry of a soccer ball or geodesic dome.

Just as periodic tilings are a powerful tool for visualizing the structure and properties of crystals, the Penrose tiling has been influential in developing intuition about quasicrystals. Levine and I had the Penrose tiling in mind when we first hypothesized the possibility of quasicrystals as a new phase of solid matter (Levine and Steinhardt 1984). The tiling is composed of two tiles, fat and thin rhombi, which repeat with incommensurate frequency. As a result, the ratio of fat and thin tiles is an irrational number, $\tau = (1 + \sqrt{5})/2$, the golden mean so beloved by the Greeks. To force the tiles to make a quasiperiodic tiling, one must introduce matching rules for how any pair of tiles can join edge-to-edge. A three-dimensional analogue was constructed (a similar three-dimensional construction had been made independently by R. Ammann), which led to the conjecture of quasicrystal solids in which each tile is replaced by a cluster of atoms. When the electron diffraction pattern for a quasicrystal was computed theoretically, it agreed qualitatively with the pattern measured for the aluminum-manganese alloy.

For an elementary introduction to quasicrystals and quasiperiodic patterns, including models of the three-dimensional icosahedral structure and colourful images that could not be reproduced in this volume see Steinhardt (1986). A recent popular review of experimental progress in the field can be found in Goldman *et al.* (1996).

In spite of these initial successes, the Penrose tiling picture has been controversial. At first, the concerns were motivated by experiments which revealed small discrepancies from the Penrose tiling picture. The Penrose tiling model predicts a diffraction pattern consisting of precise, point-like Bragg peaks, just as for ideal crystals, except arranged with forbidden symmetry. Aluminum-manganese, and other quasicrystal alloys discovered in the early years of the field, had sharp, but not perfectly sharp diffraction peaks, that were arranged very close to, but not quite perfectly at five-fold symmetric positions. One logical possibility is that the Penrose tiling picture is correct, but the samples are imperfect, just as diamonds contain flaws that disrupt their perfect crystal structure.

But, another possibility is that the Penrose structure is physically impossible to achieve, as some conjectured. It was well-known that, even using Penrose matching rules, it is impossible to construct a perfect Penrose tiling without trial and error. Adding tiles to a growing cluster according to the matching rules does not guarantee success. Often, once every ten tiles or so, a conflict arises between tiles added randomly (but according to matching rules) at different parts of the tiling. To form a perfect tiling, one must remove some tiles and start again. For a large number of atoms, the trial-and-error process seems impossibly slow. Alternative models proposing a disordered, metastable atomic structure were developed (Henley 1991).

In the late 1980s, two surprising developments, one theoretical and one experimental, began to reverse opinion. On the theoretical side, Onoda *et al.* (1988) showed that, contrary to common experience with Penrose matching rules, it is possible to find a set of local rules such that a perfect Penrose tiling can be constructed using only local decisions at each step. The key difference between the new rules and the original Penrose matching rules is that the new rules take into account all the tiles around a vertex, whereas the Penrose matching rules only depend on neighbours matching along each edge. This small difference turns out to make an enormous, qualitative difference in constructing perfect Penrose tiling. Local decisions about which tile to attach based on a vertex can produce perfect tilings of arbitrary size without trial-and-error. This theoretical discovery reinvigorated the search for better quasicrystal materials.

Within a few months, new quasicrystalline materials were found in which the quasiperiodic order is as perfect as the periodic order found in the best crystal alloys (Bancel 1991; Goldman *et al.* 1996). The peaks are so sharp that the most powerful synchrotron x-rays are unable to resolve a finite width. The peaks lie right at five-fold symmetry positions. Coherent multiple-scattering is observed as electrons pass through the lattice, which only occurs when there is negligible disorder. And, furthermore, some of the new quasicrystals are thermodynamically stable.

In spite of these successes, some skepticism concerning the Penrose tiling picture has remained. There is no firm argument so much as an intuition: an intuition that the requirements for a Penrose-tiling structure are too complex compared to the requirements for crystals (Henley 1991). The Penrose tiling picture suggests that the atoms must organize into *two* distinct clusters which act as the building blocks of the quasicrystalline structure, whereas crystals require only a single building block. The condition for crystals seems intuitively simple: it is easy to imagine a single building block arising as a low-energy atomic cluster of the given elements. For quasicrystals, the energetics must be delicately balanced to allow two distinct clusters to intermix with a specific ratio of densities. Furthermore, the atomic interactions must restrict clusters so that they join only according to the matching rules.

In the remainder of this paper, two recently discovered approaches for constructing Penrose tilings are discussed which address the criticisms of the Penrose-tiling model of quasicrystals:

- A simple proof (from Steinhardt and Jeong 1996) of the claim (Gummelt 1996) that a quasiperiodic tiling can be forced using only a single type of tile (plus matching rule).

- An independent result that matching rules can be discarded altogether (Steinhardt and Jeong 1996). Instead, maximizing the density of a chosen cluster of tiles suffices.

The results are surprising from a mathematical standpoint and suggest a new explanation of why quasicrystals form. This may enable the prediction of new quasicrystal materials, which may have practical significance since their unique symmetries result in distinctive elastic, structural, and electronic properties. Perhaps with a better understanding of why quasicrystals form, materials scientists can design quasicrystals with optimal properties.

2 New approach to Penrose tiling: single tile/matching rule

The first new result is that a quasiperiodic tiling can be forced using a single type of tile combined with a matching rule; see Fig. 1. The tiling is unconventional (perhaps a better term is a 'covering') since the decagon tiles are permitted to overlap, but only in certain discrete ways, A- or B-type. As an analogy to a real atomic structure, the overlaps should be construed as the sharing of atoms between neighbouring clusters, rather than interpenetration of two complete clusters. Realistic atomic models of known quasicrystals are known to incorporate clusters whose geometry enables sharing of atoms without distortion of the cluster shape (Henley 1991; Burkov 1992; Steurer *et al.* 1993; Aragon *et al.* 1991).

The decagon construction was originally proposed by Gummelt (1996), who presented an elaborate proof. Jeong and I have found a very simple, alternative proof outlined below which makes clear the relation to Penrose tilings and leads us to a second, novel scheme (Steinhardt and Jeong 1996).

The proof is based on inscribing each decagon with a fat Penrose rhombus tile, as illustrated in Fig. 1(c). The original Penrose tiling is constructed from fat and skinny rhombi with marked edges such that two edges may join only if the type and direction of arrows match (Penrose 1974; Steinhardt 1986). Gummelt showed that, in a perfect decagon tiling, there are exactly nine ways a decagon can be surrounded by neighbours which have A or B overlaps with it (Gummelt 1996). The allowed configurations of overlapping decagons may be mapped into configurations of inscribed rhombi. For any two overlapping decagons, the inscribed rhombi share at least one vertex and sometimes share an edge. Where the rhombi join at a vertex only, there is an open angle formed by the edges which are the location and shape where skinny rhombi can be fitted according to the Penrose matching rules. Seven of the nine decagon configurations correspond to completely surrounding a fat tile by neighbouring tiles. In the other two cases, one rhombus vertex is incompletely surrounded; but, there are only two allowed ways of adding overlapping decagons so that the inscribed rhombi complete the vertex. Counting all of these, the decagon overlap rules map into eleven ways of completely surrounding a central fat tile with fat and skinny tiles, precisely the

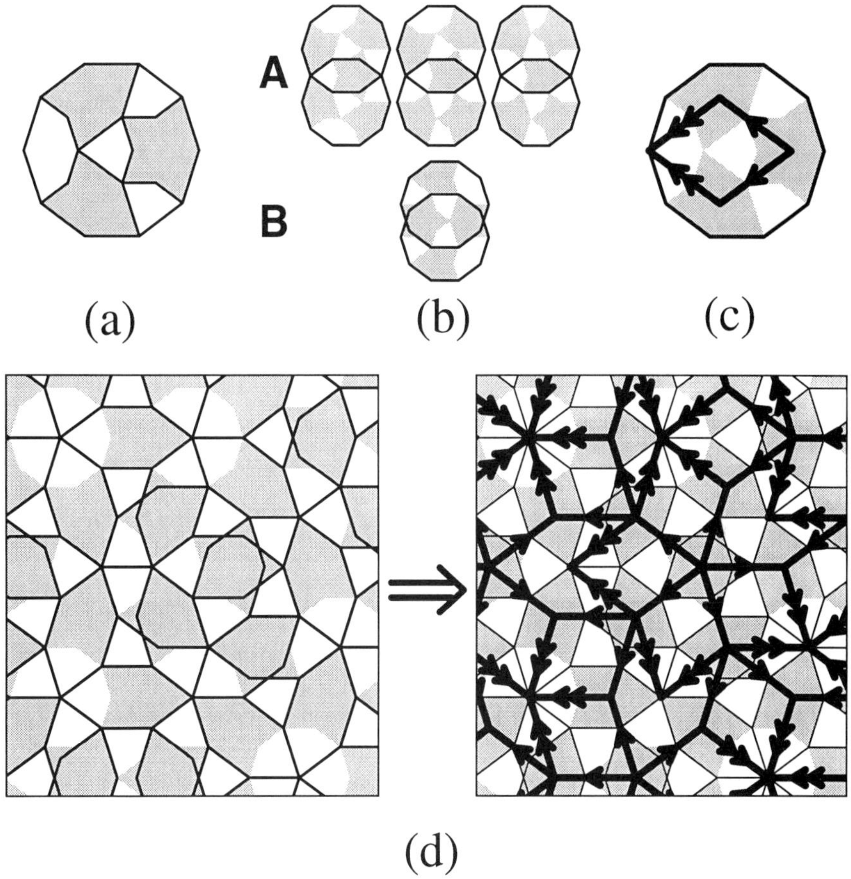

FIG. 1. A quasiperiodic tiling can be forced using a single tile, the marked decagons shown in (a). Matching rules demand that two decagons may overlap only if shaded regions overlap and the overlap area is greater than or equal to the hexagonal overlap region in A. This permits two possible types of overlap between neighbours: either small (A-type) or large (B-type), as shown in (b). If each decagon is inscribed with a fat rhombus, as shown in (c), a tiling of overlapping decagons (d, left) can be transformed into a Penrose tiling (d, right), where space for the skinny rhombi is incorporated.

number and types allowed by the Penrose arrow rules. Restricting the surroundings of every fat tile to these eleven types is equivalent to enforcing the Penrose arrow rules for fat and skinny tiles; and, thus, the proof is completed.

An important corollary is that the two-tile Penrose tiling can be reinterpreted in terms of a single, repeating motif, suggesting a similar interpretation

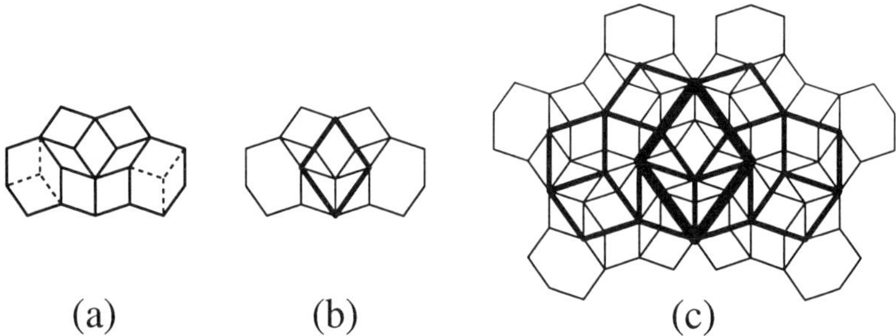

FIG. 2. The cluster C consists of 5 fat and 2 skinny rhombi with two side hexagons composed of 2 fat and 1 skinny rhombus each. There are two possible configurations for filling each side hexagon; the two possibilities are shown with dashed lines on either side in (a). Under deflation, each C-cluster can be replaced by a single 'deflated' fat rhombus, as shown in (b). There is a configuration of nine C-clusters shown in (c) (thin lines) which, under deflation, form a scaled-up C configuration (medium lines), called a DC-cluster. Under double-deflation, each DC-cluster is replaced by 'doubly-deflated' fat rhombus (thick lines).

for quasicrystals. From the point-of-view of tiling theory, this may appear to be a minor advance; but it could be an enormous simplification for the study of quasicrystals. Instead of having to simultaneously determine the decoration of two distinct tiles, we have shown that it is sufficient to identify the atomic decoration of a single cluster. Furthermore, identifying the cluster is aided by the constraint that neighbouring clusters can share atoms without significant bond distortion. Reducing the problem to the decoration of a single unit may greatly improve atomic models of quasicrystals.

3 New approach to Penrose tiling: maximizing cluster density

The second surprise is that matching rules can be avoided (Steinhardt and Jeong 1996). Instead, a perfect Penrose tiling can arise simply by maximizing the density of some chosen tile-cluster, C. Imagine all possible tilings constructed from fat and skinny rhombi with no matching rules. These include quasiperiodic, periodic, and random tilings. Then, the claim is that the Penrose tiling uniquely has the maximum density of C-clusters. (Two tilings are considered equivalent if they differ by patches whose density has zero measure.) The physical analogue is that C represents some low-energy, microscopic cluster of atoms and that minimizing the energy naturally maximizes the cluster density and forces quasiperiodicity

For the proof, we choose the cluster C shown in Fig. 2, although an infinite number of variations are possible. This choice is motivated by the fact that the

C-clusters in a Penrose tiling and the decagons in a decagon tiling are in one-to-one correspondence. Although they have different shapes, their key similarity is that two neighbouring C-clusters can share tiles in two ways isomorphic to the A- and B-overlaps of decagons. (The hexagon sidewings in Fig. 2 are introduced to prevent other kinds of overlaps.) Hence, we know the Penrose tiling has the unique property that every C-cluster has an A- or B-overlap with its neighbours. However, this does not prove that it has the maximum ρ_C, defined as the number of C-clusters per unit area in units where a skinny rhombus has area equal to unity.

Our formal proof uses the concept of 'deflation'. 'Deflation' corresponds to replacing each complete C-cluster by a larger, 'deflated' fat rhombus (see Fig. 2). The deflated rhombus has τ times the sidelength and τ^2 times the area of the original, where $\tau = (1 + \sqrt{5})/2$ is the golden ratio. Because Penrose tilings are self-similar (Penrose 1974), the density of deflated fat rhombi equals τ^{-2} times the density of original fat rhombi which equals the density of C-clusters: $\rho_C^0 = 1/(3\tau + 1)$. The deflation operation can be repeated: identify all configurations of deflated rhombi which form a scaled-up version of the C-cluster (we call this configuration a DC-cluster) and replace each with a yet-larger fat tile (see Fig. 2). Due to self-similarity, $\rho_{DC} = \rho_C^0/\tau^2$ for a Penrose tiling. For non-Penrose tilings, deflation corresponds to the same replacement wherever nine fat rhombi form a complete C-cluster, but the deflated tiling is not necessarily similar to the original and may include voids. Our proof is by contradiction: If a tiling existed with $\rho_C > 1/(3\tau + 1)$, then deflating it repeatedly increases the density without bound — an impossibility.

Because the C-clusters can overlap, a reliable scheme for assigning, or at least bounding, the area occupied by a given C-cluster is needed. A useful trick is to decorate each C-cluster as shown in Fig. 3. The kite-shaped region, which has area $3\tau + 2$, will be called the 'core-area' of the C-cluster. Although C-clusters can overlap to some degree, the only possibilities for close overlap are A-overlaps, in which the core-areas meet along an edge; or B-overlaps, in which there is a specific overlap of core-areas (Fig. 3). In a Penrose tiling, these core-areas fill the entire plane without holes. If the core-area of a C-cluster is not overlapped by any neighbouring core-areas, it can be assigned the entire core-area (at least that); for these cases, the C-clusters occupy area $\geq 3\tau + 2$, so they decrease the density relative to the Penrose value $\rho_C = 1/(3\tau + 1)$. Two C-clusters with B-overlaps are assigned area less than $3\tau + 1$ due to the overlapped core-areas. Hence, we reach an important conclusion: B-overlaps are the only mechanism for exceeding Penrose density.

To exceed the Penrose density, a tiling must have a greater density of B-overlaps than Penrose tiling. However, this condition is not sufficient. In Penrose tiling, every B-overlap of two C-clusters is surrounded by a DC-cluster (see Figs. 2 and 3). In a non-Penrose tiling, a fraction of B-overlaps may not be part of a DC-cluster (*i.e.*, one or more of the seven other C-clusters that compose a DC-cluster is not present). In these cases, it is straightforward to show by explicit

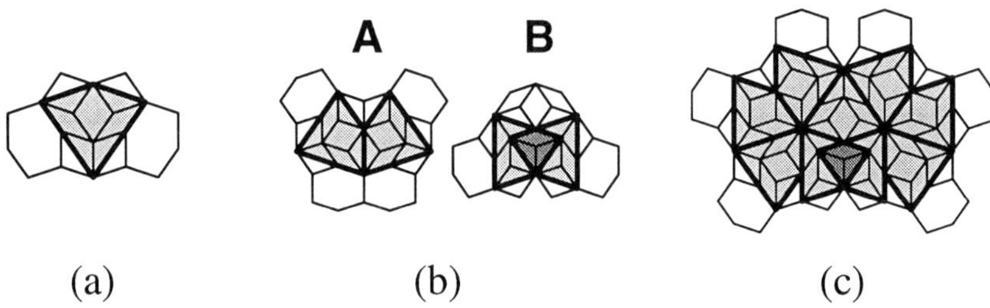

FIG. 3. Associated with each C-cluster is core-area (with area $3\tau+2$) consisting of a kite-shaped region, as shaded in (a). In a Penrose tiling, core-areas of neighbouring tiles either join edge-to-edge (A-overlap) or overlap by a fixed amount (B-overlap), as shown with dark shading in (b). Fig. (c) is a DC-cluster which illustrates the core-areas of the nine C-clusters which compose it. An isolated DC-cluster contains one B-overlap (see dark shading) and the rest A-overlaps.

constructions that one can always identify an area attached to the associated B-overlap which does not belong to the core-area of any C-cluster and is not associated with any other B-overlap (Jeong and Steinhardt 1996). This 'extra', unassigned area occupies at least as much area as saved by the B-overlap. Hence, a B-overlap which is not part of a DC-cluster does not contribute to increasing the density of C-clusters above the Penrose value.

Suppose there were a tiling with a density of C-clusters greater than the Penrose value. Then, we have just shown that it must have a higher density of DC-clusters than in a Penrose tiling, $R_{DC} > \tau^{-2}$, where R_{DC} is the number of DC-clusters divided by the number of C-clusters. Under deflation and rescaling the area by τ^2, each DC-cluster becomes a C-cluster of the deflated tiling whose density is $\tau^2 R_{DC}\rho_C$. Since $R_{DC} > \tau^{-2}$, the deflated tiling has a density of C-clusters that is strictly greater than the original tiling. Repeating the deflation *ad infinitum* would lead to an impossible tiling with an unbounded density of C-clusters.

A corollary is that, if the C-cluster density equals the Penrose value, then $R_{DC} = \tau^{-2}$ (the Penrose tiling value) and the C-cluster density in the deflated tiling must equal the Penrose value. This is useful in proving that the Penrose tiling is the *unique* tiling with $\rho_C = 1/(3\tau+1)$. Suppose there were a non-Penrose tiling with the same density. We have argued that the only local configurations which can increase the density above the Penrose value are DC-clusters, and that the increase in density is due to the B-overlap of core-areas, which is the same for each DC cluster. The corollary says that the hypothetical tiling has the same density of DC-clusters and, hence, the same density of B-overlaps surrounded by DC-clusters as Penrose tiling. However, by definition, the non-Penrose tiling must also have patches with non-zero area measure which violate the Penrose matching rules, and so cannot belong to the core-area of any C-cluster. Since the DC-cluster density is the same but there are these patches, it would appear that the average area per C-cluster must be less than the Penrose density. The only conceivable exception would be if there happen to be additional B-overlaps which do not belong to DC-clusters whose overlap area exactly compensates the area of the patches. Even this possibility can be eliminated because the corollary states that $R_{DC} = 1/\tau^2$, which means that the density of C-clusters remains unchanged under deflation and rescaling. Yet, the patches grow: a patch excluded from a C-cluster must also be excluded from a DC-cluster, but, also, some C-clusters that border the patches cannot be part of a DC-cluster and add to the patch area (Jeong and Steinhardt 1996). Since the number of C-clusters remains fixed but the patches grow, the C-cluster density in the deflated tiling must be less than the Penrose value. This contradicts the corollary; hence, uniqueness is established.

This discussion is only a brief summary of the complete proof (Jeong and Steinhardt 1996), which introduces many new aspects of Penrose tilings which experts may wish to pursue. The full proof shows how the Penrose tiling has the maximum density of C-clusters, which means that it minimizes a Hamiltonian which assigns low energy to all C-clusters. In Jeong and Steinhardt (1996), we go on to show a further new result that the Penrose tiling is the ground state of a spectrum of other Hamiltonians, including ones which assign different energies to C-clusters depending on their local environment. Hence, the Penrose tiling configuration is the ground state for a robust range of Hamiltonians. Contrary to some claims, fine-tuning of interactions is not required.

4 Implications

The two new approaches to Penrose tiling, a single tile type and maximizing cluster density, can be combined. Together, they suggest a new view of the structure of quasicrystals and why they form.

The atomic structure can be reinterpreted in terms of a single repeating cluster, rather than two different clusters. This simplifies atomic modelling since atomic decoration of only the single cluster need be considered. The modelling is further constrained since the cluster must be capable of sharing atoms in certain discrete ways with neighbours.

The results also suggest physically plausible conditions that can lead to quasicrystal formation, shedding new light on an old mystery. They imply that quasicrystals can be understood by considering the energetics of microscopic clusters and that cluster overlap is an important structural element, establishing an earlier conjecture. The simplest energetics would be assigning negative energy to the clusters and zero energy to all other local configurations, since this is sufficient to cause the minimum free energy state to be the maximum cluster density state. However, it is important that the energetics be robust. Some experiments with other energetics assignments suggest that there is a continuum

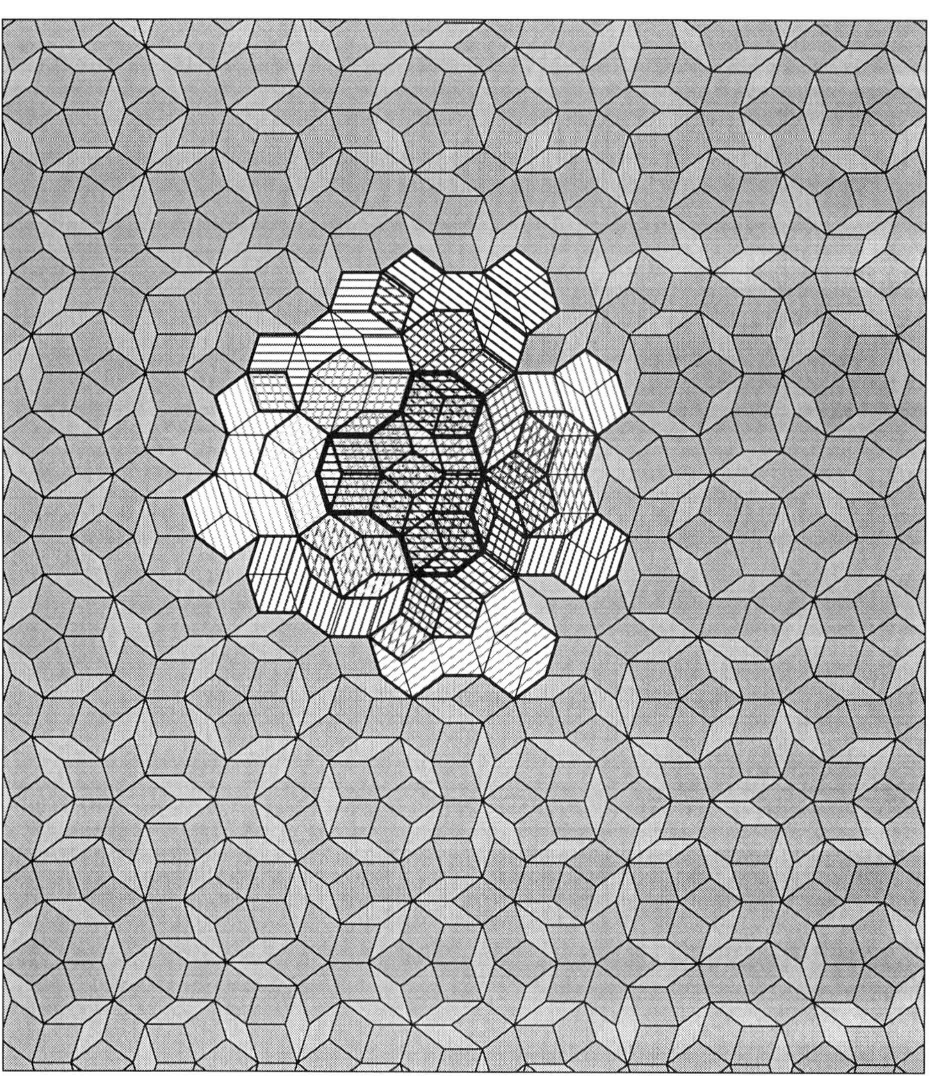

FIG. 4. A fragment of Penrose tiling in which the overlapping C-clusters are indicated with different shading in the central portion. The figure illustrates the very high C-cluster density (shown in our proof to be the maximal possible density).

of possibilities which lead to the same Penrose ground state, but this needs to be studied further.

All of these concepts can be tested using the atom clusters of known quasicrystals. Our two-dimensional tiling results can most readily be applied to decagonal quasicrystals which have periodically-spaced layers with Penrose tiling structure. The extension to three-dimensional, icosahedral symmetry is a future challenge, although past experience suggests that two-dimensional properties can be extended to three-dimensions. If these principles can be established, they may enable the reliable prediction of new quasicrystals.

As an example of the application of symmetry principles, the subject of quasicrystals is still in a primitive stage. The Bravais classification of crystal point symmetries, as well as space groups, has been achieved. But, the key confusions about the structure and formation of quasicrystals, as described in this paper, go beyond symmetry classification. Here, our concept of quasicrystals has been heavily influenced by experience with Penrose tilings and Penrose matching rules. Based on the original rules, it appeared that two or more repeating units, rigid matching rules and non-local growth rules were required to build the structure. All of these have been shown to be unnecessary, but only by further imaginative tiling constructions. Even so, it is uncertain, without more constructions, whether the results generalize to other symmetries and other dimensions.

What is missing are powerful mathematical techniques analogous to the group theoretic methods applied to the structure periodic crystals 150 years ago. Our ultimate challenge is to understand how the new results shown in this paper arise directly from quasiperiodicity and crystallographically forbidden symmetries. This demanding problem is Roger Penrose's inadvertent legacy to the world of solid state physics.

Acknowledgements I thank my collaborator, Hyeong-Chai Jeong, for his contribution to the work described herein. I also thank the organizers for the opportunity to participate in the celebration honouring Roger Penrose. Finally, I want to express my appreciation to Roger Penrose for his inspiring creativity and humanity. This work was supported in part by DOE at Penn (DOE-EY-76-C-02-3071).

Bibliography

Aragon, J. L., Romeu, D., and Gomez, A. (1991). *Phys. Rev.*, **B44**, 584–92.

Bancel, P. (1991), In: *Quasicrystals, the state of art*, eds. D. P. DiVincenzo and P. J. Steinhardt, 429–524. World Scientific, Singapore.

Burkov, S. (1992). *J. Phys.*, **2**, 695–706; *Phys. Rev. Lett.*, **67**, 614–17.

Goldman, A. *et al.* (1996). *Am. Sci.*, **84**, 230–41.

Gummelt, P. (1996). *Geometriae Dedicata*, **62**, 1-17.

Henley, C. L. (1991). In: *Quasicrystals, the state of art*, eds. D. P. DiVincenzo and P. J. Steinhardt, 17–56. World Scientific, Singapore.

Jeong, H.-C. and Steinhardt, P. J. (1994). *Phys. Rev. Lett.*, **73**, 1943–6.

Jeong, H.-C. and Steinhardt, P. J. (1996). *Phys. Rev.*, **B**, to appear.

Levine, D. and Steinhardt, P. J. (1984). *Phys. Rev. Lett.*, **53**, 2477–80.

Levine, D. and Steinhardt, P. J. (1986). *Phys. Rev.*, **B34**, 596-9.

Onoda, G., Steinhardt, P. J., DiVincenzo, D., and Socolar, J. (1988). *Phys. Rev. Lett.*, **60**, 2653–6.

Penrose, R. (1974). *Bull. Inst. Math. and Its Appl.*, **10**, 266–9.

Shechtman, D., Blech, I., Gratias, D., and Cahn, J. W. (1984). *Phys. Rev. Lett.*, **53**, 1951–4.

Steinhardt, P. J. (1986). *Am. Sci.*, **74**, 586–97.

Steinhardt. P. J. and Jeong, H. C. (1996). *Nature* **382**, 433–5.

Steurer, W., Haibach, T., Zhang, B., Kek, S., and Luck, R. (1993). *Acta Cryst.*, **B49**, 661–75.

14

Decaying Neutrinos and the Geometry of the Universe

D. W. Sciama
SISSA and ICTP, Trieste
Department of Physics, Oxford University

1 Introduction

It is a great privilege and also a great pleasure to be invited to speak at this celebration of Roger Penrose's 65th birthday. We first met about 45 years ago when, as a research student at Cambridge, I shared an office with several colleagues, one of whom was Oliver Penrose, Roger's older brother, and now a distinguished authority in statistical mechanics. My own first research interest was also in statistical mechanics, but when I changed to relativity and cosmology in the middle of my Ph.D studies, Oliver told me that I should meet his brother "who thinks geometrically".

When we did meet, we soon became close friends, and I found Roger's global way of thinking of tremendous value to me. At that time, and also over the years since, I have learnt so much from him. I am grateful for this opportunity to be able to thank him publicly for the warmth of his friendship and for the influence on me of his very special insights into the subtle relations between physics and geometry.

In thinking over what to talk about on this occasion it occurred to me that it might be appropriate to describe how the geometry of the universe is constrained by my theory that decaying dark matter neutrinos are mainly responsible for the widespread ionisation of our Galaxy (Sciama 1990, 1993). This theory is still highly speculative, but at the end of my talk I will announce a recent, still unpublished, observational result which strongly supports it (Bland–Hawthorn, Carignan and Veilleux 1996).

As you all know, the geometry of the universe is related to the still unanswered question: will the universe expand forever, or is its expansion destined to be halted by self-gravitation, then to be converted to collapse into a big crunch? If the universe just expands forever its future large-scale behaviour is described by the well-known Einstein–de Sitter model in which the 3-space orthogonal to the world-lines of substratum particles has zero curvature. If in addition the cosmological constant λ is zero the density associated with this model has the critical value ρ_c given by

$$\frac{8\pi}{3}G\rho_c = H^2,$$

where H is the Hubble constant, which determines the expansion rate of the universe.

Despite important recent observations made with the Hubble Space Telescope, the present value H_0 of H is still rather uncertain—it probably lies between 50 and 100 $\mathrm{kms^{-1}Mpc^{-1}}$. In a few years its value may be known with a precision of about 10 per cent. Already there are signs that its value might be towards the lower end of this range. A related observational parameter is the present age t_0 of the universe since the big bang. In the Einstein–de Sitter model

$$t_0 = \frac{2}{3H_0}.$$

Unfortunately, the value of t_0 is also rather uncertain. A recent detailed statistical study of the errors involved in its determination suggests a minimum acceptable value of 12×10^9 years (Chaboyer *et al.* 1996).

If we combine these results we see that a viable possibility is to have $\lambda = 0$, $\rho = \rho_c$, $t_0 = 12 \times 10^9$ years and $H_0 = 55\,\mathrm{km\,s^{-1}Mpc^{-1}}$. A number of attempts have been made to determine ρ observationally using some form of virial theorem, but again with uncertain results. The value of $\rho = \rho_c$ does lie in the permitted range, and many cosmologists hope that this is the actual value because of their preference for a flat 3-space. However, this preference would require the introduction of a dominating amount of non-baryonic dark matter in the universe, since the total baryonic contribution to ρ is known to be only a few per cent of ρ_c (from arguments involving the big bang nucleosynthesis of the light elements, D, He^3, He^4 and Li^7). There are various possibilities for the identity of this non-baryonic dark matter. In this talk we shall assume that it consists of neutrinos of non-zero rest-mass, pair-created in the hot big bang and surviving to the present day.

There is also a need to introduce dark matter into individual galaxies, like our own Milky Way, in order to account for their rotation curves, which in many cases remain flat far beyond the region containing most of the stars. In these galaxies the dark matter could be either baryonic or non-baryonic. We shall again assume in this talk that most of the dark matter in galaxies like our own consists of neutrinos with a non-zero rest-mass.

2 Relic neutrinos as dark matter

It is now established experimentally that there are three different types of neutrino, the electron-type ν_e, the muon type ν_μ, and the tau type ν_τ. Each type ν_i is pair-produced in the hot big bang via the weak interaction

$$\nu_i + \overline{\nu_i} \leftrightarrow e^- + e^+.$$

In the very early universe ν_i rapidly came into thermal equilibrium with the prevailing heat bath, but decoupled from it when the temperature dropped to about 1 Mev. A standard calculation shows that in recent times the mean number density n_{ν_i} of ν_i is given by

$$n_{\nu_i} = \frac{3}{11} n_\gamma$$

where n_γ is the number density of photons in the heat bath (corresponding today to the cosmic microwave background). According to the most recent observations made by the COBE satellite, the present temperature T_0 of this heat bath is

$$T_0 = 2.728 \pm 0.004\,\mathrm{K}.$$

Hence

$$n_\gamma = 412 \pm 2\,\mathrm{cm}^3$$

and

$$n_{\nu_i} = 112.4 \pm 0.5\,\mathrm{cm}^3.$$

It follows that, if ν_i is now moving non-relativistically,

$$m_{\nu_i} = \frac{(93.2 \pm 0.4)\rho_{\nu_i}}{\rho_c} h^2\,\mathrm{eV},$$

where

$$H_0 = 100\ \mathrm{h\,kms^{-1}\,Mpc^{-1}}.$$

We now consider the possible values of m_{ν_i}. We first emphasise that m_{ν_i} may still turn out to be zero. The present laboratory upper limits are

$$m_{\nu_e} < 10 - 15\,\mathrm{eV}$$

$$m_{\nu_\mu} < 170\,\mathrm{keV}$$

$$m_{\nu_\tau} < 24\,\mathrm{MeV}.$$

The large uncertainty in the upper limit on m_{ν_e} (which is based on observations of the end point of the beta decay spectrum of tritium) arises because a formal solution for $m^2{}_{\nu_e}$ gives a negative value, indicating that there is a significant unidentified contribution to the systematic uncertainty. There is also an upper limit on m_{ν_e} derived from the observed arrival times of ν_e from the supernova 1987A, which is about 15 eV.

We note that, if $h \leq 1$, and $\rho \sim \rho_c$, then

$$\sum m_{\nu_i} \leq 93\,\mathrm{eV}.$$

The most favoured (but not the only) particle physics models which determine m_{ν_i} lead to

$$m_{\nu_\tau} \gg m_{\nu_\mu} \gg m_{\nu_e}.$$

For simplicity, we shall assume these inequalities in what follows.

We now recall our favoured set of parameters $\lambda = 0$, $\rho = \rho_c$, $t_0 = 12-13 \times 10^9$ years and $h = 0.55$–0.50. Then if ρ_b is only a few per cent of ρ_c and the dark matter is due to neutrinos we must have

$$m_{\nu_\tau} \sim 25 \pm 3\,\text{eV}.$$

This choice would thus determine m_{ν_τ} with a precision of ~ 10 per cent.

We have now reached the starting-point of this talk, which aims to explain how the decaying neutrino theory constrains m_{ν_τ} with a precision of about 1/2 per cent. This argument is quite independent of the above considerations, but does lead to a result lying within the range 25 ± 3 eV. Thus the decaying neutrino theory would constrain the geometry of the universe.

3 Decaying neutrinos and the ionisation of the universe

If $m_{\nu_{e,\mu}} < m_{\nu_\tau}$, particle physicists tell us to expect the radiative decay

$$\nu_\tau \to \gamma + \nu_{e,\mu}.$$

Conservation of energy and momentum in the decay then imply that

$$E_\gamma = \frac{1}{2} m_{\nu_\tau} \left(1 - \left(\frac{m_{\nu_{e,\mu}}}{m_{\nu_\tau}}\right)^2\right)$$

where E_γ is the energy of the decay photon in the rest frame of the decaying ν_τ and $c = 1$. Note that E_γ is a *line*, which is very helpful in deriving stringent consequences from our assumptions. Moreover, if $m_{e,\mu} \ll m_{\nu_\tau}$ we have the simple result

$$E_\gamma \sim \frac{1}{2} m_{\nu_\tau}.$$

This result is obvious, since ν_τ is breaking up into two particles of negligible rest mass.

We now require the decay photon to be able to ionise hydrogen, so that

$$E_\gamma \geq 13.6\,\text{eV},$$

and so

$$m_{\nu_\tau} \geq 27.2\,\text{eV}.$$

The reason for making this hypothesis is that the hydrogen in the interstellar medium of our Galaxy is found to be partially ionised even in regions protected from the radiation of hot stars by opaque clouds containing neutral hydrogen. Since dark matter neutrinos would pervade the Galaxy this opacity would no longer be a problem. In fact in an opaque region every decay photon would ionise a nearby hydrogen atom which would then recombine and we would have the simple ionisation equilibrium

$$\frac{n_{\nu_\tau}}{\tau} = \alpha n_e^2,$$

where n_ν is the local number density of neutrinos, τ is their decay lifetime, α is the recombination coefficient and n_e is the number density of electrons and protons in the region concerned. Thus n_e would be independent of the density of neutral hydrogen and so of the opacity.

For this mechanism to work we need τ to have an appropriate value. In principle this value should be provided by particle physics, but as yet the correct particle physics model is not known. If we combine the observed value of n_e with the rotation curve of the Galaxy and our eventual value of m_{ν_τ} (to determine n_{ν_τ}) we find that we require

$$\tau \sim 2 \times 10^{23} \, \text{sec}.$$

This is about a million times longer than the age of the universe, so most relic neutrinos would not yet have decayed.

So far our hypothesis has provided us with only a lower limit for m_{ν_τ}. We now derive an upper limit by considering the intergalactic hydrogen-ionising flux due to the cosmological distribution of neutrinos. This flux is not yet well-measured, but an upper limit to it can be derived from the observed upper limit on the recombination radiation emitted by an opaque neutral hydrogen cloud immersed in this ionising radiation. In particular one searches for the H_α recombination line. If the flux in this line from the cloud could be measured, one could infer the flux of ionising photons incident on the cloud. We shall return to this method at the end of this talk, when we describe the new result which supports our theory. For the moment we have only an upper limit for the H_α flux from a particular intergalactic neutral hydrogen cloud (1225+01) (Vogel *et al.* 1995) which leads to the following upper limit for the intergalactic flux F of hydrogen-ionising photons at zero red shift:

$$F < 10^5 \, \text{photons cm}^{-2}\text{sec}^{-1}.$$

The contribution to F from known sources (mainly quasars) is about 2×10^4 $\text{cm}^{-2}\text{sec}^{-1}$ (Haardt and Madau 1996).

What would we expect for cosmologically produced decay photons? We must of course allow for the red shift which degrades the energy of a decay photon as it propagates through the universe so that eventually it can no longer ionise hydrogen. If we write

$$E_\gamma = 13.6 + \epsilon \, \text{eV}$$

we easily see that

$$F = \frac{n_\nu}{\tau} \frac{c}{H_0} \frac{\epsilon}{13.6},$$

where n_ν is now the number density of cosmological neutrinos at $z = 0$, and c is the velocity of light. Inserting our derived values for n_ν, τ and H_0 we find that $F < 10^5$ cm^2 sec^{-1} if

$$\epsilon < 0.1 \, \text{eV}.$$

Hence

$$E_\gamma < 13.7\,\text{eV},$$

and so

$$E_\gamma = 13.65 \pm 0.05\,\text{eV},$$

and

$$m_{\nu_\tau} = 27.2 \pm 0.1\,\text{eV}.$$

We have thus determined m_{ν_τ} with a precision of $\frac{1}{2}$ per cent. Moreover our value is within the range expected for m_{ν_τ} (25 ± 3 eV) from our argument involving the critical density, the age of the Universe and the Hubble constant. In fact we can now refine this argument in the following way. Our value of 27.2 ± 0.1 eV for m_{ν_τ} tells us that $\frac{\rho_{\nu_\tau}}{\rho_c}h^2 \sim 0.294 \pm 0.003$. If we allow for a small contribution from ρ_b, we find that if $t_0 \geq 12 \times 10^9$ years we must have $\rho \sim \rho_c$. Assuming for simplicity that $\rho = \rho_c$ exactly (flat 3-space!) we require that

$$H_0 = 55 \pm 0.5\,\text{km sec}^{-1}\text{Mpc}^{-1}.$$

Thus H_0 would be determined to within 1 per cent. It will be interesting to see whether observations of H_0 in the next few years, which may be expected to achieve a precision of 10 per cent, will be compatible with our prediction.

4 A new observational test of the decaying neutrino theory

The most decisive test of the decaying neutrino theory would be to observe the predicted decay line directly from neutrinos in the vicinity of the sun. Since the Earth's atmosphere absorbs hydrogen-ionising photons one would need to carry out such an experiment from outer space. Plans have been made to do this, and a suitable detector will soon be launched on board a Spanish mini-satellite (Bowyer *et al.* 1996). Meanwhile I have been given permission to report here the result of a recent, still unpublished, ground observation which strongly, but not decisively, supports the theory. A further measurement of the same type has been made by these observers, but the data have not yet been completely reduced.

The idea behind these measurements is to find regions of opaque gaseous hydrogen where the postulated ionising flux from neutrinos would be at least an order of magnitude greater than any ionising flux of conventional origin, and where this flux can be measured by observing the strength of the resulting H_α line emitted by this region when the ionised hydrogen recombines. Such regions can be found. They exist at the edges of some spiral galaxies, whose opaque gaseous hydrogen disks extend far beyond their distribution of stars, and whose rotation curves (as measured using the 21 cm emission line of the hydrogen discs) remain flat, so that one can be confident that the dark matter distribution extends to these outer regions.

Two galaxies are of particular interest in this connexion because their column density N of neutral hydrogen possesses a sharp edge beyond which N decreases rapidly with distance from the centre of the galaxy. This edge is thought to be an ionisation effect associated with the onset of a regime where the gas is no longer able to shield itself from the incident ionising flux. In the conventional interpretation this flux would be the extra-galactic ionising flux which, as we have seen, is less than $10^5\,\mathrm{cm}^{-2}\mathrm{sec}^{-1}$. By contrast, the ionising flux of decay photons in these edge regions, which can be derived from the rotation curves in these regions, turns out to be about $10^6\,\mathrm{cm}^{-2}\mathrm{sec}^{-1}$, in other words the desired order of magnitude greater (Sciama 1995).

The two galaxies concerned are M33, a companion of the Andromeda galaxy M31, and NGC 3198, which is called "everyone's favourite" because its rotation curve remains flat particularly far from its distribution of stars and so is the best example of the need for dark matter in galaxies. The flux in the H_α line has recently been measured at the sharp edges of both these galaxies by Bland-Hawthorn, Carignan and Veilleux (1996). Bland-Hawthorn has kindly informed me that for M33 the implied ionising flux is close to the value predicted by the decaying neutrino theory, rather than the smaller value expected in the conventional theory. This result does not provide definitive evidence in favour of the decaying neutrino theory because the disc of M33 is strongly warped, and it is conceivable that ionising photons emitted by hot stars in the central regions of the galaxy may be able to reach the sharp edge without having to penetrate the opaque gas in the disk. This problem does not arise in NGC 3198 since its disk is not appreciably warped. The results for this galaxy are therefore awaited with great interest.

If these results do turn out to support the decaying neutrino theory they would also support the suggestion that the mass density in neutrinos is close to the value needed to flatten the 3-geometry of the universe. I hope that this possible connexion between physics and geometry will give pleasure to Roger Penrose, whose penetrating insight has provided us with so many connexions between these two great concepts.

Acknowledgements This work has been supported by the Italian Ministry of Universities and Scientific and Technological Research.

Note added June 3 1997: The Spanish minisatellite was successfully launched on April 21 1997, and at the time of writing, first results should be available from it in a few weeks.

Bibliography

Bland-Hawthorn, J., Carignan, C., and Veilleux, S. (1996), private communication.

Bowyer, S. *et al.* (1996), in Proc. IAU Colloq. 152, *Astrophysics in the Extreme Ultraviolet* eds S. Bowyer and R. F. Malina (Dordrecht: Kluwer).

Chaboyer, B., Demarque, P., Kernan, P., and Krauss, L. M. (1996), *Science*, **271**, 957.

Haardt, F. and Madau, P. (1996), *Astrophysical Journal*, **461**, 20.

Sciama, D. W. (1990), *Astrophysical Journal*, **364**, 549.

Sciama, D. W. (1993), *Modern Cosmology and the Dark Matter Problem*, (Cambridge: Cambridge Unioversity Press).

Sciama, D. W. (1995), *Mon. Not. Roy. Astr. Soc.*, **276**, L1.

Vogel, S. N., Weymann, R., Rauch, M., and Hamilton, T. (1995), *Astrophysical Journal*, **441**, 162.

15

Quantum Geometric Origin of All Forces in String Theory

Gabriele Veneziano
Theory Division, CERN, 1211 Geneva 23, Switzerland

1 Introduction

I am a great admirer of Roger, but I cannot pretend to know him as well as most of you here. Let me, however, start my talk with a small episode of my life indirectly related to him. A couple of years ago I bought *The Emperor's New Mind* just before leaving for summer vacations in southern France. I took the book with me, hoping, of course, to enjoy my holiday so much as to "forget" the book in my suitcase. Things went differently: at my first tennis lesson I was struck by lumbago and finished my vacation reading the book over and over again by the hotel's swimming pool ...

This morning I went to a local bookstore and bought *Shadows of the Mind.* I am planning to take holidays in Italy in a couple of weeks and start to feel a strange ache at my left shoulder: I am not superstitious, but ...

The plan of my talk is quite simple: in the first part, after recalling how our present understanding of all forces is based on local symmetries, I will remind you of how all of them are given a geometric meaning in the Kaluza–Klein (KK) framework. Unfortunately, such an appealing idea cannot be extended beyond the semiclassical approximation because of ultraviolet infinities. In the second part of my talk I will outline how string theory, thanks to some "quantum string magic", is able to overcome the difficulties of traditional KK theory and to lead to a truly complete, quantum–geometrical unification of all forces.

2 Forces and local symmetries

Our present understanding of *all* fundamental interactions is based on the concept of *local* symmetries. In the case of "internal" symmetries such local symmetry is called gauge invariance, an invariance of the action under the transformations (I will set $c = \hbar = 1$ whenever their explicit appearance is not particularly enlightening):

$$\begin{aligned} \psi(x) &\rightarrow e^{iq\alpha(x)}\psi(x), \\ A_\mu &\rightarrow A_\mu - \partial_\mu \alpha \; . \end{aligned} \tag{2.1}$$

This leads to actions of the type:

$$S_{gauge} = -\frac{1}{4}\int d^4x F_{\mu\nu}^2 + \ldots \tag{2.2}$$

on which the standard model of strong, weak and electromagnetic interactions is based.

Similarly, in the case of "space-time" symmetries the local symmetry is called general covariance, meaning the invariance of the action with respect to the space-time diffeomorphisms:

$$\begin{aligned} x^\mu &\to \xi^\mu(x), \\ g_{\mu\nu}(x) &\to \partial_\mu \xi^\rho \partial_\nu \xi^\sigma g_{\rho\sigma}(\xi) \; . \end{aligned} \tag{2.3}$$

This leads to actions of the type:

$$S_{gravity} = \frac{1}{16\pi G_N}\int d^4x\sqrt{-g}R(g) + \ldots \tag{2.4}$$

on which general relativity (GR) is based.

Both kinds of local symmetries have been successfully tested. The former by precision tests of QED ($g-2$ being the best example), of the electroweak theory (cf. high-precision LEP experiments) and of the strong interaction sector (via jet physics, for instance). For the latter symmetry there are by now quite impressive tests of GR, last but not least the indirect observation of gravitational radiation from binary stars.

Taken all together, these observations undoubtedly represent a landmark in our understanding of physical phenomena, the situation being reminiscent of the one prevailing at the end of last century when, with Maxwell's and Newton's equations, some physicists thought that there was nothing else to be discovered. In fact some conceptual problems suggest that history may repeat itself, and that another deep revolution may be on its way in physics. I will try to explain why by taking a rather long detour.

3 Field-theoretic Kaluza–Klein

Already in the 1920s Kaluza and Klein proposed to further unify the concepts of internal and space-time symmetries by reducing the former to the latter through the introduction of some extra dimension of space. Let me briefly review the main point.

Assume that space-time contains a fifth (space-like) dimension, which has the topology of a circle, i.e. let us write:

$$x^A = (x^\mu, x^5) \tag{3.1}$$

and make the identification:

$$x^5 \equiv x^5 + 2\pi R \; . \tag{3.2}$$

Any sensible wave function will have to be periodic in x^5 and thus of the form:

$$\sum_{p_5=n/R} e^{ip_5 x^5}\,\tilde{\psi}_{p_5}(x^\mu)\;. \tag{3.3}$$

Consider now the particular coordinate transformation:

$$x^5 \to x^5 + l_P\alpha(x^\mu), \tag{3.4}$$

where, for dimensional reasons, we have introduced a length l_P. Using Eq. (3.3), this will imply:

$$\tilde{\psi}_{p_5}(x^\mu) \to e^{il_P p_5 \alpha(x)}\tilde{\psi}_{p_5}(x^\mu) \tag{3.5}$$

which looks like the gauge transformation (2.1) for a field carrying charge:

$$q = l_P p_5 = n l_P/R\;. \tag{3.6}$$

Furthermore, KK showed that the $\mu 5$ components of the five-dimensional metric transform like the gauge field in (2.1) and that the five-dimensional gravitational action generates the four-dimensional gravity–plus–gauge action ((2.4)+ (2.2)), provided l_P is identified with the so-called Planck length, $\sqrt{G_N\hbar} \sim 10^{-33}$ cm.

Besides its conceptual beauty, KK theory has two interesting consequences:

- Electric charge is automatically quantized, thanks to quantization of momentum on a circle,
- Electromagnetic and gravitational interactions get unified at energies $M_c = 1/R$ since, using (3.6) for $n = 1$, $G_N M_c^2 = l_P^2/R^2 = q^2$.

Later on the KK idea was widely generalized, e.g. to generate larger (non-Abelian) gauge groups from even higher dimensional spaces endowed with suitable isometries. Of course, even in its simplest version, the question remains of what fixes R to have the correct value $R \sim 10 l_P \sim 10^{-32}$ cm in order to explain the actual value of $\alpha \sim 1/137$.

Note, finally, that KK theory leads to a unified *classical* theory but is based, in an essential way, on quantum mechanics: the *quantization* of momentum gives the quantization of electric charge! This means that there is no way to ignore quantum mechanics within the KK theory. But are the two consistent with each other? Unfortunately, when we go from the semiclassical approximation to full-fledged quantum field theory (QFT) the problem of ultraviolet infinities immediately shows up. How do we handle that?

In $D = 4$, gauge theories can be dealt with through the process of renormalization; however, no such recipe is known for gravity. As we move to $D > 4$, both gauge and gravity become non-renormalizable. In KK theory, in particular, both diverge in a similar way in the ultraviolet, another expected consequence of KK unification.

We thus face a kind of paradoxical situation. On the one hand quantum mechanics is essential to the success of the KK idea. At the same time, QFT gives meaningless infinities and spoils the nice semiclassical results. If the beautiful KK idea is to be saved we need a better quantum theory than QFT. I will argue below that such a theory already exists: it is called (super)string theory.

4 Quantum string magic

The following is a sketchy account of what classical and quantum string theory is. Let us recall the form of the action for a system of points and for the string:

$$S_{points} = \sum_i m_i c \int d\ell_i + \text{interactions} \tag{4.1}$$

$$S_{string} = T \int d\Sigma, \tag{4.2}$$

and compare them. The first has a geometrical "free-action" term, the sum of the lengths of the paths described by the points, each one of these lengths being weighted by the corresponding mass. Interactions, however, have to be added to the action, are non-geometrical, and quite arbitrary. In contrast, S_{string} is given by a *single*, geometric term proportional to the area swept by the string. One constant appears in place of the masses (the string tension T) and, most remarkably, interactions need not be introduced by hand.

The area is swept in space-time and therefore space and time have to be measured in the same units. A constant c (later to be recognized as the speed of light) is implicitly multiplying time intervals.

In order for the action to have its usual dimensions, T has to have dimensions energy/length, as appropriate for a tension. However, as long as we work at the classical level, T is completely irrelevant (for free points, analogously, only mass ratios are important). This is because, classically, only the stationary points of S do matter, and these are invariant under a rescaling of the action.

Classically, free points move along straight lines, while the string has a rich variety of motions subject to the interesting constraint:

$$J \leq (2\pi T)^{-1} M^2 \, . \tag{4.3}$$

The mass-squared of the string is bounded from below by a multiple of its angular momentum, since a classical string with non-zero angular momentum has to have a finite size and thus, because of its finite tension, a finite energy/mass.

At the quantum level the difference between points and strings becomes even larger. Of course we know that angular momentum becomes quantized in units of $\hbar$:

$$J = n\hbar \, . \tag{4.4}$$

Not surprisingly, also $M^2(2\pi T)^{-1}$ becomes quantized in units of $\hbar$:

$$M^2(2\pi T)^{-1} = m\hbar. \tag{4.5}$$

What becomes of the classical inequality (4.3)? Here the *first* miracle occurs. One finds:

$$J \leq (2\pi T)^{-1} M^2 + a_0\hbar \text{ with } a_0 \;=\; +1\ (+2) \text{ for the open (closed) string .} \quad (4.6)$$

The origin of a_0 is that of a zero point energy (or normal-ordering constant) analogous to the famous 1/2 of the harmonic oscillator levels $(n+1/2)\hbar\omega$. This normal ordering constant is of paramount importance for string theory in that it allows the existence of classically forbidden, massless, spin 1 and spin 2 states (for open and closed strings, respectively). These particles are believed to exist in Nature, in the form of gauge bosons and gravitons respectively, and to mediate all known forces, including gravity. This is why we can dream of quantum string theory (QST) as a candidate theory of all fundamental interactions!

Quantization is responsible for a *second* miracle of QST: strings acquire a typical, better a *minimal*, size. Recall that position and momentum uncertainty behave rather symmetrically for the harmonic oscillator, both scaling like $\sqrt{\hbar}$. The same is true for the string, an infinite collection of harmonic oscillators:

$$\Delta x \;\simeq\; (\hbar/T)^{1/2} \quad ; \quad \Delta p \;\simeq\; (\hbar T)^{1/2} \;. \qquad (4.7)$$

So far we have introduced three fundamental constants:

$$c\,, \qquad T\,, \qquad \hbar\,.$$

Classically, there was no $\hbar$, of course, and T was irrelevant. Quantum mechanically, it is only a combination of $\hbar$ and T that matters. Since the quantum theory depends on S only through the pure number $S/\hbar$, the combination that survives is obviously:

$$2\pi T/\hbar \;\equiv\; \lambda_s^{-2}\,, \qquad (4.8)$$

which defines a fundamental quantization length, λ_s. It looks strange, at first sight, that we managed to get rid of $\hbar$ and T in favour of a single quantity, λ_s. Actually, in order to do so, we have implicitly changed units of energy. The natural unit of energy in string theory is *length*, with T providing the *conversion* factor to and from c.g.s. (or normal particle theory) units. Indeed, using the string tension, we can give the energy of a string by giving its length, and vice versa! If natural string units are consistently used, only two dimensionful constants are needed for the relativistic and quantum nature of the theory (Veneziano 1986). This reminds us of pure GR, where, at the classical level, one can use geometric units of energy ($G_N E$ is a length) and, at the quantum level, only the Planck length $l_P = \sqrt{G_N\hbar}$—and not G_N and $\hbar$ separately—is relevant. The possibility of using just length and time intervals to describe all physical phenomena is yet another manifestation of the geometric nature of all forces in string theory.

It looks like no achievement at all to have replaced c and $\hbar$ by c and λ_s through a change of units. But, actually, we have gained a lot (Veneziano 1986):

The quantum length λ_s is also the short-distance cut-off.

This is so because, as we have seen, strings acquire, through quantization, a finite, minimal size $O(\lambda_s)$. The finite size induces in turn a cut-off (in the way of a form factor) on large virtual momenta. Incidentally, this is what makes sense of quantum string gravity. A typical one-loop (quantum gravity) correction normalized to the tree (classical) value is of $O(G_N \Lambda_{UV}^2)$. Inserting $(\hbar T)^{1/2}$ as cut-off, one gets a finite correction of order l_P^2/λ_s^2: quantum gravity corrections are thus (typically) small if the string length parameter λ_s is somewhat larger than the Planck length l_P!

5 String-theoretic Kaluza–Klein

In order to discuss KK theory within a string context we have to mention a *third* miracle of QST, duality (for a review, see e.g. Giveon *et al.* (1994)). While in flat, non-compact space-time, λ_s is the only scale and, as such, can be defined to be 1, when we go back to the KK situation involving a circle of radius R, the dimensionless ratio:

$$r \equiv R/\lambda_s \tag{5.1}$$

becomes a physical parameter.

The third miracle of QST is that r does indeed matter but only up to a "T-duality" transformation:

$$r \to r^{-1} \tag{5.2}$$

under which physics is invariant.

Said differently, the (moduli) space of string compactifications is the interval $r = [1, \infty]$ (or $r = [0, 1]$) and thus contains a "minimal" compactification radius given by $R = \lambda_s$, the fixed point of the duality transformation (5.1).

It is not usually stressed enough that also $r \to 1/r$ duality is only there at the quantum level. This is because, under such a transformation, the roles of momentum and winding get interchanged. The energy connected to winding is an integer multiple of R even classically, since the winding of a (closed) string around a circle is a topological concept, while momentum on a circle is an integer multiple of $\hbar/R$ only in the quantum theory (as obvious from the appearance of $\hbar$).

The $r \to 1/r$ duality is very likely to stabilize the radius of internal dimensions precisely at $r = 1$. Indeed semi-realistic models show how this can occur. At the self-dual value of R, the KK gauge symmetry is enhanced. In field theory it is a $U(1)$ gauge symmetry at any R. In string theory it turns out to be $U(1) \otimes U(1)$ at generic R and $U(2)$ at $r = 1$. In the more general case of $D > 5$ dimensions (typically $D = 10$ for superstrings) even higher groups (e.g. $E_8 \otimes E_8$) emerge at special values of the compactification radii.

Furthermore, as in KK theory, gauge and gravity couplings get unified at the scale $\hbar/R$, hence, if $R = \lambda_s$, at the string scale M_s. In formulae:

$$\alpha_{GUT} = (2\pi T) G_N = l_P^2/\lambda_s^2 \; . \tag{5.3}$$

Equation (5.3) holds at tree level. However, gauge-loop corrections are controlled by α_{GUT} while, as already pointed out, gravity loops are controlled by l_P^2/λ_s^2. Thus (5.3) leads to the conclusion that also quantum corrections are unified. This is only true, actually, up to infrared effects which, being more severe for gauge interactions than for gravity (just the opposite of what is true in the ultraviolet), cause the logarithmic "running" of gauge couplings (Taylor and Veneziano 1988a) as opposed to the scale-independence of G_N.

There is one question still left unanswered: What, if anything, fixes α itself (and thus G_N)?

6 S-duality and the big fix

Once more we have to start with a miracle. At a closer look QST reveals another amazing property: it is parameter-free. What is meant by that is that any conceivable parameter gets promoted, in QST, to a field whose VEV gives the corresponding arbitrary constant of its effective QFT. Since VEVs are usually determined by the dynamics through the minimization of some potential, we may hope that the same will happen to all the constants of Nature in string theory (Veneziano 1989).

Actually, we have already encountered examples of that: the value of r, the compactification radius in string units is a field, the metric in the 55 direction, and $r \to 1/r$ duality is simply the inversion of g_{55}.

Analogously, there is a scalar field, called the dilaton, whose VEV gives the gauge-gravity coupling in string theory according to (Witten 1984)

$$\alpha_{GUT} = l_P^2/\lambda_s^2 = \exp(\langle\phi\rangle). \tag{6.1}$$

The reason why $\exp(\phi)$ becomes the loop-counting parameter is very interesting. It has to do with the peculiar way in which ϕ appears in the fundamental action of the string (it basically multiplies the Euler characteristic of the Riemann surface swept by the string). One then easily finds that there is an extra factor $\exp(\phi)$ associated with each extra "hole" (or "handle") of the Riemann surface. But each extra handle means precisely an extra loop in the field theory limit! Thus, the loop expansion of QFT is given by the "topological expansion" of QST and interactions are geometrical, in the sense that they are simply related to the possibility of splitting or recombination of strings.

We may stress again, at this point, that no concepts, other than geometrical, need to be introduced to describe interactions. Consider, for instance, two strings of size L_1, L_2 separated by a distance R and interacting with a strength given by a certain value of ϕ, a pure number. Instead of introducing the mass of the strings and the magnitude of the forces, we can look straight away to the acceleration caused on each string by the interaction with the other. Such an acceleration has dimensions l^{-1} (recall $c = 1$) and will be simply given by a combination of L_1, L_2, R and $\exp(\langle\phi\rangle)$, typically:

$$a_1 \sim \exp(\langle\phi\rangle) L_2 R^{-2}. \tag{6.2}$$

There is a bona-fide scalar particle (the dilaton) associated with ϕ. It is somewhat similar to a Jordan–Brans–Dicke scalar and, like its predecessor, can produce large violations of the equivalence principle through a composition– and momentum–dependent fifth force. This is indeed so (Taylor and Veneziano 1988b) unless the dilaton gets a mass larger than $O(10^{-5}$ eV). Unfortunately, in superstring theory, the dilaton remains massless to all orders in perturbation theory, but the same is true for the lack of supersymmetry breaking. If we want string theory to survive, we have to assume that supersymmetry gets broken at the non-perturbative level and that the dilaton gets a mass (while the cosmological constant remains small enough!).

Amazingly, if all that happens, not only QST survives, it also predicts (at least in principle) the value of α, a recurring dream of this century's physics. A simple corollary of that possibility is that the limit $\lambda_s \to 0$ of QST would *not* make sense and, in particular, a non-trivial QFT would *not* be recovered in such a limit.

This is easily understood by comparing string theory with a lattice version of QFT with lattice spacing $a \sim \lambda_s$. It is well known that, in order to (have a chance to) recover a sensible QFT at finite distances, the couplings at the cut-off have to be fine-tuned as $a \to 0$. If the coupling at the cut-off is dynamically fixed, the resulting QFT is either trivial or infinitely coupled at any finite scale. In other words, QFT would only emerge from QST as an approximate description of physics much below the latter's finite, physical cut-off, in the same way in which Fermi's theory of weak interaction is an effective description of the standard model's electroweak theory well below the W, Z masses.

Although we have no theory fixing $\phi(\alpha)$ at the moment, we can mention a new suggestive idea in that direction called S-duality. There is mounting evidence (Sen 1994) that superstring theories with extended ($N > 1$) supersymmetry have a very interesting duality of the electric-magnetic type, under which the coupling constant is inverted and charged particles are interchanged with (solitonic) magnetic monopoles, so that Dirac's quantization condition is preserved. In the supersymmetric case, ϕ and its pseudoscalar partner, the so-called axion, are combined into a complex field and the S-duality group becomes $SL(2, Z)$, a discrete group containing, in particular, the transformation $\alpha \to \alpha^{-1}$.

As with $r \to 1/r$ duality, we may expect Nature to choose the fixed point of the transformation. That would give $\alpha = 1$ which, unfortunately, is both unrealistic (experimentally $\alpha \sim 1/40$) and uncomfortably large. A way out could be that, with such a large tree-level coupling, the one-loop correction may dominate according to the famous (gauge coupling) running formula (here used with a finite UV cutoff λ_s^{-1}):

$$\alpha_{eff}^{-1}(q) = \alpha_{tree}^{-1} + \beta \ln(q\lambda_s) \sim N_{eff}, \tag{6.3}$$

where we have used the fact that β-functions are always proportional to some number N_{eff} of charged states circulating in the loop. This could very well lead to a reasonable effective $\alpha \sim O(N_{eff}^{-1})$ at the string/GUT scale.

In which sense can we then say that also the gauge–gravity coupling has a geometrical origin? Recent results (Witten 1995) indicate that the low-energy limit of superstring theory in $D = 10$ dimensions is nothing but a well-known QFT, supergravity in $D = 11$, the maximal number of dimensions in which we are able to construct supergravity. However, in order to match the number of non-compact dimensions, the latter theory has to be thought of as having one compactified spatial dimension. The radius of this (11th) dimension then gets identified with the dilaton of string theory. In other words, in this higher dimensional (so-called M) theory, of which we only know so far the low-energy limit, the gauge–gravity coupling itself acquires a geometrical meaning!

Bibliography

Giveon, A., Porrati, M. and Rabinovici, E. (1994). Target space duality in string theory. *Phys. Rep.*, **244**, 77–202.

Sen, A. (1994). Dyon-monopole bound states, selfdual harmonic forms on the multimonopole moduli space, and SL(2,2) invariance in string theory. *Phys. Lett. B*, **329**, 217.

Taylor, T.R. and Veneziano, G. (1988a). Strings and D = 4. *Phys. Lett. B*, **212**, 147.

Taylor, T.R. and Veneziano, G. (1988b). Dilaton couplings at large distances. *Phys. Lett. B*, **213**, 450–458.

Veneziano, G. (1986). A stringy nature needs just two constants. *Europhys. Lett.*, **2**, 199–204.

Veneziano, G. (1989). Quantum strings and the constants of Nature. In *The Challenging Questions*, Erice, 1989, ed. A. Zichichi. Plenum Press, New York, 1990.

Witten, E. (1984). Some properties of O(32) superstrings. *Phys. Lett. B*, **149**, 351–356.

Witten, E. (1995). String theory dynamics in various dimensions. *Nucl. Phys. B*, **443**, 85–126.

16

Space from the Point of View of Loop Groups

Graeme Segal
Department of Pure Mathematics and Mathematical Statistics
University of Cambridge, 16 Mill Lane, Cambridge, CB2 1SB

I feel honoured to be speaking at this meeting to celebrate Roger Penrose's birthday. I am sure that for all of us Penrose stands out among the people who have thought most deeply and most ingeniously about the nature of space and time. What I am going to talk about is related to ideas which he developed in the sixties, though I shall approach them from a rather different point of view. Early in life most of us fix on a number of lamp-posts under which ever after we look for things; for me these lamp-posts are algebraic topology and the theory of loop groups. My aim in this talk is very modest: I shall try to explain how the study of loop groups can shed some light on the possible properties of space.

Newtonian physics had a clear ontology: the world consisted of massive particles situated in Euclidean space. In that sense, the nature of space played a fundamental role. In the mathematical development of Newtonian mechanics, however, the role of space is not so clear. There is little fundamental difference between the description of two particles moving in $\mathbb{R}^3$ and that of a single particle moving in $\mathbb{R}^6$, nor between a pivoted rigid body and a point moving on the group-manifold SO_3. In quantum mechanics the idea of space is even more elusive, for there seems to be no ontology at all, and, whatever wave-functions are, they are certainly not functions defined in space. Still, for about seventy years we have known that elementary particles must be described not by quantum mechanics but by quantum field theory, and in field theory the role of space is quite different.

It is customary to point out that quantum field theory cannot be reconciled with general relativity. At the moment, however, I should prefer to emphasize that the two theories have a vital feature in common, for in both of them the points of space play a central and objective dynamical role. In quantum field theory two electrons are not described by a wave-function on $\mathbb{R}^6$; instead they constitute a state of a field in $\mathbb{R}^3$ which is excited in the neighbourhood of two points. The points of space *index* the observables in the theory. The mathematics of quantum field theory is an attempt to describe the nature of space, and, even after seventy years, it remains ill-understood and mysterious, as well as being incompatible with general relativity. That has led to many proposals to look at space in a completely different way. Penrose's twistor theory is a project of that kind, as is his earlier theory of spin networks. Both are radical attempts to get rid of space as a primary concept. At this meeting we have also heard about Connes's programme of non-commutative geometry, which amounts to a huge generalization of the classical notion of a manifold. Apart from that, we have heard a little about string theory, which is at present the most fashionable scheme for making space appear as an approximation to some more general kind of structure. This talk is mainly directed towards string theory.

I shall begin by explaining how the rather unexpected behaviour of space as it is described by conventional quantum field theory can be illustrated by basic facts about the representations of loop groups: the most important thing is how space seems *not* to have continuously many degrees of freedom. After that I shall try to give some idea how the more subtle properties of loop group representations lead one to the generalization of the notion of a manifold which arises in string theory. But, although it is a digression from my main theme, I shall interpose a brief account of the relation of loop group representations to two- and three-dimensional topology. I think that this helps to motivate the "stringy" constructions, and at the same time it is closely related to Penrose's theory of spin networks, as was explained in Kauffman's lecture.

1 Loop groups and quantum field theory

We start with a compact Lie group G, which for simplicity I shall take to be SU_2. The *loop group* $\mathcal{L}G$ is the group of smooth maps $\gamma : S^1 \to G$ from the circle to G. We are interested in unitary representations of $\mathcal{L}G$ on a Hilbert space $\mathcal{H}$, but to make the link with quantum field theory we think of representations of the Lie algebra

$$\mathcal{L}\mathfrak{g} = \{\text{smooth maps } \varphi : S^1 \to \mathfrak{g}\},$$

where $\mathfrak{g}$ is the Lie algebra of G. If φ had not to be smooth then $\mathcal{L}G$ would be simply the product $\prod \mathfrak{g}_\theta$ of uncountably many copies of $\mathfrak{g}$ indexed by the points θ of S^1. We can think of $\mathfrak{g}_\theta$ as an algebra of three "observables" attached to the point θ. If $\theta \neq \theta'$ then $\mathfrak{g}_\theta$ commutes with $\mathfrak{g}_{\theta'}$. The effect of restricting to the smooth loops is that the operators

$$\{\phi_1(\theta), \phi_2(\theta), \phi_3(\theta)\}$$

representing $\mathfrak{g}_\theta$ are *distributions* as functions of θ.

Our initial expectation is that an irreducible representation of $\mathcal{L}G$ should be some kind of continuous tensor product $\bigotimes V_\theta$, where V_θ is a representation of $\mathfrak{g}_\theta$. One can indeed construct representations of that kind. But if we look for so-called "*positive energy*" representations then what happens is much more subtle.

The positive energy condition is designed to model the positivity of energy in quantum field theory: it is the requirement that

(i) there is a unitary operator $R_\alpha : \mathcal{H} \to \mathcal{H}$ for each α such that

$$R_\alpha \phi_j(\theta) R_\alpha^{-1} = \phi_j(\theta + \alpha),$$

(ii) $R_\alpha = e^{i\alpha H}$, where the "Hamiltonian" H is a self-adjoint operator with *positive* spectrum.

Positive energy representations, which are completely understood and classified, have the following properties.

(a) They are necessarily *projective.* This means that the $\phi_j(\theta)$ for distinct θ cannot quite commute. Instead we must have

$$[\phi_j(\theta_1), \phi_l(\theta_2)] = \sum_m \varepsilon_{jlm}\, \delta(\theta_1 - \theta_2)\, \phi_m(\theta_2) + k\delta_{jl}\delta'(\theta_1 - \theta_2)$$

for some positive integer k which is called the *level.* The second term on the right expresses the way in which infinitesimally nearby points interact. As I shall explain in a moment, k should be regarded as the reciprocal of Planck's constant.

(b) The Hilbert space breaks up

$$\mathcal{H} = \mathcal{H}_0 \oplus \mathcal{H}_1 \oplus \mathcal{H}_2 \oplus \cdots,$$

under the action of the Hamiltonian H. The subspace $\mathcal{H}_n$ of states of energy n is *finite dimensional*, and of course G-invariant, where G is the subgroup of constant loops in $\mathcal{L}G$. The representation $\mathcal{H}$ of $\mathcal{L}G$ is completely determined by the representation $\mathcal{H}_0$ of G, and at a given level k only a finite number (in fact $k+1$) of representations $\mathcal{H}_0$ can occur. Thus although $\mathcal{H}$ is large and complicated in its structure, it is actually very rigid and tightly constrained. The loop group $\mathcal{L}G$, instead of having vastly more representations than G does, has many fewer.

(c) The number of states in $\mathcal{H}$ with energy $\leqslant E$, i.e. the dimension of

$$\bigoplus_{n \leqslant E} \mathcal{H}_n,$$

is known to grow roughly like the number of partitions of E, i.e. roughly like $e^{\sqrt{E}}$. Now for a quantum mechanical system with d degrees of freedom the number of states grows roughly like E^d, so the "system" $\mathcal{H}$ behaves like a finite dimensional system with (very roughly) $\sqrt{E}$ degrees of freedom if we are looking at states with energy of order E.

We can think about this more geometrically. The Hilbert space $\mathcal{H}$ is obtained by "geometric quantization" of an infinite dimensional classical phase space X.

In the simplest case, when $\mathcal{H}_0 = \mathbb{C}$, we have $X = \mathcal{L}G/G$. The "level" k is the class of the symplectic form in $H^2(X;\mathbb{Z}) \cong \mathbb{Z}$. When we quantize a finite dimensional symplectic manifold X with a Hamiltonian function $H_{\text{class}} : X \to \mathbb{R}$ we get a Hilbert space $\mathcal{H}$ with an operator $H : \mathcal{H} \to \mathcal{H}$, and the number of states in $\mathcal{H}$ with energy $\leqslant E$ is asymptotically equal to the Liouville volume of the region

$$X_E = \{x \in X : H_{\text{class}}(x) \leqslant E\}.$$

To analyse this when $X = \mathcal{L}G/G$, let us write elements $\gamma \in X$ in the form

$$\gamma(\theta)^{-1}\gamma'(\theta) = i \sum a_n e^{in\theta},$$

where $a_n \in \mathfrak{g} \otimes \mathbb{C}$, $a_{-n} = a_n^*$, and a_0 is determined in terms of $\{a_n\}_{n \neq 0}$ so that γ is periodic in θ. Then

$$H_{\text{class}}(\gamma) = \frac{1}{2} \int_0^{2\pi} ||\gamma(\theta)^{-1}\gamma'(\theta)||^2 \, d\theta \;=\; \pi \sum ||a_n||^2,$$

while the symplectic form is

$$\omega = k \sum_{n>0} \frac{1}{n} \operatorname{tr}\,(da_n^* \wedge da_n).$$

Thus the Liouville volume of X_E is that of an infinite dimensional ellipsoid with $2d$ axes of length $(kE/n)^{\frac{1}{2}}$ for each $n \geqslant 1$, where $d = \dim G$. We can make sense of this by counting lattice points [1] inside X_E for a lattice which is unimodular for the Liouville measure. Then at each energy all but finitely many degrees of freedom are irrelevant, and we effectively have a finite dimensional ellipsoid. A simple calculation shows that we get the expected asymptotic behaviour of the number of states.

Of course, I have done no more than restate Planck's observation about black-body radiation which was the very beginning of quantum mechanics. (The application to loop group representations was pointed out to me by my students Paul Shutler and Alex Selby.) What I want to emphasize is that the infinite dimensional manifolds which arise in quantum field theory naturally behave like nested unions of finite dimensional manifolds of increasing dimension. It is sometimes argued that this has to do with an ultimate granularity of space, but the representation theory of loop groups encourages one rather to think of a perfectly ordinary smooth manifold—the circle—in which nearby points are constrained (by the positive energy condition) to behave very coherently.

---

[1] This is very loosely expressed. "Lattice points" are not quite what is needed, as the lengths of the axes of an ellipsoid in a symplectic vector space are not invariantly defined. We must take the axes two at a time, for the invariant concept is the area of the projection of the ellipsoid on to a two-dimensional symplectic subspace, expressed as a multiple of Planck's constant.

2 Loop groups and low-dimensional topology

Let us now fix a level k, and let $\mathcal{R}$ denote the additive category of all level k representations of $\mathcal{L}G$ which are sums of finitely many irreducibles. I shall write $\mathcal{R}^{\otimes p}$ for the corresponding category of representations of

$$\underset{\leftarrow\, p\, \rightarrow}{\mathcal{L}G \times \cdots \times \mathcal{L}G}$$

These are sums of representations of the form $\mathcal{H}_1 \otimes \cdots \otimes \mathcal{H}_p$. More functorially, when S is a closed 1-manifold I shall write $\mathcal{R}_S$ for the category of representations of the group of smooth maps $S \to G$.

One of the most remarkable things about the representations of loop groups is the

Theorem *A smooth cobordism Σ from S_0 to S_1 induces an additive functor*

$$U_\Sigma : \mathcal{R}_{S_0} \to \mathcal{R}_{S_1}.$$

Oversimplifying slightly, the functor U_Σ is characterized by the following universal property: for each object $\mathcal{H}$ of $\mathcal{R}_{S_0}$, and each complex structure σ on Σ, there is a linear map

$$T_\sigma \ : \ \mathcal{H} \to U_\Sigma(\mathcal{H})$$

such that $T_\sigma \circ \gamma_0 \ = \ \gamma_1 \circ T_\sigma$ whenever γ_0 and γ_1 are the restrictions to S_0 and S_1 of a σ-holomorphic map $\gamma : \Sigma \to G_\mathbb{C}$. (This makes sense because every representation of $\mathcal{L}G$ is automatically a representation of $\mathcal{L}G_\mathbb{C}$.)

The functor $\mathcal{R} \times \mathcal{R} \to \mathcal{R}$ induced by U_Σ when Σ is a "pair of pants" is called *fusion*. I shall write it $(\mathcal{H}_1, \mathcal{H}_2) \mapsto \mathcal{H}_1 * \mathcal{H}_2$. It plays the role of a tensor product in the category $\mathcal{R}$. (The ordinary tensor product of two representations of level k is a representation of level $2k$.)

A diffeomorphism $f : \Sigma \to \Sigma'$ between cobordisms from S_0 to S_1 induces an isomorphism

$$U_f : U_\Sigma(\mathcal{H}) \to U_{\Sigma'}(\mathcal{H})$$

for any $\mathcal{H}$. This means that fusion is associative and commutative up to isomorphism, and also that the group $\mathrm{Diff}(\Sigma \operatorname{rel} \partial\Sigma)$ of diffeomorphisms of Σ which are the identity on the boundary acts on $U_\Sigma(\mathcal{H})$. It must act through the mapping-class group $\Gamma_{\Sigma,\partial\Sigma} \ = \ \pi_0 \,\mathrm{Diff}\,(\Sigma \ \operatorname{rel} \partial\Sigma)$, for the identity component of $\mathrm{Diff}(\Sigma \operatorname{rel} \partial\Sigma)$ has no finite dimensional representations. In particular, the coloured braid group on n strings, which is the mapping-class group of a disc with n holes, always acts on

$$\mathcal{H}_1 * \mathcal{H}_2 * \cdots * \mathcal{H}_n,$$

and the full braid group acts if $\mathcal{H}_1 = \mathcal{H}_2 = \cdots = \mathcal{H}_n$.

A closed surface Σ is a cobordism from the empty 1-manifold to itself, so it gives us an additive functor $U_\Sigma : \mathcal{V}ect \to \mathcal{V}ect$, where $\mathcal{V}ect$ is the category of finite dimensional vector spaces. Thus $U_\Sigma(\mathbb{C})$ is a finite dimensional representation of the mapping-class group of Σ.

But much more still is true. Not only does a diffeomorphism $f : \Sigma \to \Sigma'$ between cobordisms give us an isomorphism U_f, but a three-dimensional cobordism M between the cobordisms Σ and Σ' also induces a transformation

$$\mu_M : U_\Sigma(\mathcal{H}) \to U_{\Sigma'}(\mathcal{H}).$$

Specializing to the case when Σ and Σ' are closed surfaces, a cobordism M from Σ to Σ' gives a transformation

$$\mu_M : U_\Sigma(\mathbb{C}) \to U_{\Sigma'}(\mathbb{C})$$

of finite dimensional vector spaces. If M itself is closed then μ_M is simply a number, for $U_\varnothing(\mathbb{C}) = \mathbb{C}$.

This brings us to the final startling fact, discovered by Witten.

Theorem *The number $\mu_M \in \mathbb{C}$ associated to a closed 3-manifold is the Chern-Simons path-integral*

$$\int_{\mathcal{A}_M/\mathcal{G}_M} e^{2\pi ikCS(A)}\,\mathcal{D}A,$$

where $\mathcal{A}_M$ *is the space of* G*-connections* A *on* M,
$\mathcal{G}_M$ *is the corresponding gauge group, and*
$CS(A) = \int_M \{\langle A, dA\rangle + \frac{1}{3}\langle A, [A, A]\rangle\}$.

To be honest, the path-integral in the theorem can only be defined perturbatively, i.e. all one really has is an asymptotic expansion in $1/k$. There is also an asymptotic expansion of μ_M as the level k tends to ∞. The theorem asserts that these expansions coincide.

A whole vast theory of 3-manifold invariants and related knot invariants emerges from the facts I have listed in this section, but I believe the first inkling of it was Penrose's work on spin networks. The subject can be developed in many ways, and without mentioning loop groups. A very different treatment, closer to Penrose's, can be found in Kauffman's talk.

3 String theory

Quantum field theory on a space-time manifold M can be described as the construction of a probability measure on some space Φ_M of "fields" associated to M. Usually Φ_M is constructed by some simple process from the algebra $\Omega^\cdot(M)$ of differential forms [2] on M. For example, gauge fields are derivations of

[2] Once again this is an oversimplification. To study fermions on M we need the algebra $\Omega^\cdot(M; \wedge^\cdot\Delta)$ of differential forms with coefficients in the exterior powers of the spin bundle Δ of M.

$\Omega^{\cdot}(M) \otimes \mathrm{Mat}_N$, where Mat_N is the algebra of $N \times N$ matrices. The manifold M enters the theory through the algebra $\Omega^{\cdot}(M)$, which encodes it precisely. The way in which string theory differs from quantum field theory is that the algebra $\Omega^{\cdot}(M)$ is replaced by a much larger object $\mathcal{A}$ which I shall call a *string algebra.* (The usual name is "string background".) This is a generalized algebra in a sense I shall explain.

The first approximation to $\mathcal{A}$ starts from the loop space $\mathfrak{L}M$ of the space-time manifold M. The infinite dimensional manifold $\mathfrak{L}M$ is foliated by the orbits of the group of diffeomorphisms of the circle, which acts by reparametrizing the loops. We can consider the differential forms on $\mathfrak{L}M$ defined along the leaves of the foliation, i.e. functions of the form

$$(\gamma; \xi_1, \cdots, \xi_p) \mapsto \alpha(\gamma; \xi_1, \cdots, \xi_p),$$

where $\gamma \in \mathfrak{L}M$ and $\xi_1, \cdots, \xi_p$ are tangent vectors at γ directed along the leaf through γ. (Tangent and cotangent vectors to the leaves are traditionally called "ghost fields".) This is still not right: actually we want "semi-infinite" differential forms rather than ones of finite degree p, but I shall not explain that here. If M is a homogeneous space of a Lie group G then $\mathcal{A}$ should be a representation of the loop group $\mathfrak{L}G$, and that gives us a way of predicting its algebraic properties in general.

To state what is expected we need the concept of a *Frobenius algebra.* This is a finite dimensional commutative algebra together with a linear map $\theta : A \to \mathbb{C}$ such that $(a, b) \mapsto \theta(ab)$ is a non-degenerate quadratic form on A. The most important example for us is the cohomology algebra $A = H^*(M; \mathbb{C})$ of a compact oriented manifold, where θ is given by integration over M. (This is commutative in the *graded* sense.)

There is a folk-theorem—quasi-trivial but striking—which asserts that a Frobenius algebra is the same thing as a two-dimensional topological field theory, i.e. a rule which assigns a vector space A_S to each closed oriented 1-manifold S, and a linear map

$$U_\Sigma : A_{S_0} \to A_{S_1}$$

to each cobordism Σ from S_0 to S_1. These data are required to satisfy

(i) $A_{S_0 \amalg S_1} \cong A_{S_0} \otimes A_{S_1}$, and
(ii) $U_{\Sigma_2} \circ U_{\Sigma_1} = U_{\Sigma_2 \circ \Sigma_1}$, where $\Sigma_2 \circ \Sigma_1$ is the composite of Σ_1 and Σ_2.

Given the data $\{A_S\}$ and $\{U_\Sigma\}$ we take $A = A_{S^1}$, and define multiplication $A \otimes A \to A$ as U_Σ, where Σ is a pair of pants. The map $\theta : A \to \mathbb{C}$ is U_D, where D is a disc.

Whatever may be the merits of this paraphernalia as a definition of a Frobenius algebra, it does lend itself to useful generalizations. In the last section we saw one such generalization; string algebras are another. They are designed to resemble not the cohomology algebra $H^*(M, \mathbb{C})$ but the differential algebra $\Omega^{\cdot}(M)$ from which it comes. A string algebra gives us a cochain complex C_S for each closed 1-manifold S, and a cochain map

$$U_\Sigma : C_{S_0} \to C_{S_1}$$

for each cobordism Σ, and the properties (i) and (ii) above hold. The map induced by U_Σ on the cohomology of the complexes depends only on the topological type of Σ, and so $H^*(C_{S^1})$ is a Frobenius algebra. It should be thought of as the algebra of parametrization-invariant functions on $\mathcal{L}M$. But U_Σ itself is allowed to depend on a conformal structure σ on Σ, and when σ changes the map $U_{\Sigma,\sigma}$ changes by a cochain homotopy. The natural way to give the formal definition is as follows.

For a given cobordism Σ from S_0 to S_1 let $\mathcal{M}_\Sigma$ denote the space of isomorphism classes of conformal structures on Σ, the isomorphisms being the identity on S_0 and S_1. The structure of a string algebra is the assignment to each Σ of a cochain map

$$\hat{U}_\Sigma : C_{S_0} \to \Omega^{\cdot}(\mathcal{M}_\Sigma; C_{S_1}),$$

where the right-hand side means the total complex of the double complex. The maps $U_{\Sigma,\sigma}$ mentioned earlier are obtained from $\hat{U}_\Sigma$ by restricting to points $\sigma \in \mathcal{M}_\Sigma$. When one composes two cobordisms Σ_1 and Σ_2 the composite

$$\hat{U}_{\Sigma_2} \circ \hat{U}_{\Sigma_1} : C_{S_0} \to \Omega^{\cdot}(\mathcal{M}_{\Sigma_1} \times \mathcal{M}_{\Sigma_2}; C_{S_2})$$

must be compatible with

$$\hat{U}_{\Sigma_2 \circ \Sigma_1} : C_{S_0} \to \Omega^{\cdot}(\mathcal{M}_{\Sigma_2 \circ \Sigma_1}; C_{S_2})$$

in an obvious sense.

The definition of a string algebra is motivated by the behaviour of the differential forms on the leaves of a loop space $\mathcal{L}M$. But the hope is that the string algebra is a fundamental structure—more fundamental than space-time—and that the interpretation in terms of $\mathcal{L}M$ need not be precisely possible, or may be possible in many different ways. I should emphasize, however, that string algebras are all the same only meant to be by-products of string theory, still linked to a perturbative treatment of it: they should be one stage closer to reality than the classical algebra of functions on space-time. String theory itself, like the kingdom of Heaven, can at present be described only by poetry and metaphor.

Bibliography

A more detailed account of the material in this lecture can be found in my *Stanford Lectures on Topological Field Theory*, which are to be published by

Cambridge University Press. The most basic properties of loop-group representations are described in my book with A. Pressley (Oxford University Press 1986).

A reference for Penrose's work on spin networks is

Penrose, R. (1969). Angular momentum: an approach to combinatorial space-time. In *Quantum Theory and Beyond*, ed. T.A. Bastin. Cambridge University Press, Cambridge.

The basic work relating Chern-Simons theory to low-dimensional topology is

Witten E. (1989). Quantum field theory and the Jones polynomial. *Comm. Math. Phys.*, **121**, 351–99.

My account of string algebras roughly follows

Witten, E. (1987). Some remarks about string field theory. *Physica Scripta*, **T15**, 70–7.

PART II

Parallel Session I: Quantum Theory and Beyond

17
The Twistor Diagram Programme

Andrew P. Hodges
Wadham College, Oxford, OX1 3PN

Abstract
Recent advances in twistor diagram theory vindicate the ideas embodied in Roger Penrose's original proposals. The novel treatment of gauge fields is given particular attention.

Twistor diagrams were first written down by Roger Penrose, as an early part of the twistor programme for reformulating fundamental physics. Twistor diagrams define integrals which yield scattering amplitudes for elementary particles in flat space, and thus are roughly analogous to Feynman diagrams in standard quantum field theory (QFT). It was an essential ingredient in Penrose's programme that the divergence problems which plague QFT should be resolved in the new setting offered by twistor geometry, that twistor diagrams should be manifestly finite; and that they should supersede, rather than merely reformulate, the predictive calculus supplied by Feynman diagrams.

In this review I concentrate on just one of the diagrams first written down by Roger Penrose, to sketch the subsequent development of the theory, and to honour the prophetic power of his original intuition. This is the diagram for massless Compton scattering, as given in 1972 by Penrose (Penrose and MacCallum 1972). This is a process which in the standard treatment requires the summation of *two* Feynman diagrams, neither of them separately gauge–invariant. However Penrose saw that the amplitude could be given by just one manifestly gauge–invariant twistor diagram, which in the notation now current is written:

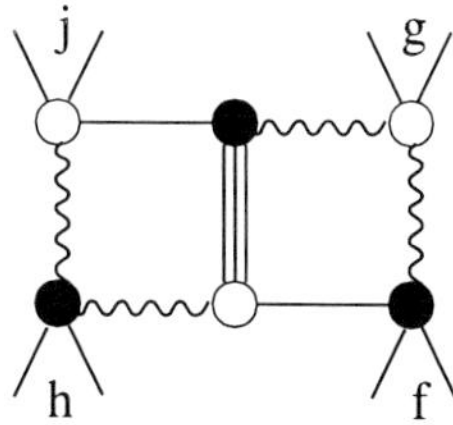

Neglecting an overall factor, this diagram specifies the integral:

$$\int_{\substack{W.Z=0,W.V=0\\U.X=0,Y.X=0}} DWXYZUV \frac{1}{U.Z}\frac{2}{(U.V)^3}\frac{1}{Y.V} f(Z^\alpha)g(W_\alpha)h(X^\alpha)j(Y_\alpha) \qquad (1.1)$$

where the $f(Z^\alpha)$, $g(W_\alpha)$, $h(X^\alpha)$, $j(Y_\alpha)$ are 1-functions of homogeneities -3, -4, -4, -3 respectively and so correspond to massless spin-$\frac{1}{2}$ and spin-1 fields via the standard Penrose correspondence. Note that the integral is defined in a product of *projective* twistor spaces. The differential form $DWXYZUV$ is the natural 18-form for this space. The prescriptions for the contour to have *boundary* on $W.Z = 0$, etc. are as significant as the pole singularities.

To arrive at this diagram, Penrose used a conjecture which amounted to drawing diagram lines for the passage of the massless electron, namely the lines corresponding to the *integrand* in (1.1). The two photons were then connected in the simplest way possible, by the lines corresponding to *boundary* prescriptions in (1.1). The diagram was only partly justified by an evaluation argument. (Indeed, in 1972 the boundary definition was still far from clear, and these elements were only treated formally.)

In fact an exact description of the 18-dimensional integral (1.1), showing agreement with the corresponding QFT amplitude, only came in the 1980s (Hodges 1989a; O'Donald 1992). It also had to wait this long for clarification of 'crossing symmetry' in twistor diagrams. In QFT, processes related by CPT symmetry have a momentum–space scattering kernel which (regarded as an analytic function of the momenta) is channel–independent. We expect some similar unifying feature in the twistor representation. In his original paper Penrose's analysis did not distinguish between the different channels, but in fact a contour for integrating (1.1) only exists if f and h are in-states and g and j out-states, or *vice versa.* Nevertheless, Penrose's original intuition was correct and there are diagrams for the crossing–related processes which also fit the idea of an electron line to which photons are attached, namely:

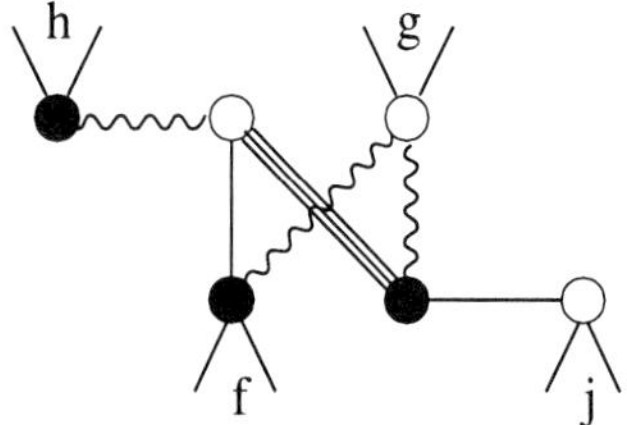

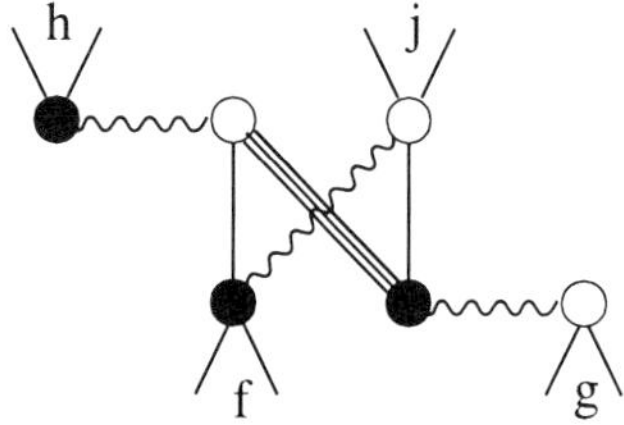

In other words, the *integrand* is the same for all three channels; only the region of integration, with its *boundaries,* is subject to change.

In the 1980s the massless QED originally considered by Penrose was extended to the massless SU(2) and Yukawa interactions in the Standard Model. Then analysis showed that this feature of a common integrand for all channels extends to all the first–order interactions of massless fields. Moreover in each case the integrand can be identified with the simple passage and interaction of spin-$\frac{1}{2}$ and spin-0 particles, a correspondence very similar to the Feynman diagram picture. However the vertices in twistor diagrams have no correspondence with the vertices in Feynman diagrams, and the boundary lines in twistor diagrams have no obvious interpretation.

Identification of this unifying feature suggested that it should hold to all orders. This hypothesis has proved the key step in the discovery of twistor diagrams for higher order processes, and a number of these are now known (Hodges 1989b, 1990; O'Donald 1992; Müller 1994). Recently, extending this work to a case of greater algebraic complexity, Johnston (1997) has established very elegant diagrams for second–order Compton scattering which, for every possible channel, also satisfy this very strong constraint.

Before illustrating the progress of the theory by writing down one of these diagrams, it is necessary to detail two features of the present diagram calculus which represent distinct changes to Penrose's original picture: (i) inhomogeneous elements and (ii) conformal symmetry breaking elements.

The original analysis of a proposed twistor diagram for massless Møller scattering by Penrose and Sparling (Penrose and MacCallum 1972; Penrose 1975; Sparling 1975) rested on an incorrectly chosen contour of integration. There is in fact an infra–red divergence in the Feynman integral, when it is evaluated for finite–normed states, and this is reflected in the impossibility of finding a corresponding diagram in projective twistor space.

In 1983, I proposed modifying the internal diagram elements, both in the integrand and in the boundary definitions, by a simple rule: every inner product such as $W.Z$ would appear instead as $W.Z - k$. The contour integration was then to be performed in non-projective space. Details were given first in Hodges (1985a). This modification strictly extended the scope of the diagram calculus, leaving unchanged all those results previously known for projective diagrams such as (1.1). (In fact, not only are the results unchanged, but the diagrams are *more consistently* defined with the new elements. Every inhomogeneous pole corresponds to an S^1 factor in the contour, and so can be interpreted as a Cauchy residue calculation; no such characterisation of the projective poles can be given.) Divergence of a Feynman diagram, such as that for Møller scattering, then corresponds to k-dependence in a corresponding twistor diagram. The extra dimension in non-projective twistor space appears to allow something analogous to dimensional regularisation.

It has more recently been realised that the number k has a natural value, up to sign, namely $e^{-\gamma}$, where γ is Euler's constant. An elementary argument can be given for this, based on the simplest twistor diagram which exhibits k-dependence:

W ∿∿∿∿∿●═══════○∿∿∿∿∿ Z

this being an integral which can be evaluated explicitly:

$$\frac{1}{(2\pi i)^4}\int_{\substack{W.V=k\\ U.Z=k}} d^4U \wedge d^4V \frac{-6}{(U.V-k)^4} = \log(W.Z/k). \qquad (1.2)$$

It is natural, given that the diagram elements are defined through derivative operations, to regard this logarithm as a term in the sequence

$$\ldots \log(W.Z/k),\ 0!(W.Z)^{-1},\ -1!(W.Z)^{-2},\ 2!(W.Z)^{-3},\ -3!(W.Z)^{-4},\ \ldots$$

Then $\log k$ is fixed as $-\gamma$ if the sequence is identified with values of

$$-\frac{d}{d\lambda}\frac{(W.Z)^{-\lambda-1}}{\Gamma(-\lambda)} \quad \text{at } \lambda = \ldots -1, 0, 1, 2, 3\ldots$$

However, an equally strong argument could be given for $k = -e^{-\gamma}$.

Secondly, Penrose's original account stressed the manifest conformal invariance of the twistor diagram representations of the first–order amplitudes for massless processes. But higher order amplitudes, even though scale–invariant, are not in general conformally invariant. Some conformal symmetry–breaking element must therefore exist in the calculus. If the integrand is to adhere to the hypothetical pattern, then it is only in the boundary prescriptions that conformal breaking can play a rôle. This means, in the inhomogeneous context, allowing boundaries of the form

$$I_{\alpha\beta}X^{\alpha}Z^{\beta} = m, \; I^{\alpha\beta}W_{\alpha}Y_{\beta} = m.$$

We can now give one of the diagrams found by Johnston for second–order Compton scattering (Johnston 1997). (The advantage of the diagram notation should now be apparent.) The lines connecting twistor vertices of the same type are conformal symmetry breaking boundaries as just defined.

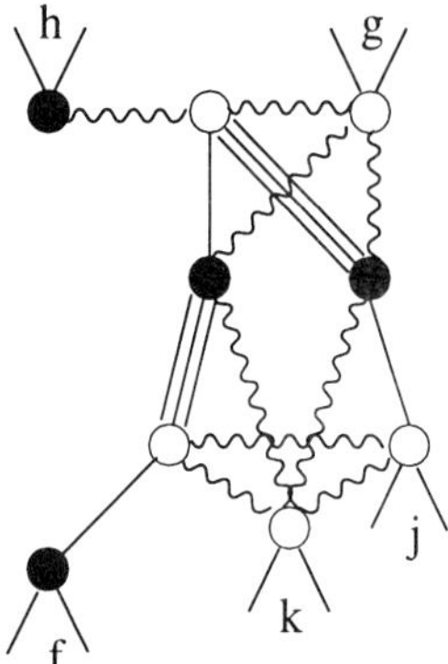

This twistor diagram notably retains Penrose's original feature of an integrand defined by the passage of the spin-$\frac{1}{2}$ field. Diagrams with the same property also exist for the other channels. This work offers yet more substantial evidence for the existence of a general twistor diagram formalism which will treat gauge fields in a simpler and more invariant manner than the Feynman calculus. (The conventional QFT calculation requires the addition of *six* Feynman diagrams, namely those with the three photons attached to the spin-$\frac{1}{2}$ line in all possible orders.)

Further recent advances have applied these inhomogeneous elements to the description of massive fields. The approach taken in Hodges (1985b) can now be improved. That first idea required the use of diagram elements with the form of integrands

$$\log(W.Z - k) + \gamma \tag{1.3}$$

rather than the more primitive form of boundaries on $W.Z = k$. These logarithmic elements were still regarded as essential at the time of the 1990 review (Hodges 1990), but it has since been seen that they are redundant, and that the more primitive poles and boundaries will suffice, though in rather an unobvious manner. First, the identification of a natural value for k allows (1.3) to be written in the form

$$\log((W.Z - \Delta)/k)$$

where $k = e^{-\gamma}$. In Hodges (1985b) the value of Δ played no essential role except that for a contour to exist, it must differ from zero. Yet its natural value, were it not for the problem of finding a contour, would be zero. This anomalous feature can now be eliminated, by noting that the logarithmic factor, with $\Delta = 0$, appears as the result of integrating the chain of primitive elements in (1.2). It has now been shown that if this chain is used to replace the logarithmic factor (1.3), a contour does emerge within the larger space thus defined. Using this idea, massive fields can then be constructed out of the the primitive pole and boundary elements, augmented by the use of just one further ingredient, simple poles of the form $(I_{\alpha\beta}X^\alpha Z^\beta - m)^{-1}$, $(I^{\alpha\beta}W_\alpha Y_\beta - m)^{-1}$ which have the effect of forcing eigenstates of mass.

If we replace these poles by boundaries on $(I_{\alpha\beta}X^\alpha Z^\beta = m)$, $(I^{\alpha\beta}W_\alpha Y_\beta = m)$, we are led to solutions of the *inhomogeneous* Klein–Gordon equation (Hodges 1991). Developments of this idea have been brought tantalisingly close to defining the Feynman propagator, but not yet quite correctly. The sign ambiguity in the k has still to be fixed, however, and this freedom may close the gap. It is possible that the sign of the k has a geometric content which only begins to play a rôle in these larger diagrams: a choice of sign is equivalent to a choice between positive and negative halves of twistor space, and so may be connected with a subtle asymmetry in time direction. A further motivation for this line of enquiry is that we may expect to find a reflection in the twistor diagram calculus of the mechanism in the Standard Model whereby Yukawa interaction with the constant part of the Higgs field acts as the origin of mass. Hence we might obtain a new geometric insight into mass. The emergent structure does indeed appear to have the potential of turning the formal Lagrangian manipulations of the Standard Model into a finite calculus.

The research described above involves difficult integrals of high dimension, but there is an underlying simplicity of purpose. Existing results suggest that the very small repertoire of inhomogeneous poles and boundaries as described, together with some further features relevant to eigenstates of mass, will suffice to represent all QFT scattering amplitudes in a manifestly finite manner.

There is much more that can yet be done to probe the structure of second–order amplitudes, but it is probably not possible to go much further with explicit calculations to yet higher orders. Rather, the ever more pressing and promising task is to use the second-order examples to identify an underlying generating principle, analogous to a Lagrangian. (A suggestion for the possible form of such a principle was given in Hodges *et al.* (1989).) This should reduce the diagram calculus to something more like a combinatorial scheme, as Penrose originally intended. The more examples of specific amplitudes known to fit a consistent and elegant pattern, the more certain it seems that such a generating principle must exist. The smaller and simpler our repertoire of elements in the calculus, the better hope of relating them to a combinatorial structure. The diagram elements are not the same as those originally suggested by Penrose, but the guiding spirit is unchanged. To an extraordinary degree, the diagrams themselves are the same in form, albeit re-interpreted, as those that he conjectured so long ago.

The twistor programme as a whole is intended for the unification of QFT and gravity through twistor geometry, and the twistor diagram programme must be seen in this light. Encouragement can be drawn from the fact that the inhomogeneous elements use twistor space in a way that has no simple space–time analogue. Further, non-projective twistor space also plays a role in Penrose's most recent ideas for the description of curved space–time. There is the possibility of a connection between these hitherto separate lines of enquiry. I end with the hope that the conjectures and formal expressions for the description of curved space–time prove to be as well–justified as those made 25 years ago for quantum field theory, but that the interpretation and vindication will not take quite so long.

Bibliography

Hodges, A. P. (1985a). A twistor approach to the regularization of divergences. *Proc. R. Soc. Lond.*, **A 397**, 341–74.

Hodges, A. P. (1985b). Mass eigenstates in twistor theory. *Proc. R. Soc. Lond.*, **A 397**, 375–96.

Hodges, A. P. (1989a). Double box diagrams. *Twistor Newsletter*, **28**.

Hodges, A. P. (1989b). Twistor translation of Feynman vertices. *Twistor Newsletter*, **29**.

Hodges, A. P. (1990). Twistor diagrams and Feynman diagrams. In *Twistors in Mathematics and Physics*, Bailey, T. N. and Baston, R. J. (eds). *Lond. Math. Soc. Lecture Note Series*, **156**. Cambridge University Press.

Hodges, A. P. (1991). Massive propagators. *Twistor Newsletter*, **32**.

Hodges, A. P., Penrose, R. and Singer, M. A. (1989). A twistor conformal field theory in four dimensions. *Phys. Letters B*, **216**, 48–52.

Johnston, D. K. (1997). D. Phil. thesis, University of Oxford.

Müller, F. X. (1994). D. Phil. thesis, University of Oxford.

O'Donald, L. J. (1992). D. Phil. thesis, University of Oxford.

Penrose, R. (1975). Twistor theory: its aims and achievements. In *Quantum Gravity.* An Oxford Symposium, Isham, C. J., Penrose, R. and Sciama, D. W.

(eds). Oxford University Press.

Penrose, R. and MacCallum, M. A. H. (1972). Twistor theory: an approach to the quantisation of fields and space–time. *Phys. Reports*, **6**, 241–315.

Sparling, G. A. J. (1975). Homology and twistor theory. In *Quantum Gravity*. An Oxford Symposium, Isham, C. J., Penrose, R. and Sciama, D. W. (eds). Oxford University Press.

Spence, S. T. (1996). D. Phil. thesis, University of Oxford.

18
Geometric Models for Quantum Statistical Inference

Dorje C. Brody
Blackett Laboratory, Imperial College, South Kensington, London SW7 2BZ

and

Lane P. Hughston
Merrill Lynch International, 25 Ropemaker Street, London EC2Y 9LY and King's College London, The Strand, London WC2R 2LS

1 Introduction

Our purpose here is to draw attention to an interesting relationship that can be shown to hold between (a) information geometry, and (b) quantum geometry. By 'information geometry', we mean the natural geometry associated with families of probability distributions. This is a subject that goes back more than half a century to the pioneering work of Rao (1945), who showed that the Fisher information matrix associated with a parametrised family of probability distributions gives rise to a natural Riemannian metric on the parameter space. This then allows one to speak in a precise way of the 'distance' between two probability distributions, an idea that turns out to be useful in problems of statistical inference. An extensive literature has thus developed, following this line of enquiry, on applications of differential geometry to statistics (see, e.g., Amari (1985), Barndorff-Nielsen *et al.* (1986), Murray and Rice (1993)). By 'quantum geometry', on the other hand, we mean the geometry of the manifold of states associated with a given quantum mechanical system. In particular, we consider the manifold of 'pure' states associated with a given complex Hilbert space. This is the space of 'rays' through the origin in the Hilbert space, which has the structure of a complex projective space. It is known that the usual rules of quantum mechanics allow one to construct a natural metric on this space, namely the Fubini–Study metric, the specification of which is equivalent to all the familiar structure associated with ordinary quantum mechanics. The significance of this geometry to the foundations of quantum theory has only begun to be appreciated relatively recently (see, e.g., Kibble (1978, 1979), Page (1987), Anandan and Aharonov (1990), Cirelli *et al.* (1990), Gibbons (1992), Ashtekar and Schilling (1995), Hughston (1995, 1996), Schilling (1996), Field (1996a, b)).

In spelling out the relationship between these two apparently rather distinct geometries, we are able to gain some new insights into the nature of statistical inference, both classical and quantum, and thus shed new light on the nature of the quantum measurement problem. Our most substantive contribution here perhaps is the establishment of a series of higher order corrections to the Heisenberg uncertainty relations for a pair of canonically conjugate operators. For example, in the case of the measurement of a 'position' parameter q, for which an unbiased estimator is given by an operator Q, with conjugate momentum operator P, where $i[P,Q] = 1$ with $\hbar = 1$, we find (cf. Brody and Hughston (1996)) that

$$\langle\tilde{P}^2\rangle\langle\tilde{Q}^2\rangle \;\geq\; \frac{1}{4}\left(1+\frac{\langle\tilde{P}^3\rangle^2}{\langle\tilde{P}^4\rangle\langle\tilde{P}^2\rangle-\langle\tilde{P}^3\rangle^2-\langle\tilde{P}^2\rangle^3}\right), \tag{1.1}$$

and also that

$$\langle\tilde{P}^2\rangle\langle\tilde{Q}^2\rangle \;\geq\; \frac{1}{4}\left(1+\frac{(\langle\tilde{P}^4\rangle-3\langle\tilde{P}^2\rangle^2)^2}{\langle\tilde{P}^6\rangle\langle\tilde{P}^2\rangle-\langle\tilde{P}^4\rangle^2}\right), \tag{1.2}$$

where $\tilde{P} = P - \langle P\rangle$ and $\tilde{Q} = Q - \langle Q\rangle$ represent the deviations of P and Q away from their means. These results act as a nice example of the strength of geometrical methods, which here are used to derive new relations of a physical or statistical character. Indeed, one of our intentions in this discussion is to emphasise the significance of the *geometric* point of view, which has been applied so successfully by Roger Penrose to numerous problems. Geometry can be used effectively at two distinct levels of scientific reasoning—first, as a mathematical tool, enabling one to arrive swiftly at results that might otherwise seem difficult or obscure; and second, as a means for directly gaining insights into phenomena.

2 Information geometry

A good example of the geometric approach, of relevance to our general argument, is found in Bhattacharyya's (1942, 1946, 1947, 1948) 'geometrisation' of the space of all probability distributions on a given configuration space. Let $p(x)$ for example be a probability density function on the real line, thus satisfying $0 \leq p(x) \leq 1$ and $\int p(x)dx = 1$. If we take the square-root density $\xi(x) \equiv \sqrt{p(x)}$, then $\int \xi^2 dx = 1$ and we can regard ξ as a point on the unit sphere $\mathcal{S}$ in a real Hilbert space $\mathcal{H}$. If $\eta(x)$ is another such square-root density function, then we can define a 'distance' function $D(\xi,\eta)$ in $\mathcal{H}$ for the two distributions corresponding to $\xi(x)$ and $\eta(x)$ by writing:

$$D^2(\xi,\eta) \;=\; \frac{1}{2}\int [\xi(x)-\eta(x)]^2\,dx\,. \tag{2.1}$$

In this case the function $D(\xi,\eta)$, known as the *Hellinger distance*, is evidently just the sine of the 'angle' made between the two Hilbert space vectors ξ and η.

At first the idea of taking the square-root of a density function may seem odd, but this turns out to be a natural construction that leads to some useful statistical insights. Physicists will already be familiar with the way in which square-root density functions arise in connection with the wave function in quantum mechanics, but may find it surprising that essentially the same construction, along with an associated Hilbert space structure, is known to statisticians.

It will be useful at this stage to formalise this geometry by the introduction of an abstract index notation for operations associated with the real Hilbert space $\mathcal{H}$. Thus, if ξ^a and η^a are typical elements of $\mathcal{H}$, we write $g_{ab}\xi^a\eta^b$ for their inner product. Those elements that satisfy the normalisation condition $g_{ab}\xi^a\xi^b = 1$ constitute the unit sphere $\mathcal{S}$ in $\mathcal{H}$. Random variables are represented by symmetric quadratic forms in $\mathcal{H}$. Thus, if X_{ab} is symmetric, then

$$E_\xi[X] = X_{ab}\xi^a\xi^b/g_{cd}\xi^c\xi^d \tag{2.2}$$

is the expectation of X_{ab} in the state ξ^a, and

$$\mathrm{Var}_\xi[X] = \tilde{X}_{ac}\tilde{X}^c_{\ b}\xi^a\xi^b/g_{ef}\xi^e\xi^f \tag{2.3}$$

is the variance, where $\tilde{X}_{ab} \equiv X_{ab} - g_{ab}E_\xi[X]$ represents the deviation of the random variable X_{ab} from its expectation. For the expected deviation we have $\tilde{X}_{ab}\xi^a\xi^b = 0$. In slightly more concrete terms, for $\mathcal{H} = L^2(R)$ the inner product $g_{ab}\xi^a\eta^b$ is the integral $\int \xi(x)\eta(x)dx$; the random variable X_{ab} corresponding to the 'position' x is the function $x\delta(x-y)$; and for the mean $X_{ab}\xi^a\xi^b$ we have the integral $\int\int x\delta(x-y)\xi(x)\xi(y)dxdy = \int xp(x)dx$; and so on. Note how effective the index notation (Penrose and Rindler 1984, 1986) is in this context. For more on the abstract index formalism in a Hilbert space setting, see, e.g., Geroch (1971a, b), Wald (1994).

Now we turn to the definition of the Fisher information and the construction of the Fisher–Rao metric. This can be carried out in a nice way in the Hilbert space setting just outlined. First, we need to introduce the idea of a 'statistical model'. By this we mean a parametric family of probability distributions $p(x|\theta^i)$, with parameters $\{\theta^i\}$, $i = 1, 2, \cdots, n$. Thus in our picture a statistical model can be viewed as a submanifold $\mathcal{M}$ of the unit sphere $\mathcal{S}$ in a real Hilbert space $\mathcal{H}$, as illustrated in Figure 1. In a problem of parametric estimation, usually one is given a statistical model $\mathcal{M} \subset \mathcal{S}$, and some data, and the idea is to infer from the data, as best as possible, which point of $\mathcal{M}$ corresponds to the 'true' distribution. We would also like to know about the errors likely to arise in making such an assessment.

Now suppose $\tau(\theta^i)$ is some specified function of the parameters $\{\theta^i\}$, and T is a random variable that is an unbiased estimator for the function $\tau(\theta^i)$; that is, the expectation of T in the state $p(x|\theta^i)$ is $\tau(\theta^i)$. From the 'measurement' (or sampling) of T we gain some information about the true distribution $p(x|\theta^i)$. The errors implicit in such a procedure are expressed in the existence of a fundamental lower bound on the variance of T that is independent of the specific choice of

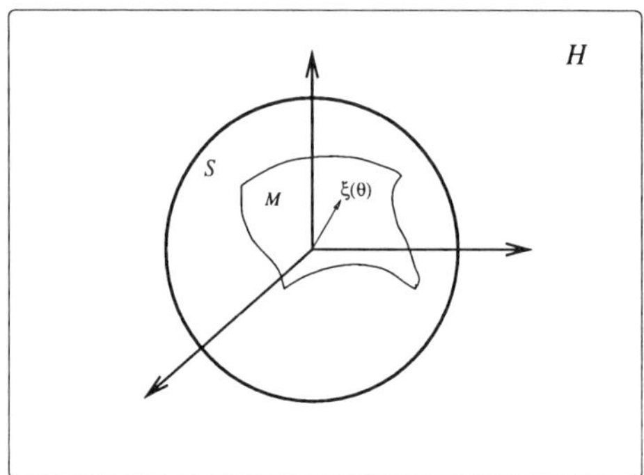

FIG. 1. **Embedding of a statistical model.** A parametric statistical model can be represented as a submanifold $\mathcal{M}$ of the unit sphere $\mathcal{S}$ in a real Hilbert space $\mathcal{H}$. A typical point in $\mathcal{M}$ can be represented by a vector $\xi^a(\theta)$. The metric geometry of $\mathcal{H}$ induces a natural Riemannian geometry on the submanifold $\mathcal{M}$, given by the Fisher–Rao metric.

estimator. This is given by the Cramér-Rao inequality

$$\mathrm{Var}_p[T] \ \geq \ \mathcal{G}^{ij}\partial_i\tau\partial_j\tau \ , \tag{2.4}$$

where $\partial_i\tau = \partial\tau/\partial\theta^i$, and $\mathcal{G}^{ij}$ is the inverse of the Fisher information matrix defined by:

$$\mathcal{G}_{ij} \ = \ \int p(x|\theta)\partial_i l(x|\theta)\partial_j l(x|\theta)dx \ , \tag{2.5}$$

where $l(x|\theta) = \ln p(x|\theta)$ is the log-likelihood function. Note that in (2.4) above, the summation convention applies to the indices i, j. In fact, the following geometrical characterisation can be established, namely that in local coordinates $\{\theta^i\}$, the Riemannian metric on $\mathcal{M}$, induced by g_{ab}, given by

$$\mathcal{G}_{ij} \ = \ 4g_{ab}\partial_i\xi^a\partial_j\xi^b \ , \tag{2.6}$$

is the Fisher information. This can be seen, for example, by substituting $p(x|\theta) = [\xi(x|\theta)]^2$ in (2.5), then converting the resulting integral expression into index notation. The factor of four is purely conventional.

Thus we see that the Fisher–Rao metric, which in terms of the 'log-likelihood' description (2.5) has a complicated, *ad hoc* character, arises naturally in a Hilbert space setting, as an induced geometry. If $\mathcal{M}$ is a curve (one-parameter statistical model), then the Fisher information

$$\mathcal{G} \ = \ 4g_{ab}\dot{\xi}^a\dot{\xi}^b \tag{2.7}$$

is four times the squared 'velocity' of $\xi^a(\theta)$ in $\mathcal{S}$. This is reminiscent of the Anadan-Aharanov (1990) characterisation of quantum mechanical energy uncertainty as the 'speed' of a Schrödinger trajectory through the state space; as we indicate later, the resemblance is not merely a coincidence, but lies at the heart of the relationship between classical and quantum statistical inference.

3 Classical estimation

Suppose we have a statistical model $\mathcal{M} = \{\xi^a(\theta)\}$, which we assume to be a curve in $\mathcal{S}$. We wish to estimate, from given data, the value of θ, or more generally a function $\tau(\theta)$ for which we have an unbiased estimator T_{ab} such that

$$\frac{T_{ab}\xi^a\xi^b}{\xi^c\xi_c} = \tau(\theta) \tag{3.1}$$

along the curve $\xi^a(\theta)$. Since τ, regarded as a function of ξ^a, is homogeneous, we can extend its definition to the whole of $\mathcal{H}$, and thus define the gradient $\nabla_a\tau$, where $\nabla_a = \partial/\partial\xi^a$. In fact, a short calculation shows that $\nabla_a\tau = 2\tilde{T}_{ab}\xi^b/g_{cd}\xi^c\xi^d$, where $\tilde{T}_{ab}$ is the deviation of T_{ab} from its mean, and as a consequence we deduce that the variance of the estimator T is given by

$$\mathrm{Var}_\xi[T] = \frac{1}{4}g^{ab}\nabla_a\tau\nabla_b\tau \tag{3.2}$$

on the unit sphere $\mathcal{S}$. This formula gives us a nice geometrical interpretation for the variance of an estimator. The unit sphere is foliated by surfaces of constant τ. For any given state ξ^a in $\mathcal{S}$, the variance of the estimator is then given by one-quarter of the squared magnitude of the gradient of τ at that point. Let us write $\dot{\xi}^a = \partial\xi^a/\partial\theta$ for the tangent vector to $\xi^a(\theta)$. Now, by the Cauchy-Schwartz inequality, clearly $(g^{ab}\nabla_a\tau\nabla_b\tau)(g_{cd}\dot{\xi}^c\dot{\xi}^d) \geq (\dot{\xi}^a\nabla_a\tau)^2$. Since $\dot{\xi}^a\nabla_a\tau = \dot{\tau} = \partial\tau/\partial\theta$ it follows immediately from (2.7) and (3.2) that $\mathrm{Var}_\xi[T] \geq \dot{\tau}^2/4\dot{\xi}^a\dot{\xi}_a$, or equivalently:

$$\mathrm{Var}_\xi[T] \geq \mathcal{G}^{-1}\left(\frac{\partial\tau}{\partial\theta}\right)^2 , \tag{3.3}$$

which is the Cramér-Rao inequality for a one dimensional statistical manifold. A similar line of reasoning in the multi-parameter setting leads to the general result (2.4). It should be apparent that the inequality (3.3) is saturated only when the tangent vector $\dot{\xi}^a(\theta)$ is parallel to the normal vector $\nabla_a\tau$. Families of probability distributions satisfying this relation are called 'exponential' families. The geometry of such distributions and their multi-parameter generalisations is of great relevance in statistical mechanics (Brody and Rivier 1995).

Once we take this geometric view of the Cramér-Rao inequality then an interesting extension of it can be obtained by use of higher derivatives of $\xi^a(\theta)$. Suppose, for example, we consider the acceleration vector $A^a(\theta)$ defined by

$$A^a \equiv \ddot{\xi}^a - \frac{\dot{\xi}^a(\dot{\xi}^b\ddot{\xi}_b)}{\dot{\xi}^c\dot{\xi}_c} - \xi^a(\xi^b\ddot{\xi}_b) , \tag{3.4}$$

so $A^a\xi_a = 0$ and $A^a\dot{\xi}_a = 0$. The squared magnitude of $\nabla_a\tau$ is clearly greater than the sum of the squared magnitudes of its projections in the directions of $\dot{\xi}^a$ and A^a. Thus, we have:

$$\mathrm{Var}_\xi[T] \geq \frac{(\dot{\xi}^a\nabla_a\tau)^2}{4\dot{\xi}^a\dot{\xi}_a} + \frac{(A^a\nabla_a\tau)^2}{4A^aA_a} . \tag{3.5}$$

The first term on the right is the Cramér-Rao bound, whereas the next term is a second-order 'correction'. Further higher-order terms can be constructed by use of higher derivatives of $\xi^a(\theta)$. We call these terms 'generalised Bhattacharyya bounds', since they formally resemble the corrections obtained by Bhattacharyya (1946) in a log-likelihood setting. In general, our higher-order bounds depend on the choice of estimator for $\tau(\theta)$. However, as we shall see in the case of quantum statistical inference, some bounds obtained in this way have the remarkable property of being independent of the choice of estimator, thus allowing us to deduce generalised quantum mechanical uncertainty relations *without having to say anything about the joint moments of the relevant conjugate observables.*

4 Quantum geometry

Quantum theory is usually approached from a 'complex' point of view, via the familiar notation of 'bras' and 'kets'. Despite its elegance there is a problem with this way of formulating the subject inasmuch as it 'builds in' the complex structure to such an extent that it is impossible to extricate it. For some purposes it is therefore advantageous to adopt a 'real' point of view in quantum theory, and introduce the complex structure as an 'extra ingredient'. This is the approach we shall take here. We start with a *real* Hilbert space $\mathcal{H}$, assumed even dimensional in the finite case, with typical elements $\xi^a, \eta^a \in \mathcal{H}$. On $\mathcal{H}$ we have a metric g_{ab} as before, and also a compatible complex structure given by a tensor $J^a_{\ b}$ satisfying $J^a_{\ b}J^b_{\ c} = -\delta^a_{\ c}$. We require that the metric is Hermitian with respect to the given complex structure, i.e., $J^a_{\ c}J^b_{\ d}g_{ab} = g_{cd}$. This implies that the tensor $\Omega_{ab} \equiv g_{ac}J^c_{\ b}$ is antisymmetric. With this structure at hand we can now define the standard 'Dirac product' $\langle\eta|\xi\rangle$ between two real vectors $\xi^a, \eta^a \in \mathcal{H}$ by

$$\langle\eta|\xi\rangle = \frac{1}{2}\eta^a(g_{ab} - i\Omega_{ab})\xi^b . \tag{4.1}$$

A real vector ξ^a can be split into a set of 'positive' and 'negative' parts with respect to the given complex structure by writing $\xi^a = \xi^a_+ + \xi^a_-$, with

$$\xi^a_\pm \equiv \frac{1}{2}(\xi^a \mp iJ^a_{\ b}\xi^b) . \tag{4.2}$$

The resulting 'positive' and 'negative' spaces can be thought of as the spaces of kets and bras, respectively, and we have $\langle\eta|\xi\rangle = \eta^a_- g_{ab}\xi^b_+$ for the Dirac product. It is worth noting that in a relativistic theory the decomposition of ξ^a corresponds to the splitting of a real field into positive and negative frequency parts.

The Schrödinger equation can also be represented neatly in real terms. This is given by $\dot{\xi}^a = J^a_{\ b}H^b_{\ c}\xi^c$, where the symmetric Hamiltonian operator H_{ab} is required to be Hermitian in the sense that $J^a_{\ c}J^b_{\ d}H_{ab} = H_{cd}$. Note that $J^a_{\ b}$ plays the role of the 'i' in the conventional formula $\partial_t|\xi\rangle = iH|\xi\rangle$. Since the actual state of the quantum system does not depend upon the phase of ξ^a, we can modify the Schrödinger equation by removing the 'dynamical' phase, and write

$$\dot{\xi}^a = J^a_{\ b}\tilde{H}^b_{\ c}\xi^c , \tag{4.3}$$

where $\tilde{H}_{ab} = H_{ab} - g_{ab}(H_{cd}\xi^c\xi^d/\xi_e\xi^e)$. Thus it is only the deviation of the Hamiltonian from its mean that affects the dynamics of the Schrödinger trajectory on the quantum mechanical state space. Two vectors $\xi^a, \eta^a \in \mathcal{S}$ are equivalent quantum states if they differ only by a phase. This means we can write, for two equivalent states, that $\eta^a = e^{i\phi}\xi^a_+ + e^{-i\phi}\xi^a_-$ for some real ϕ. At each instant of time the solutions of the linear Schrödinger equation and the modified Schrödinger equation (4.3), given an initial state $\xi^a(0)$, differ from one another in this way, where the relevant phase factor is $\phi = t\langle H\rangle$.

Another way of looking at this is as follows. Given a Schrödinger trajectory on the underlying state space (the projective Hilbert space) there is a unique lift of this trajectory to the unit sphere $\mathcal{S}$ in $\mathcal{H}$ such that, at each point along the trajectory, $\dot{\xi}^a$ is orthogonal to the direction $J^a_{\ b}\xi^b$. This lift is the trajectory determined by (4.3). It is not difficult to see then that the squared velocity of the curve $\xi^a(t)$ in $\mathcal{S}$ is

$$g_{ab}\dot{\xi}^a\dot{\xi}^b = \mathrm{Var}_\xi[H] \, , \tag{4.4}$$

where $\mathrm{Var}_\xi[H] = \tilde{H}_{ab}\tilde{H}^b_{\ c}\xi^a\xi^c/\xi_d\xi^d$ is the squared uncertainty of the energy in the given state, which is constant along a Schrödinger trajectory. Indeed, we have $g_{ab}\dot{\xi}^a\ddot{\xi}^b = 0$ along a Schrödinger trajectory, because $H_{ac}J^c_{\ b}$ is antisymmetric on account of the Hermitian condition on H_{ab}. Formula (4.4) is the Anandan-Aharanov (1990) relation, which equates the 'speed' of a quantum trajectory in its state space with the associated energy uncertainty.

5 Quantum statistical estimation

Now we are in a position to formulate the problem of parameter estimation in a quantum mechanical context. The interesting point here is that by developing quantum theory in terms of a real Hilbert space we are able to apply our 'classical' estimation theory more or less directly—the essential new features arising in the quantum mechanical case come from the fact that we supplement the real Hilbert space $\mathcal{H}$ with a compatible complex structure.

The quantum measurement problem can thus be posed as follows. Suppose we are given a number of independent, identically prepared quantum mechanical systems, and assume that the 'true' state of the system in each case belongs to a specified submanifold of the state space. The problem is by making a set of measurements on the various independent systems to determine as best as possible the true state in which these systems have been prepared.

For example, suppose we have a large number of independent identical systems, each of which evolves from some initial state $\xi^a(0)$ under the influence of the Hamiltonian H_{ab}. The problem is to estimate how much time t has elapsed since the initial preparation. Let T_{ab} be an unbiased estimator for the time parameter, so $T_{ab}\xi^a\xi^b/g_{cd}\xi^c\xi^d = t$ along the trajectory $\xi^a(t)$. This is not quite the same thing as assuming T_{ab} is a 'time observable'; we merely require a symmetric operator with the property that for any initial state $\xi^a(0)$ the expectation of T_{ab} in the state $\xi^a(t)$ is t (cf. Holevo 1982). Our strategy will be to lift the

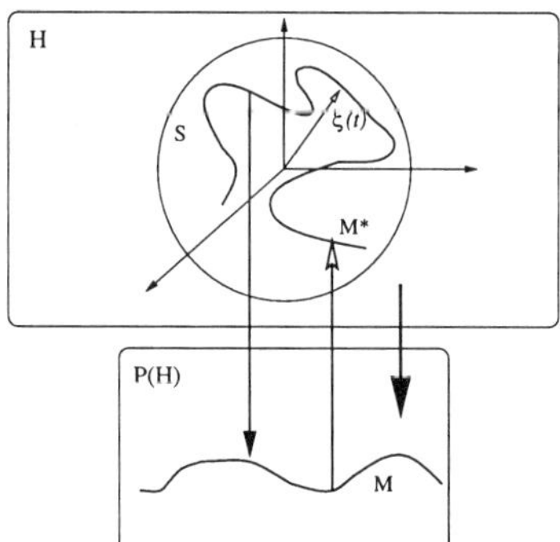

FIG. 2. **Quantum statistical estimation.** The unit sphere $\mathcal{S}$ in a real Hilbert space $\mathcal{H}$ endowed with a complex structure projects down to a complex projective Hilbert space $P(\mathcal{H})$ which is the quantum mechanical state space. A Schrödinger trajectory $\mathcal{M}$ in $P(\mathcal{H})$ lifts to a unique statistical manifold $\mathcal{M}^*$ in $\mathcal{S}$ which is everywhere orthogonal to the 'Cauchy-Riemann' direction.

Schrödinger trajectory from the state space to the unit sphere $\mathcal{S}$ in $\mathcal{H}$ via (4.3), and to regard the resulting curve as a one-dimensional statistical model in $\mathcal{H}$, as illustrated in Figure 2. Then the Cramér-Rao inequality (3.3) applies, and since by (4.4) the Fisher information in this case is $\mathcal{G} = 4\mathrm{Var}_\xi[H]$, it follows that

$$\mathrm{Var}_\xi[T]\mathrm{Var}_\xi[H] \geq \frac{1}{4}, \tag{5.1}$$

which is the Heisenberg relation for energy and time, the relevant time variable being the estimator T_{ab}. In this case the random variable H_{ab} is involved in the determination of the relevant statistical manifold, whereas the random variable T_{ab} acts as an estimator for the position on that manifold. This is different from the 'usual' interpretation of (5.1), which might, for example, make reference (albeit vaguely) to a 'simultaneous' measurement of T_{ab} and H_{ab}. The point here is that even *assuming* the state lies on the given Schrödinger trajectory determined by H_{ab}, there is still a lower bound on the variance of the estimated time elapsed, given by (5.1).

6 Higher order variance bounds

Let us now apply our theory of generalised Bhattacharyya bounds to the quantum estimation problem. First, we recall that the usual Heisenberg relation for a pair of conjugate operators is actually a special case of a more general relation

$$\mathrm{Var}_\xi[T]\mathrm{Var}_\xi[H] \geq \frac{1}{4}\left(1 + \left|\langle\xi|\{T,H\}|\xi\rangle\right|^2\right), \tag{6.1}$$

where $\{T, H\}$ is the anticommutator of T and H. This term is usually dropped, since in general it depends in a complicated way on the relationship of T and H, i.e., on their covariance, unlike the commutator term. Nevertheless, the bound

(6.1) can be reproduced satisfactorily in the present framework by noting that

$$g^{ab}\nabla_a\tau\nabla_b\tau \;\geq\; \frac{(\dot{\xi}^a\nabla_a\tau)^2}{\dot{\xi}^a\dot{\xi}_a} + \frac{(\dot{\zeta}^a\nabla_a\tau)^2}{\dot{\zeta}^a\dot{\zeta}_a}\,, \tag{6.2}$$

since $\dot{\xi}^a$ and $\dot{\zeta}^a = J^a_b\dot{\xi}^a$ are necessarily orthogonal. A short exercise then shows that the second term on the right gives rise to the familiar anticommutator term. This is a good example of a generalised bound that depends on the 'choice of estimator', and therefore is in general not so useful. That is not the end of the story, however, since we can go to higher order instead, and consider the second order correction

$$g^{ab}\nabla_a\tau\nabla_b\tau \;\geq\; \frac{(\dot{\xi}^a\nabla_a\tau)^2}{\dot{\xi}^a\dot{\xi}_a} + \frac{(\eta^a\nabla_a\tau)^2}{\eta^a\eta_a}\,, \tag{6.3}$$

where η^a is the component of $\ddot{\zeta}^a$ orthogonal to $\xi^a, \dot{\xi}^a, \zeta^a$, and $\dot{\zeta}^a$, given by

$$\eta^a \;=\; \ddot{\zeta}^a - \frac{\ddot{\zeta}^b\dot{\xi}_b}{\dot{\xi}^c\dot{\xi}_c}\dot{\xi}^a - (\ddot{\zeta}^b\zeta_b)\zeta^a\,. \tag{6.4}$$

Note that $\ddot{\zeta}^a\xi_a = 0$ and $\ddot{\zeta}^a\dot{\zeta}_a = 0$ automatically as a consequence of the modified Schrödinger equation. The second term in (6.3) can be evaluated if we make use of the identity $T_{ab}\dot{\xi}^a\dot{\xi}^b = tg_{ab}\dot{\xi}^a\dot{\xi}^b + k$, where k is a constant. Then after a short exercise we find

$$\mathrm{Var}_\xi[T]\mathrm{Var}_\xi[H] \;\geq\; \frac{1}{4}\left(1 + \frac{\langle\tilde{H}^3\rangle^2}{\langle\tilde{H}^4\rangle\langle\tilde{H}^2\rangle - \langle\tilde{H}^3\rangle^2 - \langle\tilde{H}^2\rangle^3}\right)\,, \tag{6.5}$$

where $\langle\tilde{H}^n\rangle$ denotes the n-th central moment of the Hamiltonian in the given state. The denominator in the second term is essentially the 'curvature' of the Schrödinger trajectory in the ambient space $\mathcal{S}$, which is guaranteed to be positive by a well-known statistical identity.

Another bound can be obtained by consideration of the third order Bhattacharyya style correction given by $B^a\nabla_a\tau/(B^aB_a)^{1/2}$ associated with the vector B^a defined by $B^a = \dddot{\xi}^a - (\dddot{\xi}_b\,\dot{\xi}^b/\dot{\xi}_c\dot{\xi}^c)\dot{\xi}^a$, which, as a consequence of (4.3), is orthogonal to $\dot{\xi}^a$ and ξ^a. The evaluation of the correction is facilitated if we take note of the relation $2T_{ab}\,\dddot{\xi}^a\,\xi^b = -3g_{ab}\dot{\xi}^a\dot{\xi}^b$. A short calculation making use of the identity $g_{ab}\xi^{a(n)}\xi^{b(n)} = \langle\tilde{H}^{2n}\rangle$, where $\xi^{a(n)}$ is the n-th derivative of the modified Schrödinger trajectory, then leads us to the desired result, namely:

$$\mathrm{Var}_\xi[T]\mathrm{Var}_\xi[H] \;\geq\; \frac{1}{4}\left(1 + \frac{(\langle\tilde{H}^4\rangle - 3\langle\tilde{H}^2\rangle^2)^2}{\langle\tilde{H}^6\rangle\langle\tilde{H}^2\rangle - \langle\tilde{H}^4\rangle^2}\right)\,. \tag{6.6}$$

The expression in the numerator of the correction term is the fourth cumulant of the distribution for H, which is a measure of the deviation from a normal distribution; the denominator on the other hand is guaranteed to be positive by the Cauchy-Schwartz inequality. Further higher order corrections, depending only upon moments of H, can be constructed analogously, and indeed similar results hold for any pair of conjugate operators.

7 Quantum geometry vs information geometry

The results sketched out above will indicate some of the richness of the relationship between classical and quantum statistical estimation that becomes apparent once we take a geometrical point of view, and, in particular, formulate matters in terms of an underlying real Hilbert space geometry endowed with a complex structure. The importance of complex structures in the foundations of physics is another recurrent theme in Roger Penrose's work, and it is interesting to see how this crops up here in a new approach to the quantum measurement problem. In the case of information geometry the concept of a statistical model is introduced as a submanifold of the unit sphere in a real Hilbert space, and the induced metric is the Fisher–Rao metric. When this structure is augmented with a compatible complex structure, then we can form the corresponding projective state space of quantum theory, with the Fubini–Study metric, and for a quantum statistical model we consider a submanifold of that space (see Figure 2). The approach outlined here can be contrasted with the density matrix based methodology examined by Helstrom (1976), Wootters (1981), Malley and Hornstein (1993), Braunstein and Caves (1994), and others. In the examples noted in the previous sections we have seen how the resulting theory can be applied in the case of Schrödinger type trajectories to obtain sharpened Heisenberg bounds. Another example of a natural submanifold of a quantum state space is obtained if we look at the projective Fock space associated with an ensemble of bosons—for example, the many–particle photon space of quantum electrodynamics. In that situation, the complex structure is given by the splitting of these fields into positive and negative frequency parts, and the appropriate submanifold, which is of interest in the theory of quantum measurement in electrodynamics, is the space of 'coherent' states. These states are in one-to-one correspondence with classical fields, and there is an associated probability operator valued measurement. In this case the induced metric on the submanifold is flat (Field 1996a, b), corresponding to the linearity of the associated classical theory. The statistical geometry of the resulting configuration is thus very intriguing, and also forms a natural springboard for the investigation of relativistic aspects of the theory.

Bibliography

Amari, S. (1985). *Differential–Geometrical Methods in Statistics*. Springer-Verlag, Berlin.

Anandan, J. and Aharonov, Y. (1990). Geometry of Quantum Evolution. *Phys. Rev. Lett.*, **65**, 1697–700.

Ashtekar, A. and Schilling, T. A. (1995). Geometry of Quantum Mechanics. *AIP Conference Procedings: CAM–94 Physics Meeting*, **342**, 471–8, ed. Zapeda, A. AIP Press, Woodbury, New York.

Barndorff-Nielsen, O. E., Cox, D. R., and Reid, N. (1986). The Role of Differential Geometry in Statistical Theory. *Int. Stat. Review*, **54**, 83–96.

Bhattacharyya, A. (1942). On Discrimination and Divergence. *Proc. Sc. Cong.*

Bhattacharyya, A. (1946). On Some Analogues of the Amount of Information and Their Use in Statistical Estimation. *Sankhyā*, **8**, 1–14.

Bhattacharyya, A. (1947). On Some Analogues of the Amount of Information and Their Use in Statistical Estimation. *Sankhyā*, **8**, 201–18.

Bhattacharyya, A. (1948). On Some Analogues of the Amount of Information and Their Use in Statistical Estimation. *Sankhyā*, **8**, 315–28.

Braunstein, S. L. and Caves, C. M. (1994). Statistical Distance and the Geometry of Quantum States. *Phys. Rev. Lett.*, **72**, 3439–43.

Brody, D. C. and Hughston, L. P. (1996). Geometry of Quantum Statistical Inference. *Phys. Rev. Lett.*, **77**, 2581–4.

Brody, D. and Rivier, N. (1995). Geometrical Aspects of Statistical Mechanics. *Phys. Rev.*, **E51**, 1006–11.

Cirelli, R., Mania, A., and Pizzocchero, L. (1990). Quantum mechanics as an infinite-dimensional Hamiltonian system with uncertainty structure. *J. Math. Phys.*, **31**, 2891–903.

Field, T. R. (1996a). The Quantum Complex Structure. *D.Phil. Thesis.* Oxford University, Oxford.

Field, T. R. (1996b). Coherent States and Fubini-Study Geometry. *Twistor Newsletter*, **41**, 6–14.

Gibbons, G. W. (1992). Typical States and Density Matrices. *J. Geom. and Phys.*, **8**, 147–62.

Geroch, R. (1971a). Special Topics in Particle Physics. *Unpublished lecture notes.* University of Texas, Austin.

Geroch, R. (1971b). An Approach to Quantisation of General Relativity. *Ann. Phys.*, **62**, 582–9.

Helstrom, C. W. (1976). *Quantum Detection and Estimation Theory.* Academic Press, New York.

Holevo, A. S. (1982). *Probabilistic and Statistical Aspects of Quantum Theory.* North-Holland Publishing Company, Amsterdam.

Hughston, L. P. (1995). Geometric Aspects of Quantum Mechanics. *Twistor Theory*, ed. Huggett, S. Marcel Dekker, Inc., New York.

Hughston, L. P. (1996). Geometry of Stochastic State Vector Reduction. *Proc. Roy. Soc.*, **B452**, 953–79.

Kibble, T. W. B. (1978). Relativistic Models of Nonlinear Quantum Mechanics. *Comm. Math. Phys.*, **64**, 73–82.

Kibble, T. W. B. (1979). Geometrisation of Quantum Mechanics. *Comm.*

Math. Phys., **65**, 189–201.

Malley, J. D. and Hornstein, J. (1993). Quantum Statistical Inference. *Stat. Science*, **8**, 433–57.

Murray, M. K. and Rice, J. W. (1993). *Differential Geometry and Statistics.* Chapman and Hall, London.

Page, D. A. (1987). Geometrical Description of Berry's Phase. *Phys. Rev.*, **A36**, 3479–81..

Penrose, R. and Rindler, W. (1984). *Spinors and Space-Time Volume I.* Cambridge University Press, Cambridge.

Penrose, R. and Rindler, W. (1986). *Spinors and Space-Time Volume II.* Cambridge University Press, Cambridge.

Rao, C. R. (1945). Information and the Accuracy Attainable in the Estimation of Statistical Parameters. *Bull. of Calcutta Math. Soc.*, **37**, 81–91.

Schilling, T. A. (1996). Geometry of Quantum Mechanics. *Ph.D. Thesis.* Pennsylvania State University.

Wald, R. M. (1994). *Quantum Field Theory in Curved Spacetime and Black Hole Thermodynamics.* Chicago University Press, Chicago.

Wootters, W. K. (1981). Statistical Distance and Hilbert Space. *Phys. Rev.*, **D23**, 357–62.

19
Spin Networks and Topology

Louis H. Kauffman
Department of Mathematics, Statistics and Computer Science
University of Illinois at Chicago
851 South Morgan Street
Chicago, IL, 60607–7045

Abstract

This paper discusses Penrose spin networks in relation to physics and low dimensional topology.

1 Introduction

This paper is an introduction to the relationship between Penrose spin networks and structures in physics and low dimensional topology. The paper is organised as follows. Section 2 discusses the theme of networks and discretization. Section 3 introduces the bracket model for the Jones polynomial invariant of knots and links. In Section 4 we show how the bracket state model is a natural generalization of the original Penrose spin networks, and how this model is related to the quantum group corresponding to $SL(2,C)$. In particular, we show how the binor identity of the spin networks, the skein identity of the bracket polynomial (at a special value) and the trace identity

$$\mathrm{tr}(AB) + \mathrm{tr}(AB^{-1}) = \mathrm{tr}(A)\mathrm{tr}(B)$$

of $SL(2,C)$ are really all the same. The section continues with a discussion of the role of these generalized networks in low dimensional topology. In particular we discuss the relationship of these nets with the evaluation of the Witten–Reshetikhin–Turaev invariant of 3-manifolds and corresponding relations with quantum gravity theories (Regge–Ponzano in 2+1 and Ashtekar–Smolin–Rovelli in 3+1).

2 Networks and discrete space

In topology and geometry it is common to think of spaces defined in terms of point sets or (local) coordinates. It is also possible, particularly in topology, to specify a space by purely combinatorial data, a network, a graph or a code. Knot theory provides a good example of this combinatorial approach with its systems of diagrams drawn in a plane. The knot diagrams are not knots. They are instructions for building embeddings of knots in three-dimensional space. They are a graphical notation for the topological relations in the knot. Transformations of this graphical notation can encode corresponding topological transformations on the knots, and problems in topology can be reformulated in terms of the category of diagrams.

Diagrammatic discretizations of this sort can handle non-topological properties of space as well. For example, a diagrammatic notation for scalar product and vector cross product can be used to prove identities for the triple vector cross product. This notation is based on regarding the networks as representatives for abstract tensors so that a triple vertex corresponds to a three-indexed epsilon. Here the networks can be viewed alternately as geometrical or algebraic. In formalising these notions one naturally finds that the switch of viewpoints from diagrams, to algebra, to corresponding geometry, to evaluations of contracted tensors all constitute functors among the different categories represented by these concepts. The network appears as a locus for the application of a medley of functors. The network replaces the old concept of space. That space is one of the possible targets of the many functors that extract information from the network.

In this way space, time and the idea of place and coordinate fade away to be replaced by networks and processes that unfold into both the spaces and the evaluations (partition functions, amplitudes) that elucidate the geometry and topology associated with them.

In this essay we shall see these themes illustrated in relation to knot theory, low dimensional topology and a generalization of the original Penrose theory of spin networks.

3 The bracket state summation and the Jones polynomial

In 1984 Vaughan Jones instigated a revolution in knot theory by discovering a new polynomial invariant (Jones 1985) of classical knots and links. His invariant can distinguish many knots and links from their mirror images, a capability that goes beyond the Alexander polynomial, discovered in the 1920's (Alexander 1923). The Jones polynomial was, from the outset, related to statistical mechanics and to von Neumann algebras. In 1985 the author discovered (Kauffman 1987) a model for the original Jones polynomial as a partition function defined in terms of combinatorial states of the link diagram.

This partition function will be referred to as the *bracket state model* for the Jones polynomial. It is the purpose of this section to describe the bracket model. The next section will show how this model is a perspicuous generalisation of the binor calculus of Roger Penrose (1971). The binor calculus forms the underpinning of the Penrose theory of spin networks.

In order to do this, we first digress on the combinatorial moves (the Reidemeister moves) that describe the topology of knots and links.

3.1 The Reidemeister moves

In the 1920's Reidemeister (Reidemeister 1933) discovered that three basic combinatorial moves on link diagrams suffice to capture the concept of ambient isotopy of knots and links in three-dimensional space. Two links are ambient isotopic if there is a continuous family of embeddings connecting one to the other. Reidemeister proved that two (tame or piecewise linear) links are ambient isotopic if and only if diagrams for each can be transformed into one another by his moves. See Fig. 1 for a depiction of the Reidemeister moves.

It is convenient to have terminology for the equivalence relation generated by the second and third Reidemeister moves. This equivalence is called *regular isotopy*. One reason for singling out this particular equivalence relation is that it can be used to model the topology of so-called *framed links*. A framed link is a link that is equipped with a non-vanishing normal vector field. Equivalently, a framed link can be understood as a link where each component is an embedded orientable band. By thickening (in the plane) the planar pictures into such bands, we obtain a framed link. Regular isotopy preserves the framing. Replacement of the first Reidemeister move by the move I' shown in Fig. 1 results in a diagrammatic model for the topology of framed links.

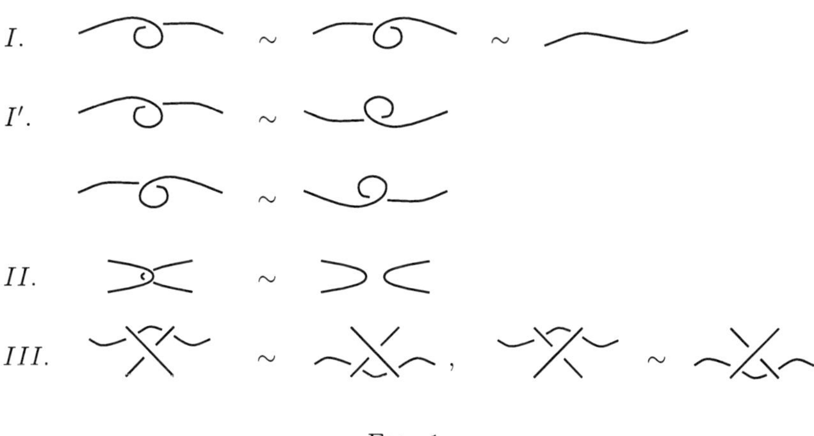

FIG. 1.

The bracket polynomial is based on the structure of the two smoothings at a crossing in a knot diagram. At a given crossing there are four local regions. Call two out of the four local regions a *pair* if they meet only at the vertex. Call the pair that are swept out by a counterclockwise turn of the overcrossing line the *A-pair*. Call the remaining pair the *B-pair*. The *A-smoothing* is the smoothing that joins the local regions of the A-pair. The *B-smoothing* is the smoothing that joins the local regions of the B-pair. See Fig. 2 for an illustration of this basic distinction.

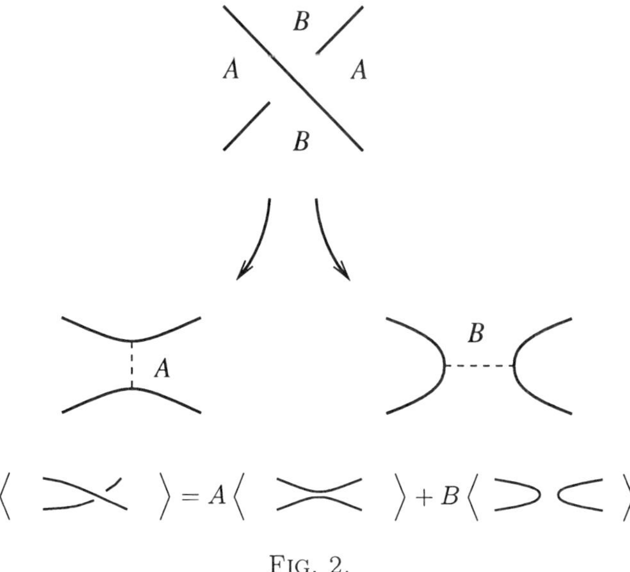

FIG. 2.

The three-variable bracket polynomial is defined on link diagrams by the following formulas:

(1) $\langle K \rangle = A\langle K_1 \rangle + B\langle K_2 \rangle$ where K_1 and K_2 are two diagrams obtained from a given crossing in K by smoothing the crossing in the two possible ways illustrated in Fig. 3 with an A-smoothing in K_1 and a B-smoothing in K_2.

(2) $\langle K \bigsqcup D \rangle = d\langle K \rangle$ where D denotes any Jordan curve that is placed into the complement of the diagram K in the plane.

(3) $\langle D \rangle = 1$ for any Jordan curve D in the plane.

Here the small diagrams stand for otherwise identical parts of larger diagrams, and the second formula means that any Jordan curve disjoint from the rest of the diagram contributes a factor of d to the polynomial. This recursive description of the bracket is well-defined so long as the variables A, B, and d commute with one another.

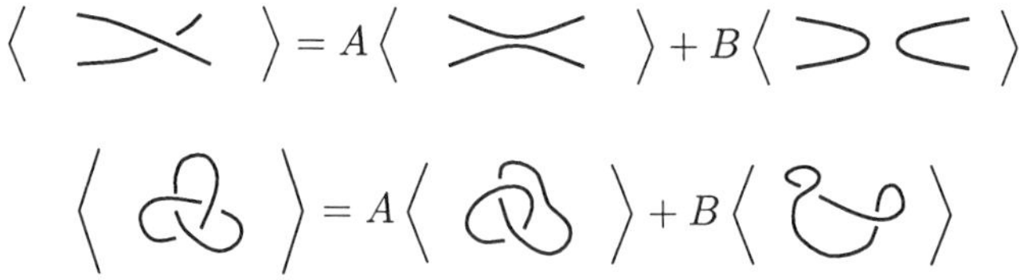

FIG. 3.

The bracket can be expressed as a state summation where the states are obtained by smoothing the link diagram in one of two ways at each crossing.

A smoothing of type A contributes a vertex weight of A to the state sum. A smoothing of type B contributes a vertex weight of B to the state sum. The norm of a state S, denoted $||S||$, is defined to be the number of Jordan curves in S. It then follows that the bracket is given by the formula

$$\langle K \rangle = \Sigma_S \langle K|S \rangle d^{||S||-1}$$

where the summation is over all states S of the diagram K, and $\langle K|S \rangle$ denotes the product of the vertex weights for the state S of K.

In the variables A, B and d the bracket polynomial is not invariant under the Reidemeister moves. However the following lemma provides the clue to finding a specialization of the polynomial that is an invariant of *regular isotopy*, the equivalence relation generated by the second and third Reidemeister moves (Reidemeister 1933).

Lemma 3.1 *Let K_{**} denote a diagram with the local configuration shown in Fig. 4. Let K_v and K_h denote the two local smoothings of this configuration as shown also in Fig. 4. Then*

$$\langle K_{**} \rangle = AB\langle K_v \rangle + (ABd + A^2 + B^2)\langle K_h \rangle.$$

Proof The proof is omitted. □

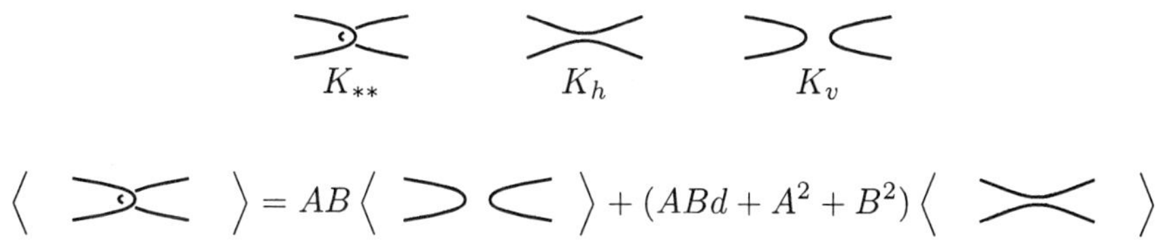

K_{**} K_h K_v

$\langle$ $\rangle = AB \langle$ $\rangle + (ABd + A^2 + B^2) \langle$ $\rangle$

FIG. 4.

Since K_v is obtained from K_{**} by a second Reidemeister move, it follows that $\langle K \rangle$ can be made invariant under the second Reidemeister move if we take $B = A^{-1}$, and $d = -A^2 - A^{-2}$. In fact, with this specialization, $\langle K \rangle$ is also invariant under the third Reidemeister move, and it behaves multiplicatively under the first Reidemeister move. This allows the normalization

$$f_K = (-A^3)^{-w(K)} \langle K \rangle.$$

f_K is an invariant of ambient isotopy for links in three-space. Here $w(K)$ is the sum of the signs of the crossings of the oriented link K.

Theorem 3.2. (Kauffman 1987) *$f_K(t^{-1/4}) = V_K(t)$ where $V_K(t)$ is the original Jones polynomial (Jones 1985).*

This completes our description of the connection of the bracket state model with the original Jones polynomial. On the knot theory side this new invariant was and is particularly exciting because of its ability to detect the topological difference between many knots and their mirror images. Furthermore, there is still no counterexample known to the conjecture that the Jones polynomial can detect knots. That is, there is no known embedded circle K in three-dimensional space with K actually knotted and $V_K(t) = 1$. On the interdisciplinary side, the connections with statistical mechanics and other areas of mathematical physics invoke an intriguing and deep mystery story that is still unfolding.

4 Spin networks

The original Penrose spin networks (Penrose 1969, 1971) were devised to create a diagrammatic and combinatorial substrate for the recoupling theory of quantum mechanical angular momentum. The key to these diagrammatics is a system of abstract tensors based on the properties of a classical epsilon (the definition will be given below) and adjusted to obtain topological invariance under planar deformations of the diagrams. It is this adjustment in the direction of topological invariance that makes these networks a special case of the bracket state model for the Jones polynomial (Kauffman 1992).

The classical epsilon is defined by the equations $\epsilon_{12} = 1$, $\epsilon_{21} = -1$ and $\epsilon_{ij} = 0$ if $i = j$. The indices are in the set $\{1, 2\}$. Note that the Kronecker delta is defined by the equation $\delta_{ab} = \delta^{ab} = \delta^b_a = 1$ if $a = b$ and 0 if $a \neq b$. Epsilons and deltas are related by the fundamental equation

$$\epsilon^{ab}\epsilon_{cd} = \delta^a_c\delta^b_d - \delta^a_d\delta^b_c.$$

We shall refer to this equation as the *epsilon identity.* It is natural to diagram this relation by letting a vertical line represent a single Kronecker delta, a cup (a local minimum, see Fig. 5) represent an epsilon with upper indices, a cap (a local maximum, see Fig. 5) represent an epsilon with lower indices. Note that in such a convention $\delta^a_c\delta^b_d$ is represented by vertical parallel lines, while $\delta^a_d\delta^b_c$ is represented by vertical crossed lines. The key to topological invariance is that the cups and the caps should satisfy the cancellation to a delta as in

$$CAP_{ai}CUP^{ib} = \delta^b_a$$

(with implicit summation over i). This is accomplished in this formalism with

$$CAP_{ab} = i\epsilon_{ab},$$

$$CUP^{ab} = i\epsilon^{ab}.$$

Here $i^2 = -1$. Finally it is convenient to introduce a minus sign at the crossing of two diagrammatic lines. With these conventions we obtain the *binor identity* as illustrated in Fig. 5. The reader will note that this identity is exactly the exchange identity for the bracket polynomial when $A = -1 = B$ and $d = -A^2 -$

$A^{-2} = -2$. Note particularly that the loop value -2 coincides with the loop value in abstract tensors as shown in Fig. 5. This loop value is $CUP^{ab}CAP_{ab}$ (summed on a and b). The binor identity is the basis of the original Penrose spin nets.

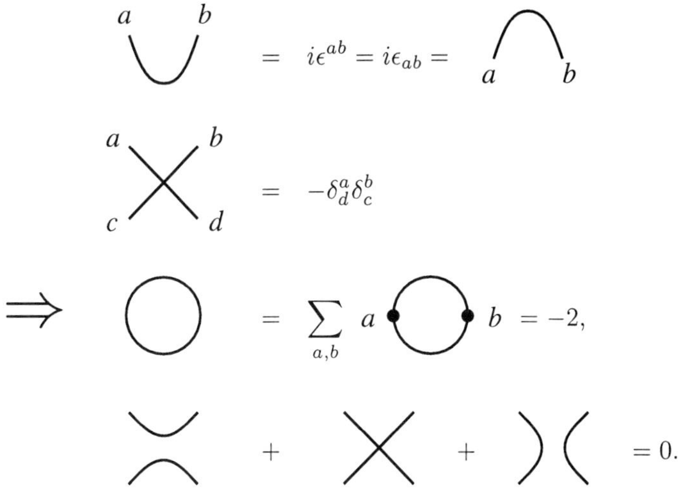

$$\bigcup_{a\ \ b} = i\epsilon^{ab} = i\epsilon_{ab} = \bigcap_{a\ \ b}$$

$$\times_{c\ \ d}^{a\ \ b} = -\delta^a_d\delta^b_c$$

$$\Rightarrow \bigcirc = \sum_{a,b} a \bullet\!\!\bigcirc\!\!\bullet\, b = -2,$$

$$\asymp + \times +)(= 0.$$

FIG. 5.

The use of the epsilon tensor in the spin networks is directly related to the group $SL(2,C)$. The algebraic reason for this is that for any 2×2 matrix P with commuting entries,

$$P\epsilon P^t = \det(P)\epsilon$$

where ϵ is regarded as a 2×2 matrix, and P^t denotes the transpose of the matrix P. Thus $SL(2,C)$ is the set of 2×2 matrices P over C such that

$$P\epsilon P^t = \epsilon.$$

The ϵ identity that is at the basis of the binors is then easily seen to be the source of the following fundamental lemma about $SL(2,C)$:

Lemma 4.1 *If A and B are matrices in $SL(2,C)$ and* tr *denotes the standard matrix trace, then*

$$\mathrm{tr}(AB) + \mathrm{tr}(AB^{-1}) = \mathrm{tr}(A)\mathrm{tr}(B).$$

Proof In this proof we will use the Einstein summation convention for repeated indices (i.e. we sum over all values in the index set $\{1,2\}$ whenever an index occurs twice in a given expression). Note that if A is in $SL(2, C)$, then $A\epsilon A^t = \epsilon$. Therefore $A = -\epsilon(A^t)^{-1}\epsilon$. Thus

$$\begin{aligned}
\mathrm{tr}(AB) &= A_{ik}B_{ki} \\
&= -A_{ir}\epsilon_{rs}((B^t)^{-1})_{sl}\epsilon_{li} \\
&= -A_{ir}((B^t)^{-1})_{sl}(\epsilon_{rs}\epsilon_{li}) \\
&= -A_{ir}((B^t)^{-1})_{sl}(\delta_{rl}\delta_{si} - \delta_{ri}\delta_{sl}) \\
&= -A_{ir}((B^t)^{-1})_{sl}\delta_{rl}\delta_{si} + A_{ir}((B^t)^{-1})_{sl}\delta_{ri}\delta_{sl} \\
&= -A_{il}((B^t)^{-1})_{il} + A_{ii}((B^t)^{-1})_{ll} \\
&= -\mathrm{tr}(AB^{-1}) + \mathrm{tr}(A)\mathrm{tr}(B^{-1}).
\end{aligned}$$

This completes the proof.

Remark This lemma shows that the SL(2) identity $\mathrm{tr}(AB) + \mathrm{tr}(AB^{-1}) = \mathrm{tr}(A)\mathrm{tr}(B)$ is essentially an matrix algebraic expression of the binor identity, and hence equivalent to the corresponding identity for the bracket polynomial at the special value $A = -1$. From this point of view, this explains the relationship of the bracket polynomial with $SL(2, C)$ that is found in the work of Bullock (1996). It also suggests the possibility of generalising Bullock's work to the corresponding quantum group, just as the introduction to the variable A into the bracket polynomial corresponds to the movement from $SL(2, C)$ to the quantum group. (See Kauffman (1991) for more details on the relationship of the bracket polynomial to the $SL(2)$ quantum group.)

The bracket polynomial provides a natural generalization or deformation of the classical spin networks where the polynomial variable A becomes a deformation parameter. In fact, this mode of generalization carries over on all levels of the structure. The group $SL(2, C)$ naturally associated with the spin networks is generalized to a corresponding quantum group (Hopf algebra). The flat networks projected into the plane become woven networks in three-dimensional space. The symmetrizers and apparatus of recoupling theory have braided analogues that can be expressed purely diagrammatically and in terms of the quantum groups. For the details of this theory the reader can consult (Kauffman 1991; Kauffman and Lins 1994; Carter *et al.* 1995; Crane *et al.* 1997; Carter *et al.* 1997).

One of the key facts about the classical spin networks is the Penrose Spin Geometry Theorem (Penrose 1969) which states that for sufficiently large and well-behaved networks, the properties of three-dimensional space (in particular properties of angles) begin to emerge of their own accord from the network structure. In this way the networks provide a combinatorial background for the emergence of properties of space and time. The emergence of space–time from such networks is not yet fully articulated.

With the Spin Geometry Theorem in mind, it is quite significant to see what is added by going to the topological spin nets with their deformation parameter

A. When A is a root of unity, the topological networks can be used to describe invariants of three-dimensional manifolds. These invariants are defined in terms of surgery on links in three-space, but the prescription of the three-manifold structure is actually purely combinatorial in terms of the link diagrams. The evaluation of these diagrams in terms of topological spin networks is accomplished via generalized recoupling theory, and it expresses everything in terms of evaluations of trivalent networks with appropriate spin assignments on their edges. In this way, certain small spin networks encode deep topological properties of three-dimensional manifolds. These results should be regarded as adding another dimension to the philosophy behind the Spin Geometry Theorem.

The first combinatorial appearance of these three-manifold invariants was in the work of Reshetikhin and Turaev (1991) who showed how to construct invariants of three-manifolds from invariants of knots and links by using the representation theory of quantum groups. In the case of the quantum group associated with $SU(2)$ (equivalently $SL(2,C)$) it is possible to rephrase the Reshetikhin–Turaev work in terms of spin networks associated with the bracket polynomial. It is to this version of the invariant that we refer when we speak of the spin network invariant of three-manifolds.

Simultaneously and independently of Reshetikhin and Turaev, Edward Witten devised a presentation of three-manifold invariants via functional integrals. Witten's work adds a significant dimension to the spin network approach. His integral formula is expressed as follows:

$$Z(M^3,K) = \int dA e^{(ik/4\pi)\int_M \mathrm{tr}(AdA+(2/3)A^3)}\mathrm{tr}(Pe^{\oint_K A}).$$

The integration is over all gauge potentials modulo gauge equivalence for $SU(2)$ gauge at the fundamental representation. A is a gauge potential on three-dimensional space with values in this representation of $SU(2)$. The Lagrangian,

$$L = \int_M \mathrm{tr}(AdA + (2/3)A^3),$$

provides the formally correct weighting factor against which to integrate the Wilson loop, $\mathrm{tr}(Pe^{\oint_K A})$, to obtain a function $Z(M^3,K)$ of a three-manifold M containing a link K. This function is, up to a normalization related to framing, an invariant of the three-manifold, link pair. We shall refer to $Z(M^3,K)$ as the Witten invariant of the three-manifold. In particular, we shall write

$$Z(M^3) = \int dA e^{(ik/4\pi)\int_M \mathrm{tr}(AdA+(2/3)A^3)}$$

for the corresponding functional, defined on a three-manifold without a specified link embedded within it.

The functional integral is, at this point in time, a purely formal approach to this invariant of links and three-manifolds. Nevertheless it does follow from Witten's work that the invariant under discussion must be identical to that invariant that we have described as computed from generalized spin networks associated with the bracket polynomial. (If $A = e^{i\pi/2r}$, then the coupling constant k in Witten's integral corresponds to r (up to a fixed finite shift).) The verification of this fact is indirect, relying on an understanding of the behaviour of the Witten invariant under surgery on links in the three-manifold. It is an open problem to derive this relationship between the Witten invariant and the spin network invariant by an appropriate discretization of the functional integral. Such a derivation would shed light on the nature of both the spin networks and the integration process.

Once the invariant has been expressed in terms of the functional integral, many conjectures and relationships about the spin network evaluations come forth. For example, the large k limit of the functional integral is asymptotically approximated by sums over flat gauge potentials, leading to specific formulas for the asymptotic approximation of the spin nets as the order of the root of unity goes to infinity. See (Freed and Gompf 1991; Jeffrey 1992; Lawrence 1995; Rozansky 1995). These conjectures have been verified in many special cases, but the general problem remains open. It is quite possible that this problem will be elucidated by an appropriate generalisation of the chromatic method of spin network evaluation (Penrose 1971; Kauffman 1992; Kauffman and Lins 1994).

Another remarkable relationship is noted by Witten (1988/89). The product of $Z(M^3)$ and its complex conjugate can be expressed as an integral that naturally interprets as a functional integral for 2+1 quantum gravity with a cosmological constant:

$$\begin{aligned} |Z(M^3)|^2 &= \int dAdBe^{(ik/4\pi)\int \operatorname{tr}(AdA+(2/3)A^3-BdB-(2/3)B^3)} \\ &= \int ded\omega e^{i\int \operatorname{tr}(eR+(\lambda/3)e^3)} \end{aligned}$$

where $e = (k/8\pi)(A-B)$, $\omega = (1/2)(A+B)$, $R = d\omega + \omega^2$ and $\lambda = (4\pi/k)^2$. Here e is interpreted as a metric, while ω is interpreted as a connection so that R is the curvature.

We know that this functional integral is a topological invariant (up to standard normalization) of the 3-manifold M, and furthermore the invariant can be computed by spin network methods from any triangulation of M via the Turaev–Viro state summation (Turaev and Viro 1992) (see also Kauffman and Lins 1994). The first hint that such a state summation might be related to gravity appeared in the work of Ponzano and Regge (1968), and Hasslacher and Perry (1981). These authors investigated analogous state sums using approximations to classical recoupling coefficients. They found that the state summations give approximations to simplicial quantum gravity in the Regge calculus. The Chern–Simons functional integral formulation shown above suggests that there should be a simplicial quantum gravity with cosmological constant that corresponds to the topological invariant $|Z(M^3)|^2$. This means that there *should* be an appropriate formulation of 2+1 quantum gravity that can be done simplicially and that involves either the deformed spin networks or (equivalently) the quantum group corresponding to $SL(2,C)$ with deformation parameter a root of unity. These are open problems.

Here we see that 2+1 quantum gravity is related to a three-dimensional topological quantum field theory. The analogous notion for 3+1 quantum gravity is equally tantalizing and even more unfulfilled due to the lack, at the present time, of combinatorial models for topological quantum field theories in dimension four. For speculations along these lines we refer the reader to Crane and Frenkel (1994), Carter *et al.* (1997) and Crane *et al.* (1997). It is very inviting to consider the possibility that amplitudes for quantum gravity in 3+1 will correspond to relative topological invariants of four-dimensional manifolds.

Finally, we must mention the long-standing work of Ashtekar, Rovelli and Smolin that starts with the loop representation of the Ashtekar formulation of quantum gravity. In this theory of quantum gravity the wave functions ψ are functions $\psi(A)$ of (the equivalent of) a gauge potential A on three-dimensional space. Smolin and Rovelli consider the *loop transform*

$$\hat{\psi}(K) = \int dA\psi(A)\mathrm{tr}(Pe^{\oint_K A}).$$

This formal transform rewrites the theory in terms of functionals on knots and links in three-dimensional space. Recent work (Smolin 1996) develops quantum gravity directly in terms of knots, links and spin networks. It remains to see how this version of quantum gravity will interface with topological quantum field theories in dimension four.

Acknowledgement This paper is an expanded version of Kauffman (1995). It gives the author pleasure to thank the National Science Foundation for support of this research under NSF Grant DMS –2528707.

Bibliography

Alexander, J. W. (1923). Topological invariants of knots and links. *Trans. Amer. Math. Soc.*, **20**, 275–306.

Bullock, D. (1996). Rings of $SL_2(C)$ characters and the Kauffman bracket skein module. Preprint.

Carter, J. S., Flath, D. E. and Saito, M. (1995). *The Classical and Quantum 6j-Symbols.* Mathematical Notes, **43**. Princeton University Press.

Carter, J.S., Kauffman, L. H. and Saito, M. (1997) *Diagrammatics, Singularities and Their Algebraic Interpretations.* To appear in Conference Proceedings of the Brasilian Mathematical Society.

Crane, L. and Frenkel, I. (1994). Four dimensional topological quantum field theory, Hopf categories and canonical bases. *J. Math. Physics*, **35**, 5136–5154.

Crane, L., Kauffman, L. H. and Yetter, D. (1997). State sum invariants of 4-manifolds. *Journal of Knot Theory and Its Ramifications.* To appear.

Freed, D. S. and Gompf, R. E. (1991). Computer calculation of Witten's three-manifold invariant. *Commun. Math. Phys.*, **141**, 79–117.

Hasslacher, B. and Perry, M.J. (1981). Spin networks are simplicial quantum gravity. *Phys. Lett.*, **103 B**.

Jeffrey, L. C. (1992). Chern–Simons–Witten invariants of lens spaces and torus bundles, and the semi-classical approximation. *Commun. Math. Phys.*, **147**, 563–604.

Jones, V. F. R. (1985). A polynomial invariant of links via von Neumann algebras. *Bull. Amer. Math. Soc.*, **129**, 103–112.

Kauffman, L. H. (1987). State models and the Jones polynomial. *Topology*, **26**, 395–407.

Kauffman, L. H. (1989). Statistical mechanics and the Jones polynomial. *AMS Contemp. Math. Series*, **78**, 263–297.

Kauffman, L. H. (1991). *Knots and Physics.* World Scientific Pub, Singapore.

Kauffman, L. H. (1992). Map coloring, q-deformed spin networks, and Turaev–Viro invariants for 3-manifolds. *Int. J. of Modern Phys. B*, **6**, Nos. 11, 12, 1765–1794.

Kauffman, L. H. (1995). Spin networks and the bracket polynomial. To appear in the Proceedings of the Summer School in Knot Theory, Warsaw.

Kauffman, L. H. and Lins, S. L. (1994). *Temperley–Lieb Recoupling Theory and Invariants of 3-Manifolds.* Annals of Mathematics Study **114**. Princeton University Press.

Lawrence, R. (1995). Asymptotic expansions of Witten–Reshetikhin–Turaev invariants for some simple 3-manifolds. *J. Math. Phys.*, **36**, (11), 6106–6129.

Penrose, R. (1969). Angular momentum: An approach to combinatorial spacetime. In *Quantum Theory and Beyond*, Bastin, T. (editor). Cambridge University Press.

Penrose, R. (1971). Applications of negative dimensional tensors. In *Combinatorial Mathematics and Its Applications*, Welsh, D. J. A. (editor). Academic Press.

Ponzano, G. and Regge, T. (1968). Semiclassical limit of Racah coefficients. In *Spectroscopic and Group Theoretical Methods in Theoretical Physics.* North Holland, Amsterdam.

Reidemeister, K. (1933). *Knotentheorie.* Springer, Berlin. Chelsea Pub. Co., N.Y. (1948).

Reshetikhin, N. Y. and Turaev, V. (1991). Invariants of three-manifolds via link polynomials and quantum groups. *Invent. Math.*, **103**, 547–597.

Rozansky, L. (1995). Witten's invariant of 3-dimensional manifolds: loop expansion and surgery calculus. In *Knots and Applications*, Kauffman, L. (editor). World Scientific Pub, Singapore.

Smolin, J. (1996). The geometry of quantum spin networks. Preprint, Center for Gravitational Physics and Geometry, Penn. State University, University Park, PA.

Turaev, V. G. and Viro, O. Y. (1992). State sum invariants of 3-manifolds and quantum $6j$ symbols. *Topology*, **31**, 865–902.

Witten, E. (1988/89). 2+1 gravity as an exactly soluble system. *Nuclear Physics* B **311**, 46–78.

Witten, E. (1989). Quantum field theory and the Jones polynomial. *Commun. Math. Phys.*, **121**, 351–399.

20
The Physics of Spin Networks

Lee Smolin
Center for Gravitational Physics and Geometry, Department of Physics, The Pennsylvania State University, University Park, PA, USA 16802 [1]

> *"A reformulation is suggested in which quantities normally requiring continuous coordinates for their description are eliminated from primary consideration. In particular, since space and time have therefore to be eliminated, what might be called a form of Mach's principle must be invoked: a relationship of an object to some background space should not be considered—only the relationships of objects to each other can have significance."*

—Roger Penrose, *Theory of quantized directions*, [1]

1 Introduction

Perhaps the most important, and most robust, result of the study of non-perturbative approaches to quantum gravity is that the space of states of a large set of theories may be described in terms of spin networks [3; 4], which were originally invented by Roger Penrose in the early 1960s [1; 2]. In this contribution I would like to honour Roger by telling the story of how it is that his spin networks, originally invented as a simple model for discrete quantum geometry, have come to occupy a central place in quantum gravity.

Briefly, a spin network is a graph, whose edges are labelled by the representations of some group, G. Its vertices are associated with finite dimensional vector spaces, whose basis elements correspond to the different ways that the incoming representations can be invariantly combined. In the case that Roger studied, and which has had the most application so far in quantum gravity, the group is $SU(2)$, but generalizations of spin networks have played a role in gauge theories and topological field theory and are under study in quantum gravity.

After telling the story of how spin networks came into quantum gravity, I will close with a few comments on how they may play a role in future developments.[2]

---

[1]smolin@phys.psu.edu

[2]This article is condensed and modified from the original version at the request of the editors. The original, called "The future of spin networks" may be found on gr-qc and will appear elsewhere. For more information about spin networks themselves, please see the article by Lou Kauffman in this volume or [1; 2; 4; 7; 17].

2 Spin networks in non-perturbative quantum gravity

The story of how spin networks entered quantum gravity is a tale that illustrates the unity and interconnections that have existed beneath the surface of contemporary theoretical physics, despite the unfortunate divisions into subfields and camps. As Roger has been involved in more than one of the strands of the story, this is a story both about his influence and the universality of some of the central ideas that have formed our understanding of gauge fields and gravity.

The first strand follows Roger's work from spin networks to twistor theory. There he discovered a very curious fact,[3] which is the importance of self-duality for an understanding of the dynamics of the gravitational field. For in spite of the fact that the Einstein's equations are duality invariant, the restriction to either the self-dual or anti-self-dual sector leads to an exactly solvable system, whose solutions can be described completely in terms of consistency conditions for the existence of certain complex manifolds. The simplification of field equations to the self-dual sector is equally profound and important for Yang-Mills theory, and indeed the self-dual or instanton solutions play an essential role now in both mathematics and physics. By another basic and profound fact all this is related to another of the strands of Roger's work, which is the expression of general relativity in terms of spinors. This is because the duality transformation in four spacetime dimensions induces chiral transformations that takes left handed spinors into right handed.

All of this suggested that the dynamics of general relativity itself might be simplified were it to be expressed in terms of chiral variables. The first concrete realization of this emerged in two crucial papers of Amitaba Sen, [24], in which he found that the constraints of the Hamiltonian formulation of general relativity took very simple polynomial forms when expressed in terms of the self-dual (or left-handed) parts of the connection and curvature. A number of us puzzled over that paper, but it took Abhay Ashtekar to realize the full import of what it implied, which was in fact the possibility of a Hamiltonian formalism for general relativity, [25], in which the configuration variable was exactly the self-dual part of the spacetime connection, A_a^{AB}, while the momentum variable is the frame field $\tilde{E}^a_{AB}$. In this formalism the constraints take the polynomial forms found by Sen; this then became the basis for much of the revival of work in quantum gravity.

Not surprisingly, this led directly to a new understanding of the self-dual sector. The already simple formulas of the canonical theory simplified still further when one restricted to the self-dual sector by setting F_{ab}^{AB}, the curvature of A_a^{AB}, to zero, as was discovered by Ted Jacobson and developed in [26] and [27].

The Ashtekar formalism is sometimes seen to be primarily a development of the Hamiltonian theory, but it led immediately to a reformulation of the Lagrangian approach as well. Indeed, it led to more than one, as the first action principle in terms of self-dual variables, [28], led the way to the discovery of

[3] Also noticed and exploited by Ted Newman and Jerzy Plebanski.

formulations in which the metric does not even appear [29].

Of course, the Ashtekar formalism had profound implications for quantum gravity, but to trace these we must return to lattice gauge theory. One of the inventors of lattice gauge theory was Polyakov, who then went on to try to express *QCD* in terms of loop variables in the continuum [30]. One of the strongest memories I have from graduate school is a seminar given by Sasha Polyakov that he began by announcing his hope to solve *QCD* exactly by expressing it purely as a theory of loops. A different approach to this idea was also developed by Migdal and Makenko [31]. While Polyakov's hope has not so far been realized,[4] these papers were the inspiration for a number of developments, no less in quantum gravity than in other areas.

Among these was an attempt to model quantum general relativity as a lattice gauge theory [32] in which the spacetime connection played the role of the gauge field. The idea of this early lattice formulation of quantum gravity was to explore non-perturbative approaches to quantum gravity. It was particularly motivated by conjectures of Wilson [33] Parisi [34] and Weinberg [35] that perturbatively non-renormalizable theories might in fact exist were there to be non-trivial fixed points of their renormalization groups.[5]

These were good ideas, but at the time they led nowhere. In fact during the early 1980s most of us working in quantum gravity were wasting our time (or at least spending it poorly, considering what we might have been doing) with various perturbative formulations. My own return to non-perturbative quantum gravity came with work with Louis Crane, most of it never published, in which we tried to develop a background independent form of string theory based on loops variables, which were dual to either the spacetime metric or connection. This work was inspired primarily by the papers of Polyakov and Migdal and Makenko. It was also motivated by work Crane and I had done on quantum gravity on fractal spacetimes, in which we understood that quantum general relativity might exist if non-perturbative effects lowered the apparent dimensionality of spacetime, as seen by the scaling behaviour of propagators above the Planck scale [37]. As a result, we were looking for a way to describe a non-smooth structure for quantum geometry, in which the effective dimension of space at Planck scales would appear to be less than three, in terms of loops.

As soon as the Ashtekar formalism appeared, it was clear that this was the way to realize these ideas. The thing to do was to construct some kind of discrete geometry from Wilson loops made from the Sen-Ashtekar connection. First, with Paul Renteln, we made a lattice formalism [38]. This has one big disadvantage, which is that it is impossible to realize diffeomorphism invariant states on it [39] but as a tool for understanding both the state space and the action of the Hamiltonian, it has been used to good effect since, most of all by

[4] Except for the case of the large N limit where progress can be made. Recently, in the supersymmetric case this has led to the very interesting developments in terms of matrix models.

[5] For an attempt to realize this in quantum gravity, see [36].

Ezawa [40], Loll [41] and Gambini and Pullin [42]. Then with Ted Jacobson we began investigating a continuum formalism. There we had a wonderful surprise, which is the discovery of an infinite class of physical states—exact solutions to the Hamiltonian constraints [43]. Moreover, we found that the Hamiltonian constraint acts in a simple way on states made from Wilson loops, with an action that is concentrated at points of intersections of the loops.

One key question we faced in this work with Jacobson was what was the actual space of states of the theory. It was clear that the states on which the Hamiltonian constraint had a simple action were not Fock states. This was good as Fock states depend on a background metric, which doesn't exist in a non-perturbative formalism. Even if we could make sense of it, the background metric would interfere with the action of the diffeomorphisms of space, which are supposed to be the gauge group, just as much as a lattice does. On the other hand, what was the alternative? We knew we didn't want to use a lattice regularization, and we weren't aware of any other choice besides the lattice regularization or Fock space to define the space of states of a quantum field theory. Something new was needed. To invent it, we were guided by simple physical ideas. First, was the physical picture going back to Penrose and spin networks, that at the Planck scale the structure of space and time should be discrete. This picture had been reinforced by the renormalization group point of view, which suggested that to realize the conjecture that the theory is defined through a non-trivial fixed point, it was necessary that quantum geometry be based on sets of lower dimensionality below the Planck scale.

Second, we took over the relationship between quantization of non-abelian electric flux and Wilson loops coming from QCD. This picture comes from ideas of Holgar Nielson and others, who had taught us to think of the vacuum of QCD as something like a dual superconductor [44]. In a superconductor the magnetic flux is quantized, so the flux through any surface comes only in integer units of a quantum of flux. In confinement we know that the non-abelian electric flux forms tubes whose energy is proportional to their length, which is also the case for the quantized magnetic flux lines in a superconductor. It is then natural to think that in the vacuum the non-abelian electric flux is quantized.

Taken together these ideas suggested that a discrete quantum geometry might be something like an idealized form of the QCD vacuum, but without any background structure, so that the flux associated with the spacetime connection would be quantized. To realize this picture, we thought about working, not with a Fock space, but with a space of states spanned by a basis, each of which was made of finite products of Wilson loops. Thus, we considered the kinematical state space $\mathcal{H}_{kinematical}$ spanned by the overcomplete basis,

$$\Psi_{\gamma_i}[A] = \prod_i T[\gamma, A] \tag{2.1}$$

where γ_i is any finite set of loops and $T[\gamma, A] = TrPe^{\int_\gamma A}$ is the Wilson loop, or traced holonomy.

States of the form (2.1) are exactly those in which the non-abelian electric field flux is quantized. That is, if we identify the operator for non-abelian electric flux through a surface S as $E(S)$, then we have

$$\hat{E}(S)\Psi_\gamma(A) = Int[\gamma, S]^2 \Psi_\gamma(A) \tag{2.2}$$

in the case that there are only simple intersections of the loop and the surface, i.e. the loop does not intersect itself at the surface. (Here $Int[\gamma, S]$ is the intersection number of the loop and the surface.)

It was thus natural to think that the discrete states (2.1) represent a discrete geometry. Furthermore, it was obvious immediately that if the diffeomorphism constraint could be solved on this space of states, the resulting set of states would be labelled by diffeomorphism classes of loops, which is to say knots, links and, most generally, networks [45]. Thus, knot theory immediately emerged as being important for understanding the state space of quantum gravity!

These ideas were later formalized by people more mathematical than ourselves, in the language of rigorous quantum field theory [46; 47; 17].[6] But I think it is important to emphasize that the roots of these constructions were in these simple physical ideas, which came from QCD, renormalization group arguments, and speculations about the Planck scale.

In fact, it took several years to realize the whole picture. For one thing one had to give a good definition of the operator $\hat{E}(S)$. Formally it looks like,

$$\hat{E}(S) = \int_S \sqrt{\tilde{E}_i^a \tilde{E}_i^b n_a n_b} \tag{2.3}$$

where n_a is the unit normal of S. The problem is how to define the operator product and square root. To do this one needs a regularization procedure, and all known regularization procedures depend on a background metric. The question is then whether one can define it through a regularization procedure such that diffeomorphism invariance is not broken, and the operator takes diffeomorphism invariant states to diffeomorphism invariant states. It took some time before a way to do this was found [23]. By this time I had realized that the non-abelian electric field flux (2.3) was none other than the area of the surface S [23]. Thus, in quantum gravity, discreteness of area corresponds exactly to confinement in QCD.[7]

---

[6]For a review of the present state of the mathematical development of these ideas, see [48]. For a demonstration of the equivalence of the formulation of Ashtekar, Lewandowski, Marlof, Mourãu and Thiemann with the earlier formulation of [22; 23; 4; 3], see [49].

[7]It is very interesting to speculate whether the correct way to formulate QCD rigorously should not be in terms of the discrete measure [47] which formalizes the notion of a discrete state space given in [43]. The problem is that, before the diffeomorphism invariance is moded out, the space of states is non-seperable. This corresponds to an unphysical situation in which any displacement at all of a loop results in an orthogonal state; as a result the theory has too many degrees of freedom. On the other hand, QCD, like quantum gravity, clearly cannot be constructed from Fock space. One interesting conjecture to consider is then that QCD cannot be defined rigorously without coupling it to quantum gravity, so that diffeomorphism invariance reduces the space of spin network states to a countable basis.

Finally, if one puts in all the constants, the Wilson loop of the Sen-Ashtekar connection is actually

$$T[\gamma, A] = e^{G \int_\gamma A} \tag{2.4}$$

where G is Newton's constant.[8] Thus, the quanta of area is proportional to $\hbar G = l^2_{Planck}$.

The second thing was to construct the diffeomorphism invariant states which required the loop representation. The idea to do this by changing to a representation in which the states were functions of loops was due to Carlo Rovelli; once we had the idea it did not take long to work it out [22].[9]

Another important operator is the volume of a region of space $V[R]$. It was immediately clear from formal expressions that if it could be defined it would be a discrete like area and the discrete eigenvalue would count something happening at points where three or more loops meet. This took a long time to work out [23] and it was this problem that led to the introduction of spin networks in quantum gravity.[10] Of course, this was known, and even mentioned at times [23] but it was unfortunately not exploited earlier. The main reason was that, unfortunately, the discovery of physical states associated with non-intersecting loops [43; 22] had pushed the question of what happened at the intersections into the background—even though it was clear, and emphasized by several people,[11] that the actions of important operators including volume, the extrinsic curvature and the Hamiltonian constraint were concentrated at the intersections.

In fact Roger had a lot to do with the realization that spin networks are important for quantum gravity. I was at a workshop in Cambridge trying to define and diagonalize the volume operator, and at some point realized that maybe the diagrammatic techniques Roger had developed to calculate with spin networks could help. I went to him and he showed me some tricks, which I used to find that the trivalent spin networks were eigenstates of the volume operator. With Carlo Rovelli we then worked out the action of all the operators we had on spin network states and found that these were even simpler than in terms of loops. Unfortunately, in the case of volume we got the eigenvalues wrong—they are zero for all the trivalent networks, as was pointed out later by Renata Loll [51].[12] In any case, we had finally realized that the central kinematical concept in

[8]Strictly speaking all we know is that G is proportional to Newton's constant. The dimensionless constant of proportionality may be thought of as a finite multiplicative renormalization factor. Such a factor is in any case expected as there ought to be a finite renormalization between the bare Planck scale that goes into the kinematics of the theory and the renormalized Planck scale that comes from the macroscopic gravitational interaction. Related issues have been raised by Imirzi and Rovelli and Thiemann.

[9]The loop representation had already been invented for QCD by Gambini and Trias [50]; a lot of time could have been saved had we been aware of it earlier.

[10]Spin networks states had previously been used in lattice gauge theory [5; 6] because they give an independent basis for the states of the form (2.1). They were also used in topological field theory [7; 8; 9] and conformal field theory [10; 11; 12; 13; 14] for the same reason.

[11]Especially Berndt Bruegmann and Jorge Pullin.

[12]And also realized independently by Georgio Imirzi and Michael Reisenberger.

quantum gravity is that the space of diffeomorphism invariant states is spanned by a basis in one to one correspondence with embeddings of spin networks.[13]

The transformation to the loop representation can be done directly in the spin network basis [4]. When one mods out by spatial diffeomorphisms one is left with a state space $\mathcal{H}_{diffeo}$ which has an independent basis in one to one correspondence with diffeomorphism classes of imbeddings of spin networks,[14] which may be labelled $|\{\Gamma\}>$. Once we had this space it was immediate that there is a space of exact solutions, given by those spin networks without nodes. There are other sets of exact solutions, which include intersections, some of which have been known for a long time [43; 52], others of which were found recently by Thiemann [53; 54]. Thus, it seemed that Polyakov's dream that reducing a theory to loops leads to its exact solutions, is to some extent realized in quantum gravity.

Is the expression of quantum gravity in terms of spin networks an important idea, or just a technical convenience? I believe it is fundamental, probably even more fundamental than the idea that the states come from applying a quantization procedure to the infinite dimensional space of connections modulo gauge transformations. There are at least four reasons to believe this. First, we have arrived at a kinematics for quantum gravity that is discrete and combinatorial, and it seems likely that such structures are fundamental, while continuum concepts such as connections are artifacts of the myth that space is continuous. In fact, at the level of spatially diffeomorphism invariant states the connections have completely disappeared. There is only a space of states spanned by a basis $|\Gamma>$, where Γ now stands for a diffeomorphism class of spin networks. The space of states has a natural inner product

$$<\Gamma|\Gamma'>= N_\Gamma \delta_{\Gamma\Gamma'} \tag{2.5}$$

where N_Γ is a norm which is related to what Penrose originally called the evaluation of the spin network [55; 1]. At this level all operators are combinatorial and topological, there is no role for a continuum concept such as a connection.

Second, the diffeomorphism classes of spin networks are somewhat more complicated than the corresponding combinatorial and topological classes. While it might seem at first that the diffeomorphism equivalence classes of networks imbedded in a spatial manifold are labelled only by their topology and connectivity, when the nodes have sufficiently high valence this is actually not the case. In the case of three dimensions, nodes with five or more incident edges require continuous parameters to label their diffeomorphism classes [62]. If one really believes that the theory is derived from a classical theory in the continuum all these should be included. But if, on the other hand, one believes that the fundamental structures are discrete, and the continuum is only an approximation, one

[13]There have been many calculations of the spectra of volume, area and length. See, for example [16; 55; 56; 57; 58; 41; 59; 60; 61].

[14]Note that even though we have gotten rid of the dependence on connections, the states still are not equivalent under the recoupling identities of ordinary spin networks.

might like to consider as meaningful only those labels of spin networks which are combinatorial or topological. Of course, the classes labelled by continuous parameters might be needed if any physically meaningful operator was known that measured those parameters, but so far none is known. Moreover, even if such an observable existed, it is likely it could be expressed as well as a slightly less local operator without the continuous parameters. For these reasons it seems best to consider the theory defined by spin networks defined only up to combinatorics and topology.[15]

Third, there are difficulties if we take too seriously the idea that the description of states and operators in terms of spin networks is in fact the result of a derivation from the continuum theory. Some of these have to do with difficulties of the diffeomorphism regularization procedures that have, so far, been developed [23; 22; 3; 64; 56; 65; 59; 60; 53]. In all of the proposals so far made, a background metric is introduced which is used to parameterize a family of point split operators. The problem is to define a diffeomorphism invariant operator, which must have no dependence on the background metric, in the limit that the regulator is removed, bringing the operators together.

There is a very nice thing about these constructions, which is that when they succeed in constructing a diffeomorphism invariant operator, that operator is necessarily finite [23]. The reason is that any divergence, if present, is measured in units of the background metric. If the result of taking the limit in which the regulator is removed is an operator that does not depend on the background metric it cannot be proportional to any divergent quantity; it must then be finite. In fact, all cases that have been worked out go exactly like this, and this may be counted as one of the successes of non-perturbative quantum gravity: diffeomorphism invariance is sufficient to guarantee finiteness of operator products, defined through such regularization procedures.

While this works simply for the case of the area operator, two kinds of problems appear when it is applied to more complicated cases including the volume, Hamiltonian constraint and Hamiltonians.[16] The first is ambiguity; different regularization procedures result in different diffeomorphism invariant operators. This is of course nothing new, it afflicts all quantum field theories. The second problem is more serious, it is that in these cases one must use highly non-trivial operator orderings in order to achieve a diffeomorphism invariant operator. For instance, in the loop or spin network representation the limit is taken along families of operators that measure various features of the loops at intersections, beyond that information which is gathered by those operators that appear in the naive transcription of the corresponding classical quantity. There can be no ob-

[15] A related argument has been raised [63] concerning even some of the topological information, that concerned with the embedding of the network in the spatial manifold. If the discrete structure is really prior to the manifold, then perhaps imbedding information ought not to play a role in the fundamental theory.

[16] In the canonical formalism a Hamiltonian is obtained whenever the time part of the gauge invariance is fixed.

jection, at least in the loop representation to the insertion of such operators, but it makes the constructions highly non-trivial.[17] As a result it is far from clear what real advantage comes from taking seriously the programme of deriving the quantum theory from the classical theory, especially as the quantum theory is believed in reality to be the exact description, while the classical description should be only an approximation. It is as if one tried to derive Newtonian mechanics by a systematic procedure from Ptolemy's astronomy.

Yet another problem is that all forms of the Hamiltonian constraint so far developed have a problem with the continuum limit, in that the physical degrees of freedom are too localized in finite regions of networks, and do not propagate in a way that can lead to long range correlations in a continuum limit [66].

The last reason to take the spin network description as more fundamental than the classical connections is that $SU(2)$ spin networks immediately generalize to a large class of cases which furthermore makes contact with conformal field theory, topological field theory and, through them, to string theory. Furthermore, the cases that make contact with conformal field theory are necessarily related to quantum groups which do not correspond in any exact sense to classical connections.

3 Future directions

In retrospect it is not surprising that spin networks have come to play a central role in quantum gravity, as this is a consequence of only the kinematical gauge invariances of the theory, which are rotations of the local frame fields and spatial diffeomorphism invariance. Because of the local frame rotations, which are manifestations of the equivalence principle, the degrees of freedom of general relativity, and a large class of generalizations and extensions, including the various versions of supergravity, are connections. But as far as we know, quantum mechanically a connection can only be in one of a few phases. The ones that are understood presently are the Coulomb phase, the Higgs phase, the confined phase and the topological phase. However, both the Coulomb and Higgs phase are incompatible with diffeomorphism invariance. This is because in these phases the connection manifests itself in terms of weakly coupled particles like the photon and these require the existence of a fixed background metric for their description. This is necessary both for the definition of the state space, which requires a splitting of the modes into positive and negative frequency components and for the inner product.

---

[17] In the connection representation [43; 47] the situation is not as good because the additional operator dependence needed cannot be expressed in terms of the basic operators involving the connection (and not the loops directly) without additional operator products which themselves need regularization. So it is not clear that an honest point split regularization can be achieved in the connection representation; one may then have no resort but to invent a category of "state dependent" regularization procedures. This problem does not occur in the loop representation because there one can construct completely well defined local operators that measure the support of the loop directly, such as $\hat{\dot{\gamma}}^a(x)|\gamma> \equiv \int ds \dot{\gamma}^a(s)\delta^3(\gamma(s),x)|\gamma>$.

The remaining phases, which are the confined and the topological, may be compatible with diffeomorphism invariance. The reason, essentially, is, as described above, the discreteness of the quantized electric field flux, which allows states to be counted, and hence an inner product to be constructed, in the absence of a background metric.

However, exactly because of this discreteness, a basis of states for either a confined or topological phase of a gauge theory may be given in terms of spin networks.

The main features of quantum general relativity, such as the discreteness of area and volume, all follow from these simple facts. As such, these are purely kinematical, and do not depend on the assumption that the dynamics of the theory is given by general relativity. At the same time, this reasoning explains why it is that spin networks also play a role in topological quantum field theory. It also explains why they are also important for understanding conformal field theories, which describe the behaviour of boundary observables in topological quantum field theory. As such, spin networks are likely to play a crucial role in future developments in quantum gravity, which may come from fitting together the different approaches of string theory, topological quantum field theory and non-perturbative quantum gravity.

Finally, it must be mentioned that the objects likely to be of importance in future developments are *quantum* spin networks, which differ from ordinary spin networks in that there are additional phase degrees of freedom associated with twisting the edges [7]. In fact, it is quantum spin networks or, as they are sometimes called, "ribbon graphs" [14; 18; 19; 20; 21] that come into topological quantum field theory and conformal field theory. They also come into quantum gravity in the presence of a cosmological constant [15; 16] and are the key to the role of Chern-Simons theory in quantum gravity [67].

Interestingly enough, quantum spin networks arise out of deep and basic mathematics, which is the theory of tensor categories [14; 18; 19; 20; 21; 68]. From this perspective, they are universal structures that arise in representation theory of rather general structures such as Hopf algebras, which encompass both ordinary Lie groups and quantum groups. However, in the absence of a classical metric or any fixed smooth structure, any non-perturbative quantum theory of gravity is likely to be based on finite combinatorics, algebra and representation theory. The fact that the quantum spin networks are basic to the mathematical vocabulary of these fields provides yet another, and perhaps the strongest, reason for believing that they are likely to continue to play a basic role in our understanding of quantum gravity.

Acknowledgements It is a pleasure to thank John Baez, Louis Crane, Ted Jacobson, Louis Kauffman, Carlo Rovelli, Roumen Borissov, Seth Major and, or course, Roger Penrose, for teaching me most of what I know about spin networks (as well as a great deal more that I forgot.) I am indebted also to Roumen Borissov, Kirill Krasnov and Fotini Markopoulou for many conversations about the future direction of this subject as well as for comments on a draft of this paper. I would also like to thank Nick Khuri and Daniel Amati for hospitality at The Rockefeller University and SISSA during the course of this work. This work was supported by the NSF grant PHY-93-96246 to The Pennsylvania State University.

Bibliography

[1] Penrose, R. Theory of quantized directions. Unpublished manuscript.

[2] Penrose, R. (1971). In *Quantum theory and beyond*, ed. T. Bastin, Cambridge University Press; Penrose, R. (1979). In *Advances in Twistor Theory*, ed. L.P. Hughston and R. S. Ward (Pitman) p. 301; Penrose, R. (1971). In *Combinatorial Mathematics and its Application*, ed. D. J. A. Welsh, Academic Press.

[3] Rovelli, C. and Smolin, L. (1995). Discreteness of area and volume in quantum gravity. *Nuclear Physics*, **B 442**, 593. Erratum: (1995) *Nucl. Phys.*, **B 456**, 734.

[4] Rovelli, C. and Smolin, L. (1995). Spin networks and quantum gravity. gr-qc/9505006, *Physical Review*, **D 52**, 5743-5759.

[5] Kogut, J. and Susskind, L. (1975). *Phys. Rev.* **D 11**, 395.

[6] Furmanski, W. and Kowala, A. (1987). *Nucl. Phys.*, **B 291**, 594.

[7] Kauffman, L. and Lins, S. (1994). *Tempereley-Lieb Recoupling Theory and Invariants of 3-Manifolds.* Princeton University Press, and references therein.

[8] Witten, E. (1989). Quantum field theory and the Jones Polynomial. *Commun. Math. Phys.*, **121**, 351.

[9] Turaev, V. and Viro, O. (1992). *Topology*, **31**, 865.

[10] Crane, L. (1991). *Commun. Math. Phys.*, **135**, 615; *Phys. Lett.*, **B 259**, 243.

[11] Verlinde, E. (1988). Fusion rules and modular transformations in 2D conformal field theory. *Nucl. Phys.*, **B 300**, 360.

[12] Moore, G. and Seiberg, N. (1988). Classical and quantum conformal field theories. *Comun. Math. Phys.*, **123**, 177.

[13] Moore, G. and Reshetikhin, N.Yu. (1989). A comment on quantum group symmetry in conformal field theory. *Nucl. Phys.*, **B 328**, 557.

[14] Reshetikhin, N. Yu. and Turaev, V. G. (1990). Ribbon graphs and their invariants derived from quantum groups. *Commun. Math. Phys.*, **127**, 1; (1991). Invariants of 3-manifolds via link polynomials and quantum groups. *Invent. Math.*, **103**, 547.

[15] Major, S. and Smolin, L. (1996). Quantum deformation of quantum gravity. gr-qc/9512020, *Nucl. Phys.*, **B 473**, 267.

[16] Borissov, R., Major, S. and Smolin, L. (1996). The geometry of quantum spin networks, gr-qc/9512043. *Class. and Quant. Grav.*, **12**, 3183.

[17] Baez, J. (1996). Spin networks in gauge theory. *Adv. Math.* **117** (1996), 253-272, gr-qc/941107; (1996). Spin networks in nonperturbative quantum gravity. In *The Interface of Knots and Physics*, ed. Louis Kauffman, American Mathematical Society, Providence, Rhode Island, gr-qc/9504036.

[18] Joyal, A. and Street, R. (1986). Braided monoidal categories. Macquarie Mathematics Reports, no. 860081; (1991). The geometry of tensor calculus I. *Adv. Math.*, **88**, 55.

[19] Yetter, D. N. (1990). Quantum groups and representations of monoidal categories. *Math. Proc. Cam. Phil. Soc.*, **108**, 261.

[20] Reshetikhin, N. Yu. (1990). Quasitriangular Hopf algebras and invariants of tangles. *Leningrad Math. J.*, **1**, 491.

[21] Chari, V. and Pressley, A. (1994). *A Guide to Quantum Groups*, Cambridge University Press. For the generalization of spin networks to monoidal categories, see Chapter 5.

[22] Rovelli, C. and Smolin, L. (1988). *Phys. Rev. Lett.*, **61**, 1155; (1990). *Nucl. Phys.*, **B133**, 80.

[23] Smolin, L. (1992). In *Quantum Gravity and Cosmology*, ed. J. Pérez-Mercader *et al.*, World Scientific, Singapore.

[24] Sen, A. (1981). On the existence of neutrino zero modes in vacuum spacetime. *J. Math. Phys.*, **22**, 1781; Gravity as a spin system. *Phys. Lett.*, **B11**, 89.

[25] Ashtekar, A. (1986). *Phys. Rev. Lett.*, **57**, 2244; *Phys. Rev.*, **D36**, 1587.

[26] Ashtekar, A., Jacobson, T. and Smolin, L. (1988). *Commun. Math. Phys.*, **115**, 631.

[27] Mason, L. and Newman, E. T. (1989). *Commun. Math. Phys.*, **121**, 659.

[28] Jacobson, T. and Smolin, L. (1988). *Phys. Lett.*, **B196**, 39; *Class. Quan. Grav.*, **5**, 583; Samuel, J. (1987). *Pramana-J Phys.*, **28**, L429.

[29] Capovillia, R., Dell, J. and Jacobson, T. (1989). *Phys. Rev. Lett.*, **63**, 2325.

[30] Polyakov, A. (1979). *Phys. Lett.*, **82B**, 247; *Nucl. Phys.*, **B 164**, 171.

[31] Makenko, Yu. M. and Migdal, A. A. (1979). *Phys. Lett.*, **88B**, 135.

[32] Smolin, L. (1979). Quantum gravity on a lattice. *Nucl. Phys.*, **B148**, 333.

[33] Wilson, K. G. (1973). *Phys. Rev.*, **D 70**, 2911.

[34] Parisi, G. (1975). *Nucl. Phys.*, **B100**, 368; On nonrenormalizable interactions. Preprint IHES/P/76/148 (1976).

[35] Weinberg, S. (1979). In *General Relativity: An Einstein Centenary Survey*, ed. S. W. Hawking and W. Israel, Cambridge University Press.

[36] Smolin, L. (1982). A fixed point for quantum gravity. *Nucl. Phys.*, **B208**, 439.

[37] Crane, L., and Smolin, L. (1986). Renormalizability of general relativity on a background of spacetime foam. *Nucl. Phys.*, **B267**, 714; (1985). Spacetime foam as a universal regulator, *Gen. Rel. and Grav.*, **17**, 1209.

[38] Renteln, P. and Smolin, L. (1989). *Class. Quant. Grav.*, **6**, 275.

[39] Renteln, P. (1990). *Class. Quantum Grav.*, **7**, 493–502.

[40] Ezawa, K. (1996). *Mod. Phys. Lett.*, **A11**, 349; gr-qc/9510019; Nonperturbative solutions for canonical quantum gravity: an overview; gr-qc/9601050.

[41] Loll, R. (1995). *Nucl. Phys.*, **B444**, 619; (1996). *Nucl. Phys.*, **B460**, 143.

[42] Fort, H., Griego, J., Gambini, R., Pullin, J. (1996). Lattice knot theory and quantum gravity in the loop representation, gr-qc/9608033; Gambini, R., and Pullin, J., gr-qc/9603019.

[43] Jacobson, T. and Smolin, L. (1988). *Nucl. Phys.*, **B 299**.

[44] Nielson, H. B. (1976). Dual strings. Published in Scottish Summer School, 465 (Bohr Inst.), NBI-HE-74-15.

[45] Smolin, L. (1988). In *New Perspectives in Canonical Gravity* by A. Ashtekar, with invited contributions, Bibliopolis, Naples.

[46] Ashtekar, A. and Isham, C. J. (1992). *Class. and Quant. Grav.*, **9**, 1069.

[47] Ashtekar, A., Lewandowski, J., Marolf, D., Mourãu, J., and Thiemann, T. (1995). Quantization of diffeomorphism invariant theories of connections with local degrees of freedom. *J. Math. Phys.*, **36**, 519, gr-qc/9504018.

[48] Ashtekar, A. chapter 11 in this volume.

[49] De Pietri, R. (1996). On the relation between the connection and the loop representation of quantum gravity, gr-qc/9605064.

[50] Gambini, R. and Trias, A. (1981). *Phys. Rev.*, **D23**, 553; (1983). *Lett. al Nuovo Cimento*, **38**, 497; (1984). *Phys. Rev. Lett.*, **53**, 2359; (1986). *Nucl. Phys.*, **B278**, 436; Gambini, R., Leal, L., and Trias, A. (1989). *Phys. Rev.*, **D39**, 3127; Gambini, R. (1991). *Phys. Lett.*, **B 255**, 180.

[51] Loll, R. (1995) *Nucl. Phys.*, **B444**, 619; (1996) *Nucl. Phys.*, **B460**, 143.

[52] Husain, V. (1989). *Nucl. Phys.*, **B313**, 711.

[53] Thiemann, T. (1996). Quantum spin dynamics I. Harvard preprint, gr-qc/9606089.

[54] Thiemann, T. (1996). Quantum spin dynamics II. Harvard preprint, gr-qc/9606090.

[55] De Pietri, R. and Rovelli, C. (1996). Geometry eigenvalues and scalar product from recoupling theory in loop quantum gravity, gr-qc/9602023. *Phys. Rev.*, **D54**, 2664; Frittelli, S., Lehner, L., and Rovelli, C. (1996). The complete spectrum of the area from recoupling theory in loop quantum gravity, gr-qc/9608043.

[56] Borissov, R. (1996). Ph.D. thesis, Temple.

[57] Thiemann, T. (1996). A length operator in canonical quantum gravity. Harvard preprint, gr-qc/9606092.

[58] Thiemann, T. (1996). Closed formula for the matrix elements of the volume operator in canonical quantum gravity. Harvard preprint, gr-qc/9606091, gr-qc/9601038.

[59] Ashtekar, A. and Lewandowski, J. (1996). Quantum geometry I: area operator, gr-qc/9602046.

[60] Lewandowski, J. (1996). Volume and quantization, gr-qc/9602035.

[61] Borissov, R., de Pietro, R., and Rovelli, C. (1997). preprint in preparation.

[62] Grott, N. and Rovelli, C. (1996). Moduli space structure of knots with intersections. *J. Math. Phys.*, **37**, 3014, gr-qc/9604010.

[63] Markopoulou, F. personal communication.

[64] Rovelli, C. and Smolin, L. (1994). *Phys. Rev. Lett.*, **72**, 446.

[65] Borissov, R. (1996). Graphical evolution of spin network states. gr-qc/9605.

[66] Smolin, L. (1996). The classical limit and the form of the hamiltonian constraint in non-pertubative quantum gravity. CGPG preprint, gr-qc/9609034.

[67] Kodama, H. (1990). *Phys. Rev.*, **D 42**, 2548.

[68] Crane, L. and Yetter, D. (1994). On algebraic structures implicit in topological quantum field theories. Kansas preprint, (1994); (1993). In *Quantum Topology*, World Scientific, p. 120; Crane, L. and Frenkel, I. B. (1994). *J. Math. Phys.*, **35**, 5136–54.

PART III

Parallel Session II: Geometry and Gravity

21

The Sen Conjecture for Distinct Fundamental Monopoles

Gary Gibbons

Departments of Applied Mathematics and Theoretical Physics, University of Cambridge, Silver St, Cambridge CB3 9EW, England

Abstract

I review some recent progress in understanding the interactions of BPS monopoles for higher rank groups. S-duality predicts the existence of a middle–dimensional square integrable harmonic form on the moduli space of distinct fundamental BPS monopoles of an arbitrary Lie group. These harmonic forms are exhibited and their uniqueness is discussed. The properties of distinct fundamental monopoles are contrasted with those of identical $SU(2)$ monopoles.

1 Introduction

It seems fitting, in a conference devoted to 'Geometrical Issues in the Foundations of Science' to illustrate by means of a simple example how the geometrical methods and ideas to which Roger Penrose has made so many contributions may be used to discuss the low energy interactions of k Bogomol'nyi monopoles and hence to check some important and far reaching conjectures concerning the S-duality of $N = 4$ supersymmetric Yang–Mills theory. This is all the more appropriate because

- The main idea is to work on the $4k$ dimensional moduli space $\mathcal{M}_k$ of classical solutions of the Bogolmol'nyi equations.
- The natural Riemannian metric on $\mathcal{M}_k$ is HyperKähler and thus satisfies the vacuum Einstein equations, i.e. it is Ricci flat.
- In the case $k = 2$ the relative moduli space is four-dimensional and self–dual, i.e. it is the real Riemannian slice of a "non-linear graviton".

Moreover, given the metric:

- Classical mechanics corresponds to geodesic motion on $\mathcal{M}_k$, and
- (Supersymmetric) Quantum mechanics corresponds to Hodge–de Rham theory on $\mathcal{M}_k$.

As we shall see, these yield to some elementary techniques familiar to anyone who has studied general relativity.

2 S-duality

We are interested in $N = 4$ SUSY Yang–Mills theory in four-dimensional Minkowski space–time. The bosonic Lagrangian is

$$-\frac{1}{4e^2}\mathrm{Tr}F\wedge\star F - \frac{\theta}{32\pi^2}F\wedge F - \frac{1}{2}|D_A\mathbf{\Phi}|^2 \tag{2.1}$$

where the Higgs field Φ is in the adjoint representation of the gauge group G. The coupling constants e and θ may be encoded in the complex number

$$\tau = \frac{\theta}{2\pi} + \frac{4\pi i}{e^2} \tag{2.2}$$

which takes values in the upper half plane. As a consequence of $N = 4$ supersymmetry, which dictates the fermion content of the theory, τ is invariant under the renormalization group flow as is the vacuum expectation value of the Higgs field, which we denote by Φ_∞. The data G, τ, and Φ_∞ should completely characterize the quantum theory. We are interested in the case of maximal symmetry breaking when the gauge group G is broken down to its maximal torus T^r, with $r =$ rank G. The quantum states of the theory are labelled by an electric charge vector $\mathbf{e}$ belonging to the root lattice Λ and a magnetic charge vector $\mathbf{g}$ belonging to the co-root lattice $\Lambda^\star$. The vacuum expectation value Φ_∞ may be regarded as defining a vector and an associated orthogonal hyperplane in the Cartan sub-algebra. Thus there is a unique set of simple positive roots β_1 and co-roots $\beta_i^\star = \frac{\beta_i}{\beta_i.\beta_i}$. One has

$$\mathbf{e} = e\sum_i n_i^{(e)}\beta_i \tag{2.3}$$

and

$$\mathbf{g} = \frac{4\pi}{g}\sum_i n_i^{(m)}\beta_i^\star \tag{2.4}$$

where $(n_i^{(m)}, n_i^{(e)})$ are integers.

The mass M of all states should satisfy the Bogomol'nyi bound

$$M \geq |(\mathbf{e}+\tau\mathbf{g}).\mathbf{\Phi}_\infty|. \tag{2.5}$$

States attaining the bound are said to be BPS states and they are invariant (in this context) under the action of half of the total supersymmetry generators.

3 Weak $\leftrightarrow$ Strong Coupling

The Olive–Montonen conjecture as extended by Sen (1994) states that the group $SL(2,\mathbb{Z})$ acts on the quantum mechanical Hilbert spaces of the set of theories parameterized by τ permuting the states but preserving the spectrum of the Hamiltonian. The action on τ is fractional linear:

$$\tau \rightarrow \frac{a\tau+b}{c\tau+d} \tag{3.1}$$

where the integers a, b, c, d satisfy $ab - cd = 1$ and the action on the electric and magnetic charges is

$$(n_i^{(m)}, n_i^{(e)}) \to (dn_i^{(m)} - bn_i^{(e)}, -cn_i^{(m)} + an_i^{(e)}). \tag{3.2}$$

In particular electric states at strong coupling are taken to magnetic states at weak coupling. Note that the Bogomol'nyi bound is invariant under this action of $SL(2, \mathbb{Z})$ and so S-duality takes BPS states to BPS states. The insight of Sen was to realize that by making a plausible assumption about the electric states at strong coupling one obtains a prediction about weak coupling which may be checked using semi-classical methods. For the case of two identical $SU(2)$ monopoles the predictions have been shown to hold true. For many $SU(2)$ monopoles the detailed metric is not known but Segal and Selby (1996) have managed, using topological arguments, to establish the existence of at least as many harmonic forms as Sen predicts. Recent work has been directed to what turns out to be a more accessible case: that of distinct fundamental monopoles for higher rank groups.

4 Bound states at threshold

The Bogomol'nyi mass formula tells us that the mass of a BPS state associated with a positive composite root α, i.e. a positive integer combination of the simple positive roots, is the same positive integer combination of the masses of the fundamental electric BPS states associated with the simple positive roots M_{β_i}. Thus for example:

$$M_{\beta_1+\beta_2} = M_{\beta_1} + M_{\beta_2}. \tag{4.1}$$

Therefore at strong coupling $k = \text{rank } G$ distinct fundamental electric states for which $n_i^{(e)} = (1, 1, \ldots, 1)$ may form an electric BPS bound state at threshold. If the S-duality hypothesis is correct then $k = \text{rank } G$ distinct fundamental BPS monopoles for which $n_i^{(m)} = (1, 1, \ldots, 1)$ should also form a bound state at threshold at weak coupling.

To see whether this is indeed the case we need to investigate supersymmetric quantum mechanics on the moduli space of classical solutions. In what follows I shall confine myself to the simplest case for which $G = SU(k+1)$. More complicated cases are treated in the papers quoted in the references, in particular three recent reviews of the subject are Lee (1996); Yi (1996); Weinberg (1996).

5 The moduli space

It is known in general for solutions of the Bogomol'nyi equations that one can take out the centre of mass motion and an overall phase, i.e. $\mathcal{M}_k$ splits isometrically as a product:

$$\mathcal{M}_k = (\mathbb{R}^3 \times \mathbb{R} \times \mathcal{M}_{k-1}^{\text{rel}})/D \tag{5.1}$$

where D is a discrete subgroup of the isometry group. The relative moduli space $\mathcal{M}_{k-1}^{\text{rel}}$ is also HyperKähler and its isometry group contains an $SO(3)$ subgroup which rotates the three complex structures I, J and K. In the case of *distinct* fundamental monopoles one has in addition a triholomorphic action of the torus group T^{k-1}. It is this fact which, by contrast with the much studied case of many identical $SU(2)$ monopoles, allows one to find the metric exactly and discuss its properties. It implies that the metrics are determined entirely by the large separation interactions of the monopoles. These are Coulomb like and contain none of the exponential terms which arise in the case of identical $SU(2)$ monopoles from the charge changing exchange of massive vector bosons. Moreover, unlike identical monopoles, the moduli space is topologically trivial. In fact, as we shall see, distinct fundamental monopoles behave with respect to one another, but not with respect to themselves, just like point particles.

General arguments show that the metric has the form

$$ds^2 = G_{ij} d\mathbf{r}_i.d\mathbf{r}_j + G_{ij}^{-1}(d\theta_i + \mathbf{W}_{im}.d\mathbf{r}_m)(d\theta_j + \mathbf{W}_{jn}.d\mathbf{r}_n) \tag{5.2}$$

where the configuration space metric G_{ij} and the one-forms $\mathbf{W}_{im}$ satisfy a set of linear Lindstrom–Rocek–Pedersen–Poon equations and $X^i = \frac{\partial}{\partial\theta_i}$ are the Killing vector fields generating the torus action. The reduced configuration space coordinates $\mathbf{r}_i$ correspond physically to the relative positions of the monopoles and mathematically to the moment maps of the torus action with respect to the three complex structures.

Because of the torus action geodesics will have conserved momenta or charges q_i associated to the Killing vectors X^i. The motion projected onto the reduced configuration space, i.e. the quotient $\mathcal{M}^{\text{rel}}_{k-1}/T^{k-1}$ is governed by the effective Lagrangian

$$L_{\text{eff}} = \frac{1}{2} G_{ij} \mathbf{v}_i.\mathbf{v}_j - \frac{1}{2} G_{ij}^{-1} q_i q_j + q_i \mathbf{W}_{im}.\mathbf{v}_m. \tag{5.3}$$

It should be clear that there is exactly as much information in the metric as there is in the effective Lagrangian. Moreover given the metric G_{ij} on the reduced configuration space the Lindstrom–Rocek–Pedersen–Poon equations enable one to deduce the connection forms $\mathbf{W}_{ij}$. Thus to specify the metric locally it suffices to give G_{ij}.

The case $k = 2$ is the simplest because the isometry group is $U(2)$. This and the fact that there is a spherically symmetric configuration with two distinct monopoles sitting on top of one another allowed Gauntlett and Lowe (1996) and independently Lee, Weinberg and Yi (1996a) to use the classification of complete self–dual Bianchi IX metrics to argue that the metric is the self–dual Taub–NUT metric (with positive mass parameter) on $\mathbb{R}^4$. The same result was already known to Connell (1996).

Next Lee, Weinberg and Yi (1996b) proposed (based on work by Manton and myself on the asymptotic form of the metric for many identical $SU(2)$ monopoles (Gibbons and Manton 1995)) an exact form for arbitrary many distinct fundamental monopoles. The basic idea is to write down the appropriate L_{eff} by considering the interactions of the monopoles at large separation and then to reconstruct the metric. In the case of identical $SU(2)$ monopoles this only gives the approximate metric at large separations but in the case of distinct fundamental monopoles Lee, Weinberg and Yi pointed out that the answer should be exact.

The essential point is to choose an appropriate basis for the relative positions. This amounts to finding a suitable basis in the root space or Cartan subalgebra t^k. A basis of roots α_a may be chosen with a labelling $a = 1, 2, \ldots, k$ and consequent ordering such that the only non-vanishing inner products

$$\lambda_i = -\alpha_a^\star . \alpha_b^\star \tag{5.4}$$

with respect to the Killing form between any two co-roots are just for those which are adjacent with respect to the ordering. For $G = SU(k+1)$ one may think of the roots associated to the vertices of the standard Dynkin diagram. The monopoles interact only if the associated vertices are connected by an edge in the Dynkin diagram. The $k-1$ relative position vectors $\mathbf{r}_i$ are associated with the $k-1$ edges of the Dynkin diagram. In fact one may identify $\mathbf{r}_i$ with $\mathbf{x}_a - \mathbf{x}_b$, where $\mathbf{x}_a$ is the physical position of the monopole associated to the root α_a. In the absence of interactions the metric G_{ij} on the spatial projection $\mathbb{R}^{3(k-1)} \equiv \mathcal{M}^{\rm rel}_{k-1}/T^{k-1}$ would be flat and given by the $(k-1)\times(k-1)$ symmetric reduced mass matrix

$$G_{ij} = \mu_{ij}. \tag{5.5}$$

Note that, for a general set of monopole masses $M_{\alpha_i} = \frac{4\pi}{e}\alpha_i.\mathbf{\Phi}_\infty$, the reduced mass matrix μ_{ij} will not be diagonal. To incorporate interactions one adds a positive diagonal contribution

$$\operatorname{diag} \frac{g^2}{8\pi}\frac{\lambda_i}{r_i}. \tag{5.6}$$

Shortly after the paper of Lee, Weinberg and Yi (1996b), Murray (1996) confirmed that the Lee,Weinberg and Yi or L-W-Y metric coincides with the natural metric on the set of Nahm data for this problem, that the manifold is topologically trivial, and that the metric is complete on $\mathbb{R}^{4(k-1)}$.

Recently Goto, Rychenkova and myself (Gibbons *et al.* 1996) have provided an elementary construction of the L-W-Y metric as a HyperKähler quotient of the flat metric on $\mathbb{H}^{k-1}\times\mathbb{H}^{k-1}$ with respect to an action of $\mathbb{R}^{k-1}$. This construction makes clear the topology and completeness and allows a simple desciption of what happens in the limit, discussed first by Lee, Weinberg and Yi (1996c) when one or more of the simple positive roots β_i becomes orthogonal to the Higgs expectation value. Physically, because the fundamental monopole masses are given by $M_{\beta_i} = \frac{4\pi}{e}\beta_i.\mathbf{\Phi}_\infty$, this corresponds to one or more fundamental monopoles becoming massless and entails a consequent enhancement of the unbroken gauge group to become non-abelian.

The HyperKähler $\mathbb{R}^{k-1} = (t_i, t_2, \ldots t_{k-1})$ action that we use is of the form:

$$q_n \to q_n \exp it_n \tag{5.7}$$

$$w_n \to w_n + \sum_m \lambda_n{}^m t_m, \tag{5.8}$$

where (q_n, w_n) are quaternionic coordinates for $\mathbb{H}^{k-1} \times \mathbb{H}^{k-1}$, i is a unit pure imaginary quaternion, and the real matrix $\lambda_n{}^m$ defines $k-1$ translation vectors in $\mathbb{R}^{k-1}$. The reduced mass matrix is given in terms of the matrix of inner products of the $k-1$ translation vectors

$$\mu = \left(\lambda^{-1}\right)^t \lambda^{-1}. \tag{5.9}$$

Thus if the translations are all finite and linearly independent then the reduced mass matrix will be non-singular. Massless monopoles arise if some of the translations tend to infinity. Infinitely heavy monopoles arise when not all the translations are linearly independent.

6 Harmonic forms

The low energy behaviour of BPS monopoles should be governed by $N = 4$ supersymmetric quantum mechanics. Restricted to the relative moduli space the Hilbert space is the direct sum of the spaces of L^2 differential forms. The four anti-commuting supersymmetry operators may be identified with d, $\overline{\partial}_I$, $\overline{\partial}_J$, and $\overline{\partial}_K$ where d is the usual exterior derivative and ∂_I is the Dolbeault operator associated to the complex structure I, etc. The Hamiltonian becomes the Hodge–de Rham operator

$$\Delta = dd^\dagger + d^\dagger d = 2(\overline{\partial}_I \overline{\partial}_I^\dagger + \overline{\partial}_I^\dagger \overline{\partial}_I) \ldots \text{etc.} \tag{6.1}$$

BPS bound states at threshold therefore correspond to L^2 harmonic forms which on a complete manifold are, by a standard theorem, closed and co-closed. Moreover they should be holomorphic with respect to all three complex structures I, J and K.

To someone brought up on general relativity it is well known that for each Killing vector field X, one may contract with a Ricci flat metric g to obtain a Killing 1-form $A = g(X, \ )$ and an exact 2-form

$$F = dA \tag{6.2}$$

which is co-closed, i.e. it satisfies Maxwell's equations. This basic result comes by contracting the identity, valid for any Killing vector field X,

$$X^{\alpha;\beta;\gamma} = R^{\alpha\beta\gamma}{}_\sigma X^\sigma \tag{6.3}$$

with the metric $g_{\beta\gamma}$ and using the Ricci flatness. For brevity I have written out this equation in the traditional (abstract) index notation used in general relativity. Thus $_;$ denotes the Levi–Civita covariant derivative and therefore, for a Killing vector field $X^{\alpha;\beta} = -X^{\beta;\alpha}$. The identity is obtained by writing out three cyclic permutations of the Ricci identity:

$$X^\alpha{}_{;\beta;\gamma} - X^\alpha{}_{;\gamma;\beta} = R^\alpha{}_{\sigma\gamma\beta} X^\sigma \tag{6.4}$$

and taking a suitable linear combination, making use of the symmetries of, and one of the Bianchi identities for, the Riemann tensor $R^{\alpha}{}_{\beta\gamma\sigma}$.

If the Killing vector field X is holomorphic with respect to some complex structure then F will also be holomorphic with respect to that complex structure because d and $\overline{\partial}$ anticommute.

In the present case we have a triholomorphic action of the torus group T^{k-1} generated by the Killing fields $X^i = \frac{\partial}{\partial\theta_i}$. Thus the relative moduli space $\mathcal{M}^{\mathrm{rel}}_{k-1}$ is naturally equipped with $k-1$ triholomorphic Maxwell 2-forms (Gibbons 1996). If $k = 2$ this construction gives the well known L^2 harmonic 2-form on the Taub-NUT metric on $\mathbb{R}^4$. Although it is exact by construction, the length of the Killing field X tends to a constant at infinity and so is not in L^2. Thus while the 2-form F is trivial from the point of view of absolute cohomology, from the point of view of L^2 cohomology it is non-trivial. However it is not detected by the Atiyah–Patodi–Singer version of the Hirzebruch Index theorem with boundary terms applied to a large ball in $\mathbb{R}^4$ (Gibbons *et al.* 1979). A similar L^2 harmonic 2-form exists on the four dimensional multi–Taub–NUT metrics.

In fact this 2-form is precisely the 2-form found earlier by a more direct calculation by Gauntlett and Lowe (1996) and independently by Lee, Weinberg and Yi (1996a) and identified by them as the bound state at threshold for two fundamental $SU(3)$ monopoles of distinct types whose existence is predicted by S-duality. The problem of finding the harmonic forms for larger values of k was left unanswered in those papers.

If $k > 2$ one has $k-1$ candidate 2-forms $F^i = d(g(X^i, \ ))$ but none is in L^2. This is because for large separations r_i the squared magnitude of the form falls off like $\frac{1}{r_i^4}$ but the volume grows like Πr_i^3. Now arbitrary wedge products of closed and co-closed forms are closed and co-closed. Thus one may construct a large number of even–dimensional harmonic forms on Lee–Weinberg–Yi space by taking the wedge products of the $k-1$ 2-forms. However it is clear that the only one which is in L^2 is the $2k-2$ form obtained by wedging together $k-1$ distinct 2-forms. Moreover this middle–dimensional form is the exterior derivative of a $2k-3$ dimensional form which is not in L^2. Thus we have an obvious candidate for the a BPS bound state at threshold predicted by S-duality (Gibbons 1996).

7 Uniqueness

Although arguments have been given for the uniqueness of the middle dimensional harmonic form on the Atiyah–Hitchin metric and on the Taub–NUT metric, there is at present no clean general argument for the uniqueness of the middle dimensional forms predicted by S-duality. As we have seen the index theorem, at least in a straightforward form, does not seem to be powerful enough to detect L^2 harmonic forms which are exterior derivatives of a form which is itself not in L^2. Note that wedging a certain number of the 2-forms coming from the torus action with various powers of the three Kähler forms $\omega_I, \omega_J, \omega_K$ will not produce L^2 forms.

One possible handle on the uniqueness, suggested to me by Nigel Hitchin, is based on the fact that on a Hyperbolic Kähler manifold, i.e. one for which the Kähler form is the exterior derivative of a bounded 1-form, all L^2 harmonic forms must be middle–dimensional (Gromov 1991). This raises the question, interesting in its own right, of whether the Lee–Weinberg–Yi metrics, or indeed the metrics on the relative moduli space of many identical $SU(2)$ monopoles are Hyperbolic Kähler with respect to one of the Kähler forms ω. They are not. If it were true that

$$\omega = db \tag{7.1}$$

with b bounded, then

$$\omega^{2s} = d(b \wedge \omega^{4s-1}), \tag{7.2}$$

where the manifold is $4s$ dimensional. Now integrate over a large ball. For large radius, when the metric behaves like the product of s copies of $\mathbb{R}^3 \times S^1$, the integral on the left hand side behaves like Πr_i^3 while the integral on the right hand side behaves like Πr_i^2. This gives a contradiction.

It seems therefore that this approach will not work in any direct way. The uniqueness of the harmonic forms therefore remains an interesting open mathematical problem.

8 Geodesics

The classical mechanics of distinct fundamental monopoles is given by geodesic motion on the Lee–Weinberg–Yi metric. It is of interest to ask whether there are analogues of the bound states at threshold at the classical level. More generally one may ask whether there are any bound geodesics. A simple argument shows that there are not. It rests on the fact that the general configuration of distinct fundamental monopoles may be obtained by starting from the spherically symmetric configuration in which all of the monopoles sit on top of one another and pulling them apart. More precisely the Lee–Weinberg–Yi metric admits a vector field

$$V = \mathbf{r}_i \frac{\partial}{\partial \mathbf{r}_i} \tag{8.1}$$

which vanishes at the spherically symmetric configuration and which is distance increasing in the sense that the Lie derivative of the metric:

$$\mathfrak{L}_V g \tag{8.2}$$

is positive definite. In fact at large separations V behaves like a homothety with respect to the spatial coordinates $\mathbf{r}_i$. Thus moving along the trajctories of V corresponds to separating the monopoles.

It follows from the geodesic equations that if T is the tangent vector of the geodesic, then

$$\frac{dg(T,V)}{dt} > 0. \tag{8.3}$$

Now if there were a bound geodesic we could average this equation over a long time interval. The left hand side would tend to zero as the time interval increased but the right hand side would tend to some finite strictly positive number. This is a contradiction. Thus there can be no bound geodesics. In fact this argument is a more geometric version of an argument presented in (Gibbons 1996) which was essentially the familiar Virial Theorem of celestial mechanics.

The physical reason that there are no bound geodesics is easy to see: the electric forces which can be read off from L_{eff}, between monopoles with distinct charges are repulsive. This is unlike the case of two identical $SU(2)$ monopoles with a non-vanishing relative electric charge. In that case the electric forces are attractive and bound geodesics and even periodic geodesics are possible.

9 Bound states in the continuum

In the case of the Atiyah–Hitchin metric it is known that in addition to the continuum which starts at zero, the scalar Laplacian admits L^2 eigenstates with positive energy. In other words there are bound states in the continuum. It seems unlikely, on the basis of the analysis of geodesics, that these exist for the Lee–Weinberg–Yi metric. One can try to apply a Virial type argument to the Laplace operator to rule them out. The idea is to study how the Rayleigh quotient

$$R[f] = \frac{\int_{\mathcal{M}^{\text{rel}}_{k-1}} (\nabla f)^2}{\int_{\mathcal{M}^{\text{rel}}_{k-1}} f^2} \tag{9.1}$$

of a putative eigenfunction f behaves as it is dragged along the integral curves of the vector field V. Although mathematically precise estimates have yet to be completed it seems rather likely that one can always alter the value of the Rayleigh quotient in this way and thus there can be no eigenfunction.

This is interesting physically because it strongly indicates that the bound state at threshold owes its existence to an effect of supersymmetry and the fermions in the theory rather than being a feature of the purely bosonic sector.

Bibliography

Connell, S.A. (1996). The Dynamics of the $SU(3)$ Charge (1,1) Magnetic Monopole. University of South Australia Preprint.

Gauntlett, J.P. and Lowe, D.A. (1996). *Nucl Phys* **B 472**, 194. *hep-th/960185.*

Gibbons, G.W. (1996). *Phys Lett* **B 382**, 53–9. *hep-th/9603317.*

Gibbons, G.W. and Manton, N.S. (1995). *Phys Lett* **B 356**, 32–8.

Gibbons, G.W., Pope, C.N., and Romer, H. (1979). *Nucl. Phys.*, **157**, 377–86.

Gibbons, G.W., Rychenkova, P., and Goto, R. (1996). HyperKähler Quotient Construction of BPS Monopole Moduli Spaces. *Commun Math Phys.* In press. *hep-th/9608085.*

Gromov, M. (1991). *J Diff Geom*, **33**, 260–92.

Lee, K. (1996). The Moduli space of BPS Monopoles. *hep-th/9608185.*

Lee, K., Weinberg E.J., and Yi, P. (1996a). *Phys Lett* **376**, 97. *hep-th/9601097.*

Lee, K., Weinberg, E.J., and Yi, P. (1996b). *Phys Rev* **D54**, 1663. *hep-th/9602167.*

Lee, K., Weinberg, E.J., and Yi, P. (1996c). Massive and Massless Monopoles with Non-abelian Magnetic Charges. *hep-th/9605229.*

Murray, M. (1996). A Note on the $(1, 1, \ldots, 1)$ Monopole. *hep-th/9605054.*

Segal, G., and Selby, A. (1996). *Comm Math Phys* **177**, 775–87.

Sen, A. (1994). *Phys Lett* **B 329**, 217–21.

Weinberg, E.J. (1996). Massless and Massive Monopoles Carrying Nonabelian Magnetic Charges. *hep-th/9610065.*

Yi, P. (1996). Duality, Multi–Monopole Dynamics and Quantum Theshold Bound States. *hep-th/9608114.*

22

An Unorthodox View of GR via Characteristic Surfaces

Simonetta Frittelli, E. T. Newman
Department of Physics and Astronomy, University of Pittsburgh, Pittsburgh, PA 15260, USA

and

Carlos Kozameh
FaMAF, Universidad Nacional de Córdoba, 5000 Córdoba, Argentina

1 Introduction

We would like to describe a novel and unorthodox approach to GR that is based on the existence of special families of 3-surfaces in a four manifold. (The details, with proofs, have been given elsewhere (Frittelli *et al.* 1995a, 1995b, 1995c, 1995d).) However, before saying what this novel approach is, we discuss the background and some issues of motivation.

We (with many others) feel that finding the relationship between GR and quantum theory, QT, is (one of) the most fundamental problems of theoretical physics and perhaps one of the most difficult.

(1) We have heard some say it is so hard "don't even bother to try".
(2) Many want to treat it "as just another field theory to be quantized a la QED".
(3) Others want to do much tinkering with GR and perhaps a bit with QT.
(4) Some (probably a minority) think that most of the difficulties lie as much with QT (measurement and meaning of QT) as with GR, and that they can and will only be resolved with some unification of the two; but not simply via the "quantization" of GR (Penrose 1994, 1996).

We are probably closest to point of view #4. Though GR is clearly a field theory, it is conceptually so very different from most others. Almost all other field theories have a background space-time—a fixed stage, so to speak—on which to "live", but GR is a theory of the stage itself. People in the field are very well aware of this but we frequently feel that after this is acknowledged, people are not quite sure how to use it; GR on many levels appears to be so similar to usual theories. For example, it is a physical theory dependent on partial differential equations, and furthermore, by giving appropriate initial data, there is a perfectly fine Cauchy development theory—the equations for the metric field (with gauge conditions) can be put into symmetric hyperbolic form and the existence of unique evolution can be proven.

We, however, want to emphasize and discuss an area of great dissimilarity. It is this dissimilarity that we feel should be stressed in a discussion of the relationship of GR with QT. It suggests that there should be something qualitatively different in the solution to this problem as compared to the problem of the quantization of Lorentz invariant field theories. We wish to point out that we are not claiming any real progress on the basic issues of GR and QT, but only that we expect that seeing GR from a completely different perspective (Frittelli *et al.* 1996) will shed light on the problem.

In almost all Lorentz invariant field theories, the (physical) characteristic surfaces of the equations, i.e., the null surfaces (which determine the domains of dependence) are fixed by the Minkowski space geometry, they do not depend on the choice or solution of the field equations.

These three-surfaces are determined as solutions to the eikonal or massless Hamilton-Jacobi equation: $S = S(x, y, z, t) =$ constant, with

$$\eta^{ab} S_{,a} S_{,b} = (\partial S/\partial t)^2 - (\partial S/\partial x)^2 - (\partial S/\partial y)^2 - (\partial S/\partial z)^2 = 0$$

and possess a variety of different types of solution; light-cones, null planes, etc. More generally, for a metric $g^{ab}(x^c)$, signature $(+ - - -)$, we have the equation for $S(x^c)$

$$g^{ab} S_{,a} S_{,b} = 0. \tag{1.1}$$

In other words, given the $g^{ab}(x^c)$ we have, again, the eikonal equation for the characteristic surfaces but now they do depend on the solution for the metric field. The Einstein equations (when augmented by appropriate gauge conditions) are, in essence, a set of symmetric hyperbolic differential equations for the components of the metric $g^{ab}(x^c)$. Quite remarkably, as the $g^{ab}(x^c)$ are evolved, they simultaneously determine the domain of dependence of the data, i.e., they determine the characteristic surfaces, $S(x^c)$ (the boundaries of the domain of dependence), though without any explicit use of (1.1). Even though the characteristic surfaces are not explicitly used in the evolution, they play an essential, but hidden, role.

The point of view we want to adopt is the reverse of the above consideration. We want to consider the characteristic surfaces, $S(x^c)$, as more fundamental than the metric $g^{ab}(x^c)$ and find equations for the surfaces (equivalent to the Einstein equations) and then, finally, use (1.1) to try to determine the $g^{ab}(x^c)$ from the surfaces. If enough surfaces (a minimum of nine are needed) are given through each space-time point, then $g^{ab}(x^c)$ will be determined up to an overall factor (the "conformal factor"). We will thus need not only equations to determine the appropriate surfaces but also equations to determine the conformal factor.

2 The null surface formulation of GR

Our unorthodox approach to GR is based on the construction of differential equations for families of surfaces and a scalar function (the conformal factor), so that the surfaces are the characteristic surfaces of the metric derived from (1.1) and, further, this metric should automatically satisfy the vacuum Einstein equations.

2.1 Kinematics

We begin with a set of N, 3-surfaces in R^4 described by

$$u_A = constant = Z_A(x^a); \qquad A = 1....N.$$

The conditions for them to be null surfaces for *some* metric is that we be able to find algebraic solutions, $g^{ab}(x^c)$, for each A, to

$$g^{ab}(x^c)\nabla_a Z_A \nabla_b Z_A = 0.$$

It is obvious that for $N \leq 9$, $g^{ab}(x^c)$ exists, and that for $N > 9$, $g^{ab}(x^c)$ does not exist unless conditions are placed on the functions $Z_A(x^a)$.

We now consider a generalization of this: we chose a sphere's worth of Z's instead of the finite number of Z's; i.e., instead of the finite index A we consider the continuous index, the points on a sphere, coordinatized by the complex stereographic coordinate, $(\zeta, \bar{\zeta}) \in S^2$ and write

$$u = Z(x^a, \zeta, \bar{\zeta}).$$

The conditions for $u = Z(x^a, \zeta, \bar{\zeta})$ to define families of null-surfaces for each constant $(\zeta, \bar{\zeta})$ is the demand that a conformal metric $g^{ab}(x^a)$ exists such that, for each $(\zeta, \bar{\zeta})$, we have

$$g^{ab}(x^c)\nabla_a Z \nabla_b Z = 0. \tag{2.1}$$

Though it is not obvious and requires considerable effort to prove (Frittelli *et al.* 1995a), this demand implies that we obtain:

(1) a unique determination of $g^{ab}(x^c)$, up to a conformal factor, $\mathfrak{w}(x^a)$; i.e.,

$$g^{ab}(x^c) = \mathfrak{w}^2 \mathfrak{g}^{ab}[Z] \tag{2.2}$$

where $\mathfrak{g}^{ab}[Z]$ is an explicit functional of Z (Frittelli *et al.* 1995a), and

(2) two differential conditions on Z and $\mathfrak{w}(x^a)$ which we refer to as the metricity conditions, m_1 and m_2. As they are relatively complicated and their details are not needed here, we omit giving them explicitly (Frittelli *et al.* 1995a).

2.2 Imposing the Einstein equations

If equation (2.2) is substituted into the vacuum Einstein equations, followed by a fair amount of manipulation (Frittelli *et al.* 1995b), we obtain a single equation for $Z(x^a, \zeta, \bar{\zeta})$ and $\mathfrak{w}(x^a)$. (The fact that the ten Einstein equations are equivalent to just one equation, though at first surprising, is easily understood since it involves the six variables $(x^a, \zeta, \bar{\zeta})$ rather than just the usual four x^a.) Before displaying it, we first introduce some needed notation. Assuming that $Z(x^a, \zeta, \bar{\zeta})$ is a known function, we can construct the four known functions

$$\theta^i \equiv (u, R, \omega, \bar{\omega}) \equiv (Z, \bar{\eth}\eth Z, \eth Z, \bar{\eth} Z) = \theta^i(x^a, \zeta, \bar{\zeta})$$

with $i = 0, 1, +, -$ and $\eth$ and $\bar{\eth}$ being (essentially) the ζ and $\bar{\zeta}$ derivatives (Newman and Penrose 1966). From the θ^i we can construct the gradient basis $\partial_a \theta^i \equiv \theta^i_a$ and the dual basis θ^a_i and the associated directional derivatives $\partial_i \equiv \theta^a_i \partial_a$. In particular, we have $D \equiv \partial_1 = \theta^a_1 \partial_a$. The one Einstein equation then has the remarkably simple form

$$D^2 \Omega = Q[Z]\Omega \tag{2.3}$$

where

$$Q \equiv -\frac{1}{4q} D^2 \Lambda D^2 \bar{\Lambda} - \frac{3}{8q^2}(Dq)^2 + \frac{1}{4q} D^2 q \ , \qquad q \equiv 1 - D\Lambda D\bar{\Lambda}$$

with $\Lambda \equiv \eth^2 Z$ and

$$\Omega \equiv \mathfrak{w}\Omega_0.$$

We have used Ω instead of $\mathfrak{w}$; Ω_0 is a *known* functional of Z which has been introduced so that Equation (2.3) does not have a first derivative term. Equation (2.3), and the two metricity conditions, constitute the vacuum Einstein equations for the variable Z and $\mathfrak{w}$.

We first illustrate these equations with the special case of the self-dual Einstein equations—the analogues of the self-dual, complex, Maxwell equations with $\mathbf{E} = \mathrm{i}\,\mathbf{B}$.

2.3 Self-dual Einstein equations

We can choose as free characteristic data an *arbitrary* complex spin-2 function of three variables, the so-called Bondi shear,

$$\sigma(u, \zeta, \bar{\zeta}).$$

The conformal factor is taken as

$$\mathfrak{w} = 1 \tag{2.4}$$

and a differential equation for Z can be found, namely

$$\eth^2 Z = \sigma(Z, \zeta, \bar{\zeta}), \tag{2.5}$$

with $\eth^2 \equiv \frac{\partial}{\partial\zeta}(1 + \zeta\bar{\zeta})^2 \frac{\partial}{\partial\zeta} \cong \partial^2/\partial\zeta^2$. Equations (2.4) and (2.5) constitute the vacuum self-dual Einstein equations (Hansen *et al.* 1978) in the following sense.

Theorem 2.1

(a) *The general regular solution of Equation (2.5) contains four parameters; i.e., it is of the form*

$$u = Z(x^a, \zeta, \bar{\zeta});$$

i.e., the space-time coordinates enter as constants of integration.

(b) *The metric, (2.2); i.e., $g^{ab}(x^c) = \mathfrak{w}^2 \mathfrak{g}^{ab}([Z])$ obtained from (2.4) and (2.5) satisfies the vacuum Einstein equations identically.*

We point out that in the important special case of vanishing data, $\sigma(u, \zeta, \bar{\zeta}) = 0$, Z has the form

$$Z = Z_0 = \sum_{l=0,1} x_{l,m} Y_{lm}(\zeta, \bar{\zeta}) \tag{2.6}$$

with

$$x_{0,0} \equiv \sqrt{2\pi} t, \qquad x_{1,1} \equiv \sqrt{\tfrac{\pi}{3}}(x - iy),$$
$$x_{1,0} \equiv -\sqrt{\tfrac{2\pi}{3}} z, \qquad x_{1,-1} \equiv -\sqrt{\tfrac{\pi}{3}}(x + iy),$$

or

$$x_{l,m} \Leftrightarrow x^a$$

and, with $\mathfrak{w} = 1$, Equation (2.2) is the Minkowski metric.

2.4 Asymptotically flat GR

In the case of asymptotically flat space-times the metricity conditions can be manipulated so that they take the form

$$\eth^2 \bar{\eth}^2 Z = \bar{\eth}^2 \sigma(Z, \zeta) + \eth^2 \bar{\sigma}(Z, \zeta) + \mathfrak{D}(Z, \Omega) \tag{2.7}$$

with characteristic data, σ and $\bar{\sigma}$, as arbitrary spin-2 functions of $(Z, \zeta, \bar{\zeta})$. $\mathfrak{D}[Z, \Omega]$ is a (not very complicated) polynomial function of the derivatives of Z and $\ln \Omega$. (For the derivation of this result in linearized GR see Frittelli *et al.* 1995c. A forthcoming paper containing the derivation of the exact result is in preparation.) Eq. (2.7) with Eq. (2.3),

$$D^2 \Omega = Q([Z])\Omega,$$

are the vacuum Einstein equations for asymptotically flat space-times.

The linearization of the vacuum Einstein equations in this version consists in choosing $\Omega = 1$ and taking $\mathfrak{D}[Z, \Omega] = 0$ obtaining (Mason 1995; Frittelli *et al.* 1995b, 1995c)

$$\eth^2 \bar{\eth}^2 Z = \bar{\eth}^2 \sigma(Z_0, \zeta) + \eth^2 \bar{\sigma}(Z_0, \zeta) \tag{2.8}$$

with σ and $\bar{\sigma}$ now depending on the flat space Z_0, Eq. (2.6). As there is a simple Green's function for (2.8), we have the immediate solution to the linearized Einstein equations

$$Z_1 = Z_0(x^a, \zeta, \bar{\zeta}) + \int_{S^2} G(\zeta, \zeta') \Big(\bar{\eth}^2 \sigma(Z_0, \zeta) + \eth^2 \bar{\sigma}(Z_0, \zeta) \Big) dS'^2. \tag{2.9}$$

This becomes the basis for, in principle, a very simple perturbation expansion (Frittelli *et al.* 1995b). Since the Q in Eq. (2.3) is quadratic in Z, it turns out that perturbatively Eqs. (2.3) and (2.7) decouple; after (2.8) is solved (linear) the Z_1 is put into (2.3) giving a quadratic correction to Ω, which in turn is substituted into (2.7) leading to a Z_2 analogous to (2.9), etc.

There is a rather surprising indication that arises from the perturbation solution: namely, the coefficients of the first four spherical harmonics (the l=0,1 harmonics) of Z can be taken (Frittelli and Newman 1996) as a set of preferred (canonical) coordinates of the space-time—i.e., there is no (apparent) gauge freedom in the solutions. This conclusion can be extended as a non-perturbative result to, sufficiently weak, asymptotically flat exact vacuum solutions.

3 Discussion and applications

We will not go into any detail here concerning our reasons for believing this approach to GR is of potential importance nor into a detailed discussion of potential applications that we foresee. We will just mention several of the ideas.

- The variables Z and $\mathfrak{w}$ have such a different meaning than the conventional GR variables that it focuses attention on different structures; null surfaces, null geodesics and a scale factor become the primary geometric objects.
- It brings us (in the case of asymptotically flat space-times) to a formulation of GR that is very close to the D'Adahmar formulation of the massless Lorentz-invariant fields on Minkowski space—with important differences.
- Via the perturbation scheme one can hope to study (and perhaps solve) to second order the problem of classical scattering; i.e., of giving data on $\mathfrak{I}^-$ and determining the resulting "data" on $\mathfrak{I}^+$. Does data of compact support go to data of non-compact support? Does Fock-space data go to Fock-space data or to infrared data?
- A main point of interest to us is: what relationship does this version of GR have with attempts to "quantize" GR? If we formally apply the Ashtekar asymptotic quantization procedure to our version then the classical data, σ and $\bar{\sigma}$, become elevated to operators with a commutator algebra between them. A surprising result (where we have the details only in linear GR, see (Frittelli *et al.* 1996)) is that the x^a (using our canonical choice of coordinates) become operators with commutator relations. We do not know the significance of this result, though we point out that the result is not obtained in an ad hoc fashion; it is obtained by simply following Ashtekar's procedure with our variables.

Acknowledgements

We would like to thank members of the mathematical physics group in Oxford for valuable discussions and NATO collaborative research grant no. 950300 for support.

Bibliography

Frittelli, S. and Newman, E.T. (1996). Pseudo-Minkowskian coordinates on asymptotically flat-space-times, *Phys. Rev. D.*

Frittelli, S., Kozameh, C.K., and Newman, E.T. (1995a). Lorentzian metrics from characteristic surfaces. *J. Math. Phys.*, **36**, 4975.

Frittelli, S., Kozameh, C. K., and Newman, E. T. (1995b). GR via characteristic surfaces. *J. Math. Phys.*, **36**, 4984.

Frittelli, S., Kozameh, C. K., and Newman, E. T. (1995c). Linearized Einstein theory via null surfaces. *J. Math. Phys.*, **36**, 5005.

Frittelli, S., Kozameh, C. K., and Newman, E. T. (1995d). On the dynamics of characteristic surfaces. *J. Math. Phys.*, **36**, 6397.

Frittelli, S., Kozameh, C. K., Newman, E. T., Rovelli, C., and Tate, R.S. (1996). Fuzzy space-times from a null-surface view of GR, gr-qc/9603061.

Hansen, R.O., Newman, E. T., Penrose, R., and Tod, K. P. (1978). The metric and curvature properties of $\mathcal{H}$-space. *Proc. R. Soc. London*, **A363**, 445.

Mason, L. J. (1995). The vacuum and Bach equations in terms of light cone cuts. *J. Math. Phys.*, **36**, 3704.

Newman, E. T. and Penrose, R. (1966). Note on the Bondi-Metzner-Sachs group. *J. Math. Phys.*, **5**, 863.

Penrose, R. (1994). *Shadows of the mind*. Oxford University Press, Oxford.

Penrose, R. (1996). On gravity's role in quantum state reduction. *Gen. Rel. Grav.*, **28**, 581.

23

Amalgamated Codazzi–Raychaudhuri Identity for Foliation

Brandon Carter
Observatoire de Paris, 92 Meudon, France.

1 Introduction

I wish to thank the editors of this volume for the opportunity of again expressing appreciation for the beautifully geometric way of perceiving the physical world that Roger Penrose communicated to so many students of my generation. In particular it was Roger's emphasis (Penrose 1964) on the importance of features that are conformally invariant that led me, with his help, to develop the systematic use (Carter 1966) of 2-dimensional *conformal projections* (the Lorentz signature analogue of Mercator type projections in ordinary terrestrial mapping) of the kind that have since become widely and appropriately known as "Penrose diagrams". An outstanding example of the kind of conformally invariant structure whose analysis was specially developed and applied under Roger's leadership is that of null-geodesic foliations: in particular, it was his derivation, with Ted Newman (Newman and Penrose 1962) of the null limit of the famous divergence identity obtained originally for a timelike flow by Raychaudhuri (1957), that provided the essential tool for deriving the singularity theorems that were subsequently developed, first by Roger himself (Penrose 1965), and later on by Stephen Hawking and others (Hawking and Penrose 1970; Hawking and Ellis 1973).

It was the work on singularity theorems (Penrose 1965; Hawking and Penrose 1970; Hawking and Ellis 1973) that first drew my attention to the original Raychaudhuri equation, whose extension to higher dimensional foliations, in a manner recently suggested by Capovilla and Guven (1995b), is described in the present article. The unified treatment provided here shows how the extended Raychaudhuri identity is fraternally related to the correspondingly extended Codazzi identity. However it is left for future work to complete the corresponding Penrose program, in the sense of treating the corresponding conformally invariant limit, meaning the case of a foliation not by surfaces with a well behaved induced metric such as will be postulated in the present work (which is physically motivated by contexts such as that of neutron star vortex congruences) but by null surfaces (of the kind whose study has been developed by Barrabès and Israel (1991)).

The present article is a sequel to a previous Penrose festschrift contribution (Carter 1992b) in which I showed how a pure background tensor formalism provides a concise but explicit and highly flexible machinery for the generalised curvature analysis of individual embedded timelike or spacelike p-dimensional surfaces in a flat or curved n-dimensional spacetime background. The relevant spacetime metric will, as usual, be denoted here by $g_{\mu\nu}$, with the understanding that the subscripts are interpretable either in the (mathematically sophisticated) sense of Roger's abstract index system (Penrose 1968) or else in the old-fashioned concrete sense (with which most physicists are still more familiar) as the labels of components with respect to some set of local coordinates x^μ, $\mu = 0, 1, n-1$. It will be shown here how this machinery can be extended in a natural way so as to treat a smooth foliation by a congruence of such surfaces, for which the complete orthonormal frame bundle characterised by the relevant group of rotations in n dimensions will have a natural reduction to the bundle of naturally adapted frames characterised by the direct product of the subgroup of tangential frame rotations in p dimensions and the complementary subgroup of orthogonal frame rotations in $(n-p)$ dimensions. This reduced frame bundle will be naturally endowed with a preferred—metric preserving but generically non-symmetric—connection, which will have an associated *foliation curvature* tensor $\mathcal{F}_{\mu\nu}{}^{\rho}{}_{\sigma}$ that is generically distinct from the ordinary Riemannian *background curvature* which will be denoted here by $B_{\mu\nu}{}^{\rho}{}_{\sigma}$. The various ways in which this foliation curvature tensor can be projected, orthogonally or tangentially, with respect to the embedded surfaces give relations of which the generalised Codazzi and Raychaudhuri identities are particular cases.

The present approach has been developed to satisfy needs arising in the context of the recent rise of interest in the theory of topological defect structures such as cosmic strings and higher dimensional cosmic membranes, as well as the related phenomenon of vortex foliations in neutron stars, which has led to the investigation of a wide range of new problems of equilibrium and dynamical evolution in a special or general relativistic framework. Various aspects of these problems (Stachel 1980; Kopczynski 1987; Hartley and Tucker 1990; Boisseau and Barrabès 1992; Boisseau and Letelier 1992; Carter 1992b, 1992a) and in recent years most particularly the requirements of general purpose perturbation analysis (Guven 1993; Carter 1993; Larsen and Frolov 1994; Capovilla and Guven 1995a; Capovilla and Guven 1995b; Battye and Carter 1995), have shown, and in some cases helped to satisfy, the need to adapt and develop pre-existing mathematical machinery for describing the relevant geometry and particularly the various kinds of curvature that are involved.

As remarked in my preceding Penrose festschrift contribution (Carter 1992b), although much of what is needed has in principle been already available in the mathematical literature (Schouten 1954; Kobyashi and Nomizu 1969; Chen 1973; Spivak 1979; Choquet-Bruhat and DeWitt-Morette 1989), it has often been in a form that is inaccessible or inconvenient for the purposes of physicists, many of whom have remained excessively dependent on obsolete sources such as Eisenhart's still very influential textbook (Eisenhart 1926) (written in ignorance of the modern concept of generalised curvature, which was at that time under development by Cartan, and which was made familiar to physicists much later by the theory of Yang and Mills). Whereas some treatments have obtained a neatly concise abstract formulation at the expense of flexibility, others have obtained general purpose adaptability at the price of using complicated and potentially confusing reference systems involving specially adapted coordinates and frames that require the simultaneous use of many different kinds of indices.

For the purpose of obtaining an optimal compromise between these two undesirable extremes, the approach (Carter 1992b, 1992a; Guven 1993; Battye and Carter 1995) used here relies as much as possible just on ordinary tensors, as defined with respect to the relevant *background* space-time with local coordinates x^μ. The advantage of avoiding explicit dependence on a specialised internal coordinate system becomes particularly clear in cases (Carter 1992a) where one is concerned with mutual contractions of tensors constructed on distinct but mutually intersecting embedded surfaces whose internal coordinates could not in general be made to be mutually compatible.

The present treatment will employ the same notation as was used in the most mathematically detailed presentation (Carter 1992b) of this approach, in which, rather than describing the induced curvature of an embedded spacelike or timelike p-surface, with internal coordinates σ^i say, in terms of the intrinsic version, with components $R_{ijk\ell}$ say, of its Riemann tensor, one prefers to describe it in terms of the corresponding background spacetime tensor $R_{\lambda\mu\nu\rho}$. The latter is definable (using the abbreviation $x^\mu_{\ ,i}$ for $\partial x^\mu/\partial\sigma^i$) as the index lowered version—obtained by contraction with the background spacetime metric $g_{\mu\nu}$—of the projection $R^{\lambda\mu\nu\rho} = R^{ijkl}\, x^\lambda_{\ ,i} x^\mu_{\ ,j} x^\nu_{\ ,k} x^\rho_{\ ,l}$ of the contravariant version of the intrinsic curvature that is obtained by internal index raising, using the inverse, with contravariant components η^{ij} (whose existence depends on the postulate that the surface is spacelike or timelike, but not null) of the induced metric with components $\eta_{ij} = g_{\mu\nu} x^\mu_{\ ,i} x^\nu_{\ ,j}$. This background spacetime representation $R_{\lambda\mu\nu\rho}$ of the internal curvature of the embedded p-surface has of course to be distinguished from the ordinary n-dimensional Riemann curvature of the background spacetime itself, whose components will be denoted simply by $B_{\lambda\mu\nu\rho}$. In the same way, rather than working with the covariant and contravariant intrinsic components η_{ij} and η^{ij} of the induced metric, one prefers to use the corresponding background coordinate components $\eta_{\mu\nu}$ and $\eta^{\mu\nu}$ of the corresponding projected tensor specified by $\eta^{\mu\nu} = \eta^{ij} x^\mu_{\ ,i} x^\nu_{\ ,j}$, which is what is referred to as the (first) *fundamental tensor* of the embedding.

To set up a systematic analysis of curvature in exclusively background tensorial terms, the natural starting point is obviously the fundamental tensor that has just been defined, whose mixed (contra/covariant) version with components $\eta^\mu_{\ \nu}$ is interpretable as a rank p *tangential projection* operator that sends a vector into the tangent subspace of the embedding. The notation $\perp^\mu_{\ \nu}$ will be used here (instead of the less suggestive symbol $\gamma^\mu_{\ \nu}$) to denote the complementary rank ($n-$

p) operator of *lateral projection* orthogonal to the surface, whose components will evidently be given in terms of those of the fundamental (tangential projection) tensor by the defining relation

$$\perp^{\mu}{}_{\nu} + \eta^{\mu}{}_{\nu} = g^{\mu}{}_{\nu} \, , \tag{1.1}$$

since the mixed version $g^{\mu}{}_{\nu}$ of the metric tensor is of course interpretable as representing the identity operator. As well as having the separate operator properties

$$\eta^{\mu}{}_{\rho}\,\eta^{\rho}{}_{\nu} = \eta^{\mu}{}_{\nu} \, , \qquad \perp^{\mu}{}_{\rho}\perp^{\rho}{}_{\nu} = \perp^{\mu}{}_{\nu} \tag{1.2}$$

the tensors thus defined will evidently be related by the conditions

$$\eta^{\mu}{}_{\rho} \perp^{\rho}{}_{\nu} = 0 = \perp^{\mu}{}_{\rho}\,\eta^{\rho}{}_{\nu} \, . \tag{1.3}$$

2 The deformation tensor

Although it is less detailed, the present article considerably extends the results of the preceding development (Carter 1992b) of the background tensorial analysis of embedding geometry, by considering not just an individual embedded non-null p-surface by itself, but the extension of any such surface to a smooth foliation by diffeomorphically similar surfaces. For such a foliation there will be corresponding *background* (not just single p-surface supported) fields of tensors $\eta^{\mu}{}_{\nu}$ and $\perp^{\mu}{}_{\nu}$, that will not only satisfy the relations (1.1), (1.2), and (1.3), but will also (unlike what was supposed in the preceding Penrose festscrift article (Carter 1992b)) have well defined (Riemannian or pseudo-Riemannian) covariant derivatives. These derivatives given will be fully determined by the specification of a certain (first) *deformation tensor*, $\mathcal{H}_{\mu}{}^{\nu}{}_{\rho}$ say, via an expression of the form

$$\nabla_{\mu}\,\eta^{\nu}{}_{\rho} = -\nabla_{\mu} \perp^{\nu}{}_{\rho} = \mathcal{H}_{\mu}{}^{\nu}{}_{\rho} + \mathcal{H}_{\mu\rho}{}^{\nu} \, . \tag{2.1}$$

It can easily be seen from (1.2) that the required deformation tensor will be given simply by

$$\mathcal{H}_{\mu}{}^{\rho}{}_{\nu} = \eta^{\rho}{}_{\sigma}\,\nabla_{\mu}\,\eta^{\sigma}{}_{\nu} = -\perp^{\sigma}{}_{\nu}\,\nabla_{\mu}\perp^{\rho}{}_{\sigma} \, . \tag{2.2}$$

The middle and last indices of this tensor will evidently have the respective properties of tangentiality and orthogonality that are expressible as

$$\perp^{\nu}{}_{\rho}\,\mathcal{H}_{\mu}{}^{\rho}{}_{\sigma} = 0 \, , \qquad \mathcal{H}_{\mu}{}^{\rho}{}_{\sigma}\,\eta^{\sigma}{}_{\nu} = 0 \, . \tag{2.3}$$

There is no automatic tangentiality or orthogonality property for the first index of the deformation tensor (2.2), which is thus reducible with respect to the tangential and orthogonally lateral projections (1.1) to a sum

$$\mathcal{H}_{\mu}{}^{\rho}{}_{\nu} = K_{\mu}{}^{\rho}{}_{\nu} - L_{\mu\nu}{}^{\rho} \tag{2.4}$$

in which such a property is obtained for each of the parts

$$K_{\mu}{}^{\rho}{}_{\nu} = \eta^{\sigma}{}_{\mu}\mathcal{H}_{\sigma}{}^{\rho}{}_{\nu} \, , \qquad L_{\mu\nu}{}^{\rho} = -\perp^{\sigma}{}_{\mu}\,\mathcal{H}_{\sigma}{}^{\rho}{}_{\nu} \, , \tag{2.5}$$

which satisfy the conditions

$$\perp^{\sigma}_{\ \mu}\, K_{\sigma\ \nu}^{\ \rho} = \perp^{\rho}_{\ \sigma}\, K_{\mu\ \nu}^{\ \sigma} = 0 = K_{\mu\ \sigma}^{\ \rho}\, \eta^{\sigma}_{\ \nu} \,. \tag{2.6}$$

and

$$\eta^{\sigma}_{\ \mu} L_{\sigma\ \nu}^{\ \rho} = \eta^{\rho}_{\ \sigma} L_{\mu\ \nu}^{\ \sigma} = 0 = L_{\mu\ \sigma}^{\ \rho} \perp^{\sigma}_{\ \nu} \ . \tag{2.7}$$

It is evident that the first of these decomposed parts is appropriately describable as the *tangential turning* tensor, since by (2.2) and (2.5) it is given by the expression

$$K_{\mu\ \nu}^{\ \rho} = \eta^{\rho}_{\ \sigma} \eta^{\tau}_{\ \mu} \nabla_{\tau} \eta^{\sigma}_{\ \nu} \,, \tag{2.8}$$

in which the only differentiation involved is contained in the tangential gradient operator $\eta^{\tau}_{\ \mu}\nabla_{\tau}$—which is well defined even for fields whose support is restricted to a single embedded surface—so that, unlike the full deformation tensor $\mathcal{H}_{\mu\ \nu}^{\ \rho}$ of the foliation, the tangential turning tensor $K_{\mu\ \nu}^{\ \rho}$ is well defined just for an individual embedded p-surface. As such, this turning tensor $K_{\mu\ \nu}^{\ \rho}$ is identifiable as what has been defined (Carter 1992b) as the ordinary *second fundamental tensor* of the particular p-surface passing through the point under consideration.

Up to this point, none of the relations formulated in this section depends on the condition that the (first) fundamental tensor field $\eta^{\mu}_{\ \nu}$ is actually tangential to well behaved p-surfaces rather than just being an arbitrary field of rank p projection tensors as characterised by the purely algebraic conditions (1.2). As pointed out in the previous analysis (Carter 1992b), the Frobenius type integrability condition that is necessary and sufficient for the local existence of well behaved p-surfaces tangential to $\eta^{\mu}_{\ \nu}$ is that the second fundamental tensor defined by (2.6) should have the generalised Weingarten property

$$K_{\mu\nu}^{\ \ \rho} = K_{(\mu\nu)}^{\ \ \ \ \rho} \qquad \Leftrightarrow \qquad K_{[\mu\nu]}^{\ \ \ \ \rho} = 0 \,, \tag{2.9}$$

(using round and square brackets to denote index symmetrisation and antisymmetrisation respectively) which means that it is symmetric with respect to its two surface tangential indices.

It is to be noticed that the second part of the decomposition (2.4), namely the *lateral turning tensor*, $L_{\mu\ \nu}^{\ \rho}$, is expressible by the formula

$$L_{\mu\ \nu}^{\ \rho} = \perp^{\rho}_{\ \sigma} \perp^{\tau}_{\ \mu} \nabla_{\tau} \perp^{\sigma}_{\ \nu} \,, \tag{2.10}$$

which differs from that in (2.8) only by the substitution of $\perp^{\nu}_{\ \mu}$ for $\eta^{\nu}_{\ \mu}$. It evidently follows that the necessary and sufficient integrability condition for the local existence of an $(n-p)$-dimensional foliation orthogonal to the p-dimensional foliation whose existence is guaranteed by (2.9) is that this lateral turning tensor should have the analogous symmetry property, which is expressible as the vanishing of the *rotation tensor*, $\omega_{\mu\nu}^{\ \ \rho}$ say, that is defined as its antisymmetric part in a decomposition of the form

$$L_{\mu\nu}^{\ \ \rho} = \omega_{\mu\nu}^{\ \ \rho} + \theta_{\mu\nu}^{\ \ \rho} \,, \qquad \omega_{(\mu\nu)}^{\ \ \ \ \rho} = 0 \,, \qquad \theta_{[\mu\nu]}^{\ \ \ \ \rho} = 0 \,. \tag{2.11}$$

The only part that remains if the foliation is $(n-p)$ surface orthogonal is the symmetric part, $\theta_{\mu\nu}{}^{\rho}$, which is the natural generalisation of the two index divergence tensor $\theta_{\mu\nu}$ whose evolution is the subject of the tensorial Raychaudhuri identity discussed by Hawking and Ellis (1973) (the original Raychaudhuri identity (Raychaudhuri 1957) being obtained by taking the scalar trace). For a 1-dimensional timelike foliation, which will have a unique future directed unit tangent vector u^{μ}, the relevant divergence tensor will be obtainable simply as $\theta_{\mu\nu} = \theta_{\mu\nu}{}^{\rho} u_{\rho}$. The generalised Raychaudhuri identity to be presented (following Capovilla and Guven 1995b) in a later section, provides an evolution equation for the three index generalised divergence tensor $\theta_{\mu\nu}{}^{\rho}$ which (unlike the ordinary divergence tensor $\theta_{\mu\nu}$) is always well defined whatever the dimension of the foliation.

3 The adapted foliation connection

Due to the existence of the decomposition whereby a background spacetime vector, with components ξ^{μ} say, is split up by the projectors (1.1) as the sum of its surface tangential part $\eta^{\mu}{}_{\nu}\,\xi^{\nu}$ and its surface orthogonal part $\perp^{\mu}_{\nu}\,\xi^{\nu}$, there will be a corresponding adaptation of the ordinary concept of parallel propagation with respect to the background connection $\Gamma_{\mu}{}^{\nu}{}_{\rho}$. The principle of the adapted propagation concept is to follow up an ordinary operation of infinitesimal parallel propagation by the projection adjustment that is needed to ensure that purely tangential vectors propagate onto purely tangential vectors while purely orthogonal vectors propagate onto purely orthogonal vectors. Thus for purely tangential vectors, the effect of the adapted propagation is equivalent to that of ordinary internal parallel propagation with respect to the induced metric in the embedded surface, while for purely orthogonal vectors it is interpretable as the natural generalisation of the standard concept of Fermi–Walker propagation. For an infinitesimal displacement dx^{μ} the deviation between the actual component variation $(dx^{\nu})\,\partial_{\nu}\xi^{\mu}$ and the variation that would be obtained by the corresponding adapted propagation law will be expressible in the form $(dx^{\nu})\,D_{\nu}\xi^{\mu}$ where D denotes the corresponding *foliation adapted differentiation* operator, whose effect will evidently be given by

$$D_{\nu}\,\xi^{\mu} = \eta^{\mu}{}_{\rho}\nabla_{\nu}\big(\eta^{\rho}{}_{\sigma}\xi^{\sigma}\big) + \perp^{\mu}_{\rho}\nabla_{\nu}\big(\perp^{\rho}_{\sigma}\,\xi^{\sigma}\big)\ . \tag{3.1}$$

It can thus be seen that this operation will be expressible in the form

$$D_{\nu}\,\xi^{\mu} = \nabla_{\nu}\xi^{\mu} + \alpha_{\nu}{}^{\mu}{}_{\sigma}\xi^{\sigma} = \partial_{\nu}\,\xi^{\mu} + \mathcal{A}_{\nu}{}^{\mu}{}_{\sigma}\xi^{\sigma}\ , \tag{3.2}$$

where the adapted *foliation connection* components $\mathcal{A}_{\mu}{}^{\nu}{}_{\rho}$ are given by the formula

$$\mathcal{A}_{\mu}{}^{\nu}{}_{\rho} = \Gamma_{\mu}{}^{\nu}{}_{\rho} + \alpha_{\mu}{}^{\nu}{}_{\rho}\ , \tag{3.3}$$

in which the $\alpha_{\mu}{}^{\nu}{}_{\rho}$ are the components of the relevant *adaptation tensor*, whose components can be seen from (2.2) to be given by

$$\alpha_{\mu\nu\rho} = 2\mathcal{H}_{\mu[\nu\rho]}\ . \tag{3.4}$$

The fact that the expression (3.4) is manifestly antisymmetric with respect to the last two indices of the adaptation tensor makes it evident that, like the usual Riemannian differentiation operator ∇, the adapted differentiation operator D will commute with index raising or lowering, since the metric itself remains invariant under adapted propagation:

$$D_\mu \, g_{\nu\rho} = 0 \ . \tag{3.5}$$

However, unlike ∇, the adapted differentiation operator has the very convenient property of also commuting with tangential and orthogonal projection, since it can be seen to follow from (2.1) and (2.3) that the corresponding operators also remain invariant under adapted propagation:

$$D_\mu \, \eta^\nu{}_\rho = 0 \ , \qquad D_\mu \perp^\nu_\rho = 0 \ . \tag{3.6}$$

There is of course a price to be paid in order to obtain this considerable advantage of D over ∇, but it is not exorbitant: all that has to be sacrificed is the analogue of the symmetry property

$$\Gamma_{[\mu}{}^\nu{}_{\rho]} = 0 \ , \tag{3.7}$$

expressing the absence of torsion in the Riemannian case. For the adapted foliation connection $\mathcal{A}_\mu{}^\nu{}_\rho$, the torsion tensor defined by

$$\Theta_\mu{}^\nu{}_\rho = 2\mathcal{A}_{[\mu}{}^\nu{}_{\rho]} = 2\alpha_{[\mu}{}^\nu{}_{\rho]} \ , \tag{3.8}$$

will not in general be zero.

4 The amalgamated foliation curvature tensor

The curvature associated with the adapted connection introduced by (4.2) in the preceding section can be read out from the ensuing commutator formula, which, for an arbitrary vector field with components ξ^μ, will take the standard form

$$2D_{[\mu} D_{\nu]} \xi^\rho = \mathcal{F}_{\mu\nu}{}^\rho{}_\sigma \xi^\sigma - \Theta_\mu{}^\sigma{}_\nu D_\sigma \xi^\rho \ , \tag{4.1}$$

in which the torsion tensor components $\Theta_\mu{}^\sigma{}_\nu$ are as defined by (3.8) while the components $\mathcal{F}_{\mu\nu}{}^\rho{}_\sigma$ are defined by a Yang–Mills type curvature formula of the form

$$\mathcal{F}_{\mu\nu}{}^\rho{}_\sigma = 2\partial_{[\mu}\mathcal{A}_{\nu]}{}^\rho{}_\sigma + 2\mathcal{A}_{[\mu}{}^{\rho\tau}\mathcal{A}_{\nu]\tau\sigma} \ . \tag{4.2}$$

Although the connection components $\mathcal{A}_\mu{}^\nu{}_\rho$ from which it is constructed are not of tensorial type, the resulting curvature components (4.2) are of course strictly tensorial. This is made evident by evaluating the components (4.2) of this *amalgamated foliation curvature* in terms of the background curvature tensor

$$B_{\mu\nu}{}^\rho{}_\sigma = 2\partial_{[\mu}\Gamma_{\nu]}{}^\rho{}_\sigma + \Gamma_\mu{}^\rho{}_\tau \Gamma_\nu{}^\tau{}_\sigma - \Gamma_\nu{}^\rho{}_\tau \Gamma_\mu{}^\tau{}_\sigma \ , \tag{4.3}$$

and the adaptation tensor $\alpha_\mu{}^\nu{}_\rho$ given by (3.4), which gives the manifestly tensorial expression

$$\mathcal{F}_{\mu\nu}{}^\rho{}_\sigma = B_{\mu\nu}{}^\rho{}_\sigma + 2\nabla_{[\mu}\alpha_{\nu]}{}^\rho{}_\sigma + 2\alpha_{[\mu}{}^{\rho\tau}\alpha_{\nu]\tau\sigma} \,. \tag{4.4}$$

Although it does not share the full set of symmetries of the Riemann tensor, the foliation curvature obtained in this way will evidently be antisymmetric in both its first and last pairs of indices:

$$\mathcal{F}_{\mu\nu\rho\sigma} = \mathcal{F}_{[\mu\nu][\rho\sigma]} \,. \tag{4.5}$$

Using the formula (3.4), it can be seen from (2.3) that the difference between this adapted curvature and the ordinary background Riemann curvature will be given by

$$\mathcal{F}_{\mu\nu}{}^{\rho\sigma} - B_{\mu\nu}{}^{\rho\sigma} = 4\mathcal{X}_{[\mu\nu]}{}^{[\rho\sigma]} + 2\mathcal{H}_{[\mu}{}^{\sigma\tau}\mathcal{H}_{\nu]}{}^\rho{}_\tau + 2\mathcal{H}_{[\mu}{}^{\tau\rho}\mathcal{H}_{\nu]\tau}{}^\sigma \,, \tag{4.6}$$

where $\mathcal{X}_{\lambda\mu}{}^\nu{}_\rho$ is what may be termed the *second deformation tensor*, which is definable by

$$\mathcal{X}_{\lambda\mu}{}^\nu{}_\rho = \eta^\nu{}_\sigma \perp^\tau{}_\rho \nabla_\lambda \mathcal{H}_\mu{}^\sigma{}_\tau \,. \tag{4.7}$$

The formula (4.6) superficially appears to depend on the higher order derivatives involved in $\mathcal{X}_{\mu\nu}{}^{\rho\sigma}$, but this is deceptive: the higher derivatives will in fact cancel, by the "amalgamated Codazzi–Raychaudhuri identity" given below.

Since the adapted derivation operator has been constructed in such a way as to map tangential vector fields into purely tangential vector fields, and lateral (surface orthogonal) vector fields into lateral vector fields, it follows that the same applies to the corresponding curvature (4.2), which will therefore consist of an additive amalgamation of two separate parts having the form

$$\mathcal{F}_{\mu\nu}{}^\rho{}_\sigma = \mathcal{P}_{\mu\nu}{}^\rho{}_\sigma + \mathcal{Q}_{\mu\nu}{}^\rho{}_\sigma \,, \tag{4.8}$$

in which the "inner" curvature acting on purely tangential vectors is given by a doubly tangential (surface parallel) projection as

$$\mathcal{P}_{\mu\nu}{}^\rho{}_\sigma = \mathcal{F}_{\mu\nu}{}^\kappa{}_\lambda \, \eta^\rho{}_\kappa \, \eta^\lambda{}_\sigma \,, \tag{4.9}$$

while the "outer" curvature acting on purely orthogonal vectors is given by a doubly lateral projection as

$$\mathcal{Q}_{\mu\nu}{}^\rho{}_\sigma = \mathcal{F}_{\mu\nu}{}^\kappa{}_\lambda \perp^\rho{}_\kappa \perp^\lambda{}_\sigma \,. \tag{4.10}$$

It is implicit in the separation expressed by (4.8) that the mixed tangential and lateral projection of the adapted curvature must vanish:

$$\mathcal{F}_{\mu\nu}{}^\kappa{}_\lambda \, \eta^\rho{}_\kappa \perp^\lambda{}_\sigma = 0 \,. \tag{4.11}$$

To get back, from the extended foliation curvature tensors that have just been introduced, to their antecedent analogues (Carter 1992b) for an individual embedded surface, the first step is to construct the *amalgamated embedding curvature* tensor, $F_{\mu\nu}{}^{\rho}{}_{\sigma}$ say, which will be obtainable from the corresponding amalgamated foliation curvature $\mathcal{F}_{\mu\nu}{}^{\rho}{}_{\sigma}$ by a doubly tangential projection having the form

$$F_{\mu\nu}{}^{\rho}{}_{\sigma} = \eta^{\alpha}{}_{\mu}\,\eta^{\beta}{}_{\nu}\mathcal{F}_{\alpha\beta}{}^{\rho}{}_{\sigma} \, . \tag{4.12}$$

As did the extended foliation curvature, so also this amalgated embedding curvature will separate as the sum of "inner" and "outer" parts in the form

$$F_{\mu\nu}{}^{\rho}{}_{\sigma} = R_{\mu\nu}{}^{\rho}{}_{\sigma} + \Omega_{\mu\nu}{}^{\rho}{}_{\sigma} \, , \tag{4.13}$$

in which the "inner" embedding curvature is given by another doubly tangential projection as

$$R_{\mu\nu}{}^{\rho}{}_{\sigma} = F_{\mu\nu}{}^{\kappa}{}_{\lambda}\,\eta^{\rho}{}_{\kappa}\,\eta^{\lambda}{}_{\sigma} = \eta^{\alpha}{}_{\mu}\eta^{\beta}{}_{\nu}\mathcal{P}_{\alpha\beta}{}^{\rho}{}_{\sigma} \, , \tag{4.14}$$

while the "outer" embedding curvature (whose noteworthy property of *conformal invariance* was pointed out in my previous Penrose festschrift contribution (Carter 1992b)) is given by the corresponding doubly lateral projection as

$$\Omega_{\mu\nu}{}^{\rho}{}_{\sigma} = F_{\mu\nu}{}^{\kappa}{}_{\lambda}\perp^{\rho}{}_{\kappa}\perp^{\lambda}{}_{\sigma} = \eta^{\alpha}{}_{\mu}\eta^{\beta}{}_{\nu}\mathcal{Q}_{\alpha\beta}{}^{\rho}{}_{\sigma} \, . \tag{4.15}$$

The formula (4.6) can be used to evaluate the "inner" tangential part of the foliation curvature tensor as

$$\mathcal{P}_{\mu\nu}{}^{\rho}{}_{\sigma} = 2\mathcal{H}_{[\nu}{}^{\rho\tau}\mathcal{H}_{\mu]\sigma\tau} + B_{\mu\nu}{}^{\kappa}{}_{\lambda}\,\eta^{\rho}{}_{\kappa}\,\eta^{\lambda}{}_{\sigma} \tag{4.16}$$

and to evaluate the "outer" orthogonal part of the foliation curvature tensor as

$$\mathcal{Q}_{\mu\nu}{}^{\rho}{}_{\sigma} = 2\mathcal{H}_{[\mu}{}^{\tau\rho}\mathcal{H}_{\nu]\tau\sigma} + B_{\mu\nu}{}^{\kappa}{}_{\lambda}\perp^{\rho}{}_{\kappa}\perp^{\lambda}{}_{\sigma} \, . \tag{4.17}$$

The formula (4.16) for the "inner" foliation curvature is evidently classifiable as a further extension of the previously derived generalisation (Carter 1992b) for $R_{\mu\nu}{}^{\rho}{}_{\sigma}$ of the historic Gauss equation. Similarly the formula (4.17) for the "outer" foliation curvature is an analogous extension of the relation (Carter 1992b) for $\Omega_{\mu\nu}{}^{\rho}{}_{\sigma}$ that corresponds to what has sometimes been referred to as the "Ricci equation" but what would seem more appropriately describable as the *Schouten equation*, with reference to the earliest relevant source with which I am familiar (Schouten 1954), since long after the time of Ricci it was not yet understood even by such a leading geometer as Eisenhart (1926).

In much the same way, the non-trivial separation identity (4.11) can be considered as a generalisation to the case of foliations of the relation that is itself interpretable as an extended generalisation to higher dimensions of the historic Codazzi equation that was originally formulated in the restricted context of 3-dimensional flat space. It can be seen from (4.6) that this extended Codazzi identity is expressible as

$$2\mathcal{X}_{[\mu\nu]}{}^{\rho}{}_{\sigma} + B_{\mu\nu}{}^{\kappa}{}_{\lambda}\,\eta^{\rho}{}_{\kappa}\perp^{\lambda}{}_{\sigma} = 0 \, , \tag{4.18}$$

which shows that the relevant higher derivatives are all determined entirely by the Riemannian background curvature so that no specific knowledge of the second deformation tensor is needed. By doubly tangential projection of the first two indices, the extended generalisation (4.18) will give back the already familiar version (Carter 1992b) of the generalised Codazzi identity for the individual embedded p-surfaces of the foliation. The corresponding doubly lateral projection would give the analogous result for the orthogonal foliation by $(n-p)$-surfaces that would exist in the irrotational case for which the lateral turning tensor given by (2.10) is symmetric. Finally the corresponding mixed tangential and lateral projection of (4.18) gives an identity that is expressible in terms of foliation adapted differentiation (3.2) as

$$\perp^{\tau}{}_{\mu} D_{\tau} K_{\nu}{}^{\rho}{}_{\sigma} + \eta^{\tau}{}_{\nu} D_{\tau} L_{\mu\sigma}{}^{\rho} = K_{\nu}{}^{\lambda}{}_{\mu} K_{\lambda}{}^{\rho}{}_{\sigma} + L_{\mu}{}^{\lambda}{}_{\nu} L_{\lambda\sigma}{}^{\rho} + \eta^{\alpha}{}_{\nu} \perp^{\beta}{}_{\mu} B_{\alpha\beta}{}^{\kappa}{}_{\lambda}\, \eta^{\rho}{}_{\kappa} \perp^{\lambda}{}_{\sigma} \,. \tag{4.19}$$

This last result is interpretable as the translation into the pure background tensorial formalism used here of the recently derived generalisation (Capovilla and Guven 1995b) to higher dimensional foliations of the well known Raychaudhuri equation (whose original scalar version (Raychaudhuri 1957), and its tensorial extension (Hawking and Ellis 1973), were formulated just for the special case of a foliation by 1-dimensional curves). The complete identity (4.18) is therefore interpretable as an amalgamated Raychaudhuri–Codazzi identity.

Bibliography

Barrabès, C. and Israel, W. (1991). *Phys. Rev.* **D43**, 1129–42.

Battye, R. A. and Carter, B. (1995). *Phys. Lett.* **B357**, 29.

Boisseau, B. and Barrabès, C. (1992). *Phys. Lett.* **B279**, 259.

Boisseau, B. and Letelier, P. S. (1992). *Phys. Rev.* **D46**, 1721.

Capovilla, R. and Guven, J. (1995a). *Phys. Rev.* **D51**, 6736.

Capovilla, R. and Guven, J. (1995b). *Phys. Rev.* **D52**, 1072.

Carter, B. (1966). *Phys. Rev.* **141**, 1242–7, and *Phys. Lett.* **21**, 423–4.

Carter, B. (1992a). *J. Class. Quantum Grav.* **9**, 19.

Carter, B. (1992b). *J. Geom. Phys.* **8**, 53.

Carter, B. (1993). *Phys. Rev.* **D48**, 4835.

Chen, B. Y. (1973). *Geometry of Submanifolds.* Dekker, New York.

Choquet-Bruhat, Y. and DeWitt-Morette C. (1989). *Analysis, Manifolds, Physics II.* North Holland, Amsterdam.

Eisenhart, L. P. (1926). *Riemannian Geometry.* Princeton University Press, reprinted 1960.

Guven, J. (1993). *Phys. Rev.* **D48**, 4464 and 5563.

Hartley, D. H. and Tucker, R. W. (1990). In *Geometry of Low Dimensional Manifolds, 1*, L.M.S. Lecture Note Series **150**, ed. S. Donaldson, C. Thomas. Cambridge University Press.

Hawking, S. W. and Ellis, G. F. R. (1973). *The Large Scale Structure of Space-Time.* Cambridge University Press.

Hawking, S. W. and Penrose R. (1970). *Proc. Roy. Soc. Lond.* **A300**, 187–201.

Kobayashi, S. and Nomizu, K. (1969). *Foundations of Differential Geometry.* Interscience, New York.

Kopczynski, W. (1987). *Phys. Rev.* **D36**, 3582 and **D36**, 3589.

Larsen, A. L. and Frolov, V. (1994). *Nucl. Phys.* **B414**, 129.

Newman, E. T. and Penrose, R. (1962). *J. Math. Phys.* **3**, 566–78.

Penrose, R. (1964). In *Relativity, Groups, and Topology,* ed. B. and C. M. DeWitt, 563–84. Gordon and Breach, New York.

Penrose, R. (1965). *Phys. Rev. Lett.* **14**, 57–9.

Penrose, R. (1968). In *Battelle Rencontres,* ed. C. M. DeWitt and J. A. Wheeler, 121–235. Benjamin, New York.

Raychaudhuri, A. (1957). *Phys. Rev.* **106**, 172.

Schouten, J. A. (1954). *Ricci Calculus.* Springer, Heidelberg.

Spivak, M. (1979). *Differential Geometry.* Publish or Perish, Berkeley.

Stachel, J. (1980). *Phys. Rev. D* **21**, 2171; **21**, 2182.

24

Abstract/Virtual/Reality/Complexity

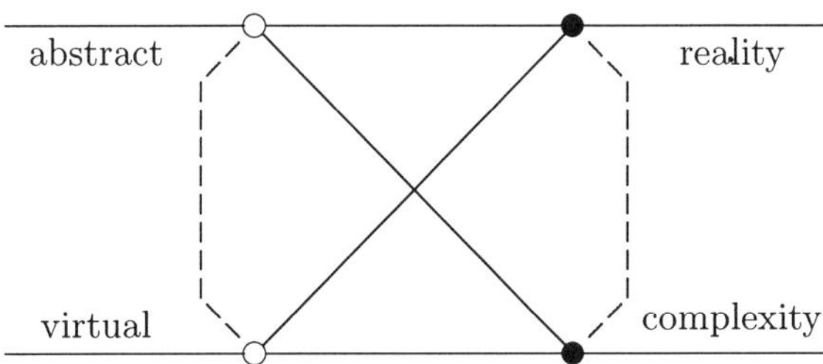

George Sparling

Department of Mathematics and Statistics, University of Pittsburgh, Pittsburgh, PA 15260[1]

Abstract

The Penrose Obstruction prevents the extension of the basic ideas of twistor theory to a curved space-time. A way to circumvent this obstacle for the case of Einstein vacuum space-times is discussed. It involves a reformulation of twistor theory using the language of abstract Grassmann algebras.

"Pluritas non est ponenda sine necessitas" Ockham (1981).

1 Introduction

This work is dedicated to Roger Penrose, mentor and friend. I thank the Conference Committee for inviting me; I especially thank Hermann Bondi, who introduced me (then a schoolboy) to relativity and Christopher Isham who gave twistor theory crucial early support. The twistor diagram of the title pays tribute to the many workers who have advanced the theory. The diagram has only single lines emerging from it indicating that it is an interaction diagram, part of a larger whole, as yet undiscovered. The dashed lines represent essential poles using infinity twistors: this reflects the idea that twistor theory should be tied to an understanding of gravity.

If twistor theory is to be a physical theory, it has to encompass the richness of Einstein's theory of gravity. Classical twistors are delicate global objects and are

[1] http://www.math.pitt.edu/ sparling.

destroyed by conformal curvature (the Penrose Obstruction, see below). Penrose has argued for their reconstitution at the quantum level, but it is hard to see exactly how to proceed. In the present work, a twistor-like theory, called abstract twistor theory, is constructed for any analytic Einstein vacuum spacetime, possibly with a cosmological constant, Sparling (1996). The Penrose Obstruction is circumvented, by suitably enlarging the Grassmann algebra of differential forms with forms of negative rank. The theory has the character of a one and one half twistor theory, as envisaged by Penrose (his self-dual theory uses only one twistor); it also controls and is compatible with the hypersurface twistor theory of Penrose, Penrose (1975), Sparling (1997), Penrose & Rindler (1986).

In the first two sections, twistor theory is reviewed, highlighting the Penrose Obstruction, the "non-linear graviton" of Penrose and hypersurface twistor theory. In the last two sections, abstract twistor theory is outlined. Starting from the twistor factorization condition, equation (4.1), repeated differentiation yields an infinite differential ideal, consistent if negative rank forms are allowed. There is gauge freedom at each iteration. The abstract twistor structure is the quotient of the ideal modulo this gauge freedom. The abstract twistor surfaces are the integral manifolds of this structure. For the prototypical case of a null constant plane wave vacuum spacetime, the four dimensional projective space of abstract twistor surfaces is shown to depend upon a third order ordinary differential equation of hypergeometric type:

$$\frac{d^3x}{du^3} = -|a|^2ux.$$

Here the complex number a determines the wave profile.

2 Complexity

The complexification and compactification of Minkowski spacetime, contains within it two families of completely null two-surfaces: the α-family, called projective twistor space, $\mathbb{PT}$, and the β-family, the dual projective twistor space, $\mathbb{PT}^*$, Mason & Woodhouse (1996), Penrose & Rindler (1984 & 1985). Each surface is a complex projective two space. Each family forms a complex projective three space. Any two surfaces of the same kind meet at a unique point. Generically an α and a β surface do not meet, but if they do meet, then they meet in a complex null geodesic and conversely, every complex null geodesic determines a unique pair of surfaces, one of each kind, that passes through it. The spaces are naturally dual, the duality corresponding to incidence of the surfaces. The spacetime is then the Grassmannian of projective lines in either twistor space.

Using spinors and abstract indices in the style of Penrose, the twistor surfaces are the integral manifolds of the differential ideal generated by factoring the canonical one-form, $\theta^a \equiv l^a n + n^a l - m^a m' - m'^a m$, where (l, n, m, m') is a null tetrad:

$$\theta^a = \alpha^{A'} \beta^A. \tag{2.1}$$

This equation describes an α-surface or a β-surface, respectively, according to whether $\alpha^{A'}$ is a spinor and β^A a spinor-valued one-form, or inversely.

If there is conformal curvature, then integrability conditions arise which generically destroy the twistor surfaces. Applying the Levi-Civita covariant exterior derivative d to equation (2.1) for the case of an α-surface, the condition of vanishing torsion, $d\theta^a = 0$, yields the equation: $\alpha^{A'} d\beta^A = -(d\alpha^{A'})\beta^A$. From this equation, we derive the equations, valid for some one-form γ: $d\beta^A = \gamma\beta^A$ and $(d\alpha^{A'} + \gamma\alpha^{A'})\beta^A{=}0$. If the twistor surface is genuinely two-dimensional, we have $\beta_C\beta^C \neq 0$ and we infer the equation: $d\alpha^{A'} = -\gamma\alpha^{A'}$. For any spinors $v^{A'}$ and v^A, we have $d^2v^{A'} = R^{A'}_{B'}v^{B'}$ and $d^2v^A = R^A_Bv^B$, where $R_{A'B'}$ and R_{AB} are the spinor curvature two-forms. Applying d to the equation for $d\alpha^{A'}$, we get: $R^{A'}_{B'}\alpha^{B'} = -(d\gamma)\alpha^{A'}$. Then contracting this equation with $\alpha_{A'}$ we get the equation: $\alpha_{A'}\alpha_{B'}R^{A'B'} = 0$. Now $R^{A'B'}$ decomposes as follows: $R^{A'B'} = \theta_{AC'}\theta^A_{D'}C^{A'B'C'D'} + \theta^{A'B}P^{B'}_B$, where $C^{A'B'C'D'}$ is the (symmetric) primed Weyl spinor and the indexed one-form P^b contains the information of the Ricci tensor. Using this decomposition with equation (2.1), we find: $0 = \alpha_{A'}\alpha_{B'}R^{A'B'} = \alpha_{A'}\alpha_{B'}\alpha_{C'}\alpha_{D'}C^{A'B'C'D'}\beta_C\beta^C$. Since $\beta_C\beta^C \neq 0$, we obtain the Penrose Obstruction in the form:

$$\alpha_{A'}\alpha_{B'}\alpha_{C'}\alpha_{D'}C^{A'B'C'D'} = 0. \tag{2.2}$$

In conformally flat spacetime, the Penrose Obstruction vanishes and there are twistor surfaces through every point, with every possible tangent spinor $\alpha^{A'}$ at that point. If we want as many twistors in the case of curved spacetime, through each point, we are forced to conclude that the primed Weyl spinor must vanish identically. If we need both twistor spaces to exist then both the primed and unprimed Weyl spinors must vanish and the spacetime is necessarily conformally flat. This is the key result of Penrose (1975), Mason & Woodhouse (1996):

Theorem 2.1 *Three dimensional projective twistor and dual twistor spaces of completely null two surfaces exist in complexified spacetime if and only the spacetime is conformally flat.*

Penrose realized that an interesting twistor theory still exists for anti-self-dual complex spacetimes: those with the primed Weyl spinor zero but not the unprimed. If $C_{A'B'C'D'}$ is zero, then the twistor space of α-surfaces exists, just as in Minkowski spacetime, but not the dual twistor space. More precisely he showed the following theorem, Penrose (1976):

Theorem 2.2 *For a spacetime with $C_{A'B'C'D'} = 0$, a three dimensional projective twistor space of completely null two surfaces exists. Each spacetime point has a Riemann sphere of twistor surfaces through it. Null separated points in spacetime have incident Riemann spheres.*

If a three dimensional manifold is given containing an embedded Riemann sphere with normal bundle the direct sum of two line bundles of Chern class one, then perturbations of the sphere give a four dimensional manifold of such spheres and that manifold then has a conformal structure, determined by incidence of the spheres and has the original three dimensional manifold as its twistor space. The conformal curvature is anti-self-dual: $C_{A'B'C'D'} = 0$.

This theorem is an application of pivotal results of Kodaira, who showed that the conditions on the normal bundle implied that there is a four parameter set of perturbed spheres, Kodaira (1962). Penrose further realized that the conformal structure is also built in and, building on work with Newman, he showed that the vacuum field equations could be axiomatized in terms of the existence of certain forms on the twistor space, Newman (1976), Penrose & Rindler (1984 & 1986). Ward showed how to allow for a cosmological constant, Ward (1980).

3 Reality

The Penrose theory of self-dual vacuum spacetimes is very beautiful and satisfactory, but his original intention was to understand real spacetimes. For real Minkowski spacetime, the twistor space $\mathbb{T}$ (the vector space whose projective space is $\mathbb{PT}$) has a pseudo-hermitian conjugation, of signature $(2, 2)$ taking the twistor space to its dual. The symmetry group is $U(2, 2)$, which induces the action of the conformal group on spacetime. The conjugation also gives a pseudo-Kähler scalar for $\mathbb{T}$. There is a preferred hypersurface $\mathbb{N}$ of $\mathbb{T}$ where the pseudo-Kähler scalar vanishes. This hypersurface acquires a $\mathcal{CR}$ structure of Levi signature $(1, 1, 0)$ from its embedding in $\mathbb{T}$. The real points of compactified Minkowski spacetime are those two dimensional subspaces of $\mathbb{T}$ that lie in $\mathbb{N}$. The projective image of $\mathbb{N}$ in $\mathbb{PT}$ is called $\mathbb{PN}$. It is a hyperquadric $\mathcal{CR}$ hypersurface in $\mathbb{PT}$ of non-degenerate Levi signature $(1, 1)$. The real spacetime points are represented in $\mathbb{PT}$ by the projective lines of $\mathbb{PT}$ that lie entirely in $\mathbb{PN}$.

Newman and Penrose found that associated to the radiation field at null infinity of an asymptotically flat analytic spacetime there is a twistor space embodying the correct structure to represent a self-dual vacuum, Newman (1976), Penrose (1976). This places the complex self-dual theory in a real context, albeit at infinity. The twistor space has a pseudo-Kähler scalar (with a Ricci flat metric), whose vanishing gives a preferred $\mathcal{CR}$ hypersurface of Levi signature $(1, 1, 0)$, Ko, Newman and Penrose (1977).

Generalizing, Penrose observed that a twistor construction exists anchored to any given complex analytic hypersurface in a complex analytic spacetime, Penrose (1975). On the hypersurface there are two three parameter families of twistor curves, the α-curves and β-curves, forming the projective twistor and dual twistor spaces of that hypersurface. These spaces coincide if the second fundamental form is pure trace. On each type of curve, the one form θ^a factorizes as $\theta^a = \alpha^{A'}\beta^A\rho$, with ρ a one-form. On an α-curve or β-curve, respectively, we have also the vanishing of the form $\alpha_{A'}d\alpha^{A'}$ or $\beta_A d\beta^A$. For each such twistor space there is an associated spacetime with self-dual or anti-self-dual Weyl curvature. If the given spacetime has anti-self-dual or self-dual Weyl curvature, then the α or β-curves are just the intersection of the hypersurface with the α or

β surfaces as appropriate. If the hypersurface is at null infinity, the hypersurface twistor spaces agree exactly with those studied by Newman and Penrose. For a general hypersurface, however, it is not known if there is a preferred metric in the conformal class of the metric of the self-dual spacetime determined by the twistor geometry.

The real version of this structure for the case of real smooth Lorentzian manifolds is as follows. To any hypersurface $\mathcal{H}$ in the manifold, there is naturally defined a seven-dimensional $\mathcal{CR}$ manifold $\mathcal{N}$, which has Levi signature $(1,1,0)$, giving a deformed analogue of $\mathbb{N}$. Here $\mathcal{N}$ is the open subset obtained from the primed co-spin bundle over $\mathcal{H}$, by deleting any point $(x, \pi_{A'})$, with $x \in \mathcal{H}$ and $\pi_{A'}$ a spinor at x, of the primed co-spin bundle, such that the real null vector $\pi^{A'}\overline{\pi}^{A}$ is tangent to $\mathcal{H}$. In particular if $\mathcal{H}$ is spacelike, then $\mathcal{N}$ is the primed co-spin bundle with only the zero section deleted. The one forms defining the $\mathcal{CR}$ structure are the restrictions to $\mathcal{N}$ of the forms $d\pi_{A'}$ and $\theta^a\pi_{A'}$ at any point $(x, \pi_{A'})$ of $\mathcal{N}$. The projective space $\mathcal{PN}$ of $\mathcal{N}$ is the quotient of $\mathcal{N}$ by identifying (non-zero) proportional spinors at any point of $\mathcal{H}$. Then $\mathcal{PN}$ has an induced non-degenerate $\mathcal{CR}$ structure of Levi signature $(1,1)$.

A main theorem here is a theorem of the author, Sparling (1997):

Theorem 3.1 *On the co-spin bundle of a real smooth Lorentzian spacetime, $\mathcal{X}$, there is defined a canonical real symmetric covariant two index tensor, the Fefferman tensor, Φ. The tensor is invariant under conformal transformations of the metric of $\mathcal{X}$. Explicitly the tensor is given by the following formula in the symmetric tensor algebra of the co-spin bundle of $\mathcal{X}$:*

$$\Phi = i\theta^a(\overline{\pi}_A d\pi_{A'} - \pi_{A'} d\overline{\pi}_A).$$

Restricted to the co-spin bundle of any hypersurface $\mathcal{H}$ of $\mathcal{X}$, Φ gives the Fefferman conformal metric of the projective twistor $\mathcal{CR}$ structure, $\mathcal{PN}$, of the projective co-spin bundle of $\mathcal{H}$ and determines that structure completely. If the spacetime is real analytic, then $\mathcal{PN}$ embeds as a $\mathcal{CR}$ hypersurface in the space of α-curves of the complexification of $\mathcal{H}$. If the spacetime is conformally flat then $\mathcal{PN}$ is naturally $\mathcal{CR}$ isomorphic to an open subset of $\mathbb{PN}$.

So the Fefferman tensor provides the glue that makes the various $\mathcal{CR}$ structures of the various hypersurfaces in spacetime cohere. It would be desirable to analyse the properties of the Fefferman tensor Φ directly, without first restricting it to a hypersurface. Unfortunately, there has been little or no progress in this direction. Also it should be pointed out that these constructions are purely kinematic, requiring no field equations. It is still unknown, in general, if any additional information is carried by the hypersurface twistor $\mathcal{CR}$ structures, or by the Fefferman tensor, when the spacetime is subject to, say, the Einstein vacuum equations.

4 Abstract

Recently the Penrose Obstruction was overthrown, Sparling (1996). An extra term is inserted in its derivation, which changes the Penrose Obstruction just to part of one equation. We first rewrite the twistor factorization equation as follows:

$$\theta^a = q^{A'}\alpha^A. \tag{4.1}$$

Here $q^{A'}$ is a spinor field and α^A is a one-form (henceforth we use Greek or Latin alphabets for Grassmann odd or even entities, respectively). As before, we get the equations: $d\alpha^A = \gamma\alpha^A$ and $(dq^{A'} + \gamma q^{A'})\alpha^A = 0$, with γ a one-form. Then $dq^{A'} + \gamma q^{A'} = \lambda_0^{A'} z$, where $z \equiv \alpha^A \alpha_A$. A grade count gives $\lambda_0^{A'}$ grade minus one, so $\lambda_0^{A'}$ vanishes. One more derivative then gives the obstruction. To circumvent the obstruction, we simply declare that minus-one-forms exist! So we enlarge the standard algebra of forms with forms of negative rank.

We continue, with $\lambda_0^{A'}$ an indexed minus-one-form. We have $d^2 = zr_0$, where r_0 is a derivation such that $r_0(v^{A'}) = q^{C'}q^{D'}C^{A'B'}_{C'D'}v_{B'} + 2Cq^{A'}q^{B'}v_{B'}$ and $r_0(v^A) = q^{C'}q^{D'}C^{AB}_{C'D'}v_B$, for any spinors $v^{A'}$ and v^A, where C and C_{ab} are proportional to the Ricci scalar and to the trace free part of the Ricci tensor, respectively. Henceforth we work in an Einstein spacetime: one for which $C_{ab} = 0$. Then d^2 and r_0 annihilate unprimed spinors. From the definition of z, we have: $dz = 2\gamma z$. Applying d to this equation and to the equation defining $\lambda_0^{A'}$, we obtain the equations: $d\gamma = x_0 z$, for some scalar x_0 and $z(d\lambda_0^{A'} + 3\gamma\lambda_0^{A'} - x_0 q^{A'} - r_0(q)^{A'}) = 0$. This last relation gives the equation:

$$d\lambda_0^{A'} + 3\gamma\lambda_0^{A'} - x_0 q^{A'} - r_0(q)^{A'} = -\alpha_A \lambda_1^{AA'} = \delta(\lambda_1^{A'}). \tag{4.2}$$

Here λ_1^a is a minus-one-form. Henceforth we work within the contravariant unprimed symmetric spinor algebra $\mathcal{S}$, omitting the unprimed spinor indices of any element of $\mathcal{S}$. Denote by $\mathcal{S}_n$ the subspace of $\mathcal{S}$ spanned by spinors of n-indices. Then δ is a one-form-valued derivation, which ignores primed spinors and which maps each $\mathcal{S}_n$ to $\mathcal{S}_{n-1}$, and whose action on any unprimed spinor $v \in \mathcal{S}_1$ is given by the formula: $\delta(v) = \alpha^A v_A$. Also define the multiplication operator α to be left multiplication by the unprimed spinor valued one-form α^A in the algebra $\mathcal{S}$; then α takes each $\mathcal{S}_n$ to $\mathcal{S}_{n+1}$.

If we contract equation (4.2) with $q_{A'}$, we recover the Penrose Obstruction, if all minus-one-forms are put to zero. Here there is no obstruction. Applying d to the defining equation for x_0, we obtain the formula, for some $x_1 \in \mathcal{S}_1$:

$$dx_0 + 2\gamma x_0 = \delta(x_1). \tag{4.3}$$

Next the Bianchi identity, $d \wedge d^2 = 0$ (where $\wedge$ denotes the commutator of derivations) yields the equation, valid for some $\mathcal{S}_1$-valued derivation r_1, which annihilates unprimed spinors:

$$d \wedge r_0 + 2\gamma r_0 = \delta \wedge r_1. \tag{4.4}$$

Repeated differentiation produces three infinite series $(x_n, r_n, \lambda_n) \in \mathcal{S}^3_n$, where we suppress the primed index of $\lambda_n^{A'}$ and each r_n is an $\mathcal{S}_n$-valued derivation which annihilates unprimed spinors. We find the following result:

Theorem 4.1 *The twistor factorization equation consistently generates an unobstructed infinite differential ideal in an analytic Einstein spacetime, using negative rank forms. The ideal is defined by the following recursions, valid for each non-negative integer n:*

$$\delta(x_{n+1}) = dx_n + (n+2)\gamma x_n + 2\alpha g_n(x, x), \tag{4.5}$$

$$\delta \wedge r_{n+1} = d \wedge r_n + (n+2)\gamma r_n + 2\alpha p_n(r, r) + 2\alpha q_n(x, r), \tag{4.6}$$

$$\delta(\lambda_{n+1}) = d\lambda_n + (n+3)\gamma\lambda_n - x_n q - r_n(q) + 2\alpha g_n(x, \lambda) + 2\alpha p_n(r, \lambda). \tag{4.7}$$

Here $g_n(x, y) \equiv p_n(x, y) + q_n(x, y)$ and we have:

$$p_n(x, y) \equiv \sum_{r+s=n-1} \frac{x_r y_s - x_s y_r}{(r+1)(r+2)}, \tag{4.8}$$

$$q_n(x, y) \equiv \sum_{r+s=n-1} \frac{2x_r y_s(r+s+3)}{(r+1)(r+2)(r+3)}. \tag{4.9}$$

In the case $n = 0$, equations (4.5), (4.6) and (4.7) exactly agree with equations (4.3), (4.4) and (4.2), respectively, given above.

For the proof of Theorem 4 see Sparling (1996). Note that for the proof it is required that, even in the enlarged Grassmann algebra, the indexed one-form α^A behaves just like a conventional pair of one forms. The necessary axiomatics are given in Sparling (1996).

A non-linear gauge freedom arises in the construction of the ideal, primarily because at each stage the operators defining the next stage have non-trivial kernels. A complete analysis of the gauge freedom has been given by Hillman and the author, Sparling and Hillman (1997).

Finally the abstract twistor structure is by definition an equivalence class consisting of the ideal summarized in this section, modulo the equivalence relation produced by the group of gauge transformations. Then we have:

Theorem 4.2 *To any real analytic Einstein spacetime, there is naturally associated an abstract twistor structure.*

5 Virtual

Here we briefly discuss the abstract twistor theory of a null constant plane wave spacetime, with global metric in complex coordinates (u, v, x, y):

$$g = 2dudv - 2dxdy - (ax^2 + cy^2)du^2. \tag{5.1}$$

Here $(a, c) \in \mathbb{C}^2$. A Lorentzian slice arises if $c = \overline{a}$, $y = \overline{x}$, $u = \overline{u}$ and $v = \overline{v}$.

We introduce a null tetrad of one-forms, such that $g = 2ln - 2mm'$:

$$l \equiv du, \quad m \equiv dx, \quad m' \equiv dy, \quad n \equiv dv - (ax^2 + cy^2)du/2. \tag{5.2}$$

The Levi-Civita covariant exterior derivative d is given as follows: $dl_a = 0$, $dm_a = cylla$, $dm'_a = axll_a$ and $dn_a = uxlm_a + cylm'_a$. The curvature, R_{ab}, such that $d^2v_a = -R_{ab}v^b$, for any co-vector v_b is: $R_{ab}{=}-2alml_{[a}m_{b]} - 2clm'l_{[a}m'_{b]}$. This is pure Weyl, so the spacetime is indeed vacuum. Passing to spinors we write $l_a = o_Ao_{A'}$, $m_a = o_A\iota_{A'}$, $m' = \iota_Ao_{A'}$ and $n_a = \iota_A\iota_{A'}$, where $o_A\iota^A = o_{A'}\iota^{A'} = 1$. Then the spin connection is as follows:

$$do_A = 0, \quad d\iota_A = axlo_A, \quad do_{A'} = 0, \quad d\iota_{A'} = cylo_{A'}. \tag{5.3}$$

The Weyl spinors are $C_{ABCD} = -ao_Ao_Bo_Co_D$ and $C_{A'B'C'D'} = -co_{A'}o_{B'}o_{C'}o_{D'}$.

Associated to the algebraic speciality of the primed Weyl spinor, some ordinary twistor two-surfaces exist, those with tangent spinor $o^{A'}$. These are the surfaces with u and y constant. We turn to the abstract twistor surfaces. We write the factorization equation in the form: $\theta^a = q^{A'}ds^A$, which amounts to taking a gauge transformation to eliminate the one-form γ of the previous section. The two components of s^A will serve as the twistor surface co-ordinates.

Taking components of the factorization equation, using equations (5.2) and (5.3), we arrive at the following system of equations:

$$\begin{aligned} du &= pds, \quad dv = qdt + (ax^2 + cy^2 - 2axsq)pds/2, \\ dx &= qds, \quad dy = pdt - axsp^2ds. \end{aligned} \tag{5.4}$$

Here we have put $p \equiv q^{A'}o_{A'}$, $q \equiv q^{A'}\iota_{A'}$, $s \equiv s^Ao_A$ and $t \equiv s^A\iota_{A'}$.

Differentiating equation (5.4), we find the following additional equations:

$$dpds = dqds = dpdt = 0, \quad dqdt = cyp^2dsdt. \tag{5.5}$$

We solve, by taking $dp = 0$ (regarding this as a choice of gauge) and $dq = cyp^2ds + \rho dsdt$, with ρ a minus-one-form. Differentiating again gives the equation: $(d\rho - cp^3)dsdt = 0$. Classically we have $\rho = 0$, so $cp^3 = 0$, which forces self-duality, $c = 0$, or $p = 0$ (the surfaces already discussed). Henceforth we take $c \neq 0$ and p generic ($p \neq 0$), so we must take ρ non-zero. Then we have $d\rho = cp^3 + \lambda ds + \mu dt$, for some minus-one-forms λ and μ. Then $d\rho$ is invertible, so we may define ω by the formula $\omega \equiv \rho(d\rho)^{-1}$. Then $d\omega = 1$ and $dq = cyp^2ds + cp^3\omega dsdt$. Summarizing, we now have the system:

$$\begin{aligned} d\omega &= 1, \quad dp = 0, \quad dq = cyp^2ds + cp^3\omega dsdt, \\ du &= pds, \quad dx = qds, \\ dy &= pdt - axsp^2ds, \quad dv = qdt + (ax^2 + cy^2 - 2axsq)pds/2. \end{aligned} \tag{5.6}$$

It is easily checked that equation (5.6) constitutes a closed differential system. To analyse it, define new variables g and z by the formulas:

$$g \equiv v - q\omega dt, \quad z \equiv y - p\omega dt. \tag{5.7}$$

Without loss of generality we may take $s \equiv up^{-1}$. Also put $r \equiv qp^{-1}$. Direct substitution in equation (5.6) gives the following list of equations:

$$dp = 0, \; dx = rdu, \; dr = czdu, \; dz = -auxdu, \tag{5.8}$$

$$dg = (ax^2 - 2arxu + cz^2)du/2. \tag{5.9}$$

This is an ordinary differential system: neither ω nor t appears explicitly, the latter in keeping with the symmetry of the problem. Equation (5.8) clearly amounts to the hypergeometric equation given in the introductory section above. Having solved this equation, equation (5.9) is a quadrature, giving a complete solution for all variables in the given gauge. For the projective twistor space there are four free parameters, so the projective twistor space has one higher dimension than the traditional twistor spaces. This extra dimension turns out to be enough to give natural projections to all the conventional hypersurface twistor spaces, as will be shown explicitly, in future work, in the case $a = 0$ (when the classical dual twistor space exists, and the hypergeometric equation degenerates, so that all the variables are polynomial in u) and somewhat less explicitly in the full case $a \neq 0$. Also there is a pseudo-Kähler scalar, which controls those of the various hypersurface twistor spaces.

Bibliography

Ko, M., Newman, E. T., and Penrose, R. (1977). "The Kähler structure of asymptotic twistor space", *Journal of Mathematical Physics* **18**, 58–64.

Kodaira, K. (1962). "A theorem of completeness of characteristic systems for analytic families of compact submanifolds of complex manifolds", *Annals of Mathematics*, **75**, 146–162.

Mason, L. J. and Woodhouse, N. M. J. (1996). *Integrability, Self-Duality, and Twistor Theory*, London Mathematical Society Monographs, New Series 15, Oxford: Clarendon Press.

Newman, E. T. (1976). "Heaven and its properties." *General Relativity and Gravitation*, **7**, 107–127.

Ockham, Guillelmi de (1981). "Quaestiones in librum tertium Sententiarum (Reportatio)", in *Opera Theologica, Vol. V*, editors Gedeon Ga'l and Rega Wood, q.18, p. 404, St. Bonaventure, New York: Franciscan Institute.

Penrose, R. (1975). "Twistor theory: its aims and achievements," in *Quantum Gravity: An Oxford Symposium*, editors C. J. Isham, R. Penrose and D. W. Sciama, Oxford: Clarendon Press.

Penrose, R. (1976). "Non-linear gravitons and curved twistor theory," *General Relativity and Gravitation*, **7**, 31–52.

Penrose, R. and MacCallum, M. A. H. (1972). "Twistor theory: an approach to the quantization of fields and space-time," *Physics Reports, Physics Letters C* **6**, 241–315.

Penrose, R. and Rindler, W. (1984). *Spinors and Space-Time. Volume 1: Two-spinor Calculus and Relativistic Fields*, Cambridge: Cambridge University Press.

Penrose R. and Rindler, W. (1986). *Spinors and Space-Time. Volume 2: Spinor and Twistor Methods in Space-Time Geometry*, Cambridge: Cambridge University Press.

Sparling, G. A. J. (1996). "An infinite differential ideal," University of Pittsburgh, preprint.

Sparling, G. A. J. (1997). "The twistor theory of hypersurfaces in space-time," to appear in *Further Advances in Twistor Theory, Vol. III*, editors L. P. Hughston and L. J. Mason, London: Pitman Press.

Sparling, G. A. J. and Hillman, D. (1997). "The gauge structure of abstract twistor theory," manuscript.

Ward, R. S. (1980). "Self-dual space-times with cosmological constant", *Communications in Mathematical Physics* **78**, 1–17.

PART IV

Parallel Session III: Fundamental Questions in Quantum Mechanics

25

Interaction-Free Measurements

Lev Vaidman
School of Physics and Astronomy, Raymond and Beverly Sackler Faculty of Exact Sciences, Tel-Aviv University, Tel-Aviv 69978, Israel

1 The Penrose bomb testing problem

I am greatly indebted to Roger Penrose. I have learned very much from his papers, from his exciting books, and from our (too short) conversations. I am most grateful to Roger for developing the idea of Avshalom Elitzur and myself on interaction-free measurements (IFM). The version of IFM Penrose described in his book (1994) is conceptually different from our original proposal, and although it is much more difficult for practical applications it has the advantage of demonstrating even more striking quantum phenomena. So I will start with presenting Penrose's version of IFM.

Suppose we have a pile of bombs equipped with super-sensitive triggers. The good bombs have a tiny mirror which is connected to a detonator such that if any particle (photon) "touches" the mirror, the mirror bounces and the bomb explodes. Some of the bombs are duds in which the mirror is rigidly connected to the massive body of the bomb. Classically, the only way to verify that a bomb is good is to touch the mirror, but then a good bomb will explode. Our task is to test a bomb without exploding it. We are not allowed to make errors in our test, i.e., to say that a bomb is good while it is a dud, but we may sometimes cause an explosion.

There cannot be a solution by weighing the bomb, or touching the mirror from the side, or any other similar way: the only observable physical difference between a good bomb and a dud is that the good bomb will explode when a single particle will touch the mirror, and the dud will not. Thus, the solution of this task seems to be logically impossible: the only difference between the bombs is that one explodes and the other does not and we are asked to test a bomb without exploding it. Nevertheless, quantum mechanics provides a solution to the problem in a surprisingly simple way.

The method uses the well known Mach-Zehnder interferometer which is shown in Fig. 1. The photons reach the first beam splitter which has transmission coefficient 1/2. The transmitted and reflected parts of their waves are then reflected by two mirrors and finally reunite at another similar beam splitter. Two detectors collect the photons after they pass through the second beam splitter. It is

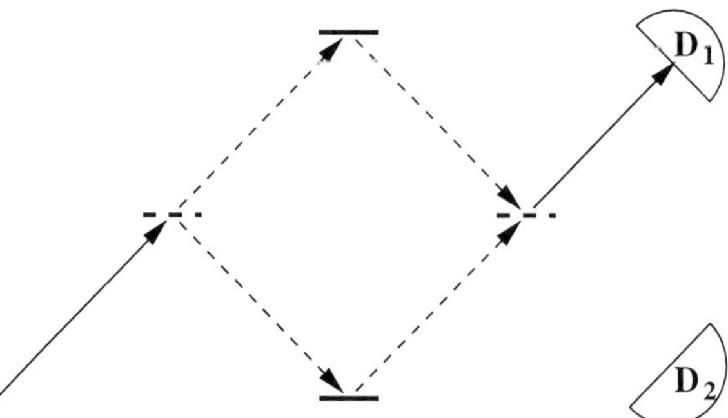

FIG. 1. **Mach-Zehnder Interferometer.** When the interferometer is properly tuned, all photons are detected by D_1 and none reach D_2. The mirrors must be massive enough and have well-defined position.

possible to arrange the positions of the beam splitters and the mirrors in such a way that due to destructive interference no photons are detected by one of the detectors, say D_2, and they all are detected by D_1.

In order to test a bomb we have to tune the interferometer in the way indicated above and replace one of its mirrors by the mirror-trigger of the bomb, see Fig. 2. We send photons through the system. If the bomb is a dud then only detector D_1 clicks. If, however the bomb is good then no interference takes

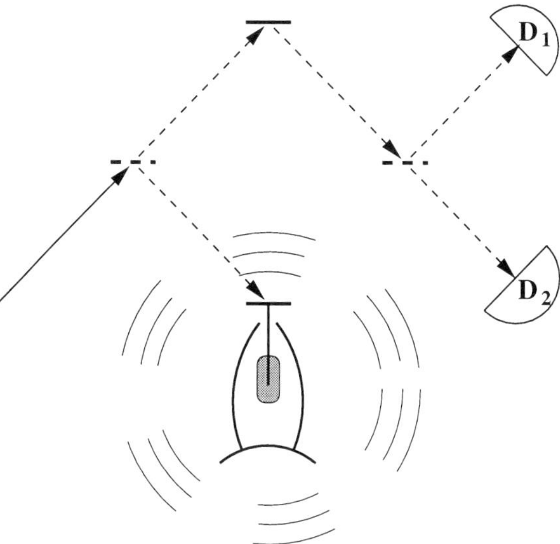

FIG. 2. **The Penrose bomb-testing device.** The mirror of the good bomb cannot reflect the photon, since the incoming photon causes an explosion. Therefore, D_2 sometimes clicks. (The mirror of a dud is connected to the massive body, and therefore the interferometer "works", i.e. D_2 never clicks.)

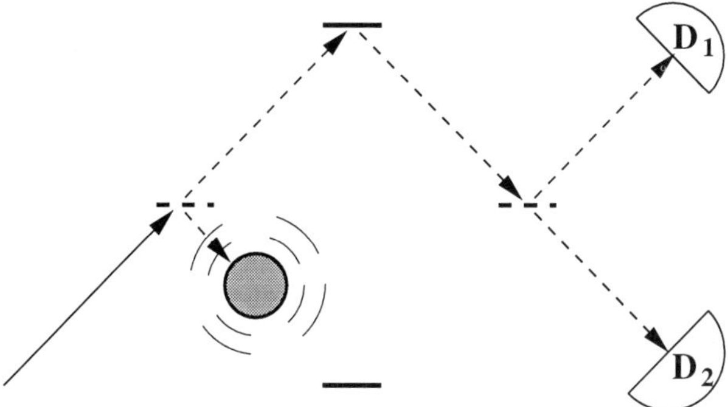

FIG. 3. **Finding an ultra-sensitive mine without exploding it.** If the mine is present, detector D_2 has probability 25% to detect the photon we send through the interferometer and in this case we know that the mine is inside the interferometer without exploding it.

place and there are three possible outcomes: the bomb might explode (probability 1/2), detector D_1 might click (probability 1/4), and detector D_2 might click (probability 1/4). In the latter case we have achieved our goal: we know that the bomb is good (otherwise D_2 could not click) without exploding it.

2 The Elitzur–Vaidman bomb testing problem

Although conceptually Penrose's proposal for testing bombs is very interesting, its implementation is very difficult. The tricky part is replacing the mirror of a tuned interferometer by the bomb. Usually, replacing a mirror requires re-tuning the interferometer, but here it is impossible to do that (the bomb might explode while we are re-tuning). The original bomb testing problem (Elitzur and Vaidman, 1993) is slightly less dramatic, but it can and has been implemented in a real experiment. This problem is equivalent to the task of finding an ultra-sensitive mine without exploding it. The solution is similar to the above: we tune the Mach-Zehnder interferometer to have no photons detected by D_2. Then we place the interferometer in such a way that the mine (if present) blocks one of the arms of the interferometer, see Fig. 3. All the discussion above about the operation of the device is the same, but the main difficulty is absent: there is no need to fix the exact position of the mine.

The efficiency of finding a good bomb (mine) without exploding it in the above procedure is only 25%. By modifying the transmission coefficients of the beam splitters and repeating the procedure in case of no explosion we can (almost) reach the efficiency of 50%. Even more surprisingly, the efficiency can be made as close as we want to 100% by integrating the idea of the IFM with the quantum Zeno effect, see Kwiat *et al.* (1995).

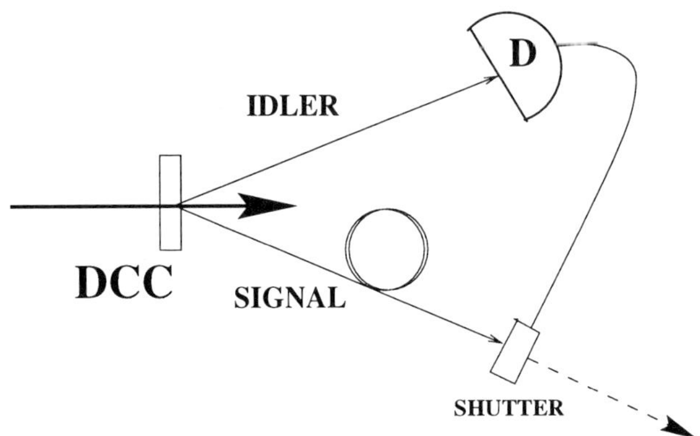

FIG. 4. **Single-photon gun.** The signal photon comes from the down-conversion crystal DCC and stored in the delay ring during the time that the idler photon activates a detector which sends a fast electronic signal to open a shutter for a short time on the way of the signal photon.

3 Experimental realization of the IFM

Two experiments verifying IFM have been performed. The first (Kwiat *et al.* 1995) used a down-conversion crystal as a photon source and state of the art optical components. The "bomb" was an efficient detector and the "explosion" was a registration of the count by a computer. A statistical analysis of thousands of counts was performed and, after normalization according to the noise and the efficiency of the detectors, the results confirmed the prediction of quantum theory with a very good precision. The second experiment was performed as a demonstration at the science fair in Groningen (du Marchie van Voorthuysen, 1997). In this experiment a standard Mach-Zehnder interferometer was operated. Occasionally a detector (connected to a loud bell in order to simulate the explosion) was inserted along one arm of the interferometer. A dimmed laser light entered the interferometer. The visitors could see that without the bell-detector usually D_1 clicked first. When the bell-detector was present, in most of the cases they first heard the bell, sometimes D_1, but also, in many cases D_2 clicked first, thus telling us that the bell-detector is inside the interferometer.

The technical achievements of the two experiments cannot be compared, but the latter had one advantage: it was really testing (although inefficiently and not very reliably) the presence of the bomb (the bell-detector). The first experiment was much more precise, but it was used more to test quantum mechanics, than to find an object without "touching" it. In the experimental run the detector which played the role of the bomb was triggered many many times. It can be considered an IFM only during the short time windows defined by the clicks of the detections of the "idler" photons coming from the down-conversion crystal.

This type of experiment with coincidence counts is commonly considered as "a single-photon" experiment. (In the Groningen experiment there was no attempt to use a single-photon source at all.) I think, however, that it is conceptually important to perform an IFM with a real single-photon source, a device which emits when commanded just one photon. I may call it a single-photon gun, see Fig. 4. In such an experiment there is another paradoxical aspect: we can get information about a region of space never visited by any particle, see Vaidman (1994a).

4 Generalized IFM

Let us consider now a more general task which can be considered as an IFM, see Vaidman (1994b). We have to verify that the system is in a certain state, say $|\Psi\rangle$. The state $|\Psi\rangle$ is such that its detection causes an explosion or destruction: destruction of a system, of a measuring device, or at least of the state $|\Psi\rangle$ itself. The states orthogonal to $|\Psi\rangle$ do not cause the destruction. Although the only physical effect of $|\Psi\rangle$ is an explosion which destroys the state, we have to detect it without any distortion.

Let us assume that if the system is in a state $|\Psi\rangle$ and a part of the measuring device is in a state $|\Phi_1\rangle$, we have an explosion. For simplicity, we will assume that if the state of the system is orthogonal to $|\Psi\rangle$ or the part of the measuring device which interacts with the system is in a state $|\Phi_2\rangle$ (which is orthogonal to $|\Phi_1\rangle$) then neither the system nor the measuring device changes their state:

$$\begin{aligned} |\Psi\rangle\ |\Phi_1\rangle &\to |expl\rangle \\ |\Psi_\perp\rangle|\Phi_1\rangle &\to |\Psi_\perp\rangle|\Phi_1\rangle \\ |\Psi\rangle\ |\Phi_2\rangle &\to |\Psi\rangle\ |\Phi_2\rangle \\ |\Psi_\perp\rangle|\Phi_2\rangle &\to |\Psi_\perp\rangle|\Phi_2\rangle. \end{aligned} \tag{4.1}$$

Now, let us start with an initial state of the measuring device $|\chi\rangle = \alpha|\Phi_1\rangle + \beta|\Phi_2\rangle$. If the initial state of the system is $|\Psi\rangle$, then the measurement interaction is:

$$|\Psi\rangle|\chi\rangle \to \alpha|expl\rangle + \beta|\Psi\rangle|\Phi_2\rangle = \alpha|expl\rangle + \beta|\Psi\rangle(\beta^*|\chi\rangle + \alpha|\chi_\perp\rangle), \tag{4.2}$$

where $|\chi_\perp\rangle = -\beta^*|\Phi_1\rangle + \alpha|\Phi_2\rangle$. If, instead, the initial state of the system is orthogonal to $|\Psi\rangle$, then the measurement interaction is:

$$|\Psi_\perp\rangle|\chi\rangle \to |\Psi_\perp\rangle|\chi\rangle. \tag{4.3}$$

To complete our measuring procedure we perform a measurement on the part of the measuring device to distinguish between $|\chi\rangle$ and $|\chi_\perp\rangle$. Since there is no component with $|\chi_\perp\rangle$ in the final state (4.3), it can be obtained only if the initial state of the system had the component $|\Psi\rangle$. This is also the final state of the system: we do not obtain $|\chi_\perp\rangle$ in the case of the explosion.

A Mach-Zehnder interferometer is a particular implementation of this scheme. Indeed, the photon entering the interferometer can be considered as a measuring device prepared by the first beam splitter in a state $|\chi\rangle = \frac{1}{\sqrt{2}}(|\Phi_1\rangle + |\Phi_2\rangle)$, where $|\Phi_1\rangle$ designates a photon moving in the lower arm of the interferometer, and $|\Phi_2\rangle$ designates a photon moving in the upper arm. Detector D_2 together with the second beam splitter tests for the state $|\chi_\perp\rangle = \frac{1}{\sqrt{2}}(|\Phi_1\rangle - |\Phi_2\rangle)$.

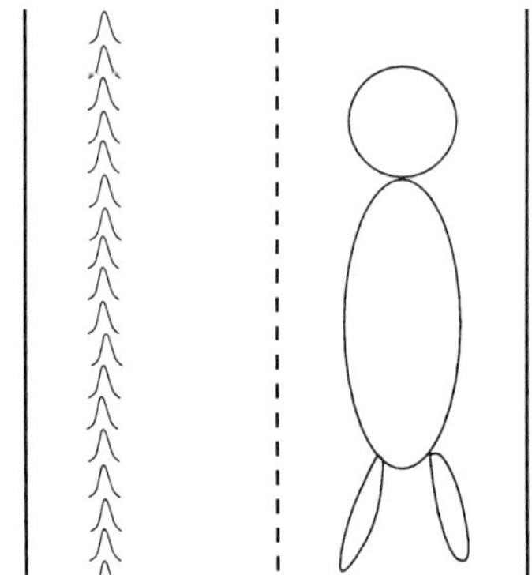

FIG. 5. **Safe X-ray photography.** Between two parallel mirrors another mirror which reflects 99.9% and transmits 0.1% is introduced. A photon starting on the left side will then end up on the right side after about 50 bounces. If, however, on the right side there is an object which can absorb the photon, then with 95% probability the photon will stay on the left. To make a photographic picture, a person enters into the right side and the photons are placed on the left side. After the time required for 50 bounces a photographic plate is introduced in the left side. In such an experiment the bones absorb only 5% of the radiation, i.e. twenty times less than in the the standard method. (It is assumed that the soft tissue does not affect the X-rays at all, while the bones are opaque.)

5 Applications of the IFM

In principle, the IFM may have many dramatic practical applications. For example, if we have a method of selecting a certain bacteria which also kills it, the IFM may allow us to select many such live bacteria. One can fantasize about safe X-ray photography, see Fig. 5. One can design a scheme of a computer which will allow knowing the result of a computation without computing, see Jozsa (1996). These are gedanken ideas, but I am optimistic about finding situations in which the IFM will have some real practical applications.

6 The IFM as counterfactuals

The IFM also has interesting philosophical implications. Let me quote Roger Penrose (1994, p.240):

> What is particularly curious about quantum theory is that there can be actual physical effects arising from what philosophers refer to as *counterfactuals* – that is, things that might have happened, although they did not happen.

For me this paradoxical situation gives another reason why we should accept the many-worlds interpretation (MWI) of quantum theory. According to the MWI, in the situations considered by Penrose, "things" did not happen in a particular world, but did happen in some other world (see Vaidman 1994a). Therefore, they did take place in the physical universe which incorporates all the worlds and thus their effect is not so surprising.

The research was supported in part by grant 614/95 of the the Basic Research Foundation (administered by the Israel Academy of Sciences and Humanities).

Bibliography

du Marchie Van Voorthuysen, E.H. (1997). Realization of an interaction-free measurement of the presence of an object in a light beam. *Am. J. Phys.*, **64**, 1504–1506.

Elitzur, A. and Vaidman, L. (1993). Quantum mechanical interaction-free measurements. *Foundation of Physics* **23**, 987–997.

Jozsa, R. (1996). Counterfactual quantum computation. University of Plymouth preprint, unpublished.

Kwiat, P., Weinfurter, H., Herzog, T., Zeilinger, A., and Kasevich, M. (1995). Interaction-free quantum measurements. *Phys. Rev. Lett.* **74**, 4763–4766.

Penrose, R. (1994). *Shadows of the Mind.* Oxford: Oxford University Press.

Vaidman, L. (1994a). On the paradoxical aspects of new quantum experiments. *Philosophy of Science Association 1994*, pp. 211–217.

Vaidman, L. (1994b). On the realization of interaction-free measurements. *Quant. Opt.* **6**, 119.

26
Quantum Measurement Problem and the Gravitational Field

Jeeva Anandan
Max-Planck-Institut für Physik
Föhringer Ring 6, D-80805 Munich, Germany
and
Department of Physics and Astronomy
University of South Carolina, Columbia, SC 20208, USA

1 Introduction

Two of the most important unsolved problems in theoretical physics are the problem of quantizing gravitation and the measurement problem in quantum theory. It is possible that the solution of each one needs the other. Since we have successful quantum theories of electroweak, and strong interactions, the solution to the problem of the collapse of the wave function, known as the measurement problem, may lie in the yet unknown quantum theory of the remaining interaction, namely quantum gravity. On the other hand, the numerous unsuccessful attempts to construct a quantum theory of gravity for more than six decades on the assumption that the present linear quantum theory is correct suggests that perhaps not only general relativity but also quantum theory should be modified in order to construct of satisfactory quantum theory of gravity.

In section 2, I shall briefly review the measurement problem and protective observations. I shall argue, in section 3, that none of the standard interpretations of quantum theory provide a solution for the measurement problem. This suggests that a modification of quantum theory may be needed, particularly since protective observation suggests that the wave function is real and therefore the reduction of the wave packet during measurement is a real objective process. I then consider, in section 4, the suggestion along the lines mentioned above due to Roger Penrose (1981, 1986, 1989, 1993, 1994, 1996). He advocated the use of the gravitational field of the wave function to explain its reduction during measurement. While several other physicists have also argued that the phenomenon of state vector reduction is an objective, real process, and not just a change in the state of knowledge of the observer (e.g. Karolyhazy 1966; Pearle 1986, 1989), an important aspect of Penrose's proposal is that he has quantitative predictions for the time of collapse of the wave function, which has the potential of being subject to experimental tests (see also Diosi 1987, 1989; Ghiradi, Grassi, and

Rimini 1990). Conversely, experiments could guide us in constructing a definite theory (which still does not exist) that would justify or modify this proposal. I then generalize this proposal for quantum superpositions of weak gravitational field states that are arbitrary in the absence of matter and stationary in the presence of matter.

2 Quantum measurement problem

The simplest, though dramatic, statement of the measurement problem in quantum theory is that *quantum theory does not explain the occurrence of events.* So, quantum theory does not explain the first thing we observe about the world around us.

To see this, consider a quantum system whose state is described by a wave function ψ just before it interacts with an apparatus, which we shall treat quantum mechanically also. Suppose $\psi = \sum_i c_i \psi_i$, where ψ_i is a state of the system, which after it interacts with the apparatus leaves it in the state that is described by the wave function $\psi'_i \alpha_i$, where ψ'_i represents the new state of the system and α_i the corresponding state of the apparatus. We represent this by

$$\psi_i \alpha \rightarrow \psi'_i \alpha_i \tag{2.1}$$

Then it follows from the linearity of quantum evolution that the interaction of ψ with the screen is represented by

$$\psi \alpha \rightarrow \sum_i c_i \psi'_i \alpha_i . \tag{2.2}$$

The resulting state is called an entangled state, meaning that it cannot be written as simple product of the form $\psi' \alpha'$.

For example, the quantum system may be a photon and the apparatus a photographic plate. Then ψ_i is a localized wave function of the photon which interacts with the plate to trigger a chemical reaction which results in a spot on the screen represented by α_i. But ψ produces a quantum superposition of many different spots on the screen that correspond to the different states ψ_i of the photon. Since the photon has now been absorbed, $\psi'_i \alpha_i$ in the right hand side of (2.1) and (2.2) may be replaced simply by α_i. We would not call the resulting state entangled. Nevertheless, the experiment (2.1) establishes a correlation between the state ψ_i and α_i. So, if we observe α_i, whether or not the quantum system is now present, we can deduce the state of the system ψ_i that would have caused this state of the apparatus. So, what is essential to measurement is this *correlation* between the system states and the corresponding apparatus states and not entanglement.

However, for a given photon in the state ψ we actually observe only one spot described, say, by α_k. This appearence of a spot may be regarded as an approximate representation of an event, because it occurs in a fairly localized region of space-time that is defined by the small spatial region on the screen and the small interval of time during which it is formed. But (2.2) by itself does not explain the appearance of this 'event'. So, we need to make an additional 'projection postulate'

$$\sum_i c_i \psi'_i \alpha_i \to \psi'_k \alpha_k, \tag{2.3}$$

where $\psi'_k \alpha_k$ represent the particular event or set of events observed. The quantum measurement problem is the problem of understanding (2.3), which is referred to as the reduction of the wave packet or collapse of the wave function. For example, is (2.3) an objective dynamical process, which we may take (2.2) to be, or is it a subjective process we make in our minds due to the additional information we obtain from the measurement? Or what determines the *preferred states* α_i into which the reduction takes place?

So, the state vector undergoes two types of changes (Wigner 1963), which using the terminology of Penrose (1989, 1996), may be called the U and R processes. (2.1) and (2.2) are U processes, whereas (2.3) is the R process. The U process is the linear unitary evolution which in the present day quantum theory is governed by Schrödinger's equation. But what causes the measurement problem is the linearity of the U process. The unitarity is really relevant to the R process. Unitarity ensures that the sum of the probabilities of the possible outcomes in any measurement, each of which is given by an R process remains constant during the U time evolution. This of course follows from the postulate that the transition probability from the initial to the final state in the R process is the square of the modulus of the inner product between normalized state vectors representing the two states.

The process of measurement, as described above, takes place in two stages: first is the entanglement (2.2) and second is the collapse (2.3). If we had no choice in preparing the initial state of the system then ψ would be in general a superposition of the ψ_is. Then the entanglement (2.2) would be the inevitable consequence of the linearity of the evolution. But if we could prepare the state then it is possible to prevent entanglement as in the case of protective observation (Aharonov, Anandan and Vaidman 1993). That is, in such an observation

$$\psi\alpha \to \psi'\alpha', \tag{2.4}$$

where the state represented by ψ' does not differ appreciably from the state ψ. The protection is usually an external interaction which puts ψ in an eigenstate of the Hamiltonian and the measurement process results in adiabatic evolution.

Then α' gives information about ψ; specifically it tells us the 'expectation value' with respect to ψ of the obervable of the system that it is coupled to an apparatus observable. By doing such experiments a large number of times it is possible to determine ψ (up to phase) even though the system is always undergoing U evolution. Consequently, the statistical interpretation of quantum mechanics is avoided during protective observations. Indeed, ψ may be determined using just one system which is subject to many experiments.

If the protection mechanism is precisely known then it would be possible to determine the ψ by means of calculation. But there is a profound difference between experimentally observing the state, which gives the manifestation of the state, and calculating it. Also, the protected state need not be in an eigenstate of the observable being measured, and yet there is no entanglement. It may appear that if the combined system evolves as (2.4), as in a protective measurement, then we cannot obtain new information about the system state, because if this state were previously unknown then there should be the possibility of the system being in more than one state with respect to the apparatus, i.e. there should be entanglement or correlation between system states and apparatus states. This is true if the system states already have a well defined meaning.

However, a state acquires meaning through its relation to other states. For example, describing a state vector $|\psi\rangle$ by means of its wave function is the same as giving the inner product of $|\psi\rangle$ with all the eigenstates of the position operator. A previously proved theorem (Anandan 1993) states that from the 'expectation values' defined as functions on the set of physical states, it is possible to construct the Hilbert space whose rays are these states. Indeed the entire machinery of quantum mechanics may be constructed from the numbers which an experimentalist obtains by protective measurements. Before the Hilbert space is reconstructed, it is not possible to calculate the wave function. However, from the information which can in principle be obtained from protective measurements, it is possible to determine the inner products between a given state vector and all other state vectors, according to the above mentioned theorem, which gives meaning to the given state vector. So if the meanings of the states are previously unknown, then in this way it is possible to obtain new information that determines the state of a system by means of protective observations, even though the evolution is according to (2.4). Also, this is done by using just one quantum system in the given state. No statistical interpretation of the wave function is needed. This suggests that the wave function may be real and objective.

3 Efforts to resolve the measurement problem

The well known Copenhagen interpretation attempts to deal with the measurement problem by introducing an artificial division between the quantum system being observed and the apparatus. The quantum system, which was assumed to be 'microscopic', is treated quantum mechanically. Its state evolves in a Hilbert space. The apparatus, assumed to be 'macroscopic', is treated classically. The discontinuous R process occurs when the microscopic system interacts with the macroscopic system. This is accounted for by supposing that the wave function represents only our knowledge of the state of the system, and this knowledge undergoes a discontinuous change when the measurement is made.

This is unsatisfactory because the apparatus is made up of electrons, protons, neutrons and photons, which are clearly quantum mechanical. At the time when the Copenhagen interpretation was formulated, it was not known that a macroscopic superconductor must be treated quantum mechanically. Also, Bose-

Einstein condensation, which provides another clearly macroscopic quantum system, was not experimentally realized. Today we know and possess macroscopic quantum systems. Moreover, there have been numerous quantum mechanical experiments on a macroscopic scale, using superconductors, electron, neutron and atomic interferometry. Another related problem is that the Copenhagen interpretation does not specify the line of division between the system and the apparatus. It does not give a number, specifying the complexity or mass of the system, which when exceeded would make the system macroscopic. Also, the early universe needs to be treated quantum mechanically because quantum gravitational effects were so important at that time. But nothing can be more macroscopic than the universe. And the universe is everything there is, so no line of division can be specified between it and the apparatus. Finally, protective observation, discussed in section 1, suggests that the wave function is real and objective, and is not just our knowledge of the system.

This brings us to two famous interpretations of quantum theory in which the wave function is regarded as real, consistent with the meaning of the wave function given by protective observation. One is the Everett interpretation (Everett 1957) in which the wave function never collapses. But this view carries with it a huge excess intellectual baggage in the form of infinitely many worlds that coexist with the world in which we observe ourselves in. Also, it does not seem to explain the 'preferred basis' or the 'interpretation basis' in which we observe the world to have a fairly classical space-time description. The latter description is very different from the Hilbert space description which is the only reality in the Everett view. Furthermore, since the Everett view gives a deterministic description of a real state vector, the only natural way of introducing probabilities is by coarse graining. But this would not agree with the probabilities determined by the inner product in Hilbert space which is well confirmed by experiment.

The Bohm interpretation (Bohm 1952) tries to overcome the old problem of wave–particle duality by asserting the simultaneous existence of both the particle and the wave. This dual ontology enables one to have the cake and eat it too. Direct experimental evidence of a particle, such as the triggering of a particle detector, or a track in a cloud chamber, etc., is explained as caused by the particle. And this is the *only* role of the particle. The motion of the particle is assumed to be guided by a 'quantum potential' determined by the wave function which explains, for example, the result of an interference experiment. Without the particle the Bohm interpretation would be like the Everett interpretation in that there is no collapse of the wave function. But the particles determine which branch of the wave function we (i.e. the particles constituting us) are in. So, there is no excess intellectual baggage of the many worlds as far as the particles are concerned. But there are the 'empty waves' of the other branches. These waves may be protectively observed and therefore may be regarded as real (Anandan and Brown 1995). This has the advantage over the previous interpretations in that there is no preferred basis problem because the particles determine 'events', e.g. spots on the photographic plate, which give the illusion of a preferred basis in the Hilbert space.

The absence of any further role for the particle is illustrated by the fact that the particle does not react back on the wave. This violates the action–reaction principle, which may be regarded as a metaphysical objection to the Bohm interpretation (Holland 1993; Anandan and Brown 1995). As mentioned above, the only reason for introducing the particle in this interpretation is in order to explain the occurrence of events, which gives the familiar space-time description. But since the space-time description is intimately associated with the gravitational field, perhaps we should explain the occurrence of these events during measurement by using the gravitational field, instead of postulating a 'particle' (whose existence may be doubted because it does not satisfy the action–reaction principle) as the 'cause' of this event.

It is also strange that in the Bohm theory the wave function plays a dual role, namely the ontological role of guiding the particle, and the epistemological role of giving initially at least the usual prescription for the probability density of finding the particle. Also, parameters such as charge, mass, etc., which are usually associated with the particle are spread out over the wave and not localized on the particle in the Bohm picture (Brown, Dewdney and Horton 1995; Anandan and Brown 1995). Finally, when one goes over to quantum field theory, the ontology undergoes a sudden change because the particle is replaced by the classical field and it is not clear what its relation is to the previous ontology.

In the Feynman path integral formalism of quantum theory, the measurement problem does not seem to occur, at least not explicitly. I asked Eugene Wigner this question in a seminar he gave in 1976, and I am sure that others must have thought about this question too. Recently, it has been advocated as a solution to the measurement problem by Kayser and Stodolsky (1995). In this approach one assumes 'events', such as the spots on a photographic plate, to be a primitive concept. Only these events are considered to be real. Given an event A caused by a system, quantum mechanics gives the probability amplitude for a subsequent event B to be caused by the same system. This is obtained by summing the probability amplitudes associated with the different paths by which the system may go from A to B. Here, as in the Copenhagen interpretation, but not the Everett nor Bohm interpretation, the wave function is not real. It is the probability amplitude for different possible events, and is therefore a prescription for the statistical prediction of these events. It may then appear that there is no measurement problem because we can deal directly with probability amplitudes without a wave function which undergoes a mysterious collapse.

However, the measurement problem may still be formulated by means of the following three questions in the amplitude language, with the translation into the wave function language given in parentheses. 1) When do we convert probability amplitudes into probabilities? (Criterion for macroscopicity of the apparatus?) 2) Why is only one of the many possible events with non-zero probability amplitude realized in a particular experiment. (Collapse problem.) 3) Why don't we see a superposition of the states that are actually observed in experiments for which there is also a non-zero probability amplitude? (The preferred basis problem.) The wave function may be regarded as the probability amplitudes to observe the particle at various points in space which then relates the above questions to the corresponding questions in the wave function language in parentheses. One cannot make the measurement problem go away by simply changing the language, which was Wigner's answer to my question.

Although I rejected the Copenhagen interpretation as unsatisfactory, it may nevertheless be telling us something important. The apparatus being 'classical' simply means that it should be given a space-time description. So, the preferred basis associated with the reduction mentioned in the previous section consists of states which appear 'classical', i.e. they have a well defined space-time representation. The quantum system, on the other hand, has its states in the Hilbert space. But the space-time geometry is very different from the natural geometry for quantum theory which is obtained from the Hilbert space (see for e.g. Anandan 1991). The gravitational field is now incorporated into the geometry of space-time. Indeed the difficulty in constructing a quantum theory of gravity may be due to these very different geometries for space-time and Hilbert space. But the R process brings these two geometries in contact with each other because of the formation of events when we attempt to observe the Hilbert space state vector (Anandan 1980). This suggests that the gravitational field may be involved in this process. If the gravitational field, which is intimately connected with space-time, causes the reduction of the wave packet, then this may explain why the states into which the collapse takes place have a well defined space-time description. Also, this argument suggests that it is not necessary to go down to the scale of Planck length for quantum gravitational effects to become important, because the above problem of relating the Hilbert space geometry to space-time geometry, which is required by the reduction of the wave packet, exists even at much bigger length scales.

4 Gravitational reduction of the wave packet

If the wave function is real, as implied by protective observation, it is likely that its collapse or reduction is also a real process. Also, as argued in the previous section, none of the interpretations of present day quantum theory advanced so far are satisfactory. We should therefore be open to the possibility of having to modify quantum theory. Several schemes have been proposed without involving the gravitational field to describe the R process, notably due to Pearle (1989), and Ghiradi, Rimini and Weber (GRW) (1986). However, as argued at the end of section 2 it is plausible that the gravitational field is involved in the reduction. Indeed several suggestions for such a reduction have been made (Karolyhazy 1966; Pearle 1981, Ghiradi, Grassi and Rimini 1990). But I shall consider here only a recent specific proposal by Penrose (1996).

To fix our ideas, consider the Stern-Gerlach experiment for a spin-half particle such as a neutron. It is well known that as the neutron passes through the inhomogeneous field of the Stern-Gerlach apparatus, its wave function splits into two, and when it interacts with a screen the combined wave function of the neutron and the screen also splits into two, as they undergo the linear U process of quantum mechanics. But the gravitational fields of the two states are different. So, if the gravitational field is to be treated quantum mechanically then the new state is the superposition

$$\Psi = \lambda|\psi_1\rangle|\alpha_1\rangle|\Gamma_1\rangle + \mu|\psi_2\rangle|\alpha_2\rangle|\Gamma_2\rangle = \lambda|\Psi_1\rangle + \mu|\Psi_2\rangle, \tag{4.1}$$

where $|\psi_1\rangle$ and $|\psi_2\rangle$ represent the states of the neutrons, $|\alpha_1\rangle$ and $|\alpha_2\rangle$ the corresponding quantum states of the screen with the different positions of the spot where the neutron strikes, $|\Gamma_1\rangle$ and $|\Gamma_2\rangle$ are the coherent states of the gravitational field, and $|\Psi_1\rangle$ and $|\Psi_2\rangle$ represent the states of the combined system. Interesting consequences of a superposition of states of a macroscopic system of the form (4.1) for a cosmic string have been obtained elsewhere (Anandan 1994, 1996).

Penrose (1996) argues that in the superposition (4.1) there must necessarily be a 'fuzziness' in the time translation operator and a corresponding 'fuzziness' in the energy. This is important for the following reason. In a dynamical collapse model, such as Penrose's being considered here, typically there is violation of conservation of energy-momentum. In the GRW scheme, this violation occurs very rarely and so, it was claimed, it cannot be detected in the usual experiments. But conservation laws are consequences of symmetries, which are to me the most fundamental aspects of physics. This is illustrated by the fact that although, as mentioned above, the Hilbert space geometry and space-time geometry are very different, they have in common the action of the Poincare symmetry group on both of them, as if this symmetry group is ontologically prior to both descriptions. I expect symmetries of laws of physics and the conservation laws which they imply to be more lasting than the laws themselves. I therefore would not like even a rare violation of the conservation of energy-momentum. The 'fuzziness' of time translation that Penrose mentions, which may be extended also to spatial translations, may change the present laws just so as to altogether prevent the violation of energy-momentum conservation.

The uncertainty of energy associated with this 'fuzziness', according to Penrose, makes superpositions of the form (4.1) unstable. This is analogous to how the uncertainty of energy ΔE of a particle makes it unstable giving it a lifetime of the order of $\frac{\hbar}{\Delta E}$. It is therefore reasonable to suppose that the superposed states in (4.1) should decay into one or other of the two states, which we observe to happen in a Stern-Gerlach experiment. The lifetime may be postulated to be

$$T = \frac{\hbar}{E}, \tag{4.2}$$

where E is to be determined. Penrose considers the special case of the state Ψ being an equal superposition of two states of a lump of mass, each of which produces a static gravitational field. In the Stern-Gerlach experiment considered above, this corresponds to the spin state being perpendicular to the inhomogeneous magnetic field. Define now a quantity which has dimension of energy

$$\Delta = \frac{1}{G}\int (\nabla\Phi_1 - \nabla\Phi_2)^2 d^3x, \tag{4.3}$$

where Φ_1 and Φ_2 are the Newtonian gravitational potentials of the two lump states, and G is Newton's gravitational constant. Penrose postulates that E is some numerical multiple of Δ.

Two questions which arise now are whether Penrose's postulate can be obtained in some natural way and how it could be generalized. I shall try to answer both questions. Note first that the classical gravitational field corresponds to the mean value of the metric operator $\hat{g}_{\mu\nu}$ and the connection operators $\hat{\Gamma}^{\mu}{}_{\nu\rho}$. Quantum gravitational effects, however, depend on the fluctuation of the gravitational field. The *fluctuation* of a component of the connection $\Delta\Gamma$ is given by

$$\Delta\Gamma^2 = \sum \int \langle\Psi|(\hat{\Gamma}^{\rho}_{\mu\nu} - \langle\Psi|\hat{\Gamma}^{\rho}_{\mu\nu}|\Psi\rangle)^2|\Psi\rangle d^3x \tag{4.4}$$

where the sum is over a set of independent components of the connection which is explained below.

If Ψ is an eigenstate of $\hat{\Gamma}$ then (4.4) vanishes, and the geometry is essentially classical. So, we would not expect it to decay, i.e. T is infinite. It is reasonable therefore to take E to be proportional to some positive power of $\Delta\Gamma$. Since $\frac{1}{G}\Delta\Gamma^2$ has the dimension of energy, I postulate that

$$E \sim \frac{1}{G}\Delta\Gamma^2. \tag{4.5}$$

Thus E is a measure of the fluctuation of the quantum gravitational field.

Of course, (4.4) and therefore (4.5) depend on the gauge in which the gravitational field is quantized. So, to make (4.5) physically meaningful, it is necessary to eliminate the gauge degrees of freedom by quantizing the connection coefficients in an appropriate gauge in which these coefficients are unique. Consider the linearized approximation so that the gravitational field is treated as a spin 2 field $h_{\mu\nu} = g_{\mu\nu} - \eta_{\mu\nu}$, where $\eta_{\mu\nu}$ is the Minkowski metric. I shall assume the usual 'Lorentz gauge' condition $\bar{h}^{\mu\nu}_{\ ,\nu} = 0$, where $\bar{h}^{\mu\nu} = h^{\mu\nu} - \frac{1}{2}\eta^{\alpha\beta}h_{\alpha\beta}\eta^{\mu\nu}$. The latter condition does not fix the gauge uniquely. In the absence of matter further specialization gives the transverse, traceless gauge, which is unique (see e.g. Misner, Thorne and Wheeler 1973). I shall take the quantized metric components in this gauge to be the physical degrees of freedom of the gravitational field, and therefore make the prediction (4.2) with (4.5) expressed in terms of the fluctuations of the corresponding connection coefficients. In the presence of matter, the choice of a natural gauge requires further investigation. But in the special case of a stationary situation, we can fix the gauge uniquely by supplementing the above 'Lorentz gauge' condition by the reasonable requirement that the metric is time independent. Then, (4.4) and therefore (4.5) may be defined in this gauge. Once the prediction (4.2) is made in a particular gauge, it is unique. One may then transform the expression of this prediction to one's favourite gauge.

Consider now the superposition of two gravitational fields of the form (4.1), where $|\Psi_1\rangle$ and $|\Psi_2\rangle$ are approximate eigenstates of $\hat{\Gamma}$ with eigenvalues Γ_1 and Γ_2. Then, from (4.4),

$$\Delta\Gamma^2 = \sum \int \{|\lambda|^2(1-|\lambda|^2){\Gamma_1}^2 + |\mu|^2(1-|\mu|^2){\Gamma_2}^2 - 2|\lambda|^2|\mu|^2\Gamma_1\Gamma_2\} d^3x. \quad (4.6)$$

Consider first, for simplicity, the quantized Newtonian gravitational field. Then, in the Newton-Cartan theory, I shall take a Galilean coordinate system (Misner, Thorne and Wheeler 1973) to be the gauge in which (4.5) is defined. The only non-vanishing components of the connection for the two superposed gravitational fields in this gauge are ${\Gamma_1}^i{}_{00} = -\frac{\partial\Phi_1}{\partial x^i}$ and ${\Gamma_2}^i{}_{00} = -\frac{\partial\Phi_2}{\partial x^i}$. Therefore, in the special case considered by Penrose for which $\lambda = \mu = \frac{1}{\sqrt{2}}$, from (4.5) and (4.6),

$$E \sim \frac{1}{4G}\int (\nabla\Phi_1 - \nabla\Phi_2)^2 d^3x. \quad (4.7)$$

This E is proportional to Δ given by (4.3) in agreement with Penrose's result.

In the linearized limit mentioned above, the gauge or coordinate system may be chosen such that each of the two superposed stationary fields is time independent. Then, in this case, there are also the following other non-zero connection coefficients: ${\Gamma_1}^i{}_{jj} = -\frac{\partial\Phi_1}{\partial x^i}, {\Gamma_2}^i{}_{jj} = -\frac{\partial\Phi_2}{\partial x^i}, j = 1,2,3$. ${\Gamma_1}^\mu{}_{i\mu} = \frac{\partial\Phi_1}{\partial x^i}, {\Gamma_2}^\mu{}_{i\mu} = \frac{\partial\Phi_2}{\partial x^i}, \mu = 0,1,2,3$. Therefore, the order of magnitude of $\Delta\Gamma^2$ is still the same as before. So, we still obtain (4.7) for the special case considered by Penrose.

Hence, (4.5) generalizes Penrose's ansatz in three ways. We can now predict the order of magnitude of T for arbitrary coefficients λ and μ in (4.1). Second, (4.2) is valid for superpositions of more than two lump states. For example, if we do a Stern-Gerlach experiment for a spin 3/2 particle, there would be a superposition of *four* spots on the screen according to Schrödinger's equation. The order of magnitude of the time of collapse of this superposition may be obtained from (4.2) and (4.5). Finally, we can now obtain T not only for arbitrary superpositions of static gravitational fields but also more generally for stationary gravitational fields and non-stationary fields in the absence of matter for which there are other components of $\Gamma^\mu{}_{\nu\rho}$ besides the ones mentioned above. For example, the above results may be applied to Leggett's proposed experiment to realize the quantum superposition of two currents in a SQUID (Leggett 1980; Chakravarty and Leggett 1984; Leggett and Garg 1985). In this case, owing to the Lense-Thirring fields of these currents, $\Gamma^i{}_{j0}$ components are also non-zero, and their fluctuation should contribute to the reduction of the superposition of the wave functions corresponding to these currents.

The question of how well the above predictions agree with experiment, for example the superposition of two currents mentioned above, will be investigated in a future paper.

Acknowledgements

I thank Abhay Ashtekar and Roger Penrose for useful discussions. This work was supported by ONR grant R&T 3124141 and NSF grant PHY-9307708.

Bibliography

Aharonov, Y., Anandan, J. and Vaidman, L. (1993). Meaning of the wavefunction. *Phys. Rev. A*, **47,** no. 6, 4616–26; Aharonov, Y. and Vaidman, L. (1993). Schrödinger waves are observable after all. *Phys. Lett. A*, **178,** 38–42.

Anandan, J. (1980). On the hypotheses underlying physical geometry. *Found. of Phys.* **10,** 601–29.

Anandan, J. (1991). A geometric approach to quantum mechanics. *Found. of Phys.*, **21,** 1265–84.

Anandan, J. (1993). Protective measurement and quantum reality. *Found. of Phys. Lett.*, **6,** no. 6, 503–32.

Anandan, J. (1994). Interference of geometries in quantum gravity. *Gen. Rel. and Grav.* **26,** 125–33.

Anandan, J. (1996). Gravitational phase operator and the cosmic string. *Phys. Rev. D*, **53,** 779–86.

Anandan, J. and Brown, H. R. (1995). On the reality of space-time geometry and the wavefunction. *Found. of Phys.*, **25,** 349–60.

Bohm, D. (1952). A suggested interpretation of the quantum theory in terms of "hidden" variables, I and II. *Phys. Rev.*, **85,** 166–93.

Brown, H. R., Dewdney, C. and Horton, G. (1995). Bohm particles and their detection in the light of neutron interferometry. *Found. of Phys.*, **25,** 329.

Chakravarty, S. and Leggett, A. J. (1984). Dynamics of the two–state system with ohmic dissipation. *Phys. Rev. Lett.*, **52,** 5.

Diosi, L. (1987). A universal master equation for the gravitational violation of quantum mechanics. *Phys. Lett. A*, **120,** 377–81.

Diosi, L. (1989). Models for universal reduction of macroscopic quantum fluctuations. *Phys. Rev. A* **40,** 1165–74.

Everett, H. (1957). Relative state formulation of quantum mechanics. *Rev. Mod. Phys.*, **29,** 454–62.

Ghiradi, G. C., Rimini, A. and Weber, T. (1986). Unified dynamics for microscopic and macroscopic systems. *Phys. Rev. D*, **34,** 470–91.

Ghiradi, G. C., Grassi R. and Rimini, A. (1990). Continuous–spontaneous–reduction model involving gravity. *Phys. Rev. D*, **42,** 1057–64.

Holland, P. R. (1993). *The Quantum Theory of Motion.* Cambridge University Press, Cambridge.

Karolyhazy, F. (1966). Gravitation and Quantum Mechanics of Macroscopic Objects. *Nuovo Cim. A*, **42,** 390.

Kayser, B. and Stodolsky, L. (1995). EPR experiments without collapse of the wavefunction. *Phys. Lett. B*, **359,** 343–50.

Leggett, A. J. (1980). Macroscopic quantum systems and the quantum theory of measurement. *Prog. Theor. Phys. Suppl.*, **69,** 80–100.

Leggett, A. J. and Garg, A. (1985). Quantum mechanics versus macroscopic realism. *Phys. Rev. Lett.*, **54,** 857–60.

Misner, C. W., Thorne, K. S. and Wheeler, J. A. (1973). *Gravitation.* W.H. Freeman, New York.

Pearle, P. (1986). Models of reduction. In *Quantum Concepts in Space and Time,* eds. R. Penrose and C. J. Isham. Clarendon Press, Oxford.

Pearle, P. (1989). Combining stochastic dynamical state-vector reduction with spontaneous localization. *Phys. Rev. A* **39,** 2277–89.

Penrose, R. (1981). Time asymmetry and quantum gravity. In *Quantum Gravity 2: A Second Oxford Symposium,* eds. C. J. Isham, R. Penrose, and D. W. Sciama. Oxford University Press, Oxford.

Penrose, R. (1986). Gravity and state vector reduction. In *Quantum Concepts in Space and Time,* eds. R. Penrose and C. J. Isham. Clarendon Press, Oxford.

Penrose, R. (1989). *Emperor's New Mind,* Oxford University Press, Oxford.

Penrose, R. (1993). Gravity and quantum mechanics. In *General Relativity and Gravitation 1992*: Part 1, Plenary Lectures eds. R. J. Gleiser, C. Kozameh and O. M. Moreschi. IOP Publ.

Penrose, R. (1994). *Shadows of the Mind.* Oxford University Press, Oxford.

Penrose, R. (1996). On gravity's role in quantum state reduction. *Gen. Rel. and Grav.*, **28,** 581–600.

Wigner, Eugene (1963). The problem of measurement. *Am. J. Phys.*, **31,** 6.

27
Entanglement and Quantum Computation

Richard Jozsa
School of Mathematics and Statistics, University of Plymouth, Plymouth, PL4 8AA, England. Email: rjozsa@plymouth.ac.uk

1 Introduction

The phenomenon of quantum entanglement is perhaps the most enigmatic feature of the formalism of quantum theory. It underlies many of the most curious and controversial aspects of the quantum mechanical description of the world. Penrose (1994) gives a delightful and accessible account of entanglement illustrated by some of its extraordinary manifestations. Many of the best known features depend on issues of *non-locality*. These include the seminal work of Einstein, Podolsky and Rosen (1935), Bell's work (1964) on the EPR singlet state, properties of the GHZ state (Greenberger *et al.* 1989; Mermin 1990) and Penrose's dodecahedra (Penrose 1994). In this paper we describe a new feature of entanglement which is entirely independent of the auxiliary notion of non-locality.

We will argue that the phenomenon of entanglement is responsible for an essential difference in the *complexity* (as quantified below) of physical evolution allowed by the laws of quantum physics in contrast to that allowed by the laws of classical physics. This distinction was perhaps first explicitly realised by Feynman (1982) when he noted that the simulation of a quantum evolution on a classical computer appears to involve an unavoidable exponential slowdown in running time. Subsequently in the development of the subject of quantum computation – which represents a hybrid of quantum physics and theorctical computer science – it was realised that quantum systems could be harnessed to perform useful computations more efficiently than any classical device. Below we will examine some of the basic ingredients of quantum computations and relate their singular efficacy to the existence of entanglement.

The perspective of information theory also provides further new insights into the relationship between entanglement and non-locality, beyond the well studied mediation of non-local correlations between local measurement outcomes. To outline some of these effects we first introduce the notion of "quantum information".

A fundamental differencc between quantum and classical physics is that the state of an unknown quantum system is in principle unmeasurable, e.g. given an unknown state $|\psi\rangle$ of a 2-level system it is not possible to identify it. In fact any measurement on $|\psi\rangle$ will reveal at most one bit of information about its identity whereas the full description of $|\psi\rangle$ requires the specification of two real numbers. In terms of binary expansions this corresponds to a double infinity of bits of information. We refer to the full (largely inaccessible) information represented by a quantum state as quantum information in contrast to the more familiar notion of classical information such as the outcome of a measurement which is in principle fully accessible.

The inaccessibility of quantum information is closely related to the possibility of non-local influences which do not violate classical causality, as necessitated by Bell's (1964) analysis of local measurements on an EPR pair – the non-local influences are simply restricted to a level which is inaccessible to any local observations. However it has recently been shown (Bennett *et al.* 1993, 1996) that entanglement plays a more subtle role here than just mediating correlations between the classical information of local measurement outcomes. According to the process known as quantum teleportation (Bennett *et al.* 1993) entanglement acts as a *channel* for the transmision of quantum information: an unknown quantum state of a 2-level system may be transferred intact from one location to another by sending only two bits of classical information, if the locations are also linked by entanglement in the form of a shared EPR pair. The remaining quantum information can be interpreted as having flown across the entanglement which is destroyed by the process.

Another novel application of entanglement and non-locality inspired by concepts from theoretical computer science is the existence and construction of quantum error correcting codes, first introduced by Shor (1995). Entanglement provides a way of delocalising quantum information in a system composed of several subsytems. For example if $|0\rangle$ and $|1\rangle$ are orthogonal states then the entangled states $\frac{1}{\sqrt{2}}(|0\rangle\,|0\rangle + |1\rangle\,|1\rangle)$ and $\frac{1}{\sqrt{2}}(|0\rangle\,|1\rangle + |1\rangle\,|0\rangle)$ are identical in terms of local quantum information (each subsystem being in the maximally mixed state in each case) whereas they differ in terms of their global quantum information content. By an ingenious extension of this idea (Shor 1995) one may encode an unknown state of a 2-level system into an entangled state of several 2-level systems in such a way that if any (unknown) one of the subsystems is arbitrarily corrupted then the original state may still be perfectly reconstructed, i.e. none of the information of the original state resides locally in the encoding. In this way a state may be protected against the effects of spurious environmental interactions if (as is generally a good approximation) these interactions act by local means.

Thus quantum computation and quantum information theory provide a rich variety of new applications of entanglement and we now turn to a more detailed discussion of issues in quantum computation in particular.

2 Quantum computation and complexity

The theory of computation and computational complexity (Papadimitriou 1994) is normally presented as an entirely mathematical theory with no reference to considerations of physics. However any actual computation is a physical process involving the physical evolution of selected properties of a physical system. Consequently the issues of "what is computable" and "what is the complexity of a computation" must depend essentially on the laws of physics and cannot be characterised by mathematics alone. This fundamental point was emphasised by Deutsch (1985) and it is dramatically confirmed by the recent discoveries (Deutsch and Jozsa 1992; Bernstein and Vazirani 1993; Simon 1994; Shor 1994; Grover 1996; Kitaev 1995) that the formalism of quantum physics allows one to transgress some of the boundaries of the classical theory of computational complexity, whose formulation was based on classical intuitions. Penrose (1994) proposes the possible introduction of non-computable elements into physics (i.e. non-computable within the standard existing theory of computability). This however requires going beyond the existing formalism of quantum theory since the latter lies entirely within the bounds of classical computability. Our considerations here lie entirely within the standard framework so that, for us, quantum processes cannot result in any computation which is not already possible by classical means. This notwithstanding, there does appear to be a significant difference in the *efficiency* of computation as noted in Feynman's remark.

A fundamental notion of the theory of computational complexity is the distinction between polynomial and exponential use of resources in a computation. This will provide a quantitative measure of our essential distinction between quantum and classical computation. Consider a computational task such as the following: given an integer N, decide whether N is a prime number or not. We wish to assess the resources required for this task as a function of the size of the input which is measured by $n = \log_2 N$, the number of bits needed to store N. If $T(n)$ denotes the number of steps (on a standard universal computer) needed to solve the problem, we ask whether $T(n)$ can be bounded by some polynomial function in n or whether $T(n)$ grows faster than any polynomial. More generally we may consider any language $\mathcal{L}$ – a language being a subset of the set of all finite strings of 0's and 1's – and consider the computational task of recognising the language i.e. given a string σ of length n the computation outputs 0 if $\sigma \in \mathcal{L}$ and outputs 1 if $\sigma \notin \mathcal{L}$. (In our example above $\mathcal{L}$ is the set of all prime numbers written in binary.) The language $\mathcal{L}$ is said to be in complexity class $\mathcal{P}$ ("polynomial time") if there exists an algorithm which recognises $\mathcal{L}$ and runs in time $T(n)$ bounded by a polynomial function. Otherwise the recognition of $\mathcal{L}$ is said to require exponential time. More generally it is useful to consider algorithms which involve probabilistic choices ("coin tosses") (Papadimitriou 1994; Ekert and Jozsa 1996). $\mathcal{L}$ is said to be in the class $\mathcal{B}PP$ ("bounded-error probabilistic polynomial time") if there is a polynomial time algorithm which correctly classifies the input string σ with probability at least 2/3 (or equivalently, any other value strictly between 1/2 and 1). Thus a $\mathcal{B}PP$ algorithm may give the wrong answer but by simple repetition and taking the majority answer, we can amplify the probability of success as close to 1 as desired while retaining a polynomial time for the whole process (Papadimitriou 1994; Ekert and Jozsa 1996).

The class $\mathcal{BPP}$ of algorithms which run in polynomial time but allow for "small imperfections", is often regarded as the class of computational tasks which are "feasible in practice" (or at least a first mathematical approximation to this notion). We also use the term "efficient computation" for a computation which runs in polynomial time.

In the above considerations the exact number of steps $T(n)$ will generally depend on the choice of underlying computer and the model of computation adopted. However if we stay within the confines of classical physics, the distinction between polynomial and exponential time i.e. between efficient and non-efficient computation, appears to be robust, being independent of these choices. It is thus a distinction with *physical* significance. In the discussion above we have illustrated it in its most familiar form – in terms of the resource of time. From the physical point of view it is natural to extend the notion of efficient computation to require the efficient use of *all* possible physical resources.

Indeed in the following discussion we will be led to consider other resources such as energy. The absolute significance of the distinction between efficient and non-efficient computation provides an extension of the classical Church-Turing thesis (Deutsch 1985; Shor 1994) which refers to a similar distinction between computability and non-computability. The fundamental *raison d'être* of quantum computation is the fact that quantum physics appears to allow one to transgress this classical boundary between polynomial and exponential computations.

The concept of quantum computation may be rigorously formalised as a natural extension of classical mathematical models of computation (Deutsch 1985; Bernstein and Vazirani 1993; Shor 1994; Yao 1993; Ekert and Jozsa 1996) in which the computational steps are allowed to be quantum processes restricted by a suitable notion of locality. For our purposes it will suffice to envisage a quantum computer as a standard universal computer in which the memory bits are 2–level quantum systems instead of 2–state classical systems. The quantum systems are each endowed with a preferred basis $\{|0\rangle, |1\rangle\}$ corresponding to the classical bit values of 0 and 1. We refer to these 2–level systems as qubits, a term first introduced by Schumacher (1995).

The computer is able to support arbitrary superpositions of the values 0 and 1 within each qubit and also entanglements of many qubits. Furthermore the computer may be programmed to perform unitary transformations of any number of qubits. It is important however that large unitary transformations be constructed or "programmed" from a finite set of fixed given unitary transformations. In this way we can assess the complexity of unitary transformations by the length of their programs. There are many known examples of small finite sets of transformations, out of which one can program any unitary transformation of any number of qubits (to arbitrary precision) (Deutsch 1989; Barenco *et al.* 1995a; Barenco 1995). Indeed it is known (Barenco *et al.* 1995b) that almost any single transformation of two qubits by itself suffices. The distinction between polynomial and exponential time does not depend on the choice of these basic transformations as any one such set can first be used to build the members of any other set leading to only an overall constant expansion in the number of elementary steps in any computation.

3 Superposition and entanglement in quantum computation

There are several quantum algorithms known (Deutsch and Jozsa 1992; Bernstein and Vazirani 1993; Simon 1994; Shor 1994) which strongly support the view that a quantum computer can perform some computational tasks exponentially faster than any classical device. The most significant of these is Shor's polynomial time algorithm for integer factorisation (Shor 1994; Ekert and Jozsa 1996), a problem which is believed to lie outside the class $\mathcal{BPP}$ of classical computation. We then ask: what is the essential quantum effect that gives rise to this exponential increase in computing power?

All of the quantum algorithms utilise the process of computation by quantum parallelism (Deutsch 1985). This refers to a quantum computer's capability to carry out many computations simultaneously in superposition if the input is set up in a suitable superposition of classically distinct inputs. Thus one might conclude that it is *superposition* that is at the root of the quantum computational speedup. However closer examination will show that *entanglement* is the essential feature rather than just superposition itself. Note that entanglement may be viewed as a special kind of superposition – superposition in the presence of a product structure on the state space – which arises from the system being made up of several subsystems. In our considerations these are the qubits comprising the computer. An entangled state is then a superposition of product states which cannot be expressed as a single product state.

To see that superposition itself is not the essential feature we need only note that classical waves also exhibit superposition. Any effect depending on quantum interference alone can be readily mimicked by classical waves. However in other respects quantum states and classical waves differ considerably e.g. the measurement theories are very different (being far more favourable for computation with classical waves than with quantum states!) and *there is no classical analogue of the phenomenon of entanglement.*

To illustrate the above remarks and highlight the role of entanglement consider the following example of computation by quantum parallelism. Let $B = \{0, 1\}$ and consider any (non-trivial) function $f : B^n \to B$. Suppose that we have a quantum computer programmed to compute f in polynomial time. The computer has n input qubits and one output qubit and its operation corresponds to a unitary transformation $\mathcal{U}_f$ on $n+1$ qubits which effects the evolution:

$$\mathcal{U}_f : \underbrace{|i_1\rangle\,|i_2\rangle\ldots|i_n\rangle}_{\text{input}}\,|0\rangle \longrightarrow |i_1\rangle\,|i_2\rangle\ldots|i_n\rangle\,|f(i_1,\ldots,i_n)\rangle\,.$$

Here each i_k is either 0 or 1. The output register is initially in state $|0\rangle$ and at the end of the computation it contains the basis state corresponding to the value of the function. Consider the one-qubit operation:

$$H = \frac{1}{\sqrt{2}} \begin{pmatrix} 1 & 1 \\ 1 & -1 \end{pmatrix}.$$

If the input qubits are all initially in state $|0\rangle$ then applying H to each one successively gives an equal superposition of all 2^n values in B^n:

$$H \otimes \cdots \otimes H \; |0\rangle \cdots |0\rangle = \frac{1}{2^{n/2}} \left(|0\rangle + |1\rangle\right)^n = \frac{1}{2^{n/2}} \sum_{i=0}^{2^n-1} |i\rangle \tag{3.1}$$

(where we have identified $(i_1, \ldots, i_n)$ with the binary number $i_1 \ldots i_n < 2^n$). Note that this state is prepared in polynomial (in fact linear) time. Running the computer with (3.1) as input yields the final state

$$|f\rangle = \sum_{i=0}^{2^n-1} |i\rangle \, |f(i)\rangle \, . \tag{3.2}$$

Thus by quantum parallelism we have computed exponentially many values of f in superposition with only polynomial computing effort.

Can we mimick the above with classical waves? We represent each qubit by a classical wave system and select two modes of vibration to represent the states $|0\rangle$ and $|1\rangle$ e.g. $|0\rangle$ and $|1\rangle$ may be the two lowest energy modes of a vibrating elastic string with fixed endpoints. It is then straightforward to construct the superposition corresponding to $|0\rangle + |1\rangle$ and performing this separately on n pieces of string we obtain the product state (3.1). However, regardless of how much the strings interact with each other in their subsequent (externally driven) vibrational evolution, their joint state is always a *product* state of n separate vibrations. The total state space of the total classical system is the *Cartesian* product of the individual state spaces of the subsystems whereas quantum-mechanically, it is the *tensor* product. This crucial distinction between Cartesian and tensor products is precisely the phenomenon of quantum entanglement. The state (3.2) is generally entangled (for non-trivial f's) so that the transition from (3.1) to (3.2) cannot be achieved in a classical scenario.

We may attempt to represent entanglement using classical waves in the following manner. The state of n qubits is a 2^n dimensional space and can be isomorphically viewed as the state space of a *single* particle with 2^n levels. Thus we simply interpret certain states of a single 2^n level particle as "entangled" via their correspondence under a chosen isomorphism between $\bigotimes^n \mathcal{H}_2$ and $\mathcal{H}_{2^n}$ (where $\mathcal{H}_k$ denotes a Hilbert space of dimension k.) In this way, 2^n modes of a classical vibrating system can apparently be used to mimick general entanglements of n qubits. However the physical implementation of this correspondence appears always to involve an exponential overhead in some physical resource so that the isomorphism is *not* a valid correspondence for considerations of complexity. For example suppose that the 2^n levels of the one-particle quantum system are equally spaced energy levels. A general state of n qubits requires an amount of energy that grows *linearly* with n whereas a general state of this 2^n level system (and also the corresponding classical wave system) requires an amount of energy that grows *exponentially* with n. To physically realise a system in a general superposition of 2^n modes we need exponential resources classically and linear resources quantum mechanically *because of the existence of entanglement.*

This comparison is reminiscent of the representation of whole numbers in unary (i.e. representing k as a string of k 1's analogous to k equally spaced levels) versus the binary representation of k which requires a string of length $\log_2 k$ and is therefore exponentially more efficient. Note that n classical bits can also accommodate 2^n possible alternatives but only one such alternative can be present at any time, even if probabilistically determined. In contrast n qubits can accommodate 2^n possible alternatives which may all be *simultaneously* present in superposition. As a quantum computation proceeds a new qubit may be incorporated into the overall processed state at each step leading to an exponential growth in time of the quantum information, because of entanglement. If we were to compute this quantum evolution "by hand" from the laws of quantum mechanics then we would suffer an exponential slowdown in handling an exponentially growing amount of information. In contrast, Nature manages to process this growing information in linear time! This is an example of Feynman's remark mentioned in the introduction. Note that if the state of the accumulating qubits were always a product state (i.e. no entanglement) then the quantum information would accumulate only linearly.

There exist physical systems with infinitely many discrete energy levels which accumulate below a finite bound E_0. Thus we may use these levels to represent general superpositions of exponentially many modes with only a constant cost in energy and apparently circumvent the above objections! However in this case the levels will crowd together exponentially closely and we will need to build our instruments with exponentially finer precision. This again will presumably require exponentially increasing physical resources.

It has been occasionally suggested that the interferences in Shor's efficient quantum factoring algorithm and other quantum algorithms, can be readily represented by classical wave interferences but on closer inspection all such proposals involve an exponential overhead in some physical resource as illustrated in our discussion above.

The standard mathematical theory of computational complexity (Papadimitriou 1994) assesses the complexity of a computation in terms of the resources of time (number of steps needed) and space (amount of memory required). In the above we have been led to consider the accounting of other physical resources such as energy and precision (Shor 1994). This reinforces our earlier remark that the notion of computational complexity must rest on the laws of physics and consequently the proper assessment of complexity will need to take into account all possible varieties of physical resource. A theory of computational complexity based on such general physical foundations remains to be formulated.

4 Entanglement and the super-fast quantum Fourier transform

The efficiency of Shor's factoring algorithm rests largely on the fact that the discrete Fourier transform (Ekert and Jozsa 1996) DFT_{2^n} in dimension 2^n (a particular unitary transformation in 2^n dimensions) may be implemented on a quantum computer in time polynomial in n (in fact quadratic in n). Similarly the efficiency of Deutsch's and Simon's algorithms (Deutsch and Jozsa 1992; Simon 1994) rest on the polynomial-time computability of the Fourier transform over the additive group B^n. The standard classical Fourier transform algorithm implements DFT_{2^n} in time $O((2^n)^2)$ and the classical *fast* Fourier transform algorithm improves this to $O(n2^n)$ which is an exponential saving but still remains exponential in n. We will argue that the extra quantum mechanical speedup to $O(n^2)$ resulting in a genuinely polynomial-time algorithm, is due to the effects of entanglement. We will describe a simplified example which illustrates the essential principle involved.

Let M be a unitary matrix of size 2^n and v a column vector of length 2^n. To compute $w = Mv$ by direct matrix multiplication requires $O((2^n)^2)$ operations, each entry of the result requiring $O(2^n)$ operations. Suppose now that the space of v is the tensor product of n two dimensional spaces $V^{(1)} \otimes \ldots \otimes V^{(n)}$ and that M decomposes as a simple tensor product

$$M = S^{(1)} \otimes \ldots \otimes S^{(n)} \tag{4.1}$$

where each $S^{(i)}$ is a 2 by 2 matrix acting on the respective component space $V^{(i)}$. Thus we can label the components of v by indices $i_1 \ldots i_n \in B^n$ and the computation of w becomes

$$w_{j_1 \ldots j_n} = \sum_{i_1 \ldots i_n} S^{(1)}_{j_1 i_1} \ldots S^{(n)}_{j_n i_n} v_{i_1 \ldots i_n}. \tag{4.2}$$

Now consider first $S^{(1)}$. Each application of this 2 by 2 matrix requires some fixed (i.e. independent of n) number of operations and $S^{(1)}$ needs to be applied 2^{n-1} times, once for each choice of the indices $i_2 \ldots i_n \in B^{n-1}$. The same accounting applies to each of the n matrices $S^{(i)}$ giving a total number of operations $O(n2^{n-1}) = O(n2^n)$. Thus the tensor product factorisation (4.1) leads to an exponential speed-up compared to straightforward matrix multiplication for a general M. A similar argument will apply if M decomposes more generally into the successive application of a polynomial number of matrices, each of which applies to a bounded number b of the component spaces (not necessarily disjoint) and b is independent of n. (4.1) is merely the simplest example of such a decomposition. The classical fast Fourier transform algorithm is based on the fact that the Fourier transform in dimension 2^n decomposes in just this way (although not as simply as (4.1) c.f. (Ekert and Jozsa 1996) for an explicit description of the decomposition).

Consider finally the implementation of M as given in (4.1) on a quantum computer. The data given by the components of v is represented by the amplitudes of a general state of n qubits. Each of the n operations $S^{(i)}$ is a one-qubit operation and needs to be applied *only once* to its respective qubit i.e. the 2^{n-1} repetition of the classical calculation is eliminated! This is due to the rules of formation of the tensor product (i.e. entanglement) requiring for example that $S^{(1)}$ applied to the first qubit automatically carries through to all possible values of the indices $i_2 \dots i_n$ in (4.2) i.e. the global operation $S^{(1)} \otimes I \otimes \dots \otimes I$ is implemented on the whole space. Thus M is implemented in time $O(n)$. In a similar way, the more complicated decomposition of the Fourier transform can be seen (Ekert and Jozsa 1996) to lead to a time $O(n^2)$. This comparison of classical and quantum implementations of M is yet another illustration of Feynman's (1982) remark that the simulation of a quantum process on a classical computer generally involves an exponential slowdown.

The classical and quantum scenarios for the above computation of Mv differ significantly in the following respect. After the classical computation we are able to read off all 2^n components of w presented as classical information whereas the quantum computation results in the quantum information of one copy of the state $|w\rangle = \sum w_{i_1 \dots i_n} |i_1\rangle \dots |i_n\rangle$ from which we are unable to extract the individual values of $w_{i_i \dots i_n}$. This is the issue of inaccessibility of quantum information that was mentioned in the introduction. Nevertheless we are able to extract classical information from $|w\rangle$ that depends on exponentially many of the values and classically this information would require a preliminary exponential computational effort. This phenomenon also occurs in computation by quantum parallelism. From the quantum information $|f\rangle$ in (3.2) we are unable to extract all the individual values $f(i_1 \dots i_n)$ but we can obtain certain global properties of the function. Indeed in Shor's factoring algorithm (Shor 1994; Ekert and Jozsa 1996), the analogue of f is a periodic function and after applying the Fourier transform we are able to extract the period.

5 Concluding remarks

In summary the effects of quantum entanglement enable certain large unitary transformations to be implemented exponentially more efficiently on a quantum computer than on any classical computer. However after quantum computation the full results are coded in a largely inaccessible form. Remarkably this inaccessibility, dictated by the principles of quantum measurement theory, does *not* serve to cancel out the exponential gain in computing effort. Limited information may be obtained about the transformed data which, although small, would nevertheless require an exponential amount of computing effort to obtain by classical computation.

Another fundamental issue intimately related to entanglement is the so-called measurement problem of quantum mechanics. This refers to the reconciliation of the apparent "collapse of the wave function" in a measurement with the unitary evolution of quantum mechanics and to the explanation of why, after a measurement, we see merely one of the possible outcomes rather than experiencing some weird reality in entanglement with all possible outcomes. Penrose (1994) discusses several of the best known interpretations of quantum mechanics and it appears that none of them provides a resolution of this phenomenon. With extraordinarily innovative and broadranging arguments, Penrose suggests that the resolution might involve non-computational ingredients. The fact that quantum theory has resisted unification with the theory of gravitation suggests that the essentially linear concept of entanglement may not persist at a macroscopic level. Indeed it is difficult to imagine that the linearity of quantum theory could survive a unification with the non-linear foundations of the general theory of relativity. The algorithms of quantum computation such as Shor's algorithm depend critically for their efficiency and validity on effects of increasingly large scale entanglements with increasing input size. Thus efforts to experimentally implement these algorithms may provide particularly acute opportunities to witness a possible breakdown of the current conventional quantum formalism.

Bibliography

Barenco, A. (1995) *Proc. Roy. Soc. London Ser. A* **449**, 679.

Barenco, A., Bennett, C., Cleve, R., DiVincenzo, D., Margolus, N., Shor, P., Sleator, T., Smolin, J. and Weinfurter, H. (1995a) *Phys. Rev. A* **52**, 3457.

Barenco, A., Deutsch, D. and Ekert, A. (1995b) *Proc. Roy. Soc. London Ser. A* **449**, 669.

Bell, J. S. (1964) *Physics* **1**, 195.

Bennett, C., Brassard, G., Crepeau, C., Jozsa, R., Peres, A. and Wootters, W. (1993) *Phys. Rev. Lett.* **70**, 1895.

Bennett, C., DiVincenzo, D., Smolin, J. and Wootters, W. (1996) *Phys. Rev. A* **54**, 3824.

Bernstein, E. and Vazirani, U. (1993) *Proc. 25th Annual ACM Symposium on the Theory of Computing* (ACM Press, New York), pp. 11–20 (Extended Abstract). Full version of this paper appears in *S. I. A. M. Journal on Computing* **26** (Oct 1997).

Deutsch, D. (1985) *Proc. Roy. Soc. London Ser. A* **400**, 97.

Deutsch. D. (1989) *Proc. Roy. Soc. London Ser. A* **425**, 73.

Deutsch, D. and Jozsa, R. (1992) *Proc. Roy. Soc. London Ser. A* **439**, 553–558.

Einstein, A., Podolsky, B. and Rosen, N. (1935) *Phys. Rev.* **47**, 777.

Ekert, A. and Jozsa, R. (1996) *Rev. Mod. Phys.* **68**, 733.

Feynman, R. (1982) *Int. J. Theor. Phys.* **21**, 467.

Greenberger, D., Horne, M. and Zeilinger, A. (1989) in *Bell's Theorem, Quantum Theory and Conceptions of the Universe* (ed. M. Kafatos) (Kluwer Academic, Dordrecht, The Netherlands), p. 73.

Grover, L. (1996) *Proc. 28th Annual ACM Symposium on the Theory of Computing* (ACM Press, New York), pp. 212–219.

Kitaev, A. (1995) *Quantum Measurements and the Abelian Stabiliser Problem*, preprint available at LANL quant-ph preprint archive 9511026.

Mermin, N. D. (1990) *Am. J. Phys.* **58**, 731.

Papadimitriou, C. H. (1994) *Computational Complexity* (Addison-Wesley, Reading, MA).

Penrose, R. (1994) *Shadows of the Mind* (Oxford University Press).

Schumacher, B. (1995) *Phys. Rev. A* **52**, 2738.

Shor, P. (1994) *Proc. of 35th Annual Symposium on the Foundations of Computer Science* (IEEE Computer Society, Los Alamitos), p. 124 (Extended Abstract). Full version of this paper appears in *S. I. A. M. Journal on Computing* **26** (Oct 1997) and is also available at LANL quant-ph preprint archive 9508027.

Shor, P. (1995) *Phys. Rev. A* **52**, R2493.

Simon, D. (1994) *Proc. of 35th Annual Symposium on the Foundations of Computer Science* (IEEE Computer Society, Los Alamitos), p. 116 (Extended Abstract). Full version of this paper appears in *S. I. A. M. Journal on Computing* **26** (Oct 1997).

Yao, A. (1993) *Proceedings of the 34th Annual Symposium on the Foundations of Computer Science* (ed. S. Goldwasser) (IEEE Computer Society, Los Alamitos, CA), p. 352.

PART V

Parallel Session IV: Mathematical Aspects of Twistor Theory

28

Penrose Transform for Flag Domains

Simon Gindikin

Department of Mathematics, Rutgers University, New Brunswick, NJ 08903[1]

It is a great pleasure for me to talk at this conference in Roger's honor on the Penrose transform with which I have been friends for the last 18 years of my mathematical life. I chose to talk about a generalized Penrose transform. There are two sides of the Penrose transform which are at the focus of possible generalizations: cohomological representations of solutions of a class of differential equations and representations of higher $\bar{\partial}$-cohomology in a holomorphic language. We need to understand on which equations and complex manifolds it is possible to generalize Penrose's constructions. It is not necessary to connect them with group actions, but it is reasonable to start experimental work with homogeneous manifolds, and flag domains (homogeneous domains on flag manifolds) are an appropriate class for such a consideration. Some results of W. Schmid on realizations of the discrete series of representations are already fragments of the theory. We will not review here known results on the Penrose transform on some flag domains. I prefer to describe what the final theory is supposed to look like and to formulate a chain of conjectures without technical details. In this theory the central role must be played by explicit formulas for some intertwining operators and that is why I will start off the introduction with a short review of formulas for the usual Penrose transform.

1 The Penrose transform in formulas

1.1 Twistor geometry

Let us consider the projective space $\mathbb{C}P^3$ with homogeneous coordinates $z = (z_0, z_1, z_2, z_3)$ and the domains $D_\pm$:

$$zHz^* \gtrless 0, \tag{1.1}$$

where H is a Hermitian form of signature (2,2) e.g.

$$H_1 = \frac{1}{i}\begin{pmatrix} 0 & I \\ -I & 0 \end{pmatrix}, \quad H_2 = \begin{pmatrix} I & 0 \\ 0 & -I \end{pmatrix},$$

[1]email: gindikin@math.rutgers.edu

I being the unit matrix. Let S be the Stiefel manifold of 2×4 matrices

$$Z = \begin{pmatrix} u \\ v \end{pmatrix}, \qquad rank\, Z = 2,$$

and S_+ be the domain

$$ZHZ^* \gg 0. \tag{1.2}$$

We consider the fibering over the Grassmannian

$$\pi : S \to M = Gr(2;4) \tag{1.3}$$

where fibers are equivalence classes

$$Z \sim gZ, \qquad g \in GL(2;\mathbb{C})$$

and $\pi(Z)$ is the line(u,v) passing through u, v.

The domain $M_+ = \pi(S_+)$ consists of lines inside D_+ and it is holomorphically equivalent to the future tube (because we can take $H = H_1$ and the coordinate chart $Z = (I, z)$, $z \in M(2,\mathbb{C})$).

We have the double fibering

$$\begin{array}{ccc} & S_+ & \\ \rho \swarrow & & \searrow \pi \\ D_+ & & M_+ \end{array}, \qquad \rho\begin{pmatrix} u \\ v \end{pmatrix} = u, \tag{1.4}$$

where $D_+ \subset \mathbb{C}P^3$ is a 1-linear concave manifold and S_+, M_+ are Stein manifolds.

1.2 Penrose transform

Let $\omega(z, \overline{z}; d\overline{z}) \in Z^{(0,1)}(D_+, \mathcal{O}(-2))$ be a $\overline{\partial}$-closed (0,1)-form on D_+ with coefficients in $\mathcal{O}(-2)$ (for simplicity we will consider only such coefficients). This means that

$$\omega(\lambda z, \overline{\lambda z}; d\overline{(\lambda z)}) = \lambda^{-2}\omega(z, \overline{z}; d\overline{z}), \qquad \lambda \in \mathbb{C}.$$

The Penrose transform is

$$\mathcal{P}\omega(Z) = \int_{\mathbb{C}P^1_\tau} \omega\Big|_{z=\tau_0 u + \tau_1 v} \wedge (-\tau_0 d\tau_1 + \tau_1 d\tau_0), \qquad Z = \begin{pmatrix} u \\ v \end{pmatrix} \in S_+. \tag{1.5}$$

We can interpret $\mathcal{P}\omega$ as a section of a line bundle on M_+ :

$$\mathcal{P}\omega(gZ) = (\det g)^{-1}\mathcal{P}\omega(Z).$$

We have

$$\Box_{ij}(\mathcal{P}\omega) = 0,$$

$$\Box_{ij} = \frac{\partial^2}{\partial u_i \partial v_j} - \frac{\partial^2}{\partial u_j \partial v_i}. \tag{1.6}$$

If we restrict ω to the coordinate chart $Z = (I, z)$, then we will have only one (wave) equation. The operator $\mathcal{P}$ can be pushed down on cohomology $H^{(0,1)}(D_+, \mathcal{O}(-2))$.

Proposition 1.1 *(Eastwood et al. (1981); Gindikin and Henkin (1978b, 1981) The operator $\mathcal{P}$ is an isomorphism*

$$\mathcal{P} : H^{(0,1)}(D_+, \mathcal{O}(-2)) \to \mathrm{Sol}(\Box_{ij} F(Z) = 0, Z \in S_+). \tag{1.7}$$

1.3 Inverse Penrose transform

It is remarkable that it is possible to write the inverse operator explicitly and moreover this operator is, in a sense, differential.

Proposition 1.2. (Gindikin and Henkin 1981) *Let F be a solution of the system*

$$\Box_{ij} F(Z) = 0 \tag{1.8}$$

on S_+ and

$$\kappa F = \sum \frac{\partial F(u, v)}{\partial u_j} dv_j. \tag{1.9}$$

If γ is a section of the fibering $\rho : S_+ \to D_+$, then

$$\omega - c\{\kappa(\mathcal{P}\omega)|_\gamma\}^{(0,1)} \tag{1.10}$$

for a constant c is a $\bar{\partial}$-exact form.

In (1.10) we take the $(0,1)$-part of the restriction of $\mathcal{P}\omega$ to the section γ. In this way we have reconstructed a form from the same cohomology class as ω.

1.4 Holomorphic cohomology

We will reformulate the last result as follows. Let us consider the complex of holomorphic differential forms on S_+ with differentials only along fibers: $\varphi(u, v|dv)$, $(u, v) \in S_+$, $u \in D_+$ ($\kappa\mathcal{P}_\omega$ has such a structure) and corresponding cohomology (the differential acts along fibers) $H^{(q)}_{\mathrm{hol}}(D_+ \mathcal{O}(-2))$.

Proposition 1.3 *There is an isomorphism*

$$H^{(1)}_{\mathrm{hol}}(D_+, \mathcal{O}(-2)) \cong H^{(0,1)}(D_+, \mathcal{O}(-2))$$

and a (chain) morphism from the first space to the second is

$$\varphi \mapsto \{\varphi|_\gamma\}^{(0,1)} \tag{1.11}$$

(cf. (1.10)). From the other side, the operator $\kappa \circ \mathcal{P}$ is the morphism from the Dolbeault cohomology $H^{(0,1)}$ to the holomorphic cohomology $H^{(1)}_{hol}$.

Thus the Penrose transform gives a possibility for a holomorphic realization of the Dolbeault cohomology. Moreover the operator $\kappa \circ \mathcal{P}$ picks up in each (holomorphic) cohomology class a unique representative satisfying the condition

$$\varphi(u + \lambda v, v|dv) = \varphi(u, v|dv)$$

(holomorphic Hodge theorem).

1.5 Boundary integral formula

It is possible to reconstruct $\mathcal{P}\omega$ only using boundary values of ω on ∂D_+ (Gindikin 1984):

$$\mathcal{P}\omega(u,v) = c\int_{\partial D_+} \frac{\omega \wedge [z,w,dz,dz] \wedge [u,v,dz,dw]}{[z,w,u,v]^2} \tag{1.12}$$

where $[a,b,c,d] = \det(a,b,c,d)$ relative to exterior multiplication of 1-forms and $w = f(z), z \in \partial D_+$, such that $\mathrm{line}(z, f(z)) \cap D_+ = \emptyset$. The integrands for different f differ on exact forms.

What is important in this formula is the structure of the denominator: it is equal to zero if

$$\mathrm{line}(z,w) \cap \mathrm{line}(u,v) \neq \emptyset.$$

So we have an analog of the Cauchy kernel for lines. If we apply κ to both parts we will obtain a boundary integral formula for $\kappa \circ \mathcal{P}$.

2 Generalized Penrose transform (geometrical problems)

The basic problem is to understand on which class of complex manifolds it is possible to generalize all the components of the theory that we have discussed above. Let us recall that Martineau (1966) considered the analog of the Penrose transform for concave domains in $\mathbb{C}P^n$ in 1962. In this case there are no differential equations on the image (like for the Radon transform). The first class of manifolds on which it is possible to generalize the Penrose transform is the class of q-linear concave domains in $\mathbb{C}P^n$ (Gindikin and Henkin 1978a, 1978b). We are sure that the Penrose transform is essentially connected with the possibility of describing analytic cohomology in holomorphic language. Therefore we will start to do so as of this point.

2.1 Holomorphic cohomology

It turns out that such a possibility exists in a quite general situation (Gindikin 1993; Eastwood *et al.* 1995, 1996). Namely let us take a fibering

$$\rho : S \to D, \tag{2.1}$$

where S is a Stein manifold and the fibers are contractible. Such a fibering exists for any complex manifold D (Eastwood *et al.* 1996). Let us consider the complex of holomorphic forms on S with differentials only along fibers (and coefficients pulled up from a bundle on D). Let $H^{(q)}_{\mathrm{hol}}(D)$ be the cohomology for this complex. Then

$$H^{(q)}_{\mathrm{hol}}(D) \cong H^{(0,q)}(D) \tag{2.2}$$

where we take on D the Dolbeault cohomology with the corresponding coefficients. A natural morphism of the holomorphic cohomology to the Dolbeault one can be constructed in the same way as in (1.11): we take $(0,q)$-*parts of restrictions of holomorphic forms on a section of* ρ.

In our example the inverse operator had the structure $\kappa \circ \mathcal{P}$ and it gives the canonical (Hodge) representatives of the holomorphic cohomology classes. I believe that this part of the program can be realized only for a special class of complex manifolds D where there are enough compact complex submanifolds with some special properties of incidence allowing us to proceed with the Penrose transform. It is the reason why we restrict ourselves below to the class of homogeneous manifolds.

2.2 Flag domains

The domain $D_+ \subset \mathbb{C}P^3$ (1.1) is an example of a pseudo-Hermitian symmetric space and the manifold of cycles M_+ (1.2) is the Hermitian symmetric space with the same group $SU(2,2)$. So it is natural to consider the class of pseudo-Hermitian symmetric spaces, or more generally, the class of flag domains. Let us state the definitions.

Flag manifolds are compact complex homogeneous manifolds

$$F = G/P \cong G_u/C \tag{2.3}$$

where G is a complex semisimple Lie group, P is a parabolic subgroup, G_u is the maximal compact subgroup of G, and C is the centralizer of a torus in G_u. The case when $P = B$ is the Borelian subgroup and C is the torus corresponds to the manifold of complete flags. Here G is the group of complex automorphisms of F and G_u consists of automorphisms of the canonical Kählerian metric on F.

Flag domains are open orbits of real forms $G_{\mathbb{R}}$ of G. Special cases of flag manifolds are compact Hermitian symmetric manifolds; special cases of flag domains are pseudo-Hermitian symmetric manifolds (then F must be symmetric, however a flag domain in a symmetric compact manifold can be nonsymmetric, cf. example 5 below).

Examples

(1) Pseudo-Hermitian symmetric domains $D_\pm \subset F = \mathbb{C}P^3$ (1.1). Here $G = SL(3;\mathbb{C})$, $G_{\mathbb{R}} = SU(2,2)$.

(2) Hermitian symmetric domain $M_+ \subset F = Gr(2;4)$ (1.2).

(3) Pseudo-Hermitian domain $D \subset Gr(2,4)$ which consists of lines intersecting both $D_\pm$(1.1).

(4) Let $G = SL(3,\mathbb{C})$, $\tilde{F} = \mathbb{C}P^2$, and F be the manifold of flags (z,l), $z \in \mathbb{C}P^2$, a line $l \ni z$. For F the parabolic group is Borelian. Let $G_{\mathbb{R}} = SU(2,1)$.

   (a) On $\tilde{F}$ we have 2 flag domains $\tilde{D}_\pm$:

$$|z_0|^2 - |z_1|^2 - |z_2|^2 \gtrless 0$$

   where the complex ball $\tilde{D}_+$ is the Hermitian symmetric domain and $\tilde{D}_-$ is pseudo-Hermitian.

   (b) On F there are 3 flag domains:

$$(z,l) \in D_1 \quad \text{if} \quad z \in \tilde{D}_+;$$

$$(z,l) \in D_2 \quad \text{if} \quad l \subset \tilde{D}_-;$$
$$(z,l) \in D_3 \quad \text{if} \quad z \in \tilde{D}_-, \quad l \cap \tilde{D}_+ \neq \emptyset.$$

(5) Flag domains corresponding to $G = SL(n+1;\mathbb{C})$, $G_{\mathbb{R}} = SL(n+1;\mathbb{R})$:

(a) $F = \mathbb{C}P^n, D = \mathbb{C}P^n \backslash \mathbb{R}P^n$. This flag domain in the symmetric manifold $\mathbb{C}P^n$ is not symmetric.

(b) $n = 2m-1$, $F = Gr(m, 2m)$ (Grassmanian of $(m-1)$-planes in $\mathbb{C}P^{2m-1}$). The domain $D = SL(2m;\mathbb{R})/SL(m;\mathbb{C}) \times T$ is symmetric (one of two components of the set of $(m-1)$-planes without real points), T is the circle.

2.3 Dual manifolds

The next step is to consider manifolds of maximal compact submanifolds (analogs of M_+). The naive idea that manifolds of cycles are the corresponding Hermitian symmetric domains works only for a few examples. The reason is that the Riemann symmetric space $M_{\mathbb{R}} = G_{\mathbb{R}}/K_{\mathbb{R}}$ cannot be Hermitian but there are not only flag domains with the group $G_{\mathbb{R}}$ but pseudo-Hermitian spaces also ($G_{\mathbb{R}}$ is not a group of Hermitian type). Moreover, for groups of Hermitian type the manifold of cycles does not as a rule coincide with the corresponding Hermitian symmetric domain.

Let us describe the generic situation. We start from an object dual to a flag manifold $F = G/P$ associated with $G_{\mathbb{R}}$. Let $K_{\mathbb{R}}$ be a maximal compact subgroup of $G_{\mathbb{R}}$ and K be its complexification. Let us consider

$$M = G/K. \tag{2.4}$$

The intrinsic description of the class of such manifolds is that they are *symmetric Stein manifolds* (K corresponds to a holomorphic involution). Let K_m be the isotropy subgroup of a point $m \in M$. Then minimal orbits $\Omega_1(m), \ldots, \Omega_j(m)$ of K_m on F are compact complex submanifolds: they are isomorphic to some flag manifolds with the group K.

If $D \subset F$ is a flag domain then there is one type of orbit ($\Omega(m)$ say) that is contained in D for some $m \in M$. If H is the isotropy subgroup of $z \in D$ in $G_{\mathbb{R}}$ and $K_{\mathbb{R}}$ is maximal compact in $G_{\mathbb{R}}$ such that $H \cap K_{\mathbb{R}}$ is maximal compact in H then $K_{\mathbb{R}}/K_{\mathbb{R}} \cap H$ is one such manifolds as well as its translation by $G_{\mathbb{R}}$. In this way we construct a family of $\Omega(m)$ that are parametrized by points of a Riemann symmetric manifold $m \in M_{\mathbb{R}} \subset M$. Let us take *the set of such* $m \in M$ *such that* $\Omega(m) \subset D$ *and let* $M(D)$ *be its connected component containing* $M_{\mathbb{R}}$.

The manifold $M(D)$ is supposed to be the basic geometrical object for the Penrose transform for flag domains. The problem is that so far we have no final description of $M(D)$. Wolf (1992) proved that $M(D)$ is a Stein manifold. Wolf and Zierau are probably close to proving that for the groups of Hermitian type $M(D)$ either coincides with corresponding Hermitian symmetric space $M_{\mathbb{R}}$ or with $M_{\mathbb{R}} \times M_{\mathbb{R}}$ for the generic situation. For groups of non-Hermitian type

$M(D)$ is apparently never homogeneous and it is the focus of the problem. Let us give the description of $M(D)$ for our examples.

Examples of $M(D)$. If $G_{\mathbb{R}}$ is a group of Hermitian type then the manifold of cycles in D can be isomorphic to $M_{\mathbb{R}}$ and as a result will not be a domain in M. We will continue nevertheless to use the notation $M(D)$ for this domain and call such a situation degenerate. We follow below the previous enumeration of our examples:

(1) $M(D_+) = M_+$ of (2) (degenerate case)

(2) $M(M_+) = M_+$

(3) $G_{\mathbb{R}} = SU(2,2)$, $K_{\mathbb{R}} = S(U(2) \times U(2))$, $\Omega(m) = \mathbb{C}P^1 \times \mathbb{C}P^1$, $K = S(L(2) \times L(2))$, and $M = SL(4;\mathbb{C})/S(L(2;\mathbb{C}) \times L(2;\mathbb{C}))$ is the manifold of generic pairs m of lines (l_1, l_2) in $\mathbb{C}P^3$ $(l_1 \cap l_2 = \emptyset)$; $\Omega(m) \subset Gr(2,4)$ is realized as the set of lines in $\mathbb{C}P^2$ which intersect both lines l_1, l_2. Finally,

$$M(D) = \{m = (l_1, l_2), \quad l_1 \subset D_+, \quad l_2 \subset D_-\}$$

where $D_\pm$ are from (1).

(4a) $\tilde{F} = \mathbb{C}P^2$, $M(\tilde{D}_+) = \tilde{D}_+$, $M(\tilde{D}_-) = \tilde{D}'_+$ (domain $(\tilde{D}_+)$ in the dual projective plane $(\mathbb{C}P^2)'$). These examples are degenerate.

(4b) For $D_j \subset F$ we have $M = SL(3;\mathbb{C})/GL(2;\mathbb{C})$ consisting of generic pairs $m = (z, l)$, $z \in \mathbb{C}P^2$, line $l \not\ni z$. D_1, D_2 are fibering over $\tilde{D}_+$, $\tilde{D}'_+$ correspondingly and $M(D_1) = \tilde{D}_+$, $M(D_2) = \tilde{D}'_+$. For all these examples we have degenerate $M(D)$. Now

$$M(D_3) = \tilde{D}_+ \times \tilde{D}'_+ = \{(z,l), z \in \tilde{D}_+, l \subset \tilde{D}_-\}$$

(5) For both examples $K_{\mathbb{R}} = SO(n+1;\mathbb{R})$, $K = SO(n+1;\mathbb{C})$, and

$$\begin{aligned} M &= SL(n+1;\mathbb{C})/SO(n+1;\mathbb{C}) \\ &= \{Q \in M(n+1;\mathbb{C}), Q = Q^T, \det Q = 1\}, \\ M_D &= \{Q : \text{quadric } zQz^T = 0 \text{ in } \mathbb{C}P^n \text{ has no real points}\}_0. \end{aligned}$$

We take here the connected component $\{\quad\}_0$ containing

$$M_{\mathbb{R}} = \{Q \in M(n+1;\mathbb{R}), Q = Q^T, \det Q = 1,\ Q \gg 0\}.$$

The last example illustrates the situation in the non-Hermitian case: $M(D)$ is $G_{\mathbb{R}}$-invariant, but is not homogeneous. Let us formulate a few conjectures for this case.

2.4 Basic conjectures

Conjecture 2.1 *If $G_{\mathbb{R}}$ is a group of non-Hermitian type then $M(D)$ will be the same for all flag domains with the group $G_{\mathbb{R}}$.*

This conjecture is true for groups of Hermitian type in the generic situation. To make this conjecture explicit we need to find a language to describe $G_{\mathbb{R}}$-invariant nonhomogeneous manifolds at M. One possibility is to parameterize $G_{\mathbb{R}}$-orbits close to $M_{\mathbb{R}}$ and to find parameters corresponding to orbits at the $G_{\mathbb{R}}$-invariant manifold $M(D)$. In Akhiezer and Gindikin (1990) there was described in such a language a Stein neighborhood of $M_{\mathbb{R}}$ at M, which I believe coincides with $M(D)$.

Let us discuss another possibility. We will define special functions—determinant functions—which I believe play a central role in analysis on flag domains (joint project with H.-W.Wong). We consider again the action of an isotropy subgroup K_m on F. There is one open orbit, but we are interested in orbits of codimension 1. Their closures are algebraic varieties which are defined by the equations

$$\Delta_j(m|z) = 0, \qquad m \in M, \quad z \in F.$$

These algebraic functions Δ_j on $M \times F$ are called *determinant functions.* They can be explicitly computed in the language of roots in the general case. For example 5a

$$\Delta(Q|z) = zQz^T.$$

All flag domains D_j on F have as a joint edge of boundaries a compact homogeneous (CR) manifold Ξ with group $K_{\mathbb{R}}$ (it is also the minimal orbit of $G_{\mathbb{R}}$).

Conjecture 2.2 *For groups $G_{\mathbb{R}}$ of non-Hermitian type $M(D)$ coincides with the connected component of the set*

$$\{m \in M : \Delta_j(m|z) \neq 0 \quad \text{for all } z \in \Xi\}$$

containing $M_{\mathbb{R}}$. For its description it is enough to take any one of Δ_j.

We can see that in example 5a, we had just such a description of $M(D)$. Again this conjecture is true for a group of Hermitian type in the generic case.

3 Generalized Penrose transform (analytic problems)

Let $D \subset F$ be a flag domain, $\Omega(m)$, $m \in M(D) \subset M$, be maximal compact submanifolds in D, $q = \dim_{\mathbb{C}} \Omega(m)$. Then for appropriate invariant vector bundles V

$$\dim H^{(q)}(D, V) = \infty$$

and a representation of $G_{\mathbb{R}}$ is realized in this cohomology. In particular, Schmid proved the Kostant–Langlands conjecture that discrete series of representations are realized in such a way for complete flags and some line bundles. In example 4 the domains D_1, D_2 correspond to holomorphic and antiholomorphic series of $SU(2,1)$ and D_3 to nonholomorphic discrete series. In the first two cases cohomology is integrated along fibers of $D_1 \to \tilde{D}_+$, $D_2 \to \tilde{D}'_+$ and we obtain sections of holomorphic or antiholomorphic bundles.

We obtain the Penrose transform if we integrate the cohomology $H^{(q)}(D)$ along cycles $\Omega(m)$, $m \in M(D)$. Results will be holomorphic sections of some vector bundles on $M(D)$ satisfying some systems of differential equations. Under some restrictions such systems of first order were introduced by Schmid (1989). More precisely, he considered these equations only on $M_{\mathbb{R}} \subset M(D)$ where they are elliptic and he realized representations of discrete series in their solutions. The intertwining operator from cohomology on D to solutions of the Schmid equations on $M_{\mathbb{R}}$—a variant of the Penrose transform—was considered by Schmid. This transform is sometimes called the real Penrose–Schmid transform. I want to emphasize that this "real transform" is not something specifically real: it was just not extended to the complex domain. I believe that the construction of the extended complex Penrose transform is very essential. As a consequence we must obtain

Conjecture 3.1 *All solutions of Schmid equations can be simultaneously holomorphically extended on $M(D)$.*

Such a phenomenon, of the simultaneous extension of solutions, is known for some elliptic equations starting from the Laplace equations on the sphere (Aronshain). Conjecture 3 is proved for $SU(p,q)$ in Barchini *et al.* (1997).

For the usual Penrose transform analogs of the Schmid equations are massless equations which are conformally invariant. It is interesting to find a geometrical interpretation of the Schmid equations. Examples show that probably there is a possibility of connecting the Schmid equations with a generalized conformal structure (Gindikin 1990). Let us consider on tangent spaces $T_m M$ conic varieties V_m, which are minimal orbits of M_m (cf. examples in Gindikin 1995). The field $\{V_m, m \in M\}$ defines on M a generalized conformal structure (which is not flat).

Among conformally invariant equations there are not only massless equations of first order, but also the wave equation of second order. It turns out that in the general case the class of Schmid equations can be extended to systems of equations of second order. If $G_{\mathbb{R}}$ is a group of Hermitian type, then such systems are Hua equations which, at first glance, had appeared in absolutely different problems. Analogs of Hua systems exist also for groups of non-Hermitian type.

Problem 3.2 *To develop a unified geometrical theory of Hua and Schmid equations which probably must be connected with the generalized conformal structure on M.*

It is natural to connect the construction of the inverse Penrose transform with the consideration of holomorphic cohomology on D. For flag domains D there are very explicit (but nonhomogeneous) Stein coverings S and we do not need to use here the general construction of Eastwood *et al.* (1996). Since the operator from the holomorphic cohomology to Dolbeault cohomology has the standard structure (like (1.11)) we need only to construct an explicit intertwining operator from solutions to holomorphic cohomology.

Conjecture 3.3 *Such an operator is a differential operator (similar to the operator κ of Gelfand–Graev–Shapiro in integral geometry).*

Finally we will say a few words about analogs of the boundary representation of the Penrose transform (1.12). Such an operator was defined by Knapp and Wallach (1976): their Szegö operator acts from sections of bundles on $\Xi(D)$ to solutions of Schmid equations on $M_{\mathbb{R}}$. Again the crucial and nontrivial problem is to extend the kernel to the complex domain and to find its singularities.

Conjecture 3.4 *The Szegö–Knapp–Wallach kernel is a rational function whose denominator is a product of degrees of determinant functions.*

Out of this conjecture and conjecture 2.2 follows conjecture 3.1. Conjecture 3.4 was proved in Barchini *et al.* (1997) for $SU(p,q)$.

There are several other intertwining operators in this structure which it is interesting to compute explicitly. The most interesting operator is from cohomology on D to sections on $\Xi(D)$ which complete the Penrose and Szegö transforms to form a commutative diagram. It must be a nonstandard operator of boundary values for $\overline{\partial}$-cohomology. It is essential that in generic situations the edge Ξ is only a part of the boundary and for complete flags they are totally real (then D looks like a curved nonconvex tube). For such domains cohomology can have functions or distributions or hyperfunctions as boundary values on edges (Gindikin 1994b). In Gindikin (1994a, 1996) these conjectures were investigated for $SU(2,1)$, and in Gindikin (1997) for $SO(1,n)$.

Bibliography

Akhiezer, D. N. and Gindikin, S. G. (1990). On the Stein extension of real symmetric spaces. *Math. Ann.*, **286**, 1–12.

Barchini, L., Gindikin, S. G. and Wong H.-W. (1997). Determinant functions and holomorphic extension of Szegö kernels for $SU(p,q)$. *Pacific J. Math.* To appear.

Eastwood, M. G., Penrose, R. and Wells, Jr., R. O. (1981). Cohomology and Massless fields. *Comm. Math. Phys.*, **78**, 305–51.

Eastwood, M. G., Gindikin, S. G. and Wong H.-W. (1995). Holomorphic realization of $\overline{\partial}$-cohomology and constructions of representations. *J. Geom. Phys.*, **17**, 231–244.

Eastwood, M. G., Gindikin, S. G. and Wong H.-W. (1996). A holomorphic realization of analytic cohomology. *C. R. Acad. Sci.*, **322**, Serie 1, 529–534.

Gindikin, S. G. (1984). Integral formulas and integral geometry for $\overline{\partial}$-cohomology at $\mathbb{C}P^n$. *Funk. Anal. Priloz.*, **18**, 2, 26–33. Transl.: *Func. Anal. Appl.*, **18**, 108–118 (1984).

Gindikin, S. G. (1990). Generalized conformal structures. In *Twistors in Mathematics and Physics*, Bailey, T. N. and Baston, R. J. (eds). *London Math. Soc. Lect. Notes Ser.* **156**, 36–52. London.

Gindikin, S. G. (1993). Holomorphic language for $\overline{\partial}$-cohomology and representations of real semisimple Lie groups. *Contemporary Math.*, **154**, 103–115.

Gindikin, S. (1994a). The holomorphic Cauchy–Szegö kernel for nonholomorphic representations of $SU(2,1)$. In *Representation Theory and Harmonic Analysis. Contemp. Math. AMS*, **191**, 75–82.

Gindikin, S. (1994b). The Radon transform from cohomological point of view. In *75 Years of Radon Transform*, Gindikin, S. and Michor, P. (eds). International Press, 123–128.

Gindikin, S. (1995). Fubini–Study structures on Grassmanians. IHES Preprint.

Gindikin, S. (1996). The Penrose transform on flag domains in $F(\mathbb{C}P^2)$. *Contemporary Mathematical Physics*, F.A. Berezin memorial volume. *AMS Translations* Series 2, **175**, 49–56.

Gindikin, S. (1997). $SO(1,n)$-twistors. *J. Geom. Phys.* To appear.

Gindikin, S. G. and Henkin, G. M. (1978a). Integral geometry for $\overline{\partial}$-cohomology in q-linear concave domains in $\mathbb{C}P^n$. *Funk. Anal. Prilozh.*, **12**, 4, 6–23. Engl.transl.: *Func. Anal. Appl.*, **12**, 247–261 (1978).

Gindikin S. and Henkin, G. (1978b). Transformation de Radon pour la d''-cohomologie des domains q-linéar linéarement concaves. *C. R. Acad. Sci.*, **AB297**, 4, 209–212.

Gindikin, S. G. and Henkin, G. M. (1981). The Penrose transform and complex integral geometry. *Modern Problems of Mathematics*, VINITI, Moscow, **17**, 57–112.

Knapp, A. W. and Wallach, N. R. (1976). Szegö kernels associated with discrete series. *Inv. Mat.*, **34** , 163–200.

Martineau, A. (1966). Sur la topologie des espaces de fonction holomorphes. *Math. Ann.*, **163**, 62–88.

Schmid, W. (1989). Homogeneous complex manifolds and representations of semisimple Lie groups. In *Representation Theory and Harmonic Analysis on Semisimple Lie Groups*, Math. Surveys and Monographs, AMS, **31**, 223–286.

Wolf, J. A. (1992). The Stein condition for cycle spaces of open orbits on complex flag manifolds. *Ann. Math.*, **136**, 541–555.

29

Twistor Solution of the Holonomy Problem

S.A. Merkulov
University of Glasgow

and

L.J. Schwachhöfer
Universität Leipzig

1 Introduction

The holonomy group is one of the most informative characteristics of an affine connection. The problem of classifying all possible holonomy groups has a long history, beginning in the 1920s with the works of Cartan (1926a, b) who used this notion to classify locally symmetric Riemannian manifolds. Berger (1955) showed that the list of irreducibly acting matrix Lie groups which can, in principle, occur as the holonomy of a torsion-free affine connection must be very restrictive. This sharply contrasts a result of Hano and Ozeki (1956) which says that there is no interesting holonomy classification in the class of arbitrary affine connections — *any* closed subgroup of $\mathrm{GL}(n, \mathbb{R})$ can be realized as the holonomy of an affine connection (with torsion, in general).

Berger presented his classification list of all possible candidates for irreducible holonomies[1] in two parts—the first part was claimed to contain all possible groups which preserve a non-degenerate symmetric bilinear form, while the second part was conjectured to contain all the rest, *up to a finite number of missing terms* which Bryant (1991) suggested calling the *exotic holonomies*.

2 Main result

The classification of all metric holonomies has been recently completed (Bryant 1995). The resulting Table 1 is a culmination of efforts of many people to show that most entries of Berger's original metric list do occur as holonomies of Levi Civita connections and that just a few of them are superfluous (see, e.g., Alek-

[1] From now on, by a holonomy group we always understand the irreducibly acting holonomy of a *torsion-free* affine connection which is *not* locally symmetric. The second assumption is motivated by the fact that, due to Cartan (1926) and Berger (1957), the list of locally symmetric affine spaces is completely known.

Table 1 Complete list of metric holonomies

Group G	Representation space	Group G	Representation space
$\mathrm{SO}(p,q)$	$\mathbb{R}^{p+q}$, $(p+q) \geqslant 2$	G_2	$\mathbb{R}^7$
$\mathrm{SO}(n,\mathbb{C})$	$\mathbb{R}^{2n}$, $n \geqslant 2$	G_2'	$\mathbb{R}^7$
$\mathrm{U}(p,q)$	$\mathbb{R}^{2(p+q)}$, $(p+q) \geqslant 2$	$\mathrm{G}_2^{\mathbb{C}}$	$\mathbb{R}^{14}$
$\mathrm{SU}(p,q)$	$\mathbb{R}^{2(p+q)}$, $(p+q) \geqslant 2$	$\mathrm{Spin}(7)$	$\mathbb{R}^8$
$\mathrm{Sp}(p,q) \cdot \mathrm{Sp}(1)$	$\mathbb{R}^{4(p+q)}$, $(p+q) \geqslant 2$	$\mathrm{Spin}(4,3)$	$\mathbb{R}^8$
$\mathrm{Sp}(p,q)$	$\mathbb{R}^{4(p+q)}$, $(p+q) \geqslant 2$	$\mathrm{Spin}(7,\mathbb{C})$	$\mathbb{R}^{16}$
$\mathrm{Sp}(n,\mathbb{R}) \cdot \mathrm{SL}(2,\mathbb{R})$	$\mathbb{R}^{4n}$, $n \geqslant 2$		
$\mathrm{Sp}(n,\mathbb{C}) \cdot \mathrm{SL}(2,\mathbb{C})$	$\mathbb{R}^{8n}$, $n \geqslant 2$		

seevski 1968; Besse 1987; Bryant 1987, 1991, 1995; Salamon 1989, and the references cited therein).

Berger's second list of non-metric holonomies, refined and extended, is given in Tables 1 and 2. The 4-dimensional representations of $\mathrm{T}_{\mathbb{R}} \cdot \mathrm{SL}(2,\mathbb{R})$, $\mathrm{T}_{\mathbb{C}} \cdot \mathrm{SL}(2,\mathbb{C})$, $\mathbb{R}^* \cdot \mathrm{SO}(2) \cdot \mathrm{SL}(2,\mathbb{R})$ and $\mathbb{C}^* \cdot \mathrm{SU}(2)$, and the fundamental representations of various real forms of $\mathrm{T}_{\mathbb{C}} \cdot \mathrm{Spin}(10,\mathbb{C})$ and $\mathrm{T}_{\mathbb{C}} \cdot \mathrm{E}_6^{\mathbb{C}}$ have been added to the list of non-metric holonomies by Bryant (1991, 1995). He also conjectured that the 4-dimensional representations of $H_\lambda \cdot \mathrm{SU}(2)$ and $H_\lambda \cdot \mathrm{SL}(2,\mathbb{R})$ may occur as holonomies. The *infinite* series $\mathrm{SL}(2,\mathbb{R}) \cdot \mathrm{SO}(p,q)$ and $\mathrm{SL}(2,\mathbb{C}) \cdot \mathrm{SO}(n,\mathbb{C})$ as well as exceptional representations E_7^5, E_7^7, $\mathrm{E}_7^{\mathbb{C}}$ have been found by Chi *et al.* (1996). Finally, the representations of the groups $\mathrm{Sp}(3,\mathbb{R})$, $\mathrm{Sp}(3,\mathbb{C})$, $\mathrm{SU}(1,5)$, $\mathrm{SU}(3,3)$, $\mathrm{SL}(6,\mathbb{R})$, $\mathrm{SL}(6,\mathbb{C})$, $\mathrm{Spin}(2,10)$, $\mathrm{Spin}(6,6)$, $\mathrm{Spin}(12,\mathbb{C})$ have been added to the list by the authors (1996) who showed that the moduli space of torsion-free affine connections with these holonomies is non-empty and finite dimensional. The point of the present article is that these representations are the last ones which are missing from the original Berger lists:

Main Theorem *If G is an irreducible representation of a reductive Lie group which occurs as the holonomy of a non-locally symmetric torsion-free affine connection, then G is one of the entries in Tables 1–3.*

Moreover, due to the efforts of a number of people over the last 40 years all entries of Tables 1–3 are known to occur as holonomies except the 4-dimensional representations of $H_\lambda \cdot \mathrm{SU}(2)$ and $H_\lambda \cdot \mathrm{SL}(2,\mathbb{R})$ which, as candidates for holonomies, have been suggested by Bryant (1995).

Table 2 Berger's list of non-metric holonomies

group G	representation V	restrictions
$T_{\mathbb{R}} \cdot SL(n,\mathbb{R})$	$\mathbb{R}^n$	$n \geqslant 2$
	$\odot^2\mathbb{R}^n \simeq \mathbb{R}^{n(n+1)/2}$	$n \geqslant 3$
	$\Lambda^2\mathbb{R}^n \simeq \mathbb{R}^{n(n-1)/2}$	$n \geqslant 5$
$T_{\mathbb{C}} \cdot SL(n,\mathbb{C})$	$\mathbb{C}^n \simeq \mathbb{R}^{2n}$	$n \geqslant 1$
$T^*_{\mathbb{C}} \cdot SL(n,\mathbb{C})$	$\odot^2\mathbb{C}^n \simeq \mathbb{R}^{n(n+1)}$	$n \geqslant 3$
	$\Lambda^2\mathbb{C}^n \simeq \mathbb{R}^{n(n-1)}$	$n \geqslant 5$
$\mathbb{R}^* \cdot SL(n,\mathbb{C})$	$\{A \in M_n(\mathbb{C}) : \bar{A} = A^t\} \simeq \mathbb{R}^{n^2}$	$n \geqslant 3$
$T_{\mathbb{R}} \cdot SL(n,\mathbb{H})$	$\mathbb{H}^n \simeq \mathbb{R}^{4n}$	$n \geqslant 1$
	$\{A \in M_n(\mathbb{H}) : A^* = -A^t\} \simeq \mathbb{R}^{n(2n+1)}$	$n \geqslant 2$
	$\{A \in M_n(\mathbb{H}) : A^* = A^t\} \simeq \mathbb{R}^{n(2n-1)}$	$n \geqslant 3$
$T_{\mathbb{R}} \cdot Sp(n,\mathbb{R})$	$\mathbb{R}^{2n}$	$n \geqslant 2$
$T_{\mathbb{C}} \cdot Sp(n,\mathbb{C})$	$\mathbb{C}^{2n} \simeq \mathbb{R}^{4n}$	$n \geqslant 2$
$\mathbb{R}^* \cdot SO(p,q)$	$\mathbb{R}^{p+q}$	$p+q \geqslant 3$
$T^*_{\mathbb{C}} \cdot SO(n,\mathbb{C})$	$\mathbb{C}^n \simeq \mathbb{R}^{2n}$	$n \geqslant 3$
$T_{\mathbb{R}} \cdot SL(m,\mathbb{R}) \cdot SL(n,\mathbb{R})$	$\mathbb{R}^{mn}$	$m > n \geqslant 2$ or $m \geqslant n > 2$
$T_{\mathbb{C}} \cdot SL(m,\mathbb{C}) \cdot SL(n,\mathbb{C})$	$\mathbb{C}^m \otimes \mathbb{C}^n \simeq \mathbb{R}^{2mn}$	$m > n \geqslant$ or $m \geqslant n > 2$
$T_{\mathbb{R}} \cdot SL(m,\mathbb{H}) \cdot SL(n,\mathbb{H})$	$\mathbb{R}^{16mn}$	$m > n \geqslant 1$ or $m \geqslant n > 1$
$SU(2) \cdot SO(n,\mathbb{H})$	$\mathbb{R}^2 \otimes \mathbb{R}^{4n} \simeq \mathbb{R}^{8n}$	$n \geqslant 2$

NOTATIONS: $T_{\mathbb{F}}$ denotes any connected Lie subgroup of $\mathbb{F}^*$,
$T^*_{\mathbb{F}}$ denotes any non-trivial connected Lie subgroup of $\mathbb{F}^*$,
$M_n(\mathbb{F})$ denotes the algebra of $n \times n$ matrices over $\mathbb{F}$.

3 Twistor theory of holonomy groups

Let V be a vector space and $\mathfrak{g}$ an irreducible Lie subalgebra of $\mathfrak{gl}(V) \simeq V \otimes V^*$. In the holonomy group context, one is interested in the following three $\mathfrak{g}$-modules:

(i) $\mathfrak{g}^{(1)} := (\mathfrak{g} \otimes V^*) \cap (V \otimes \odot^2 V^*)$,

(ii) the *curvature space* $K(\mathfrak{g}) := \ker i_1$, where i_1 is the composition

$$i_1 : \mathfrak{g} \otimes \Lambda^2 V^* \longrightarrow V \otimes V^* \otimes \Lambda^2 V^* \longrightarrow V \otimes \Lambda^3 V^*,$$

(iii) the *2nd curvature space* $K^1(\mathfrak{g}) := \ker i_2$, where i_2 is the composition

$$i_2 : K(\mathfrak{g}) \otimes V^* \longrightarrow \mathfrak{g} \otimes \Lambda^2 V^* \longrightarrow \mathfrak{g} \otimes \Lambda^3 V^*.$$

Note that if ∂ is the composition

$$\mathfrak{g}^{(1)} \otimes V^* \to \mathfrak{g} \otimes V^* \otimes V^* \to \mathfrak{g} \otimes \Lambda^2 V^*$$

Table 3 List of exotic holonomies

group G	representation V	restrictions
$T_{\mathbb{R}} \cdot \mathrm{Spin}(5,5)$	$\mathbb{R}^{16}$	
$T_{\mathbb{R}} \cdot \mathrm{Spin}(1,9)$	$\mathbb{R}^{16}$	
$T_{\mathbb{C}} \cdot \mathrm{Spin}(10,\mathbb{C})$	$\mathbb{C}^{16} \simeq \mathbb{R}^{32}$	
$T_{\mathbb{R}} \cdot E_6^1$	$\mathbb{R}^{27}$	
$T_{\mathbb{R}} \cdot E_6^4$	$\mathbb{R}^{27}$	
$T_{\mathbb{C}} \cdot E_6^{\mathbb{C}}$	$\mathbb{C}^{27} \simeq \mathbb{R}^{54}$	
$T_{\mathbb{R}} \cdot \mathrm{SL}(2,\mathbb{R})$	$\odot^3\mathbb{R}^2 \simeq \mathbb{R}^4$	
$T_{\mathbb{C}} \cdot \mathrm{SL}(2,\mathbb{C})$	$\odot^3\mathbb{C}^2 \simeq \mathbb{R}^8$	
$\mathbb{R}^* \cdot \mathrm{SO}(2) \cdot \mathrm{SL}(2,\mathbb{R})$	$\mathbb{R}^2 \otimes \mathbb{R}^2 \simeq \mathbb{R}^4$	
$\mathbb{C}^* \cdot \mathrm{SU}(2)$	$\mathbb{C}^2 \simeq \mathbb{R}^4$	
$H_\lambda \cdot \mathrm{SU}(2)$	$\mathbb{R}^4$	
$H_\lambda \cdot \mathrm{SL}(2,\mathbb{R})$	$\mathbb{R}^4$	
$\mathrm{SL}(2,\mathbb{R}) \cdot \mathrm{SO}(p,q)$	$\mathbb{R}^2 \otimes \mathbb{R}^{p+q} \simeq \mathbb{R}^{2p+2q}$	$p+q>2$
$\mathrm{SL}(2,\mathbb{C}) \cdot \mathrm{SO}(n,\mathbb{C})$	$\mathbb{C}^2 \otimes \mathbb{C}^n \simeq \mathbb{R}^{4n}$	$n \geqslant 3$
E_7^5	$\mathbb{R}^{56}$	
E_7^7	$\mathbb{R}^{56}$	
$E_7^{\mathbb{C}}$	$\mathbb{R}^{112} \simeq \mathbb{C}^{56}$	
$\mathrm{Sp}(3,\mathbb{R})$	$\mathbb{R}^{14} \subset \Lambda^3\mathbb{R}^6$	
$\mathrm{Sp}(3,\mathbb{C})$	$\mathbb{R}^{28} \simeq \mathbb{C}^{14} \subset \Lambda^3\mathbb{C}^6$	
$\mathrm{SU}(1,5)$	$\mathbb{R}^{28}$	
$\mathrm{SU}(3,3)$	$\mathbb{R}^{28}$	
$\mathrm{SL}(6,\mathbb{R})$	$\mathbb{R}^{20} \simeq \Lambda^3\mathbb{R}^6$	
$\mathrm{SL}(6,\mathbb{C})$	$\mathbb{R}^{40} \simeq \Lambda^3\mathbb{C}^6$	
$\mathrm{Spin}(2,10)$	$\mathbb{R}^{32}$	
$\mathrm{Spin}(6,6)$	$\mathbb{R}^{32}$	
$\mathrm{Spin}(12,\mathbb{C})$	$\mathbb{R}^{64} \simeq \mathbb{C}^{32}$	

NOTATIONS: $T_{\mathbb{F}}$ denotes any connected Lie subgroup of $\mathbb{F}^*$,

$$H_\lambda = \left\{ e^{\lambda t} \begin{pmatrix} \cos t & -\sin t \\ \sin t & \cos t \end{pmatrix} : t \in \mathbb{R} \right\}, \quad \lambda > 0.$$

then $\partial(\mathfrak{g}^{(1)} \otimes V^*) \subseteq K(\mathfrak{g})$.

The geometric meaning of $\mathfrak{g}^{(1)}$ is that if there exists a (local) torsion-free affine connection ∇ on a manifold M with holonomy algebra $\mathfrak{g}$ then, for any (local) function $\Gamma : M \to \mathfrak{g}^{(1)}$, the affine connection $\nabla + \Gamma$ is again torsion-free and has holonomy algebra $\mathfrak{g}$; put another way, $\mathfrak{g}^{(1)}$ measures non-uniqueness of torsion-free affine connections with holonomy $\mathfrak{g}$ on a fixed manifold.

The meaning of $K(\mathfrak{g})$ and $K^1(\mathfrak{g})$ is that the curvature tensor of a torsion-free affine connection ∇ on a manifold M with holonomy in $\mathfrak{g}$ has the curvature tensor at each $x \in M$ isomorphic to an element of $K(\mathfrak{g})$ while the non-vanishing of $K^1(\mathfrak{g})$ is a necessary condition for ∇ *not* to be locally symmetric.

Therefore, $\mathfrak{g}$ can be a candidate to the holonomy algebra of a torsion-free affine connection only if $K(\mathfrak{g}) \neq 0$. Then the question is how to compute $K(\mathfrak{g})$?

With any real irreducible representation of a real reductive Lie algebra one may associate an irreducible complex representation of a complex reductive Lie algebra. Since all the above $\mathfrak{g}$-modules behave reasonably well under this association, we may assume from now on that V is a finite dimensional complex vector space and $\mathfrak{g} \subseteq \mathfrak{gl}(V)$ is an irreducible representation of a complex reductive Lie algebra. Clearly, $G = \exp(\mathfrak{g})$ acts irreducibly in V^* via the dual representation. Let $\tilde{X}$ be the G-orbit of a highest weight vector in $V^* \setminus 0$. Then the quotient $X := \tilde{X}/\mathbb{C}^*$ is a compact complex homogeneous-rational manifold canonically embedded into $\mathbb{P}(V^*)$, and there is a commutative diagram

$$\begin{array}{ccc} \tilde{X} & \hookrightarrow & V^* \setminus 0 \\ \downarrow & & \downarrow \\ X & \hookrightarrow & \mathbb{P}(V^*) \end{array}$$

In fact, $X = G_s/P$, where G_s is the semisimple part of G and P is the parabolic subgroup of G_s leaving a highest weight vector in V^* invariant up to a scale. Let L be the restriction of the hyperplane section bundle $\mathcal{O}(1)$ on $\mathbb{P}(V^*)$ to the submanifold X. Clearly, L is an ample homogeneous line bundle on X. We call (X, L) the *Borel-Weil data* associated with $(\mathfrak{g}, V)$.

According to Borel-Weil, the representation space V can be easily reconstructed from (X, L) as $V = \mathrm{H}^0(X, L)$. What about $\mathfrak{g}$? The Lie algebra of the Lie group of all global biholomorphisms of the line bundle L which commute with the projection $L \to X$ is isomorphic to $H^0(X, L \otimes (J^1 L)^*)$—a central extension of the Lie algebra $\mathrm{H}^0(X, TX)$. Whence, as a complex Lie algebra, $\mathrm{H}^0(X, L \otimes (J^1 L)^*)$ has a natural complex irreducible representation in $\mathrm{H}^0(X, L) = V$; with very few (and well studied in the holonomy context) exceptions, this representation is isomorphic, up to a central extension, to the original $\mathfrak{g}$.

Remarkably enough, the basic $\mathfrak{g}$-modules defined above fit nicely the Borel-Weil paradigm as well. The resulting twistor formulae were among our basic instruments in solving the holonomy problem.

Proposition 3.1 *For a compact complex homogeneous-rational manifold X and an ample line bundle $L \to X$, there is an isomorphism*

$$\mathfrak{g}^{(1)} = H^0\left(X, L \otimes \odot^2 N^*\right),$$

and an exact sequence of $\mathfrak{g}$-modules,

$$0 \longrightarrow \frac{K(\mathfrak{g})}{\partial(\mathfrak{g}^{(1)} \otimes V^*)} \longrightarrow H^1\left(X, L \otimes \odot^3 N^*\right) \longrightarrow H^1\left(X, L \otimes \odot^2 N^*\right) \otimes V^*,$$

where $\mathfrak{g}$ is $H^0(X, L \otimes N^*)$ represented in $V = H^0(X, L)$, and where $N := J^1 L$.

Proof. The result follows easily from the exact sequences

$$0 \longrightarrow L \otimes \odot^2 N^* \longrightarrow L \otimes N^* \otimes V^* \longrightarrow L \otimes N^* \otimes \Lambda^2 V^*$$

and

$$0 \longrightarrow L \otimes \odot^3 N^* \longrightarrow L \otimes \odot^2 N^* \otimes V^* \longrightarrow L \otimes N^* \otimes \Lambda^2 V^* \longrightarrow L \otimes \Lambda^3 V^*,$$

where arrows are a combination of a natural monomorphism $N^* \longrightarrow V^* \otimes \mathcal{O}_X$ (which holds due to ampleness of L) with the antisymmetrization. □

This proposition is a group-theoretic manifestation of the fact (Merkulov 1994) that *any* torsion-free affine connection with irreducibly acting holonomy can, in principle, be constructed by twistor methods. This universal twistor construction can be formulated in the language of G-structures as follows.

Theorem 3.2 *Let X be a generalised flag variety embedded as a Legendre submanifold into a complex contact manifold Y with contact line bundle L such that L_X is very ample on X. Then*

(i) *There exists a complete analytic family $F \hookrightarrow Y \times M$ of compact Legendre submanifolds with moduli space M being an $h^0(X, L_X)$-dimensional complex manifold. For each $t \in M$, the associated Legendre submanifold X_t is isomorphic to X.*

(ii) *The Legendre moduli space M comes equipped with an induced irreducible G-structure, $\mathcal{G}_{ind} \to M$, with G isomorphic to the connected component of the identity of the group of all global biholomorphisms $\phi : L_X \to L_X$ which commute with the projection $\pi : L_X \to X$. The Lie algebra of G is isomorphic to $H^0\left(X, L_X \otimes (J^1 L_X)^*\right)$.*

(iii) *If $\mathcal{G}_{ind}$ is k-flat, $k \geqslant 0$, then the obstruction for $\mathcal{G}_{ind}$ to be $(k+1)$-flat is given by a tensor field on M whose value at each $t \in M$ is represented by a cohomology class $\rho_t^{[k+1]} \in \tilde{H}^1\left(X_t, L_{X_t} \otimes S^{k+2}(J^1 L_{X_t})^*\right)$.*

(iv) *If $\mathcal{G}_{ind}$ is 1-flat, then the bundle of all torsion-free connections in $\mathcal{G}_{ind}$ has as the typical fiber an affine space modeled on $H^0\left(X, L_X \otimes S^2(J^1 L_X)^*\right)$.*

The geometric meaning of cohomology classes

$$\rho_t^{[k+1]} \in \tilde{H}^1\left(X_t, L_{X_t} \otimes S^{k+2}(J^1 L_{X_t})^*\right)$$

is very simple—they compare to $(k+2)$th order the germ of the Legendre embedding $X_t \hookrightarrow Y$ with the "flat" model, $X_t \hookrightarrow J^1L_{X_t}$, where the ambient contact manifold is just the total space of the vector bundle $J^1L_{X_t}$ together with its canonical contact structure and the Legendre submanifold X_t is realized as a zero section of $J^1L_{X_t} \to X_t$. Therefore, the cohomology class $\rho_t^{[k]}$ can be called the kth Legendre jet of X_t in Y. Then it is natural to call a Legendre submanifold $X_t \hookrightarrow Y$ *k-flat* if $\rho_t^{[k]} = 0$. With this terminology, the item (iii) of the above theorem acquires a rather symmetric form: *the induced G-structure on the moduli space M of a complete analytic family of compact Legendre submanifolds is k-flat if and only if the family consists of k-flat Legendre submanifolds.*

A very intriguing aspect of the holonomy list of the Main Theorem is that all holonomy groups share one and the same property—the associated Borel-Weil data (X, L) have X biholomorphic to a compact Hermitian symmetric manifold. Put another way, twistor theory of holonomy groups reveals a surprising pattern in the classification list of holonomy groups which is not visible in the standard $(\mathfrak{g}, V)$-description. The list of compact Hermitian symmetric manifolds is very short, and it is desirable to get an independent proof (explanation) of this phenomenon. Such an explanation may result in a much shorter proof of the Main Theorem than the one given in Merkulov and Schwachhöfer (1996).

Bibliography

Alekseevski, D. V. (1968). Riemannian spaces with unusual holonomy groups. *Funct. Anal. Appl.* **2**, 97–105.

Berger, M. (1955). Sur les groupes d'holonomie des variétés à connexion affine et des variétés Riemanniennes. *Bull. Soc. Math. France* **83**, 279–330.

Berger, M. (1957). Les espaces symétriques noncompacts. *Ann. Sci. École Norm. Sup.* **74**, 85-177.

Besse, A. (1987). *Einstein Manifolds.* Springer.

Bryant, R. (1987). Metrics with exceptional holonomy. *Ann. of Math.* **126**, 525–576.

Bryant, R. (1991). Two exotic holonomies in dimension four, path geometries, and twistor theory. *Proc. Symposia in Pure Mathematics* **83**, 33–88.

Bryant, R. (1995). Classical, exceptional, and exotic holonomies: a status report. Preprint.

Cartan, É. (1909). Les groupes de transformations continus, infinis, simples. *Ann. Éc. Norm.* **26**, 93–161.

Cartan, É. (1926a). Les groups d'holonomie des espaces généralisés. *Acta Math.* **48**, 1–40.

Cartan, É. (1926b). Sur une classe remarquable d'espaces de Riemann. *Bull. Soc. Math. France.* **54**, 214–264; **55**, 114–134.

Chi, Q.-S., Merkulov, S. A., and Schwachhöfer, L. J. (1996). On the existence of an infinite series of exotic holonomies. *Inv. Math.* **126**, 391–411.

Hano, J. and Ozeki, H. (1956). On the holonomy groups of linear connections. *Nagoya Math. J.* **10**, 97–100.

Merkulov, S. A. (1994). Existence and geometry of Legendre moduli spaces. *Math.Z.*, to appear.

Merkulov, S. A. and Schwachhöfer, L. J. (1996). Classification of irreducible holonomies of torsion-free affine connections, preprint.

Penrose, R. (1976). Non-linear gravitons and curved twistor theory. *Gen. Rel. Grav.* **7**, 31–52.

Salamon, S. M. (1989). *Riemannian Geometry and Holonomy Groups.* Longman.

30

The Penrose Transform and Real Integral Geometry

Toby N. Bailey
Department of Mathematics, University of Edinburgh

1 Introduction

I remember, some years ago, Roger Penrose drawing a picture of twistor theory as a tree, which he was trying to train mainly towards fundamental physics, but which was always sending off vigorous shoots in the direction of pure mathematics. Twistor theory has certainly been a source of an astonishing amount of mathematics. One example is what we now call the Penrose transform, and in this note I want to report on some joint research with Michael Eastwood (Bailey and Eastwood 1997), whereby methods involving holomorphic geometry inspired by the Penrose transform are used to obtain results on transforms in real integral geometry. (This follows on from joint work with Eastwood, Rod Gover and Lionel Mason (1997). It was Mason's idea originally that one ought to be able to obtain results of this sort.) A full account of our results appears in (Bailey and Eastwood 1997), and here I want just to outline the methods and make more precise the connection with the Penrose transform which is largely suppressed in (Bailey and Eastwood 1997).

2 The twistor programme

Roger Penrose's Twistor Programme aims to obtain an alternative formulation of basic physics, with the hope that it will cast light on the the fundamental problem of unifying general relativity with quantum theory. From a mathematical point of view, it is based on the following geometry: the Grassmanian $\mathbb{M} = \mathrm{Gr}_2(\mathbb{C}^4)$ of 2-dimensional subspaces of $\mathbb{C}^4$ comes equipped with a natural holomorphic conformal structure, and it can be regarded as a compactification of complexified Minkowski space. One then takes (projective) *twistor space* $\mathbb{P} = \mathbb{CP}_3$ to be the projective space of $\mathbb{C}^4$, so that by the classical Klein correspondence $\mathbb{M}$ parameterises the (complex projective) lines in $\mathbb{P}$. Penrose's suggestion is that the space $\mathbb{P}$ may be more fundamental than $\mathbb{M}$ itself, and that if one could describe known physics intrinsically on $\mathbb{P}$, then it might be possible to see a way forward.

The basic conformal geometry of Minkowski space is easily seen to be encoded in incidence relations in $\mathbb{P}$, and it was soon discovered that solutions of many important conformally invariant linear equations (the wave equation, Weyl neutrino equation, source-free Maxwell equations, etc.) could be generated by certain contour integrals of sections of line bundles over appropriate regions of $\mathbb{P}$. (For a review of the twistor programme as it stood at this time, see (Penrose and MacCallum 1972).) The freedom in the choice of f and its domain of definition remained a mystery until it was realised that one should think of it as a Cech representative of an element of a first sheaf cohomology group. At this point, it was possible to use the machinery of sheaf cohomology to obtain results, and the procedure adopted in (Eastwood *et al.* 1981), applied in more general situations, is what is now often called the (holomorphic) Penrose transform.

3 The holomorphic Penrose transform

The general method of the holomorphic Penrose transform in (Eastwood *et al.* 1981) can be described quite quickly. The starting data is that one has a complex manifold Z and a family $\{Y_x | x \in X\}$ of compact complex submanifolds of Z parameterised holomorphically by a complex manifold X. One requires that the *correspondence space*

$$F = \{(z, x) | x \in X, z \in Y_x\}$$

is a holomorphic submanifold of $Z \times X$, and that the obvious maps

$$Z \overset{\mu}{\leftarrow} F \overset{\nu}{\rightarrow} X \tag{3.1}$$

are holomorphic surjections of maximal rank. We assume also that the fibres of μ are contractible. In the context of Penrose's original twistor picture, such a situation arises by taking X to be a "sufficiently convex" open subset of $\mathbb{M}$ and Z to be the subset of $\mathbb{P}$ swept out by all the complex projective lines corresponding to points in X.

Let $E \to Z$ be a holomorphic vector bundle. Then it follows from a result of Buchdahl (1983) that pull-back provides an isomorphism

$$H^p(Z, \mathcal{O}(E)) \to H^p(F, \mu^{-1}\mathcal{O}(E))$$

where μ^{-1} denotes the topological inverse image sheaf and $\mathcal{O}(E)$ denotes the sheaf of holomorphic local sections of E. Local sections of $\mu^{-1}\mathcal{O}(E)$ are holomorphic sections of the pull-back bundle μ^*E constant along the fibres of μ, and it is then not hard to show that there is an exact sequence of sheaves on F,

$$0 \to \mu^{-1}\mathcal{O}(E) \to \mathcal{O}(\mu^*E) \to \Omega^1_\mu(E) \to \Omega^2_\mu(E) \to \ldots, \tag{3.2}$$

where $\Omega^p_\mu(E)$ is the sheaf of holomorphic relative p-forms (i.e. forms in the fibres of μ, parameterised over Z), with values in $\mu^*(E)$. The differential is essentially the holomorphic de Rham operator in the fibres of μ.

For simplicity, let us assume also that X is Stein. Then the Leray spectral sequence collapses and one concludes that

$$H^q(F, \Omega^p_\mu(E)) = \Gamma(X, \nu^q_*(\Omega^p_\mu(E))),$$

where ν_*^q denotes the q-th direct image. Now we make this substitution in the hypercohomology spectral sequence of (3.2) to obtain a spectral sequence

$$E_1^{p,q} = \Gamma(X, \nu_*^q(\Omega_\mu^p(E))) \Rightarrow H^{p+q}(F, \mu^{-1}\mathcal{O}(E)) = H^{p+q}(Z, \mathcal{O}(E)). \qquad (3.3)$$

Generically, the direct images here can be identified with sheaves of holomorphic sections of holomorphic vector bundles on X, and the maps are holomorphic differential operators between them. By the Penrose transform of $H^{p+q}(Z, \mathcal{O}(E))$, I mean the output from this spectral sequence.

4 The connection with integral geometry

A typical output from the Penrose transform is an isomorphism

$$\mathcal{P} : H^p(Z, \mathcal{O}(E)) \to \{\phi \in \Gamma(X, \mathcal{O}(V)) | D\phi = 0\},$$

where V is a holomorphic vector bundle over X and D is some holomorphic differential operator taking values in another holomorphic vector bundle. In many cases where p is the complex dimension of the fibres of ν, one can evaluate $(\mathcal{P}(f))(x)$ by taking a Dolbeault representative for $f \in H^p(Z, \mathcal{O}(E))$ and integrating this over Y_x. (Strictly, one can only do this directly when E is a line bundle that restricts to the canonical line bundle on each Y_x.) This places the Penrose transform squarely in the realm of integral geometry: the transform consists of integrating $\bar{\partial}$-closed differential forms over families of cycles. This is usually the view being taken when one hears the term "Radon-Penrose transform" being used. It is worth pointing out though that the Penrose transform often produces results of interest when the transform map is not so directly realisable.

The work I am reporting on here exploits a different, and less precise, connection between the Penrose transform and integral geometry. To start from the integral geometry end, the real version of the Klein correspondence is the fact that $\mathrm{Gr}_2(\mathbb{R}^4)$ parameterises the real projective lines in $\mathbb{RP}_3$. Consider the transform from smooth 1-forms on $\mathbb{RP}_3$ to twisted functions on $\mathrm{Gr}_2(\mathbb{R}^4)$ obtained by integrating over real projective lines. (The twist arises because an orientation must be chosen.)

The connection between this and the Penrose transform is as follows. The inclusion $\mathbb{R}^4 \subset \mathbb{C}^4$ induces inclusions $\mathrm{Gr}_2(\mathbb{R}^4) \subset \mathrm{Gr}_2(\mathbb{C}^4) = \mathbb{M}$ and $\mathbb{RP}_3 \subset \mathbb{CP}_3 = \mathbb{P}$. Now consider a holomorphic double fibration (3.1) obtained as described immediately below (3.1) from Penrose's original twistor correspondence so that X is some "sufficiently convex" open subset of $\mathbb{M}$. Suppose that $X_R = X \cap \mathrm{Gr}_2(\mathbb{R}^4)$ is non-empty. For $x \in X_R$, the corresponding complex projective line $Y_x \subset Z$ intersects $\mathbb{RP}_3$ in a real projective line, and if one takes a suitable Cech representative for $f \in H^1(Z, \Omega^1)$ (where Ω^1 is the sheaf of holomorphic 1-forms), one component of the Penrose transform $\mathcal{P}f(x)$ can be obtained by integrating the Cech representative over this real projective line. Thus, the Penrose transform is obtained for these "real" points precisely by performing the real integral transform we have just discussed.

This fact suggests that one might be able to use some variant on the holomorphic Penrose transform arguments to obtain results on the real integral transform. One obstruction to doing so is an interesting difference between the holomorphic Penrose transform and most classical transforms in real integral geometry: the Penrose transform is essentially local, in that, for example, one takes X to be any "sufficiently convex" open set in $\mathbb{M}$ (indeed, one cannot take X to be all of $\mathbb{M}$ because the correspondence would not then satisfy the required conditions); on the other hand, most real integral geometry transforms do not localise well in the sense that although the transform may be defined, one does not get clean results about the range.

5 A new method in real integral geometry

It turns out that one can adapt the arguments of the holomorphic Penrose transform to analyse transforms in real integral geometry. This programme is continuing, and I will just outline the methods for one particular example. (See Bailey and Eastwood 1997 for details.)

Let us take the real integral transform mentioned in the previous section. If we try and regard it as a Penrose transform, it suggests that we should take $X = \mathrm{Gr}_2(\mathbb{R}^4) \subset \mathbb{M}$ in place of the usual "sufficiently convex" open subset. The complex projective lines corresponding to points in $\mathrm{Gr}_2(\mathbb{R}^4)$ sweep out the whole of $\mathbb{CP}_3$. Defining a correspondence space as usual by

$$F = \{(z, x) | z \in \mathbb{CP}_3, x \in \mathrm{Gr}_2(\mathbb{R}^4), z \in Y_x\},$$

we obtain a diagram

$$\mathbb{CP}_3 \overset{\eta}{\leftarrow} F \overset{\tau}{\rightarrow} \mathrm{Gr}_2(\mathbb{R}^4).$$

The fibres of τ are $\mathbb{CP}_1$ (as they are in the original holomorphic transform), but the map η is not a fibre bundle projection: in fact, it is precisely the blow up (in the real category) of $\mathbb{CP}_3$ along $\mathbb{RP}_3$.

Let E be a holomorphic vector bundle on $\mathbb{CP}_3$. Our analysis proceeds by relating smooth sections of E over $\mathbb{RP}_3$ to a certain cohomology group $H^1_{\mathrm{d}}(F, \widetilde{\eta^* E})$ which I will define below, and then computing this cohomology by means of a spectral sequence very analogous to the hypercohomology sequence appearing in the holomorphic Penrose transform, and whose terms are smooth sections of vector bundles over $\mathrm{Gr}_2(\mathbb{R}^4)$.

5.1 Pull-back from $\mathbb{RP}_3$ to F

The cohomology we consider on F is constructed as follows. Define a complex sub-bundle $\mathcal{A}^{1,0} \subset \mathcal{E}^1$ of the bundle of smooth complex-valued 1-forms on F to be the image of the canonical map

$$\eta^* \mathcal{E}^{1,0}_{\mathbb{CP}_3} \to \mathcal{E}^1.$$

(It is easy to check this is injective, and so does define a sub-bundle.) By properties of pull-back, $\mathcal{A}^{1,0}$ generates a differentially closed ideal in the (complexified) de Rham complex on F, and thus we obtain a quotient differential complex which we denote

$$\Gamma(F, \mathcal{E}) \xrightarrow{\bar{\mathrm{d}}} \Gamma(F, \mathcal{A}^{0,1}) \xrightarrow{\bar{\mathrm{d}}} \Gamma(F, \mathcal{A}^{0,2}) \xrightarrow{\bar{\mathrm{d}}} \cdots. \tag{5.1}$$

(The structure we are defining on F is called an "involutive structure", or a "formally integrable structure". See (Treves 1992).)

Let $\widetilde{\mathbb{C}}$ denote the pull-back by τ of the locally constant non-trivial line bundle on $\mathrm{Gr}_2(\mathbb{R}^4)$. Then it is easy to see that if E is a holomorphic vector bundle over $\mathbb{CP}_3$, then one can tensor (5.1) through by $\eta^*E \otimes \widetilde{\mathbb{C}}$ to obtain a differential complex, and we denote by

$$H^1_{\bar{\mathrm{d}}}(F, \widetilde{\eta^*E})$$

its first cohomology group. The transfer of information from $\mathbb{RP}_3$ to F is accomplished by showing that there is an exact sequence

$$0 \to H^0(\mathbb{CP}_3, \mathcal{O}(E)) \to \Gamma(\mathbb{RP}_3, \mathcal{E}(E)) \xrightarrow{\alpha} H^1_{\bar{\mathrm{d}}}(F, \widetilde{\eta^*E}) \to H^1(\mathbb{CP}_3, \mathcal{O}(E)) \to 0. \tag{5.2}$$

The map α is obtained as follows: take $f \in \Gamma(\mathbb{RP}_3, \mathcal{E}(E))$ and pull back to obtain a section of η^*E over $\eta^{-1}\mathbb{RP}_3$. Now let $\tilde{f}$ denote a smooth section of η^*E over F, agreeing with f on $\eta^{-1}\mathbb{RP}_3$ and such that $\bar{\mathrm{d}}\tilde{f}$ vanishes to infinite order on $\eta^{-1}\mathbb{RP}_3$. (Such a thing exists by formal power series calculations and partition of unity arguments.) There is a canonical locally constant section H of $\widetilde{\mathbb{C}}$ with a unit jump discontinuity across the hypersurface $\eta^{-1}\mathbb{RP}_3$. Then αf is the equivalence class of $H\bar{\mathrm{d}}f$ in $H^1_{\bar{\mathrm{d}}}(F, \widetilde{\eta^*E})$. The exactness of (5.2) is established by formal power series calculations. Note that the map α is an isomorphism up to finite-dimensional errors.

5.2 Push-down from F to $\mathrm{Gr}_2(\mathbb{R}^4)$

Recall that the fibres of τ are $\mathbb{CP}_1$. For a vector bundle V with a complex structure on the fibres of τ, denote by Œ(V) the sheaf of smooth sections of V which are holomorphic on the fibres of τ. Denote by $\mathcal{E}^{0,1}$ the sheaf of $(0,1)$-forms in the fibres of τ. It is easy to check that there is a natural inclusion $\mathcal{E}^{0,1} \hookrightarrow \mathcal{A}^{0,1}$, and we define $\mathcal{E}^1_\eta$ by the exactness of

$$0 \to \mathcal{E}^{0,1} \to \mathcal{A}^{0,1} \to \mathcal{E}^1_\eta \to 0.$$

This short exact sequence induces a filtration on the complex which computes $H^*_{\bar{\mathrm{d}}}(F, \widetilde{\eta^*E})$, and results in a spectral sequence

$$E_1^{p,q} = \Gamma(\mathrm{Gr}_2(\mathbb{R}^4), \tau_*^q \text{Œ}(\widetilde{\mathcal{E}^p_\eta \otimes \eta^*E})) \Rightarrow H^{p+q}_{\bar{\mathrm{d}}}(F, \widetilde{\eta^*E}), \tag{5.3}$$

where we write $\mathcal{E}^p_\eta = \bigwedge^p \mathcal{E}^1_\eta$.

If E is an $\mathrm{SL}(4,\mathbb{C})$-homogeneous bundle, then the direct image sheaves appearing here are sheaves of smooth sections of complex vector bundles over $\mathrm{Gr}_2(\mathbb{R}^4)$. The maps are differential operators, and so combined with the map α

of §5.2, we are able to relate $\Gamma(\mathbb{RP}_3, E)$ to kernels and cokernels of differential operators on $\mathrm{Gr}_2(\mathbb{R}^4)$.

An important point is that the E_1 terms in the spectral sequence (5.3) consist of smooth sections over $\mathrm{Gr}_2(\mathbb{R}^4)$ of the restrictions of the vector bundles whose holomorphic sections appear in the spectral sequence (3.3) for the holomorphic Penrose transform for $\mathbb{CP}_3$, and the maps are also "the same" in this sense. Thus, once the basic geometry has been established and the "pull-back" results of the previous section proved, one can more or less import results wholesale from the holomorphic Penrose transform. For a wide class of holomorphic double fibrations where the relevant spaces are of the form G/P where G is a holomorphic semisimple Lie group and P is a parabolic subgroup, the resulting Penrose transform has been extensively studied (Baston and Eastwood 1989), and the calculation of the vector bundles appearing in (3.3) has been reduced to a simple algorithm. Consequently, for real integral geometry transforms obtained from these by the process we have been discussing, the considerations of this section are purely routine, and the only real work is in the "pull-back mechanism".

5.3 Results

It is now possible to state the main result of (Bailey and Eastwood 1997). First note that everything we have said generalises easily to the family of integral transforms where $\mathbb{RP}_3$ and $\mathrm{Gr}_2(\mathbb{R}^4)$ are replaced by $\mathbb{RP}_n$ and $\mathrm{Gr}_2(\mathbb{R}^{n+1})$, the integral transform still being integration over real projective lines. We regard $\mathbb{RP}_n$ as a homogeneous space for $\mathrm{SL}(n+1,\mathbb{R})$. Given a complex, finite-dimensional irreducible representation $\mathbb{E}$ of $\mathrm{SL}(n+1,\mathbb{C})$, we have the Bernstein-Gelfand-Gelfand resolution on $\mathbb{RP}_n$,

$$0 \to \mathbb{E} \to \mathcal{E}(\mathbb{E}^0) \xrightarrow{\delta} \mathcal{E}(\mathbb{E}^1) \xrightarrow{\delta} \mathcal{E}(\mathbb{E}^2) \xrightarrow{\delta} \cdots \xrightarrow{\delta} \mathcal{E}(\mathbb{E}^n) \to 0.$$

Here, we are writing $\mathbb{E}$ also for the sheaf of locally constant $\mathbb{E}$-valued functions, the remaining terms are sheaves of smooth sections of certain irreducible complex homogeneous vector bundles and the maps δ are differential operators, not necessarily first-order. This generalises the de Rham resolution, which is the case where $\mathbb{E}$ is the trivial 1-dimensional representation.

There is an $\mathrm{SL}(n+1,\mathbb{R})$-equivariant "integration" of $\omega \in \Gamma(\mathbb{RP}_n, \mathcal{E}(\mathbb{E}^1))$ over a real projective line. We say that ω has *zero energy* if this integral vanishes on every projective line.

Theorem 5.1 *The field $\omega \in \Gamma(\mathbb{RP}_n, \mathcal{E}(\mathbb{E}^1))$ is zero energy if and only if there exists $f \in \Gamma(\mathbb{RP}_n, \mathcal{E}(\mathbb{E}^0))$ with $\omega = \delta f$.*

The proof of the theorem is straightforward once the spectral sequence (5.3) is understood in the cases $E = \mathbb{E}^0$ and $E = \mathbb{E}^1$. A particular family of examples of the above is where ω is a symmetric covariant k-tensor field. In this case, the invariant integral over a real projective line γ is given by integrating $\omega(\dot{\gamma}, \ldots, \dot{\gamma})$. The theorem gives that ω is zero energy if and only if it is the symmetrised covariant derivative of a symmetric covariant $(k-1)$-tensor field on $\mathbb{RP}_n$. The case $k = 1$ of this (which says that for a smooth 1-form on $\mathbb{RP}_n$, its integral over every projective line vanishes if and only if it is the exterior derivative of a smooth function) is due to Michel (1978), as is the case $k = 2$ (Michel 1973). (This last result is equivalent to Blaschke rigidity of $\mathbb{RP}_n$, i.e. the fact that one cannot infinitesimally perturb the standard metric on this space while keeping all geodesics closed and the same fixed length.) The case $k = 3$ is due to Estezet (1988) (see Goldschmidt 1990). Another case, not in this family, gives the analogue of Blaschke rigidity in the category of manifolds with projective structure.

Bibliography

Buchdahl, N. P. (1983). On the relative De Rham sequence. *Proc. A.M.S.*, **87**, 363–6.

Bailey, T. N. and Eastwood, M. G. (1997). Zero-energy fields on real projective space. *Geom. Dedicata,* to appear.

Bailey, T. N., Eastwood, M. G., Gover, A. R., and Mason, L. J. (1997). The Funk transform as a Penrose transform. *Math. Proc. Camb. Phil. Soc.*, to appear.

Baston, R. J. and Eastwood, M. G. (1989). *The Penrose Transform; its Interaction with Representation Theory*, Oxford University Press.

Eastwood, M. G., Penrose, R. and Wells, R. O., Jr. (1981). Cohomology and massless fields. *Commun. Math. Phys.*, **78**, 305–51.

Estezet, P. (1988). *Tenseurs symétriques à énergie nulle sur les variétés à courbure constante,* Thèse de doctorat de troisième cycle, Université de Grenoble I.

Goldschmidt, H. (1990). The Radon transform for symmetric forms on real projective spaces, in *Integral Geometry and Tomography, Cont. Math.* **113**, Amer. Math. Soc., 81–96.

Michel, R. (1973). Problèmes d'analyse géométriques liés à la conjecture de Blaschke. *Bull. Soc. Math. France,* **101**, 17–69.

Michel, R. (1978). Sur quelques problèmes de géométrie globale des géodésiques. *Bol. Soc. Bras. Mat.*, **9**, 19–38.

Penrose, R. and MacCallum, M. A. H. (1972). Twistor theory: an approach to the quantisation of fields and space-time, *Phys Reports.*, **6C**, 241–315.

Treves, F. (1992). *Hypo-analytic Structures*, Princeton University Press.

31
Pythagorean Spinors and Penrose Twistors

Andrzej Trautman
Instytut Fizyki Teoretycznej, Uniwersytet Warszawski
ul. Hoża 69, PL–00681 Warszawa, Poland[1]

1 Introduction

Besides the major applications of spinors and twistors to equations of mathematical physics, there are minor results, where these objects play an auxiliary role or bring a new light on otherwise well-known facts. One example of such a result is the twistor-inspired derivation, based on the use of the conformal compactification of a (pseudo-) Euclidean space, of the fractional-linear form of Möbius transformations. Even simpler is the remark that the solution, attributed to Euclid, of the Pythagorean equation, has a spinorial interpretation: it is equivalent to the statement that a null vector in $\mathbb{Z}^3$, considered as a subset of $\mathbb{R}^3$ with a scalar product of signature $(1,2)$, is the (tensor) square of an integer-valued spinor. In this short article, I expand the latter observation and give a summary of the rudiments of twistor notions associated with higher-dimensional spaces; this account is included here not because of its novelty, but in the hope that there may be some interest in a presentation by an outsider. Only 'global' twistors are considered here; I gave a brief account of my view of 'local' twistors in (Trautman 1993).

Originally, Roger Penrose intended twistor spaces to be associated with, or serve as replacements for, the 4-dimensional, Lorentzian space-times. His belief in the privileged and exceptional role of four dimensions, apparent in twistor theory, was strikingly confirmed by the discoveries of exotic differential structures on $\mathbb{R}^4$, and of the Donaldson and Seiberg–Witten invariants. There are, however, interesting generalizations of twistor ideas to other dimensions and signatures; especially to proper Riemannian 3- and 4-manifolds. As is often the case with important ideas, the original notion of twistor has been generalized in many ways; only some of them are briefly presented below.

2 Pythagorean spinors

If p and q are integers, then the triple of integers (x, y, z), given by

$$x = p^2 - q^2, \quad y = 2pq, \quad z = p^2 + q^2, \tag{2.1}$$

[1]email: amt@fuw.edu.pl

is *Pythagorean*: it satisfies the equation $x^2+y^2=z^2$. If (x,y,z) is Pythagorean, then at least one of the numbers x and y is even; moreover, if $t \in \mathbb{Z}$, then (y,x,z) and (tx,ty,tz) are also Pythagorean. I say that a Pythagorean triple (x,y,z) is *standard* if $z>0$ and either the triple (x,y,z) is relatively prime (*rp*) and y is even or $(x/2,y/2,z/2)$ is a triple of *rp* integers and $y/2$ is odd. For example, the triples $(-1,0,1)$ and $(8,6,10)$ are standard, but $(4,3,5)$ and $(6,8,10)$ are not. Every Pythagorean triple can be written as (tx,ty,tz), where $t\in\mathbb{Z}$, the integers (x,y,z) are *rp* and $z>0$; if y is even, then (x,y,z) is standard; if y is odd, then $(2x,2y,2z)$ is standard.

Proposition 2.1 *If (x,y,z) is a standard Pythagorean triple, then there is a pair (p,q) of relatively prime integers such that* (2.1) *holds.*

In other words: there is a bijection between the set of directions in $\mathbb{Z}^2$ and the set of 'null directions' in $\mathbb{Z}^3$.

Proof Note that $z>0$, y even and $y^2=(z+x)(z-x)$ imply $z+x=2m\geq 0$ and $z-x=2n\geq 0$, where m and n are integers. If $y=2r$, then the Pythagorean equation is equivalent to $r^2=mn$. If the triple (x,y,z) is *rp*, then so is the triple (m,n,r). If r is odd and the integers x, y and z are all even, but have no divisor >2, then the triple (m,n,r) is also *rp*. If (m,n,r) is *rp*, then $r^2=mn$ implies that both m and n are squares. □

Recall the classical lemma (Sierpiński 1987):

Lemma 2.2 *If p and q are integers, then a necessary and sufficient condition for the existence of integers u and v, such that $pu+qv=1$, is that p and q be relatively prime.*

It leads to

Proposition 2.3 *The group $\mathsf{SL}_2(\mathbb{Z})$ acts transitively on the set $P\subset\mathbb{Z}^2$ of integer-valued 'spinors' with relatively prime components.*

In other words: the group $\mathsf{SL}_2(\mathbb{Z})$ acts transitively on the set of directions in $\mathbb{Z}^2$.

Proof Indeed, let $a,b,c,d\in\mathbb{Z}$ and consider

$$A=\begin{pmatrix} a & b \\ c & d \end{pmatrix}. \tag{2.2}$$

The matrix A is in $\mathsf{SL}_2(\mathbb{Z})$ iff $ad-bc=1$; it acts in $\mathbb{Z}^2$ by sending $\begin{pmatrix} p \\ q \end{pmatrix}$ to $\begin{pmatrix} p' \\ q' \end{pmatrix}=A\begin{pmatrix} p \\ q \end{pmatrix}$. If $\begin{pmatrix} p \\ q \end{pmatrix}\in P$, then there are integers u and v such that $pu+qv=1$. Putting $(u',v')=(u,v)A^{-1}$ one obtains $p'u'+q'v'=1$; therefore, $\mathsf{SL}_2(\mathbb{Z})$ acts in P. This action is transitive: if $\begin{pmatrix} p \\ q \end{pmatrix}\in P$ and $pu+qv=1$, then the matrix $\begin{pmatrix} p & -v \\ q & u \end{pmatrix}$ sends $\begin{pmatrix} 1 \\ 0 \end{pmatrix}$ to $\begin{pmatrix} p \\ q \end{pmatrix}$. □

The stabilizer of an element of P is a subgroup of $\mathsf{SL}_2(\mathbb{Z})$ isomorphic to $\mathbb{Z}$.

Recall that the group $\mathsf{SL}_2(\mathbb{R})$ is the connected component of the group $\mathsf{Spin}_{1,2}(\mathbb{R})$: there is the exact sequence of homomorphisms of groups,

$$1 \to \mathbb{Z}_2 \to \mathsf{SL}_2(\mathbb{R}) \xrightarrow{\rho} \mathsf{SO}^{\circ}_{1,2}(\mathbb{R}) \to 1.$$

If $(x, y, z) \in \mathbb{R}^3$ is represented by the matrix $\begin{pmatrix} z+x & y \\ y & z-x \end{pmatrix} = 2\begin{pmatrix} p \\ q \end{pmatrix}(p \;\; q)$ and

$$v = \begin{pmatrix} z+x & y \\ y & z-x \end{pmatrix} J, \text{ where } J = \begin{pmatrix} 0 & -1 \\ 1 & 0 \end{pmatrix}, \tag{2.3}$$

so that $v^2 = (x^2 + y^2 - z^2)I$, then $\rho(A)v = AvA^{-1}$. By restriction, one obtains the exact sequence

$$1 \to \mathbb{Z}_2 \to \mathsf{SL}_2(\mathbb{Z}) \xrightarrow{\rho} \mathsf{G} \to 1.$$

The group $\mathsf{G} \subset \mathsf{SO}^{\circ}_{1,2}(\mathbb{R})$ which, by definition, is the image of $\mathsf{SL}_2(\mathbb{Z})$ by ρ, is a group of matrices with entries that are either integer or half-integer. For example,

$$\rho\begin{pmatrix} 1 & 1 \\ 0 & 1 \end{pmatrix} = \begin{pmatrix} \frac{1}{2} & 1 & \frac{1}{2} \\ -1 & 1 & 1 \\ -\frac{1}{2} & 1 & \frac{3}{2} \end{pmatrix}.$$

The group G acts on the set $\{(x, y, z) \in \mathbb{Z}^3 : x + z \text{ is even}\}$. It is an easy exercise to find the subgroup of $\mathsf{SL}_2(\mathbb{Z})$ that covers the subgroup of G containing all matrices with integer elements.

3 Projective quadrics and twistors

Global twistors, described in the first paper on the subject (Penrose 1967), are associated with *projective quadrics*, i.e. with conformal compactifications of (pseudo-) Euclidean spaces. Most of the time, the name 'projective quadric' is shortened to 'quadric'. Let V be an m-dimensional vector space over $K = \mathbb{R}$ or $\mathbb{C}$ with a non-degenerate quadratic form g. In the vector space $W = V \oplus K^2$ one introduces the quadratic form h by putting $h(w) = g(v) + \lambda\mu$, where $w = (v, \lambda, \mu) \in W$, and defines the quadric $\mathsf{Q}(h)$ to be the subset of the projective space $\mathsf{P}(W)$ consisting of all null directions, $\mathsf{Q}(h) = \{\operatorname{dir} w : w \in W, w \neq 0, h(w) = 0\}$. The quadric inherits from the quadratic space (W, h) a (locally flat) *conformal structure*. The map $V \to \mathsf{Q}(h)$, given by $v \mapsto \operatorname{dir}(v, -g(v), 1)$, is a conformal embedding with an image open and dense in $\mathsf{Q}(h)$; the complement of the image is the 'null cone at infinity'. The group $\mathsf{Spin}(h)$ acts on $\mathsf{Q}(h)$ by sending $\operatorname{dir} w$ to $\operatorname{dir}(AwA^{-1})$, where $w \in W$ and $A \in \mathsf{Spin}(h)$. This action is conformal and transitive. The Clifford algebra $\mathsf{Cl}(h)$, associated with the quadratic space (W, h), is isomorphic to the algebra $\mathsf{Cl}(g) \otimes \mathsf{End}K^2$; an isomorphism is induced by the Clifford map

$$(v, \lambda, \mu) \mapsto \begin{pmatrix} v & \lambda \\ \mu & -v \end{pmatrix} = X. \tag{3.1}$$

Therefore, every $A \in \mathsf{Spin}(h) \subset \mathsf{Cl}(h)$ can be written in the form (2.2), where a, b, c and d are now suitable elements of $\mathsf{Cl}(g)$. Let $a \mapsto {}^t a$ be the antiautomorphism of the algebra $\mathsf{Cl}(g)$ such that ${}^t 1 = 1$ and ${}^t v = v$ for every $v \in V$. There holds

Proposition 3.1 *If $A \in \mathsf{Spin}(h)$, then the space*

$$V_A = \{v \in V : cv + d \text{ is invertible in } \mathsf{Cl}(g)\}$$

is open and dense in V; the map

$$f_A : V_A \to V_{A^{-1}} \text{ defined by } f_A(v) = (av + b)(cv + d)^{-1}$$

is a conformal diffeomorphism, $g(\mathrm{d}v) = {}^t(cv + d)(cv + d)g(\mathrm{d}f_A(v))$.

This result goes back to Th. Vahlen; see (Robinson and Trautman 1993) and the references given there. The Clifford algebra $\mathsf{Cl}(g)$ has an irreducible Dirac (m even) or Pauli (m odd) representation γ in a complex, $2^{[\frac{1}{2}m]}$-dimensional vector space S of spinors. The Dirac representation, restricted to the even Clifford algebra, decomposes into the sum of two Weyl representations in the spaces of spinors of opposite 'chirality'. A representation γ of $\mathsf{Cl}(g)$ in S extends to a representation δ of $\mathsf{Cl}(h)$ in $S \oplus S$. Namely, if $a \in \mathsf{Cl}(g)$ and $b \in \mathsf{End}K^2 \subseteq \mathsf{End}\mathbb{C}^2$, then

$$\delta : \mathsf{Cl}(h) \to \mathsf{End}(S \oplus S) = (\mathsf{End}S) \otimes \mathsf{End}\mathbb{C}^2 \text{ is given by } \delta(a \otimes b) = \gamma(a) \otimes b. \tag{3.2}$$

3.1 The complex case

3.1.1 *Complex projective quadrics of dimension m*

Assume $K = \mathbb{C}$ so that $V = \mathbb{C}^m$. The corresponding complex quadric Q_m is, in the words of Kobayashi and Ochiai (1982), 'a holomorphic analogue of a sphere'. It has no complex-bilinear Riemannian structure; its complex conformal structure supports a unique conformal spin structure which can be described as follows. Let Cl_m denote the Clifford algebra associated with $(\mathbb{C}^m, g)$. The conformal spin (Clifford) group is defined here as the subset Cpin_m of Cl_m consisting of products of all *even* sequences of non-null vectors in V. This group acts on vectors by sending, for every $a \in \mathsf{Cpin}_m$, the vector v to $\rho(a)v = av\,{}^t a$. There is the exact sequence of group homomorphisms,

$$1 \to \mathsf{K}_m \to \mathsf{Cpin}_m \xrightarrow{\rho} \mathsf{CO}_m \to 1,$$

where CO_m is the connected component of the conformal group (understood here as the group of rotations and dilations). If m is odd, then the kernel K_m of ρ is $\mathbb{Z}_2 = \{1, -1\}$. For m even, ρ gives a four-fold cover and $\mathsf{K}_m = \{1, -1, \eta, -\eta\}$, where $\eta \in \mathsf{Cl}_m$ is a volume element normalized so that ${}^t\eta\eta = -1$. The group K_m is isomorphic to $\mathbb{Z}_4$ for $m \equiv 0 \bmod 4$ and to $\mathbb{Z}_2 \times \mathbb{Z}_2$ for $m \equiv 2 \bmod 4$. The spin group is $\mathsf{Spin}(g) = \mathsf{Spin}_m = \{a \in \mathsf{Cpin}_m : {}^t aa = 1\}$. The group Spin_{m+2} acts transitively on Q_m. The image of the null vector $w_\infty = (0, 1, 0) \in W$ by (3.1) is $\begin{pmatrix} 0 & 1 \\ 0 & 0 \end{pmatrix} \in \mathsf{Cl}_{m+2}$. The stabilizer of $\operatorname{dir} w_\infty \in \mathsf{Q}_m$ is the semi-direct product H_m of Cpin_m by $\mathbb{C}^m$ given explicitly by

$$\mathsf{H}_m = \{\begin{pmatrix} a & v \\ 0 & {}^t a^{-1} \end{pmatrix} : a \in \mathsf{Cpin}_m,\ v \in \mathbb{C}^m\} \subset \mathsf{Spin}_{m+2}\,.$$

The groups Cpin_m and $\mathbb{C}^m$ are thus made into subgroups of Spin_{m+2}. Let PO_m be the quotient of the complex special orthogonal group SO_m by its centre: $\mathsf{PO}_{2n+1} = \mathsf{SO}_{2n+1}$ and $\mathsf{PO}_{2n} = \mathsf{SO}_{2n}/\mathbb{Z}_2$. There is the commutative diagram

$$\begin{array}{ccc} \mathsf{Cpin}_m & \longrightarrow & \mathsf{CO}_m \\ \downarrow & & \downarrow \\ \mathsf{Spin}_{m+2} & \longrightarrow & \mathsf{PO}_{m+2} \end{array}$$

of group homomorphisms: the vertical arrows are injective and the horizontal ones are 4:1 or 2:1 depending on whether m is even or odd. The map $\mathbb{C}^m \to \mathsf{Spin}_{m+2}$ descends to a monomorphism of groups, $\mathbb{C}^m \to \mathsf{PO}_{m+2}$. With these observations in mind, one can formulate

Proposition 3.2 (i) *The conformal spin structure on* Q_m *is given by the principal bundle maps*

$$\begin{array}{cccc} \mathsf{Cpin}_m & \longrightarrow & \mathsf{CO}_m & \\ \downarrow & & \downarrow & \\ \mathsf{Spin}_{m+2}/\mathbb{C}^m & \to & \mathsf{PO}_{m+2}/\mathbb{C}^m & \to \mathsf{Q}_m\,. \end{array}$$

(ii) *The associated bundle of spinors,*

$$(\mathsf{Spin}_{m+2}/\mathbb{C}^m) \times_{\mathsf{Cpin}_m} S \to \mathsf{Q}_m,$$

corresponding to the representation $\gamma : \mathsf{Cpin}_m \to \mathsf{GL}(S)$, *is isomorphic to the bundle* $\Sigma \to \mathsf{Q}_m$, *where*

$$\Sigma = \{(\operatorname{dir} w, \Phi) \in \mathsf{Q}_m \times S \oplus S : \delta(w)\Phi = 0\}$$

and δ *is as in* (3.2).
(iii) *The Maurer–Cartan form* $A^{-1}\mathrm{d}A$ *defines a flat Cartan connection on the* H_m*-bundle* $\varpi : \mathsf{Spin}_{m+2} \to \mathsf{Q}_m$.

A proof of (i) is in (Robinson and Trautman 1993). The map $\varpi : \mathsf{Spin}_{m+2} \to \mathsf{Q}_m$ is given by $\varpi(A) = \mathrm{dir}(Aw_\infty A^{-1})$. Part (ii) generalizes a similar observation made by Manin (1981) for $m = 4$. I learned of this generalization from Harnad; the isomorphism in question is given by $[(A\mathbb{C}^m, \varphi)] \mapsto (\varpi(A), \delta(A)(\varphi, 0))$. Part (iii) follows from the definition of a Cartan connection; see (Friedrich 1977) and the references given there.

If $\Phi \in S \oplus S$ is a non-zero spinor, then the vector space $\{w \in W : \delta(w)\Phi = 0\}$ is totally null; if it is maximal (*mtn*), then Φ is said to be *pure*. If $m = 2n$ is even, then a pure spinor is Weyl (chiral) and the $(n+1)$-vector formed from a linear basis spanning the corresponding *mtn* is either self-dual or antiself-dual. The projective *twistor space* T_m for Q_m is the manifold of directions of pure spinors associated with (W, h). For m even, it has two components, T_m^+ and T_m^-. If one puts $m = 2n$ (m even) or $m = 2n-1$ (m odd), then $\dim \mathsf{T}_m = \frac{1}{2}n(n+1)$. In particular, each of the spaces T_3, T_4^+ and T_4^- is diffeomorphic to $\mathbb{C}\mathrm{P}_3$. A global twistor $\mathrm{dir}\,\Phi \in \mathsf{T}_m$ is identified with the *mtn* space of vectors annihilating Φ. This space descends to a totally null geodesic submanifold of Q_m of the maximal dimension $[\frac{1}{2}m]$. The dimensions of Q_m and T_m coincide only for $m = 3$ (minitwistors; cf. the papers by K. P. Tod in (Mason *et al.* 1995); see also (Ward 1996) and the papers by N. J. Hitchin referred to there) and $m = 6$ (in this case, the three spaces Q_6, T_6^+ and T_6^- are diffeomorphic to each other; this coincidence reflects triality; cf. the papers by L. P. Hughston in (Mason *et al.* 1995)). The flag manifold for Q_m is defined as the 'projectivized' bundle of pure spinors, $\mathsf{F}_m = \{(\mathrm{dir}\, w, \mathrm{dir}\,\Phi) \in \mathsf{Q}_m \times \mathsf{T}_m : \delta(w)\Phi = 0\}$. The two natural projections define the double fibration $\mathsf{Q}_m \leftarrow \mathsf{F}_m \rightarrow \mathsf{T}_m$ which underlies the *Penrose correspondence* (Wells 1979). For m even, F_m has two connected components and there are two such double fibrations.

3.1.2 *The case of four dimensions*

Instead of representing W as $V \oplus \mathbb{C}^2$, one uses, in this case, the identification of $\mathbb{C}^6$ with $\wedge^2\mathbb{C}^4$. Let $\mathbb{T}$ be the complex, four-dimensional vector space of *Penrose twistors*; $\mathbb{T}$ is assumed to be endowed with a volume element $\varepsilon \in \wedge^4\mathbb{T}^*$, $\varepsilon \neq 0$. A frame $(e_\alpha)_{\alpha=1,\ldots,4}$ in $\mathbb{T}$ is said to be *unimodular* if $\varepsilon = e^1 \wedge e^2 \wedge e^3 \wedge e^4$, where (e^α) is the frame in $\mathbb{T}^*$, dual to (e_α). From now on, only unimodular frames are used. The six-dimensional vector space $W = \wedge^2\mathbb{T}$ has a quadratic form h—the *Pfaffian*—defined by $\frac{1}{2}w \wedge w = h(w) e_1 \wedge e_2 \wedge e_3 \wedge e_4$. The volume element defines also the Hodge map $\star : \wedge\mathbb{T} \to \wedge\mathbb{T}^*$, such that $\star(1_{\wedge\mathbb{T}}) = \varepsilon$. If $w = \frac{1}{2}w^{\alpha\beta} e_\alpha \wedge e_\beta$, then $\star w = \frac{1}{2} \star w_{\alpha\beta} e^\alpha \wedge e^\beta$, where $\star w_{12} = w^{34}$, etc. If $w \in W$ is considered as a linear map $\mathbb{T}^* \to \mathbb{T}$ and $\star w$ as a linear map $\mathbb{T} \to \mathbb{T}^*$, then

$$w \circ \star w = -h(w)\mathrm{id}_{\mathbb{T}} \quad \text{and} \quad \star\, w \circ w = -h(w)\mathrm{id}_{\mathbb{T}^*}. \tag{3.3}$$

In the notation with indices, these equations read $w^{\alpha\gamma} \star w_{\gamma\beta} = -\delta^\alpha_\beta(w^{12}w^{34} + w^{13}w^{42} + w^{14}w^{23})$. The *Klein quadric* is $\mathsf{Q}_4 = \{\mathrm{dir}\, w : w \in W, w \neq 0, w \wedge w = 0\}$.

By (3.3), the linear map $W \to \mathsf{End}(\mathbb{T} \oplus \mathbb{T}^*)$ given by

$$w \mapsto \begin{pmatrix} 0 & w \\ \star w & 0 \end{pmatrix} \tag{3.4}$$

has the Clifford property and yields a faithful and irreducible representation of $\mathsf{Cl}(h) = \mathsf{Cl}_6$ in $\mathbb{T} \oplus \mathbb{T}^*$. With respect to this representation, the elements of $\mathbb{T}$ and $\mathbb{T}^*$ are Weyl spinors of opposite chirality; using the notation of §3.1.1 one can put $\mathsf{T}_4^+ = \mathsf{P}(\mathbb{T})$ and $\mathsf{T}_4^- = \mathsf{P}(\mathbb{T}^*)$. The projective twistor $\operatorname{dir}\Phi$, $0 \neq \Phi \in \mathbb{T}$, is identified with the *mtn* 3-space $\{w \in W : w \wedge \Phi = 0\}$; this space projects to a totally null, geodesic, self-dual 2-dimensional submanifold of Q_4: $\alpha(\Phi) = \{\operatorname{dir}(\Phi \wedge \Phi') : \Phi' \in \mathbb{T}, \Phi \wedge \Phi' \neq 0\}$. As a complex manifold, $\alpha(\Phi)$ is $\mathbb{C}\mathsf{P}_2$. If $\Phi, \Phi' \in \mathbb{T}$ and $\Phi \wedge \Phi' \neq 0$, then $\operatorname{dir}(\Phi \wedge \Phi') \in \mathsf{Q}_4$ is the intersection of $\alpha(\Phi)$ and $\alpha(\Phi')$. Similarly, if $\Psi \in \mathbb{T}^*$, $\Psi \neq 0$, then there is the submanifold of Q_4: $\beta(\Psi) = \{\operatorname{dir}(\Phi \wedge \Phi') : \Phi, \Phi' \in \mathbb{T}, \Phi \wedge \Phi' \neq 0, \langle \Phi, \Psi \rangle = \langle \Phi', \Psi \rangle = 0\}$. The submanifolds $\alpha(\Phi)$ and $\beta(\Psi)$ intersect along a null geodesic iff $\langle \Phi, \Psi \rangle = 0$; as a complex manifold, such a null geodesic is $\mathbb{C}\mathsf{P}_1$; two distinct points $\operatorname{dir} w$, $\operatorname{dir} w' \in \mathsf{Q}_4$ lie on such a null geodesic iff $w \wedge w' = 0$; see §9.3 in (Penrose and Rindler 1986) and (Penrose 1996).

The group $\mathsf{Spin}(h) = \mathsf{Spin}_6$ is isomorphic to $\mathsf{SL}_4 = \mathsf{SL}(\mathbb{T})$ embedded in $\mathsf{Cl}(h)$ by

$$A \mapsto \begin{pmatrix} A & 0 \\ 0 & A^{*-1} \end{pmatrix} \tag{3.5}$$

where $A^* \in \mathsf{SL}(\mathbb{T}^*)$ is the transpose of $A \in \mathsf{SL}(\mathbb{T})$. The element A acts in W by sending w to AwA^*, as may be checked from (3.4), (3.5) and the equation $\star(AwA^*) = (\det A)(A^{*-1} \star wA^{-1})$ valid for every $w \in W$ and $A \in \mathsf{GL}(\mathbb{T})$. A frame (e_α) in $\mathbb{T}$ can be used to construct a 'null frame' $(w_a)_{a=0,1,\ldots,4,\infty}$ in W by putting (say): $w_0 = e_3 \wedge e_4$, $w_1 = e_1 \wedge e_3$, $w_2 = e_1 \wedge e_4$, $w_3 = e_2 \wedge e_3$, $w_4 = e_2 \wedge e_4$ and $w_\infty = e_1 \wedge e_2$. For $z = (z^\mu) \in \mathbb{C}^4$, put $w(z) = w_0 + z^\mu w_\mu + (z_1 z_4 - z_2 z_3) w_\infty$; then for every z one has $w(z) \neq 0$ and $w(z) \wedge w(z) = 0$; the map $z \mapsto \operatorname{dir} w(z)$ is a conformal embedding of $V = \mathbb{C}^4$ in Q_4. Put $S = \operatorname{span}\{e_1, e_2\}$ and $S' = \operatorname{span}\{e_3, e_4\}$; the direction of w_∞ (equivalently: the plane S) is preserved by the subgroup

$$\mathsf{H}_4 = \{\begin{pmatrix} a & v \\ 0 & b \end{pmatrix} : a \in \mathsf{GL}(S),\ b \in \mathsf{GL}(S'),\ \det a \det b = 1 \ \text{ and } \ v \in \mathsf{Hom}(S', S)\}$$

of SL_4 so that Cpin_4 is isomorphic to $\{(a, b) \in \mathsf{GL}_2 \times \mathsf{GL}_2 : \det a \det b = 1\}$.

3.2 The real case

Assume now $K = \mathbb{R}$ and let (k, l), $k + l = m$, be the signature of g. The real quadric $\mathsf{Q}_{k,l}$ is diffeomorphic to $(\mathbb{S}_k \times \mathbb{S}_l)/\mathbb{Z}_2$. In particular, $\mathsf{Q}_{k,0} = \mathbb{S}_k$; a proper real quadric, i.e. one with $kl \neq 0$, is orientable iff $k + l$ is even (Cahen *et al.*

1993). An essential difference between the complex and the real case is that, in the latter, the conformal structure on the quadric is generated by a pseudo-Riemannian metric. One can consider spin or pin structures corresponding to such a metric. (S)pin structures on real quadrics have been determined and a method for finding the spectrum of the Dirac operator given in (Cahen *et al.* 1995). There is neither room nor need to describe here the construction of the twistor spaces associated with the real quadrics. The most important case of $\mathrm{Q}_{1,3}$ is fully treated in the works of Penrose and his school. Instead, I describe here the *real twistors* on $\mathrm{Q}_{1,2}$ that could have been discovered by Euclid, had he followed the 'spinorial method' of solving the Pythagorean equation, outlined in §2.

3.2.1 *Real twistors on* $\mathrm{Q}_{1,2}$

Let $\lambda, \mu \in \mathbb{R}$ and let v be as in (2.3). The matrix X, given by (3.1), can be now considered as an endomorphism of $\mathbb{U}$, a four-dimensional vector space of *real* twistors. The antisymmetric matrix

$$\omega = \begin{pmatrix} 0 & J \\ J & 0 \end{pmatrix} : \mathbb{U} \to \mathbb{U}^*$$

is a symplectic 2-form on $\mathbb{U}$ and $\varepsilon = \frac{1}{2}\omega \wedge \omega$ is the corresponding volume 4-form. It follows from $X^* = \omega \circ X \circ \omega^{-1}$ that the map $X \circ \omega^{-1} : \mathbb{U}^* \to \mathbb{U}$ is antisymmetric. Since $X^2 = (x^2 + y^2 - z^2 + \lambda\mu)\mathrm{id}_{\mathbb{U}}$, if the vector $(x, y, z, \lambda, \mu) \in \mathbb{R}^6$ is null, then the bivector $X \circ \omega^{-1}$ is of rank ≤ 2 and there are twistors $\Phi, \Psi \in \mathbb{U}$ such that $X \circ \omega^{-1} = \Phi \wedge \Psi$. Moreover, $\mathrm{tr} X = 0$ implies $\omega(\Phi, \Psi) = 0$. Conversely, given a four-dimensional real symplectic space $(\mathbb{U}, \omega)$, the vector space $W = \{w \in \wedge^2\mathbb{U} : \mathrm{tr}(w \circ \omega) = 0\}$ is five-dimensional and the restriction of the Pfaffian to W is a quadratic form of signature (2,3). Therefore, the quadric $\mathrm{Q}_{1,2}$ can be identified with the set of null directions in W, or, equivalently, with the set of *lagrangian planes* in $\mathbb{U}$. A real twistor $\Phi \in \mathbb{U}$ defines the null geodesic $\gamma(\Phi) = \{\mathrm{dir}(\Phi \wedge \Psi) : \Psi \in \mathbb{U}, \omega(\Phi, \Psi) = 0\}$ on $\mathrm{Q}_{1,2}$. If $\Phi \wedge \Psi \neq 0$ and $\omega(\Phi, \Psi) = 0$, then $\gamma(\Phi) \cap \gamma(\Psi) = \mathrm{dir}(\Phi \wedge \Psi)$. Two distinct points of $\mathrm{Q}_{1,2}$ lie on one null geodesic iff the corresponding lagrangian planes intersect along a line. For the material of this paragraph, see Note 1 to Chapter 6 in (Woodhouse 1980) and §7.2 in (Penrose and Rindler 1986).

Acknowledgements This research was sponsored by the Polish Committee on Scientific Research (KBN) under Grant No. 2 P302 112 7. The author is grateful to the Mathematical Institute at Oxford for the hospitality and financial support extended to him during the conference on *Geometric issues in the foundations of science.*

Bibliography

Cahen, M., Gutt, S., and Trautman, A. (1993). Spin structures on real projective quadrics. *J. Geom. Phys.*, **10**, 127–154.

Cahen, M., Gutt, S., and Trautman, A. (1995). Pin structures and the modified Dirac operator. *J. Geom. Phys.*, **17**, 283–297.

Friedrich, H. (1977). Twistor connection and normal conformal Cartan connection, *Gen. Rel. Grav.*, **8**, 303–312.

Kobayashi, S. and Ochiai, T. (1982). Holomorphic structures modeled after hyperquadrics, *Tôhoku Math. J.*, **34**, 587–629.

Manin, Yu. I. (1981). Gauge fields and holomorphic geometry (in Russian), *Current Problems in Mathematics*, **17**, 3–55, Akad. Nauk USSR, Moscow.

Mason, L.J., Hughston, L.P., and Kobak, P.Z. (eds). (1995). *Further Advances in Twistor Theory: Volume II.* (Pitman Research Notes in Mathematics Series 232). Longman and Wiley, Harlow and New York.

Penrose, R. (1967). Twistor algebra. *J. Math. Phys.*, **8**, 345–366.

Penrose, R. (1996). Incidence between complex null rays. *Twistor Newsletter*, **41**, 1–5.

Penrose, R. and Rindler, W. (1986). *Spinors and Space-Time*, vol. 2: *Spinor and Twistor Methods in Space-Time Geometry.* Cambridge University Press, Cambridge.

Robinson, I. and Trautman, A. (1993). The conformal geometry of complex quadrics and the fractional-linear form of Möbius transformations. *J. Math. Phys.*, **34**, 5391–5406.

Sierpiński, W. (1987). *Elementary Theory of Numbers.* PWN and North-Holland, Warszawa and Amsterdam.

Trautman, A. (1993). Geometric aspects of spinors. In *Proceedings of the Third International Conference on Clifford Algebras and their Applications in Mathematical Physics*, ed. R. Delanghe, F. Brackx and H. Serras Kluwer, Dordrecht.

Ward, R.S. (1996). Twistors, geometry and integrable systems, chapter 6 in this volume.

Wells, R.O. Jr (1979). Complex manifolds and mathematical physics. *Bull. (N. S.) Amer. Math. Soc.*, **1**, 296–336.

Woodhouse, N. (1980). *Geometric Quantization.* Clarendon Press, Oxford.

PART VI

Afterword

32
Afterword

Roger Penrose
Mathematical Institute
24–29 St. Giles, Oxford OX1 3LB

1 Geometry, and the roots and aims of twistor theory

Geometry, in its various forms, has always been one of my greatest delights—and even obsessions. In addition to the intrinsic beauty of the subject, there is the fact that geometry can also shed profound light on the deepest issues of science and philosophy—a wondrous fact that is indeed one of Nature's miracles. One such a miracle was Einstein's general relativity. But quantum theory, also, can in many respects be vastly illuminated by the geometrical perspective.

The prospect of a new kind of geometry for physics, namely *twistor theory*, which attempts to unite space–time structure with quantum–mechanical principles, has indeed become one of my obsessions. It is something to which I have now devoted over thirty-three of my years—which is more than half my life. This subject has provided me with great satisfaction, but also much frustration. In some respects it has developed in ways that I had vaguely anticipated, but it has also moved in directions that I had not expected at all. Among the things that I had not fully expected about twistor theory was the mathematical difficulty of the subject, as it was to develop. Another was the slowness of its maturing into a serious physical theory. It has not yet really done so even after all these years, despite impressive and determined efforts by a number of different researchers. But equally, I had little inkling of the vast array of purely mathematical uses to which twistor theory and its numerous variants might be put. A number of elegant articles in this book amply illustrate the scope of the mathematical applications of twistor theory and of twistor–related ideas.

Whilst general relativity has been my main professional concern, whether addressed within the framework of twistor theory or by more conventional means, quantum theory also has long held a particular fascination for me. Not only does quantum mechanics have some striking geometrical manifestations, but its descriptions of the world also reveal a wealth of deep underlying mysteries—even bordering on paradox. I have long held the view that not all the seeming paradoxes of quantum theory can simply be argued away in accordance with a belief that they arise merely from an inadequate human understanding of the implications of the theory's mathematical formalism. Instead, at some level, there must be a deviation from purely unitary evolution, so that state–vector reduction can become a *real* phenomenon. Accordingly, I believe that a major revolution in our physical theory must be waiting in the wings. Moreover, because of the (mysterious) non-local nature of quantum entanglement (EPR effects), whatever the nature of this revolution might be, the final theory that emerges must have a fundamentally *non-local* character. The desire for such a non-local scheme was one of the important motivations behind the original formulation of twistor theory.

Among the other inputs into twistor theory coming from quantum mechanics is *spin* and its mathematical representation in terms of *spinors*. Indeed, the geometry and algebra of 2-spinors, and their relation to the intrinsic structure of space–time light cones (most specifically, the holomorphic geometry of the celestial sphere), have been particularly influential for me. These ingredients, taken in conjunction with the basic role of complex numbers and complex analysis in quantum theory, provided additional important initial motivations behind twistor theory. Spin–networks, themselves partly motivated by the EPR phenomenon, also had a significant early influence, and it is noteworthy that spin–networks (and their q-deformed variants) have acquired a renewed interest in relation to quantum gravity. Other motivations came explicitly from the role of complex functions in general relativity. Twistor theory is indeed fundamentally complex and spinorial; moreover it provides a non-local description of space–time.

It is still not completely clear what is the relationship between twistor non-locality and the specific non-locality of quantum entanglement. However, a clue is undoubtedly to be found in the non-local effects that are described (in the weak–field limit) by the use of sheaf cohomology, this being the means whereby twistor theory describes (linear) massless fields. Moreover, the known non-linear versions of these cohomological descriptions—namely the 'non-linear (leg–break) graviton construction', to describe anti–self–dual vacuum space–times, and the Ward construction, to describe anti–self–dual gauge fields—are also fundamentally non-local. The (local) information of the space–time fields is stored in global structure in the twistor picture. There is no local information in twistor space to specify what specific space–time field is being referred to.

2 Towards a twistor description of Einsteinian physics

It has been clear to me for some time that in order for twistor theory to make major progress towards becoming a genuine physical theory, it is necessary that general relativity proper (in respect of the vacuum equations at least) be brought within the theory's compass. Thus, not only the (anti–)self–dual field, but the properly combined self–dual and anti–self–dual parts of the full Einstein field must have a natural twistor (and not merely an 'ambitwistor') interpretation. There now is a reasonably clear programme for a full twistor description of Einstein vacuum fields, although definitive results are not yet to hand. This programme is based on the following two facts:

(a) the Einstein vacuum equations constitute the consistency condition for the equations of a massless field of helicity $\frac{3}{2}$ (expressed in terms of a potential);

(b) in Minkowski space $\mathbb{M}$, the space of charges for massless fields of helicity $\frac{3}{2}$ is twistor space $\mathbb{T}$.

Thus, the notion of a massless field of helicity $\frac{3}{2}$ mediates between the Einstein vacuum equations (i.e., the condition of Ricci–flatness in four dimensions) and the concept of twistor space.

This basic programme for finding a twistorial formulation of the Einstein vacuum equations takes the following overall form. For a given vacuum space–time $\mathcal{M}$, we study the massless fields, of helicity $\frac{3}{2}$, defined within (but not necessarily globally within) $\mathcal{M}$. We then try to find the relevant notion of 'charge' for the fields of helicity $\frac{3}{2}$, in the background of $\mathcal{M}$, the *twistor space* $\mathcal{T}$ for $\mathcal{M}$ being the space of such charges. This space $\mathcal{T}$ is to be 'curved' in some appropriate way that encodes the geometrical structure of $\mathcal{M}$. The hope would be that the structure of $\mathcal{T}$ that arises can be given in terms of *free functions* (as is the case with the specific non-linear graviton and Ward constructions referred to above) and that this structure indeed determines the geometry of $\mathcal{M}$, as required. For this, some twistorial notion of 'space–time point' is needed which generalizes the holomorphic curves (Riemann spheres) of the above non-linear graviton construction.

As a half–way stage towards understanding the structure of $\mathcal{T}$, the gravitational *googly* problem must, it seems, be solved. This problem is to find a twistor construction for the *self–dual* (as opposed to anti–self–dual) solutions of the Einstein vacuum equations (or Yang–Mills equations, in the case of the googly problem for gauge fields). Having the standard non-linear graviton construction for *anti*–self–dual Einstein fields, it would be a triviality (a mere redefinition of terms) to find a construction for self–dual Einstein fields in terms of a deformed *dual* twistor space; what is required, instead, is a construction for *self*–dual Einstein fields in terms of *twistor* space. The 'googly' terminology is borrowed from the game of cricket; a googly is a cricket ball bowled with a *right*–handed helicity, but by means of a bowling action which appears to be that which normally imparts a *left*–handed helicity—called a *leg–break*. In the conventional description of gravitons as solutions of the *linear* spin 2 massless equations, left–handed gravitons (helicity -2) would indeed be described by (positive–frequency) anti–self–dual fields and right–handed ones (helicity $+2$) by self–dual fields, so the terminology is apposite. In accordance with this, I shall henceforth refer to the original non-linear graviton construction (in terms of twistor, rather than dual twistor space) as the *leg–break* graviton. For the opposite helicity $+2$ (again in terms of twistor space), we require a corresponding *googly* graviton construction.

To get some more understanding of what is involved here, we recall how linear massless fields of spin 2 are described in twistor terms. The left–handed fields are obtained from twistor functions (representatives of first cohomology) which are homogeneous of degree $+2$, and it is not hard to see how to provide an infinitesimal leg–break deformation of (an appropriate region of) twistor space by use of such a twistor function. The right–handed fields, on the other hand, are obtained from twistor functions homogeneous of degree -6, and it has always seemed to be a considerable puzzle how such functions might effect a deformation of some appropriate kind.

The most promising role for such -6 functions arises in connection with the *blown-up* twistor space. In the flat case, the blown-up space $\mathbb{T}^{\#}$ is obtained from the ordinary twistor space $\mathbb{T}$, in terms of standard spinor parts $Z^\alpha = (\omega^A, \pi_{A'})$, by

$$(\omega^A, \pi_{A'}) \mapsto (\omega^A \pi_{B'}, \pi_{A'}\pi_{B'}).$$

In terms of projective twistor space $\mathbb{PT}$, the line $\mathbb{P}\mathbf{I}$, which represents infinity for Minkowski space $\mathbb{M}$, is replaced by a quadric $\mathbb{P}\mathbf{I}^{\#}$ in $\mathbb{PT}^{\#}$. If a sequence of points $\mathbf{p}_1, \mathbf{p}_2, \mathbf{p}_3, \ldots$ in $\mathbb{PT} - \mathbb{P}\mathbf{I}$ approaches a point $\mathbf{p} \in \mathbb{P}\mathbf{I}$, where all the points $\mathbf{p}_i$ lie in a particular plane $\mathbf{W}$ through $\mathbb{P}\mathbf{I}$, then, in $\mathbb{PT}^{\#}$, the points $\mathbf{p}_i$ reach a point $\mathbf{p_W}$ of $\mathbb{P}\mathbf{I}^{\#}$ which depends on the choice of $\mathbf{W}$. If $\mathbf{p}$ is held fixed on $\mathbb{P}\mathbf{I}$, but the plane $\mathbf{W}$ varies, then the corresponding points $\mathbf{p_W}$ sweep out a generator (determined by $\mathbf{p}$) of the quadric $\mathbb{P}\mathbf{I}^{\#}$. If, on the other hand, the plane $\mathbf{W}$ through $\mathbb{P}\mathbf{I}$ is held fixed but the point $\mathbf{p}$ on $\mathbb{P}\mathbf{I}$ varies, then the points $\mathbf{p_W}$ sweep out a generator of the opposite system on $\mathbb{P}\mathbf{I}^{\#}$ determined by (and determining) $\mathbf{W}$.

This picture is not sufficient for our purposes, however. We require, also, the non-projective space $\mathbb{T}^{\#}$. For example, if the twistors

$$(\omega^A, \lambda^2\pi_{A'}), \quad \text{for} \quad \lambda = \lambda_1,\ \lambda_2,\ \lambda_3, \ldots$$

represent the points $\mathbf{p}_1, \mathbf{p}_2, \mathbf{p}_3, \ldots$ in $\mathbb{PT}$, respectively, then the limit point $\mathbf{p} \in \mathbb{P}\mathbf{I}$, attained as $\lambda \to 0$, is given by the point $(\omega^A, 0)$ of $\mathbb{T}$. But if we take the scalings for the twistors in the form

$$(\lambda^{-1}\omega^A, \lambda\pi_{A'}),$$

then in the limit $\lambda \to 0$, we obtain a well–defined point $\mathbf{p_W} \in \mathbb{P}\mathbf{I}^{\#}$ given by the point $(\omega^A\pi_{B'}, 0)$ of $\mathbb{T}^{\#}$. Note that whereas in the case of $\mathbb{PT}^{\#}$, the quadric $\mathbb{P}\mathbf{I}^{\#}$ simply *replaces* the line $\mathbb{P}\mathbf{I}$ (their neighbourhoods in $\mathbb{PT} - \mathbb{P}\mathbf{I}$ being identical), in the non-projective case, $\mathbf{I}^{\#}$ and $\mathbf{I}$ are attached to $\mathbb{T} - \mathbf{I}$ at two quite different places (so that there would be no loss of Hausdorffness if both were to be attached together).

We now examine the behaviour of the 1-form δ and 3-form θ, given by

$$\delta = \epsilon^{A'B'}\pi_{A'}d\pi_{B'}, \qquad \theta = \frac{1}{6}\epsilon_{\alpha\beta\gamma\delta}Z^\alpha dZ^\beta \wedge dZ^\gamma \wedge dZ^\delta,$$

which satisfy

$$\delta \wedge \theta = 0,$$

and find that δ and θ have, respectively, a simple zero and a simple pole at $\mathbf{I}^{\#}$. Hence, the object

$$\mathbb{I}\Gamma = \delta \otimes \theta$$

extends smoothly to $\mathbf{I}^{\#}$. A differential operator $\mathbf{D}$ exists which acts on quantities of this kind and is invariant under the rescalings $\delta \mapsto \nu\delta$, $\theta \mapsto \nu^{-1}\theta$. This is defined by

$$\mathbf{D}(\alpha \otimes \beta) = \mathbf{d}\alpha \oslash \beta - \alpha \otimes \mathbf{d}\beta,$$

α being a 1-form and β a 3-form subject to $\alpha \wedge \beta = 0$, and where the bilinear operation '$\oslash$' defines a product between 2-forms and 3-forms according to

$$(dp \wedge dq) \oslash \beta = dp \otimes (dq \wedge \beta) - dq \otimes (dp \wedge \beta).$$

The significance of this is that the kernel of the operator $\mathbf{D}$ consists of expressions of the form $f\mathbb{I}\Gamma$, where the twistor function $f(Z^\alpha)$ has homogeneity degree -6 in Z^α. Thus, we can think of our required *googly deformations* of $\mathbb{T}^{\#}$ as being in some sense generated by quantities which are locally like some multiple of $\mathbf{D}\mathbb{I}\Gamma$, but which are not globally of this form.

The blown-up space $\mathbb{PT}^{\#}$ and its non-projective version $\mathbb{T}^{\#}$ have direct relevance to the asymptotic structure of Minkowski space $\mathbb{M}$. Applying a conformal rescaling to the Lorentzian metric of $\mathbb{M}$, we can smoothly adjoin a (null) hypersurface boundary $\mathcal{I}$ to $\mathbb{M}$. The complexification $\mathbb{C}\mathcal{I}$ of $\mathcal{I}$ contains null geodesics of three kinds: there are α-*lines*, lying in β-planes but not α-planes within $\mathbb{C}\mathcal{I}$; there are β-*lines*, lying in α-planes but not β-planes within $\mathbb{C}\mathcal{I}$; finally, there are the *generators* of $\mathbb{C}\mathcal{I}$, lying both in α-planes and β-planes in $\mathbb{C}\mathcal{I}$. The points of $\mathbb{PT}^{\#}$ correspond precisely to the α-lines together with the generators, the generators themselves being represented by the points of $\mathbb{P}\mathbf{I}^{\#}$.

Under appropriate circumstances, an *asymptotically flat* curved space–time $\mathcal{M}$ has a complexified null infinity of a very similar nature, and the complex manifold representing the α-lines is essentially a leg–break graviton $\mathcal{T}$, as described above. The googly problem then takes the following form: how do we code the information contained in the location of the β-lines in the way that $\mathbf{I}^{\#}$ is attached to $\mathcal{T}$? At the time of writing, it appears that this information can be coded in roughly the way indicated above, in terms of locally defined 1-forms $\otimes$ 3-forms, with a structure resembling that of $\mathbb{I}\Gamma$. According to this proposal, the conformal factor which is needed in order to make (null) infinity into a finite hypersurface $\mathcal{I}$ splits into two complex factors Ω and $\widetilde{\Omega}$, one of which rescales ϵ_{AB} and the other rescales $\epsilon_{A'B'}$:

$$\epsilon_{AB} \mapsto \Omega\epsilon_{AB}, \quad \epsilon_{A'B'} \mapsto \widetilde{\Omega}\epsilon_{A'B'},$$

where the metric rescales according to

$$g_{ab} \mapsto \Omega\widetilde{\Omega}g_{ab}.$$

In the real case, we have $\widetilde{\Omega} = \overline{\Omega}$, but for a complex space–time we can have $\widetilde{\Omega}$ and Ω as independent quantities. To accompany these rescalings is a *torsion*, which takes the form of the dual of the 1-form $\mathrm{i}\, d\,[\log(\Omega/\widetilde{\Omega})]$. The idea is to use a patchwork of rescaling (and consequent) torsion to encode the information of the β-lines. The rescaling also affects the quantity $\mathbb{\Gamma}$, thereby relating all this to the earlier discussion.

The next part of the googly problem would be to find a notion of 'space–time point' which is, in the appropriate sense 'dual' to that which is used in the leg–break construction. The essential clue to this lies in the nature of the Poincaré–invariant exact sequence for flat twistor space

$$0 \to \mathbb{S} \to \mathbb{T} \to \overline{\mathbb{S}}^* \to 0.$$

Here, $\mathbb{S}$ stands for spin–space (the space of ω^A) and $\overline{\mathbb{S}}^*$ stands for the conjugate dual spin–space (the space of $\pi_{A'}$). The two middle maps are given, respectively, by

$$\omega^A \mapsto (\omega^A, 0) \quad \text{and} \quad (\omega^A, \pi_{A'}) \mapsto \pi_{A'}.$$

A space–time point x can be interpreted as a 'splitting' of the sequence, according to which we have a map in the opposite direction from each of these two, given, respectively, by

$$(\omega^A, \pi_{A'}) \mapsto \omega^A - \mathrm{i}x^{AA'}\pi_{A'} \quad \text{and} \quad \pi_{A'} \mapsto (\mathrm{i}x^{AA'}\pi_{A'}, \pi_{A'}).$$

In the leg–break construction, we have a deformed twistor space $\mathcal{T}$ taking the place of $\mathbb{T}$, and the projection $\mathcal{T} \to \overline{\mathbb{S}}^* - \{0\}$ takes the place of the canonical map $\mathbb{T} \to \overline{\mathbb{S}}^*$. In this construction, the 'space–time points' arise as cross-sections of the fibration $\mathcal{T} \to \overline{\mathbb{S}}^* - \{0\}$, generalizing the map $\pi_{A'} \mapsto (\mathrm{i}\, x^{AA'}\pi_{A'}, \pi_{A'})$, described above. For the googly construction, we require a deformed version $\mathbb{S} - \{0\} \to \mathcal{T}$ of the injection $\mathbb{S} \to \mathbb{T}$, where the (singular) way in which $\mathbb{S}$ is attached to $\mathcal{T}$ is to characterize the required reverse maps that correspondingly generalize $(\omega^A, \pi_{A'}) \mapsto \omega^A - \mathrm{i}x^{AA'}\pi_{A'}$. The specific mode of attachment of $\mathbb{S}$ to $\mathcal{T}$ is to be determined by the 'googly' information that is given on the 'blown-up' space $\mathcal{T}^{\#}$ as indicated above, the specific singularity at $\mathbb{S} - \{0\}$ ($\subset \mathcal{T}$) that corresponds to this googly information being obtained when various 'blow-downs', applied to $\mathcal{T}^{\#}$, are compared. This supplies a notion of *'co-patching'* that appears to dualize the normal patching of the leg–break manifold $\mathcal{M}$.

There is a formal procedure which, in principle, gives rise to this required 'dualization' of the leg–break construction. We imagine proceeding in the following way. First of all we 'cheat' by simply taking the complex conjugate $\overline{Z}_\alpha$ of each twistor quantity Z^α. This converts a left–handed field, such as a leg–break graviton, into a right–handed one. However, holomorphicity has been lost because of the appearance of complex–conjugate quantities. To remedy this, we 'uncheat' by applying the twistor quantization rule whereby each occurrence of $\overline{Z}_\alpha$ is replaced by the operator $-\partial/\partial Z^\alpha$, thus restoring holomorphicity. We think of these operators as applying to the appropriate sheaves, finding that this gives us 'dual' operations also in the 'logical' sense of the *arrow reversal* that is involved in category theory. To illustrate this, consider the example of functions of a single complex variable. If $\mathcal{O}$ denotes the sheaf of holomorphic functions on $\mathbb{C}$, we have the two exact sequences

$$0 \longrightarrow \mathbb{C} \longrightarrow \mathcal{O} \xrightarrow{\partial/\partial z} \mathcal{O} \longrightarrow 0$$

and

$$0 \longleftarrow \mathbb{C} \longleftarrow \mathcal{O} \xrightarrow{\times z} \mathcal{O} \longleftarrow 0$$

being, in an appropriate sense *dual* to one another. In the first case, the '$\mathbb{C}$' stands for the sheaf of *constant* complex functions, and in the second case, it stands for the sheaf of complex functions which are zero except at the origin 0. Thus, there is an 'arrow reversal' accompanying the interchange of z with $\partial/\partial z$. This phenomenon occurs quite generally, and it relates in a consistent way with the kind of duality that is relevant to twistor theory, as indicated above.

In principle, this should enable us to provide a direct dualization of the 'patching' that leads to the leg–break graviton construction, so that an appropriate notion of 'co-patching' for the googly graviton can indeed be obtained. However, in practice, this leads to power series in $\partial/\partial Z^\alpha$ which may be rather formal, where questions of convergence, etc. present difficult issues of principle. This appears to lead one into the area of pseudo–differential operators rather than geometry, and it is hard to see how to make genuine progress unless a clear–cut geometrical interpretation is presented also. The purpose of the above 'googly geometry' is to present such a geometrical interpretation, but all this is 'work in progress' at the present stage.

The present status of the twistor description of the (vacuum) Einstein equations is somewhat uncertain, as things stand. It appears to be a genuinely difficult problem, but its solution is perhaps not too enormously far off, in my opinion. I believe, also, that the satisfactory solution of this problem will serve as a springboard for the twistor understanding of a number of other important physical problems. As was the case for the leg–break construction, the solution of the Einstein (vacuum) problem led, shortly thereafter, to a corresponding solution to the anti–self–dual Yang–Mills equations, and it is to be expected that the same ought also to be true in the case of the full Yang–Mills equations. As another point, the role of *mass* in twistor theory ought to be substantially illuminated when it is fully understood how gravitation proper is given a full twistor interpretation. This should have a significant impact on the twistor particle programme and on the problems of twistor diagram theory.

3 Further issues of physics and biology

It is to be hoped, also, that some insights will be provided towards an understanding of the very issue for which twistor theory was originally devised, namely the finding of the appropriate union of space–time structure with quantum mechanics. As was stated above, it is my own strong opinion that the very rules of quantum mechanics must become modified in order for this appropriate union to take place, the phenomenon of quantum state–vector reduction being a quantum–gravity effect. Does twistor theory shed any light on the nature of this phenomenon? It would be hard to maintain that there is anything in our present understanding of the theory which has much of a clear–cut nature to say on this issue. But the hope would be that when the correct twistor notion of 'space–time point' (at least in appropriate approximation) is to hand, for which both left–handed and right–handed parts of the graviton are accommodated, then some suggestion will be provided as to how the (non-local) 'jumps' which are characteristic of quantum mechanics can come about.

It is also my opinion that the accommodating of quantum state reduction into a satisfactory fundamental theory will have profound implications in a great many different areas. Unlike the main other suggested applications of quantum gravity (such as the taming or elimination of the space–time singularities at the big bang and in black holes, and in the regularization of infinities in quantum field theory), which refer to circumstances where dimensions of the order of the Planck length (10^{-33} cm) or the Planck time (10^{-43} sec) have relevance, this phenomenon would have implications for many other areas of much more direct relation to 'everyday' experience. A question arises, for example, in connection with crystal growth and quasicrystal growth. Are these things that can be fully understood without the non-local effects of quantum state reduction? When a crystal (or quasicrystal) forms, a solid substance is produced, in which the pattern of individual atoms forms a well–defined structure where the positions of the individual constituents are reasonably definite. This substance has formed, however, from a quite different sort of structure—a gas or a liquid—in which the quantum state is a highly entangled one where the individual atoms have no specific locations. To get from the one kind of state to the other, the process of quantum state reduction has to take place. In my view, the standard 'decoherence' point of view does not, without further ingredients, provide a satisfactory answer to this problem. Does it have a bearing on the issue of whether there is any significant non-locality in the growth of actual quasicrystals? The problem of what is *really* going on in quantum state reduction could well have relevance to many different physical processes.

If this is indeed the case, there must be important relevance to numerous *biological* processes also. Can one really hope to obtain a proper understanding of the processes that control the behaviour of cells without knowing how to apply quantum mechanics to them in a consistent way? Without a proper *objective* theory of quantum state reduction, it is hard to see how this can be achieved. If it is true that quantum state–vector reduction is a gravitational effect, then (at least for the kind of model that I have in mind) it is possible to estimate the level at which spontaneous state reduction will take place, and its associated time–scale. This would have particular relevance in those circumstances where the system is able to maintain an internal quantum coherence for a sufficient length of time that environmental decoherence does not mask whatever specific characteristic effects objective reduction (as a gravitational phenomenon) ought to have.

Are there likely to be biological circumstances in which this is the case? This brings me to my final topic: the question of whether the phenomenon of *consciousness* is related to the occurrence of such spontaneous objective reduction effects. Does the cytoskeleton, with its system of interconnected microtubules provide a plausible environment, first for quantum computation, and then for quantum coherence, especially within neurons? Could the effects of such objective quantum state reduction influence the strengths of synapses and the consequent firing of neurons—in a way which is consistent with what we know of the circumstances in which consciousness arises and with our feelings of 'free will'? It may be that much will be revealed to us concerning these issues in biological studies over the next several years. But without a further revolution in physics itself, I do not see the phenomenon of consciousness finding a proper home within our scientific world–view. An important question will be whether *non-computability* will be a feature of this revolution. If the holomorphic notions that are central to twistor theory find themselves to be part of such a revolution, then there must be some fundamental link between the notions of computability and holomorphicity. I am not aware of such a link, but according to the ideas that I am expressing, such a link ought to reveal itself at some stage.

It is clear that we are very far from satisfactorily addressing all the difficult issues that I have raised here. Nevertheless, a great deal of remarkable progress has been made on many of the topics that I have touched upon above. This volume provides excellent accounts of a good many of these relevant advances. I very much welcome and greatly appreciate them all.